D1571130

STOCHASTIC EQUATIONS IN INFINITE DIMENSIONS

Now in its second edition, this book gives a systematic and self-contained presentation of basic results on stochastic evolution equations in infinite dimensional spaces, typically Hilbert and Banach spaces. In the first part the authors give an exposition of the main properties of probability measures on separable Banach and Hilbert spaces, as required later; they assume a reasonable background in probability theory and finite dimensional stochastic processes. The second part is devoted to the existence and uniqueness of solutions of a general stochastic evolution equation, and the third concerns the qualitative properties of those solutions. Appendices gather together background results from analysis that are otherwise hard to find under one roof.

This revised edition includes two brand new chapters surveying recent developments in the field and an even more comprehensive bibliography, making this book an essential and up-to-date resource for all those working in stochastic differential equations.

Encyclopedia of Mathematics and Its Applications

This series is devoted to significant topics or themes that have wide application in mathematics or mathematical science and for which a detailed development of the abstract theory is less important than a thorough and concrete exploration of the implications and applications.

Books in the **Encyclopedia of Mathematics and Its Applications** cover their subjects comprehensively. Less important results may be summarized as exercises at the ends of chapters. For technicalities, readers can be referred to the bibliography, which is expected to be comprehensive. As a result, volumes are encyclopedic references or manageable guides to major subjects.

ENCYCLOPEDIA OF MATHEMATICS AND ITS APPLICATIONS

All the titles listed below can be obtained from good booksellers or from Cambridge University Press. For a complete series listing visit www.cambridge.org/mathematics.

106 A. Markoe *Analytic Tomography*
107 P. A. Martin *Multiple Scattering*
108 R. A. Brualdi *Combinatorial Matrix Classes*
109 J. M. Borwein and J. D. Vanderwerff *Convex Functions*
110 M.-J. Lai and L. L. Schumaker *Spline Functions on Triangulations*
111 R. T. Curtis *Symmetric Generation of Groups*
112 H. Salzmann *et al. The Classical Fields*
113 S. Peszat and J. Zabczyk *Stochastic Partial Differential Equations with Lévy Noise*
114 J. Beck *Combinatorial Games*
115 L. Barreira and Y. Pesin *Nonuniform Hyperbolicity*
116 D. Z. Arov and H. Dym *J-Contractive Matrix Valued Functions and Related Topics*
117 R. Glowinski, J.-L. Lions and J. He *Exact and Approximate Controllability for Distributed Parameter Systems*
118 A. A. Borovkov and K. A. Borovkov *Asymptotic Analysis of Random Walks*
119 M. Deza and M. Dutour Sikirić *Geometry of Chemical Graphs*
120 T. Nishiura *Absolute Measurable Spaces*
121 M. Prest *Purity, Spectra and Localisation*
122 S. Khrushchev *Orthogonal Polynomials and Continued Fractions*
123 H. Nagamochi and T. Ibaraki *Algorithmic Aspects of Graph Connectivity*
124 F. W. King *Hilbert Transforms I*
125 F. W. King *Hilbert Transforms II*
126 O. Calin and D.-C. Chang *Sub-Riemannian Geometry*
127 M. Grabisch *et al. Aggregation Functions*
128 L. W. Beineke and R. J. Wilson (eds.) with J. L. Gross and T. W. Tucker *Topics in Topological Graph Theory*
129 J. Berstel, D. Perrin and C. Reutenauer *Codes and Automata*
130 T. G. Faticoni *Modules over Endomorphism Rings*
131 H. Morimoto *Stochastic Control and Mathematical Modeling*
132 G. Schmidt *Relational Mathematics*
133 P. Kornerup and D. W. Matula *Finite Precision Number Systems and Arithmetic*
134 Y. Crama and P. L. Hammer (eds.) *Boolean Models and Methods in Mathematics, Computer Science, and Engineering*
135 V. Berthé and M. Rigo (eds.) *Combinatorics, Automata and Number Theory*
136 A. Kristály, V. D. Rădulescu and C. Varga *Variational Principles in Mathematical Physics, Geometry, and Economics*
137 J. Berstel and C. Reutenauer *Noncommutative Rational Series with Applications*
138 B. Courcelle and J. Engelfriet *Graph Structure and Monadic Second-Order Logic*
139 M. Fiedler *Matrices and Graphs in Geometry*
140 N. Vakil *Real Analysis through Modern Infinitesimals*
141 R. B. Paris *Hadamard Expansions and Hyperasymptotic Evaluation*
142 Y. Crama and P. L. Hammer *Boolean Functions*
143 A. Arapostathis, V. S. Borkar and M. K. Ghosh *Ergodic Control of Diffusion Processes*
144 N. Caspard, B. Leclerc and B. Monjardet *Finite Ordered Sets*
145 D. Z. Arov and H. Dym *Bitangential Direct and Inverse Problems for Systems of Integral and Differential Equations*
146 G. Dassios *Ellipsoidal Harmonics*
147 L. W. Beineke and R. J. Wilson (eds.) with O. R. Oellermann *Topics in Structural Graph Theory*
148 L. Berlyand, A. G. Kolpakov and A. Novikov *Introduction to the Network Approximation Method for Materials Modeling*
149 M. Baake and U. Grimm *Aperiodic Order I: A Mathematical Invitation*
150 J. Borwein *et al. Lattice Sums Then and Now*
151 R. Schneider *Convex Bodies: The Brunn–Minkowski Theory (Second Edition)*
152 G. Da Prato and J. Zabczyk *Stochastic Equations in Infinite Dimensions (Second Edition)*
153 D. Hofmann, G. J. Seal and W. Tholen (eds.) *Monoidal Topology*
154 M. Cabrera-García and Á. Rodriguez Palacios *Non-Associative Normed Algebras I: The Vidav–Palmer and Gelfand–Naimark Theorems*
155 C. F. Dunkl and Y. Xu *Orthogonal Polynomials of Several Variables (Second Edition)*

ENCYCLOPEDIA OF MATHEMATICS AND ITS APPLICATIONS

Stochastic Equations in Infinite Dimensions

Second Edition

GIUSEPPE DA PRATO

Scuola Normale Superiore, Pisa

JERZY ZABCZYK

Polish Academy of Sciences

CAMBRIDGE
UNIVERSITY PRESS

University Printing House, Cambridge CB2 8BS, United Kingdom

Cambridge University Press is part of the University of Cambridge.

It furthers the University's mission by disseminating knowledge in the pursuit of education, learning and research at the highest international levels of excellence.

www.cambridge.org
Information on this title: www.cambridge.org/9781107055841

First published 1992
Reprinted with corrections 2008
Second edition 2014

Printed in the United Kingdom by CPI Group Ltd, Croydon CR0 4YY

A catalog record for this publication is available from the British Library

Library of Congress Cataloging in Publication data
Da Prato, Giuseppe, author.
Stochastic equations in infinite dimensions / Giuseppe Da Prato, Scuola Normale Superiore, Pisa, Jerzy Zabczyk, Polish Academy of Sciences. – Second edition.
pages cm. – (Encyclopedia of mathematics and its applications)
Includes bibliographical references and index.
ISBN 978-1-107-05584-1 (hardback)
1. Stochastic partial differential equations. I. Zabczyk, Jerzy, author. II. Title.
QA274.25.D4 2014
519.2′2 – dc23 2013049903

ISBN 978-1-107-05584-1 Hardback

Contents

Preface

This book is devoted to stochastic evolution equations on infinite dimensional spaces, mainly Hilbert and Banach spaces. These equations are generalizations of Itô stochastic equations introduced in the 1940s by Itô [423] and in a different form by Gikhman [347].

First results on infinite dimensional Itô equations started to appear in the mid-1960s and were motivated by the internal development of analysis and the theory of stochastic processes on the one hand, and by a need to describe random phenomena studied in the natural sciences like physics, chemistry, biology, engineering as well as in finance, on the other hand.

Hilbert space valued Wiener processes and, more generally, Hilbert space valued diffusion processes, were introduced by Gross [363] and Daleckii [183] as a tool to investigate the Dirichlet problem and some classes of parabolic equations for functions of infinitely many variables. An infinite dimensional version of an Ornstein–Uhlenbeck process was introduced by Malliavin [518, 519] as a tool for stochastic study of the regularity of fundamental solutions of deterministic parabolic equations.

Stochastic parabolic type equations appeared naturally in the study of conditional distributions of finite dimensional processes in the form of the so called nonlinear filtering equation derived by Fujisaki, Kallianpur and Kunita [330] and Liptser and Shiryayev [501] or as a linear stochastic equation introduced by Zakaï [737]. Another source of inspiration was provided by the study of stochastic flows defined by ordinary stochastic equations. Such flows are in fact processes with values in an infinite dimensional space of continuous or even more regular mappings acting in a Euclidean space. They are solutions of the corresponding backward and forward stochastic Kolmogorov like equations; see Krylov and Rozovskii [469], Carverhill and Elworthy [146], Kunita [476] and Pardoux [577]. Stochastic partial differential equations constitute an important tool in the analysis of partial differential equations on Hilbert spaces. Such studies are the subject of the monograph [220] by the authors. Let us also mention that the idea of treating delay equations as infinite dimensional processes, systematically used for deterministic equations [258, 392], also proved to be useful for stochastic delay equations (Vinter [698], Chojnowska-Michalik [164]).

As far as applications are concerned, stochastic evolution equations have been motivated by such phenomena as wave propagation in random media (Keller [445], Frish [318]) and turbulence (Novikov [560], Chow [166]). Important motivations came also from biological sciences, in particular from population biology (Dawson [225], Fleming [311]). One has to mention also the early control theoretic applications of Wang [707], Kushner [481] and Bensoussan and Viot [60]. Since the early days, the number of specific equations studied in the literature has increased considerably and we could even say that we are witnessing an explosion of interest in the subject. In particular, stochastic versions of various equations such as reaction-diffusion, wave, beam, Burgers, Musiela, Navier–Stokes, Kardar–Parisi–Zhang, Kuramoto–Sivashinsky, Cahn–Hilliard, Korteweg–de Vries, Schrödinger, Landau–Lifshitz–Gilbert, to mention only a few, have been the subject of numerous studies. In this book we treat only some of them. However, descriptions and bibliographical comments on most of them are given in Chapters 13 and 14 and in the Introduction which is devoted to motivating examples.

Basic theoretical questions on *existence and uniqueness* of solutions were asked and answered, under various sets of conditions, in the 1970s and 1980s and are still of great interest today. An important contribution is due to Pardoux, who, in his thesis [576], obtained fundamental results on stochastic nonlinear partial differential equations (PDEs) of monotone type; see also Krylov and Rozovskii [469]. Basic results on weak solutions are due to Viot; see his thesis [695], and papers [696, 697]. Early important contributions are also due to Bensoussan and Temam [58, 59] and Dawson [226]. More recently, interesting results have been obtained on SPDEs with *random boundary conditions*. For first contributions see Sowers [654] and Da Prato and Zabczyk [218]. Early papers used the Wiener process as a model of noise and stochastic perturbations. The number of studies devoted to equations with *different noise processes* is increasing. In particular, equations with *fractional Brownian motion* and with *Lévy processes* are attracting much attention; see Duncan, Maslowski and Duncan-Pasik [269, 270], Maslowski and Nualart [534] and the recent monograph by Peszat and Zabczyk [596]. Important contributions on numerical solutions have been published. New approaches and original points of view like *Hida's white noise* approach, the *rough paths* approach or *Wiener chaos expansions*, are appearing. They are all discussed in Chapter 14.

The aim of this book is to present basic results on stochastic evolution equations in a rather systematic and self-contained way. We discuss topics covered traditionally by books on ordinary stochastic differential equations: stochastic calculus, existence and uniqueness results, continuous and regular dependence on initial data, Markov property, equations for transition probabilities of Kolmogorov type, absolute continuity of laws induced by solutions on the spaces of trajectories, and asymptotic properties.

The book systematically uses the theory of linear semigroups. Semigroup theory is an important part of mathematics, having several connections with the theory of partial differential equations. Semigroups have been successfully applied to treat semilinear equations; see for instance [401, 512]. The assumption, which we will often make in this book, that the linear part of the equation is the infinitesimal generator of a linear semigroup, is equivalent to the minimal requirement that the equation under study, in its simplest form, has a unique solution continuously depending on the initial data. The semigroup formulation allows a uniform treatment of parabolic, hyperbolic and delay equations. In numerous situations results obtained by more specialized PDE methods can be recovered by the semigroup approach. Early contributions using that approach include some sections of the book by Balakrishnan [32], papers by Curtain and Falb [176] and by Métivier and Pistone [543], and the thesis by Chojnowska-Michalik [163].

A different method for studying stochastic partial differential equations, the so called *variational approach*, was introduced by Pardoux [575] and Krylov and Rozovskii [469]. We do not treat this method in this book. For a recent presentation see the monograph by Prévot and Röckner [602].

In several parts of the book an important role is played by control theory. In particular, control theoretic results are used in the study of transition semigroups, invariant measures and large deviations.

The book is divided into three main parts devoted respectively to foundations of the theory, existence and uniqueness results, and properties of solutions. Analytical results needed in the book, not always easily available in the literature, are gathered in the appendices. Appendix A is devoted to the semigroup treatment of linear deterministic evolutionary problems, so can be regarded as a kind of introduction to the book. Appendices B, C and D concern, respectively, control theory, nuclear and Hilbert–Schmidt operators, and dissipative mappings.

In Part I we recall the measure theoretic foundations of probability theory and give a self-contained exposition of the basic properties of probability measures on separable Banach and Hilbert spaces, needed in what follows. In particular, we prove the Fernique theorem on exponential moments of Gaussian measures, the Bochner characterization of measures on Hilbert spaces, and the Feldman–Hajek theorem on absolute continuity of Gaussian measures. We also introduce reproducing kernels of Gaussian measures and apply this concept to expansions of white noise. In Chapter 3 we list commonly used concepts and theorems from the theory of stochastic processes. We take for granted several results on finite dimensional stochastic processes, in particular classical martingale inequalities. We introduce infinite dimensional Wiener processes and analyze a specific case including spatially homogeneous ones. Finally, we construct the stochastic integral with respect to infinite dimensional Wiener processes and establish Itô's formula and the stochastic Fubini theorem. Maximal inequalities for stochastic integrals are treated in a detailed way.

In Part II we proceed to the main subject of the book, stochastic equations of the form

$$dX = (AX + F(X))dt + B(X)dW(t), \quad x(0) = x, \tag{1}$$

where A is a linear operator, generally unbounded, acting on a Hilbert space H, and F and B are nonlinear, in general discontinuous mappings acting on appropriate spaces. Moreover W is a Wiener process on a Hilbert space U and $x \in H$.

Part II is devoted to existence and uniqueness of solutions. In Chapter 5 we set $F = 0$, B constant and establish existence of weak solutions. We elaborate on the factorization method introduced in [205] and use it to prove the time continuity of the weak solution under broad conditions. Continuity with respect to spatial variables is treated as well. Distribution valued Ornstein–Uhlenbeck processes are also investigated. We give more refined regularity results in the case when A generates an analytic semigroup. In Chapter 6, F is again 0 but B is linear. We first derive sharp estimates for stochastic convolution

$$W_A^{\Phi}(t) = \int_0^t S(t-s)\Phi(s)dW(s), \quad t \geq 0,$$

where Φ is an operator valued process and $S(\cdot)$ is the contraction semigroup generated by A. We deal with estimates in a wide class of Banach spaces and also include some maximal regularity results from [206]. With good estimates in hand, we establish existence of solutions to (1) by a fixed point argument. We also present a method, applicable only in special situations, of transforming the Itô equation into a deterministic one with random coefficients, which can be treated pathwise. Chapter 7 is devoted to nonlinear equations. We first prove existence and uniqueness when F and B are Lipschitz continuous in H, and then turn to more general B to cover equations with a Nemytskiĭ nonlinearity. The case when F and B are locally Lipschitz continuous or dissipative on a suitable Banach space $E \subset H$ is treated as well. Chapter 8 is devoted to martingale solutions solving the martingale problem, also called weak solutions. We give a proof of the Viot theorem, in the so called compact case, based on the above mentioned factorization method.

Part III of the book is devoted to qualitative properties of solutions. In Chapter 9 we establish continuous dependence of solutions on the initial data and the Feller and Markov properties by an adaptation of finite dimensional methods. We indicate a large class of equations for which the transition semigroup is strongly Feller and for which the Kolmogorov equation can be solved for an arbitrary bounded Borel initial function. The important Bismut–Elworthy–Li formula is derived here and it is used to establish differentiability of solutions to Kolmogorov equations for a wide family of equations. Chapter 10 is on absolute continuity of laws corresponding to solutions of two different equations. We first give a detailed treatment of linear equations based on the Feldman–Hajek theorem. Next we prove the Girsanov theorem and give sufficient conditions for absolute continuity for nonlinear equations. We also establish existence of martingale solutions to equations with irregular drifts by Girsanov's method. Two following chapters concern the asymptotic properties of solutions. Existence and uniqueness of invariant measures and mean square stability are treated first. A careful analysis is carried out for linear equations with an additive and/or multiplicative noise. Nonlinear equations are treated under two types of hypothesis: dissipativity and compactness. Chapter 12 examines the asymptotic properties of solutions when $B(X) = \varepsilon B$, where B is a fixed bounded operator from U to H and ε is small. We establish the large deviation principle for laws of solutions and apply the resulting estimates to the so called exit problem. We generalize finite dimensional results of Freidlin–Wentzell and derive specific asymptotic formulae for so called gradient systems.

As already mentioned, the book covers only basic results of the theory and a number of specific equations are not treated. A comprehensive discussion of the literature and new developments is postponed to Chapters 13 and 14.

In particular we have not covered stochastic equations in nuclear spaces, which have appeared in the study of fluctuation limits of infinite particle systems. They are discussed by Itô [425]; see also Kallianpur and Pérez-Abreu [435]. We do not consider time dependent systems although several extensions to this case are possible. Nor do we discuss quasi-linear equations or equations with stochastic boundary conditions and variational inequalities. Each of those subjects would require several additional chapters. For the same reason we do not report on recent results on the corresponding Fokker–Planck equations, the theory of Dirichlet forms and its applications for solving equations with very irregular coefficients (see Ma and Röckner [516] and the references therein) or on potential theoretic concepts like the Martin boundary (see Föllmer [313]). We do not treat stochastic equations with Lévy noise (see [599] and references therein).

The present book is the second edition of *Stochastic Equations in Infinite Dimensions* published in 1992. We now describe the changes incorporated in the new edition.

There are no major changes in Chapter 1 on random variables or in Chapter 2 on probability measures. We have improved a theorem on white noise expansions, stating it for an arbitrary complete basis in the reproducing kernel. Estimates on the moments of Gaussian measures are derived in more detail and the proof of the Feldman–Hajek theorem is presented, we believe, in a clearer way.

In Chapter 3 we have added the Kolmogorov–Loève–Chentsov theorem on existence of a Hölder continuous version of a random field on bounded open subsets of $\mathbb{R}^d$ with a proof based on the Garsia, Rademich and Rumsay lemma.

In Chapter 4 on stochastic processes we have expanded sections on infinite dimensional Wiener processes. We devote more space to Wiener processes with general, non-trace-class, covariances and discuss specific examples of Wiener processess in $L^2(\mathscr{O})$. We also elaborate an important case of spatially homogeneous Wiener processes. Maximal inequalities for stochastic integrals are treated in a systematic and complete way.

In Chapter 5 we describe in more detail the so called factorization method, a tool to establish time regularity of the solution and existence of invariant measures. At the moment of writing the first edition the method, introduced in the paper [205] by Da Prato, Kwapień and Zabczyk, was fairly new. Since then it has found many applications, some explained in the present edition. In this chapter we also give a new proof of convergence of solutions to equations with Yosida approximations of the linear part of the drift. Existence of regular solutions to equations with higher order operators is treated as well. A section on strong solutions is thoroughly reworked.

A novelty in Chapter 6 is a section on maximal regularity for stochastic convolutions in L^p and $W^{k,p}$ spaces based on a paper by Da Prato and Lunardi [206].

In Chapter 7 on existence and uniqueness of solutions, we add general results which allow us to treat stochastic parabolic equations with Nemytskiĭ diffusion operators. The section on dissipative nonlinearities is extended as well.

Chapter 8 on martingale solutions remains basically as it was.

In Chapter 9 on Markov properties and Kolmogorov equations, we have essentially simplified the proof of differentiability of solutions with respect to initial data. We have also included explicit formulae for higher derivatives of Ornstein–Uhlenbeck transition semigroups. We also expand the section on mild Kolmogorov equations.

Chapters 10–12 are essentially unchanged, although we have tried to simplify the presentation and eliminate some misprints. For more recent results on large time behavior of solutions we refer to our book [220]. Additional results on large deviations can be found in the monograph by Feng and Kurtz [289].

Chapters 13 and 14 are new and are devoted respectively to a survey of results on specific equations and to a description of new developments.

There exist several excellent books on ordinary stochastic differential equations, which provide inspiration for infinite dimensional theory. Earlier books include those by Gikhman and Skorokhod [348], Has'minskii [394], Ikeda and Watanabe [418], Elworthy [280] and more recently Protter [609], Øksendal [567] and Applebaum [23].

Several books on infinite dimensional theory were published before 1992, for example Walsh [702], Belopolskaya and Daleckij [52], Rozovskii [633] and Métivier [542]. More recent are books by Chow [169], Grecksch and Tudor [360], Prévot and Röckner [602], Sanz-Solé [637], Dalang, Khoshnevisan, Mueller, Nualart and Xiao [180], Holden, Øksendal, Ubøe and Zhang [407], Peszat and Zabczyk [599], Kotelenez [458], Veraar [693], internet lecture notes by Hairer [385] and van Neerven's Internet seminar [686]. They all emphasize different aspects of the theory.

It is our pleasure to thank our colleagues and collaborators, Z. Brzeźniak, S. Cerrai, A. Chojnowska-Michalik, A. Debussche, F. Flandoli, M. Fuhrman, B. Goldys, D. Gątarek, M. Gubinelli, S. Peszat, E. Priola, G. Tessitore, L. Tubaro and L. Zambotti for reading some parts of the book and for useful comments. Our special thanks go to Professor Kai Liu from the University of Liverpool, who sent us a very long list of misprints, mistakes and suggestions which helped us to improve the presentation in a considerable way.

Finally we would like to thank the team of the Cambridge University Press, in particular Vania Cunha, for help and understanding.

Introduction: motivating examples

As mentioned in the Preface, stochastic evolution equations in infinite dimensions are natural generalizations of stochastic ordinary differential equations and their theory has motivations coming from both mathematics and natural sciences: physics, chemistry, biology and mathematical finance.

We present here several examples of stochastic equations of the form

$$dX = (AX + F(X))dt + B(X)dW(t), \quad x(0) = x, \tag{1}$$

together with some comments concerning their derivation. Examples 0.1–0.3 have purely mathematical motivations, examples 0.4–0.6 come from physics, 0.7 from chemistry, 0.8–0.9 from biology and 0.10 from finance.

0.1 Lifts of diffusion processes

Consider an ordinary stochastic differential equation on $\mathbb{R}^d$ of the form

$$\begin{cases} dy = f(y)dt + \sum_{j=1}^{N} b_j(y)d\beta_j, \\ y(0) = \xi \in \mathbb{R}^d, \end{cases} \tag{2}$$

where f and $b_1, \ldots, b_N$ are continuous mappings from $\mathbb{R}^d$ into $\mathbb{R}^d$. Let us fix a closed subset $K \subset \mathbb{R}^d$ and let E be a Hilbert space of mappings from K into $\mathbb{R}^d$ contained in the space $C(K; \mathbb{R}^d)$ of continuous mappings from K into $\mathbb{R}^d$. The following equation on E

$$\begin{cases} dX = F(X)dt + \sum_{j=1}^{N} B_j(X)d\beta_j \\ X(0) = x \in E \end{cases} \tag{3}$$

in which

$$F(x)(\xi) = f(x(\xi)), \quad B_j(x)(\xi) = b_j(x(\xi)), \quad \xi \in K, \tag{4}$$

is called the *lift* of (2) to E.

Note that if the identity mapping $I_d(\xi) : I_d(\xi) = \xi \in K$ belongs to E and there exists a solution to (3) with $x = I_d$ then the formula

$$y(t, \xi) = X(t, I_d)(\xi), \quad \xi \in K \tag{5}$$

defines a version of (2) depending continuously on the initial condition ξ. Such versions are called *stochastic flows*. If in addition the space E consists of diffeomorphisms, then the stochastic flow (4) is the flow of diffeomorphisms. This way one can obtain basic results about stochastic flows from elementary facts on stochastic equations with values in infinite dimensional spaces and known results about Sobolev spaces. See [146] for a detailed exposition of the subject.

0.2 Markovian lifting of stochastic delay equations

A different type of lifting proved to be very useful in the study and applications of stochastic delay equations of the form

$$\begin{cases} dy(t) = \left(\int_{-r}^{0} a(d\theta)y(t+\theta) + f(y(t))\right) dt + \sum_{j=1}^{N} b_j(y(t))d\beta_j(t), \\ y(0) = x_0 \in \mathbb{R}^n, \quad y(\theta) = x_1(\theta), \quad \theta \in [-r, 0], \end{cases} \tag{6}$$

where $a(\cdot)$ is an $n \times n$ matrix valued finite measure on $[-r, 0]$ and f and b_j are as in the preceeding example. It turns out that if we define $H = \mathbb{R}^n \times L^2(-r, 0, \mathbb{R}^n)$ then the H-valued process $X(\cdot)$

$$X(t) = \begin{pmatrix} y(t) \\ y_t(\cdot) \end{pmatrix}$$

where $y_t(\theta) = y(t+\theta),\ t \geq 0,\ \theta \in [-r, 0]$, is a solution of the equation

$$\begin{cases} dX = (AX + F(X))dt + \sum_{j=1}^{N} B_j(X)d\beta_j, \\ X(0) = \begin{pmatrix} x_0 \\ x_1(\cdot) \end{pmatrix} \in H, \end{cases} \tag{7}$$

with operators A, F, B_j defined as follows.

The operator A is linear and unbounded with the domain

$$D(A) = \left\{ \begin{pmatrix} x \\ y \end{pmatrix} \in H : x = y(0),\ y \in W^{1,2}(-r, 0; \mathbb{R}^n) \right\}$$

and is given by the formula

$$A \begin{pmatrix} x \\ y \end{pmatrix} = \begin{pmatrix} \int_{-r}^{0} a(d\theta)y(\theta) \\ \frac{dy}{d\theta} \end{pmatrix}, \quad \begin{pmatrix} x \\ y \end{pmatrix} \in D(A).$$

Moreover

$$F\begin{pmatrix}x\\y\end{pmatrix}=\begin{pmatrix}f(x)\\0\end{pmatrix},\quad \begin{pmatrix}x\\y\end{pmatrix}\in H$$

$$B_j\begin{pmatrix}x\\y\end{pmatrix}u=\begin{pmatrix}b_j(x)\\0\end{pmatrix}u,\quad \begin{pmatrix}x\\y\end{pmatrix}\in H,\quad u\in\mathbb{R}^1,\ j=1,\dots,N.$$

Conversely, under fairly general conditions, the $\mathbb{R}^n$-dimensional coordinate of the solution X of equation (7) is a solution of the stochastic equation (6). The main advantage of dealing with equation (7) rather than with (6) is that the solution of (7) is Markovian and the solution of (6) is not. For more details and applications we refer to [164, 258, 698].

0.3 Zakaï's equation

Let y be the solution of the equation in $\mathbb{R}^n$

$$dy(t)=f(y(t))dt+dW(t)+BdV(t),\quad y(0)=\xi\in\mathbb{R}^n$$

and let γ be the "observation"

$$\gamma(t)=\int_0^t g(y(s))ds+V(t),\quad t\ge 0, \tag{8}$$

where $f:\mathbb{R}^n\to\mathbb{R}^n$, $g:\mathbb{R}^n\to\mathbb{R}^m$ are given mappings, $B=(b_{j,k})$ is an $n\times m$ matrix and W,V are independent Wiener processes of dimensions n and m respectively.

It is of great interest to describe the evolution in time of the conditional distribution μ_t of $y(t)$ with respect to the σ-field generated by $\gamma(s),\ s\le t,\ t\ge 0$. One approach to the problem was proposed by Zakaï [737] and is related to the so called *Zakaï equation*:

$$\begin{cases} dX(t,x)=\left[\dfrac{1}{2}\displaystyle\sum_{j=1}^n\frac{\partial^2 X}{\partial x_j^2}(t,x)-\sum_{j=1}^n\frac{\partial X}{\partial x_j}(t,x)f_j(x)\right]dt\\ \qquad\qquad +X(t,x)\displaystyle\sum_{k=1}^m g_k(x)d\gamma_k(t)-\sum_{k=1}^m\sum_{j=1}^n\frac{\partial X}{\partial x_j}(t,x)b_{jk}d\gamma_k(t)\\ X(0,x)\ =x\in E \end{cases} \tag{9}$$

on a space of real valued functions defined on $\mathbb{R}^k$. Equation (9) is of the form (1), with differential operators A and B of, respectively, second and first order, and $F=0$. Moreover $\gamma_1,\dots,\gamma_k$ are coordinates of the process γ, which is a Wiener process, under an equivalent probability measure. If there exists a solution $X(\cdot)$ of (9) then, under rather general conditions, the function valued process

$p(t) := X(t)/\langle X(t), 1\rangle,\ t \geq 0$, where

$$\langle X(t), 1\rangle = \int_{\mathbb{R}^k} X(t, \xi)d\xi, \quad t \geq 0,$$

is identical with the process of the densities of the conditional laws $\mu_t,\ t \geq 0$. Since the law $\mathscr{L}(\gamma(\cdot))$ of the process $\gamma(\cdot)$ on $C([0, T]; \mathbb{R}^k)$, T arbitrary positive constant, is equivalent to the law of $\mathscr{L}(V(\cdot))$ so, to study the problem of existence and uniqueness of the solutions to (9) or path properties of $X(\cdot)$, one can assume that processes $\gamma_1, \dots, \gamma_k$ are independent real valued Wiener processes. In this sense Zakai's equation is of the type (1). For more details about the equation see [434] and [577].

We pass now to examples arising in the natural sciences.

0.4 Random motion of a string

The following model of motion of an elastic string in a viscous random environment was proposed by Funaki [331]. There is a vast literature on related models so we will be more detailed here.

Let us start by remarking that the motion of a particle in a viscous environnement in $\mathbb{R}^d$ under a forcing field $f(y)$, $y \in \mathbb{R}^d$, can be described by the first order equation

$$y' = f(y), \quad y(0) \in \mathbb{R}^d.$$

Fix a natural number $N > 1$ and a sequence $W_1, \dots, W_N$ of independent d-dimensional Wiener processes and consider a system of N particles that move under the influence of three kinds of forces: *elastic forces* acting between neighboring particles, proportional to the distance between particles, the *external force* f and the *random forces* of white noise type.

The movement of the kth particle is then described by a properly normalized stochastic ordinary differential equation:

$$\begin{aligned} dy_k = &\left[(\kappa/2)N^2(y_{k+1} + y_{k-1} - 2y_k) + f(y_k)\right] dt \\ &+ \sqrt{N}\, b(y_k)dW_k, \quad k = 1, \dots, N, \end{aligned} \tag{10}$$

where κ is the modulus of the elastic forces and for each y, $b(y)$ is a matrix describing the intensities of the random forces. Assume that transformations f and b are Lipschitz continuous, then the system (10) determines uniquely processes $y_k(t)$, $k = 1, \dots, N$, $t \geq 0$, provided the initial conditions $y_k(0)$, $k = 1, \dots, N$ are given as well as the processes $y_0(t)$, $t \geq 0$, and $y_{N+1}(t)$, $t \geq 0$, which appear in the equations describing motion of the first and the Nth particles respectively. Let $\xi_k = \frac{k-1}{N-1}$, $k = 1, \dots, N$, and let $x(\xi)$, $\xi \in [0, 1]$, be a fixed continuous function with values in $\mathbb{R}^d$. We fix the initial conditions $y_1(0), \ldots, y_N(0)$ by requiring that

$$y_k(0) = x(\xi_k), \quad k = 1, \ldots, N.$$

The processes y_0 and y_{N+1} are determined by one of the following three boundary conditions

$$y_0(t) = y_1(t), \quad y_{N+1}(t) = y_N(t), \quad t \geq 0, \tag{11}$$

or

$$y_0(t) = y_1(t) = 0, \quad y_{N+1}(t) = y_N(t), \quad t \geq 0, \tag{12}$$

or

$$y_0(t) = y_1(t) = 0, \quad y_{N+1}(t) = y_N(t) = 0, \quad t \geq 0. \tag{13}$$

When considering cases (12) and (13) we will require also that $x_0(0) = 0$ and $x_0(0) = x_0(1) = 0$. The following process $X_N(t, \cdot)$, $t \geq 0$, with values in the function space $E = L^2(0, 1; \mathbb{R}^d)$ or in $E = C([0, 1]; \mathbb{R}^d)$

$$X_N(t, \xi) = y_k(t) + \frac{\xi - \xi_k}{\xi_{k+1} - \xi_k} y_{k+1}(t), \quad \xi \subset [\xi_k, \xi_{k+1}],\ k = 1, 2, \ldots, N-1,$$

can be regarded as a discrete approximation of the moving string. Let $\mathscr{L}(X_N)$ be its distribution on $C([0, T]; E)$ ($T > 0$ a fixed number) of the process X_N, $N = 2, 3, \ldots$ The following result is due to Funaki [331]. In its formulation

$$E^2 = \left\{ z \in W^{2,2}([0, 1]; \mathbb{R}^d) : \frac{d^2 z}{d\xi^2} \in E \right\},$$

E^2 stands for the Sobolev space of functions $x \in W^{2,2}([0, 1]; \mathbb{R}^d)$.

Theorem 0.1 (see [331]) *The sequence $\{\mathscr{L}(X_N)\}$ converges weakly on $C([0, T]; E)$ to $\mathscr{L}(X)$, where the process X is a solution of equation* (1) *and the operator $A = d^2/d\xi^2$ with the domain $D(A)$ is equal respectively to*

$$D(A) = E^2 \cap \left\{ z : \frac{dz}{d\xi}(0) = \frac{dz}{d\xi}(1) = 0 \right\},$$

or

$$D(A) = E^2 \cap \left\{ z : z(0) = \frac{dz}{d\xi}(1) = 0 \right\},$$

or

$$D(A) = E^2 \cap \{z : z(0) = z(1) = 0\},$$

according to which of the conditions (11), (12), (13) *are considered in the definition of X_N. The process $W(\cdot)$ is a cylindrical Wiener process on $U := L^2(0, 1; \mathbb{R}^1)$ with* *the identity covariance operator. Moreover*

$$F(z)(\xi) = f(z(\xi)), \quad z \in E, \ \xi \in [0,1],$$
$$(B(z)u)(\xi) = b(z(\xi))u(\xi), \quad z \in E, \ u \in U, \ \xi \in [0,1].$$

0.5 Stochastic equation of the free field

Hida and Streit showed in [403] that the equation

$$dX = -(\lambda - \Delta)^{1/2} X dt + dW,$$

where W is again a cylindrical Wiener process on the Hilbert space $E = U = L^2(D)$ with the covariance operator I, has a stationary solution corresponding to the Gaussian invariant measure with the covariance

$$C = (\lambda - \Delta)^{-1/2}.$$

This stationary solution can be interpreted as the Euclidean free field.

0.6 Equation of stochastic quantization

Let D be an arbitrary open subset of $\mathbb{R}^n$ and $r(u, v), \ u, v \in D$, a *positive definite*, continuous function. This means that r is a continuous function such that

$$\sum_{j,k=1}^{n} r(u_j, u_k)\lambda_j\lambda_k \geq 0, \quad \forall\, u_j, u_k \in D, \ \lambda_j, \lambda_k \in \mathbb{R}^1. \tag{14}$$

It is well known that positive definite, continuous functions are precisely the covariance functions of the mean-square continuous, *Gaussian random fields* $\{\xi_u, u \in D\}$, i.e.,

$$r(u, v) = \mathbb{E}(\xi_u \xi_v), \quad u, v \in D. \tag{15}$$

In (15), $\mathbb{E}$ stands for the expectation with respect to the probability measure $\mathbb{P}$ given on $(\Omega, \mathscr{F})$; we assume that $\mathbb{E}(\xi_u) = 0$, $u \in U$. On the other hand, let $H_m, \ m \in \mathbb{N}$, be Hermite polynomials

$$H_m(z) = m! \sum_{n+2k=m} (-1)^k \frac{z^n}{n!k!2^k}, \quad m \in \mathbb{N}, \ z \in \mathbb{R}^1 \tag{16}$$

given by the *generating function* formula

$$e^{\lambda z - \frac{1}{2}\lambda^2} = \sum_{m=0}^{\infty} \frac{\lambda^m}{m!} H_m(z), \quad \lambda, z \in \mathbb{R}^1.$$

Let E be a space of continuous functions defined on D. Given a positive definite, continuous function $r(\cdot,\cdot)$, the so called mth-*Wick power*: $x^m(z) : z \in D$, of an arbitrary function $x \in E$, is defined by the formula

$$: x^m(z) := \sqrt{r^m(z,z)}H_m\left(\frac{x(z)}{\sqrt{r(z,z)}}\right), \quad z \in D,\ x \in E. \tag{17}$$

It follows from (16) that

$$: x^m(z) := m! \sum_{n+2k=m} (-1)^k \frac{x^n(z) r^k(z,z)}{n!k!2^k}, \quad z \in D,\ x \in E. \tag{18}$$

It is therefore clear that for $x \in E$ the formulae (17) and (18) define continuous functions.

The *stochastic quantization equation* is of the form

$$dX = [AX- :X^m:]dt + dW, \tag{19}$$

where the term $:X^m:$, m an odd number, is a fairly irregular drift, called the "Wick power." The equation is important in statistical physics as the invariant measure ν for the solution is, up to a multiplicative constant, of the Gibbs form

$$\nu(dx) = e^{-\frac{1}{m+1}\int_D :x^{m+1}:d\xi}\mu(dx),$$

where μ is the invariant measure for the Ornstein–Uhlenbeck equation

$$dX = AX + dW.$$

W is a cylindrical Wiener process on $U = L^2(D)$, with the covariance I, A is a self-adjoint negative definite operator, generally unbounded on U, and m is an odd number.

To throw some light on the very definition of the Wick power and its connection with Gaussian random fields, assume that γ is a symmetric Gaussian measure on a Hilbert space H with the covariance operator $C = \frac{1}{2}A^{-1}$ of the integral type with the kernel function $r(\cdot,\cdot)$. The following two observations are essential.

Fact 1 Let x be a real valued, symmetric, Gaussian random variable defined on a probability space $(\Omega, \mathscr{F}, \mathbb{P})$ and let G be the σ- algebra generated by x. Then the random variables $(m!)^{-1/2}H_m(x)$, $m \in \mathbb{N}$, constitute a complete and orthonormal basis in $L^2(\Omega, \mathscr{F}, \mathbb{P})$. They are obtained from x^m, $m \in \mathbb{N}$, by the Hilbert–Schmidt orthogonalization procedure.

Fact 2 Let us assume that the random field $x(u), u \in D$, has continuous realizations belonging the the space E almost surely. Then the law γ of the random field $x(u), u \in D$, is a symmetric Gaussian measure on E equipped with its Borel σ-algebra B. Moreover the functionals $x(z)$, $z \in D$, defined for x in E, can be regarded as random variables on (E, B, γ). Therefore the family $x(z), z \in D$, constitutes a random field with the law γ and the covariance function $r(\cdot,\cdot)$. Consequently the families

$\{(x(z)^m, z \in D\}$, $m = 0, 1, \ldots$ and $\{: x(z)^m :, z \in D\}$ are random fields as well and the latter can be obtained from the former by the orthogonalization procedure sometimes called *renormalization*.

In applications, H is an appropriate Sobolev space and the covariance operator is $C = (\lambda - \Delta)^{-1}$, $\lambda \in \mathbb{R}^1$, where λ is a fixed number (in the definition of the Laplace operator one has to take into account boundary conditions). If the operator C obtained this way is nuclear then it determines a Gaussian, symmetric measure γ on H. If the measure is concentrated on the space E of continuous functions then the Wick powers have, according to the previous considerations, well defined meaning as well as the stochastic equations (19). However, in several situations of physical interest the measure γ is not concentrated on a space of continuous functions but on some spaces of distributions on D. Then additional care must be given to the nonlinear term in (19). For more details we refer to the papers [92, 288, 432].

In particular, let $D = [0, \pi]$, $H = L^2(D)$. The operator $\Delta = d^2/dz^2$, with the domain $D(\Delta) = H_0^1(0, \pi) \cap H^2(0, \pi)$, is negative and self-adjoint and the operators $R_\lambda = (\lambda - \Delta)^{-1}$, $\lambda \geq 0$, are nuclear, integral operators with kernels:

$$r_\lambda(u, v) = r_\lambda(v, u) = \frac{\sinh(\lambda^{1/2} v) \sinh(\lambda^{1/2}(\pi - u))}{\lambda^{1/2} \sinh(\lambda^{1/2}\pi)}, \qquad 0 \leq u \leq v \leq \pi, \lambda > 0$$

$$r_0(u, v) = r_0(v, u) = \frac{1}{\pi}(v(\pi - u)), \quad 0 \leq u \leq v \leq \pi.$$

One can easily show that stochastic Gaussian processes with covariance functions $r_\lambda(\cdot, \cdot)$, $\lambda > 0$, have continuous versions with values 0 at the boundary points 0 and π. Therefore their laws are Gaussian measures γ_λ, concentrated on the space $E = C_0([0, \pi])$. The measure γ_0 corresponding to $\lambda = 0$ is the law of a *Brownian bridge process*. In this case equation (19) is of the form

$$\begin{cases} dX(t, \xi) = \dfrac{d^2}{dz^2} X(t, \xi) - [X^3(t, \xi) - \dfrac{3}{\pi}(\xi(\pi - \xi))X(t, \xi)]dt + dW, \\ \xi \in [0, \pi], \quad t \geq 0 \\ X(t, 0) = X(t, \pi) = 0, \quad X(0, \cdot) = x \in C_0([0, \pi]). \end{cases} \tag{20}$$

Equations similar to (20) appear in the following two different contexts.

0.7 Reaction-diffusion equation

The deterministic reaction-diffusion equations are of the form

$$\frac{\partial u}{\partial t}(t, \xi) = \sigma^2 \frac{\partial^2 u}{\partial \xi^2}(t, \xi) + f(u(t, \xi)), \quad t \geq 0, \ \xi \in [0, T], \tag{21}$$

with appropriate boundary conditions. The function $(\partial u/\partial t)(t, \xi)$ is the so called *rate function* consisting of the gain term l and the loss term m. In the case of two initial and one final products: $l(u) = au^2 + g, m(u) = u^3 + bu, u \in \mathbb{R}^1$ and a, b and g are positive constants. The many-particle nature of a real system results in internal fluctuations. This leads to

$$\frac{\partial u}{\partial t}(t, \xi) = \sigma^2 \frac{\partial^2 u}{\partial \xi^2}(t, \xi) + f(u(t, \xi)) + \dot{W}(t, \xi), \quad t \geq 0, \ \xi \in [0, T], \tag{22}$$

where $\dot{W}$ is a temporal and spatial white noise. Equation (22) can be formulated as an equation of type (1). More information about the model can be found in [26] and [520].

0.8 An example arising in neurophysiology

Neurons or nerve cells can be regarded as long thin cylinders, which act much like electrical cables. Let us identify such a cylinder with the interval $[0, L]$. Let $V(t, \xi)$ be the electrical potential at the point ξ at time t. The potential satisfies a nonlinear parabolic equation coupled with a system of ordinary differential equations, called Hodgkin–Huxley equations. We will not write those equations here but indicate only two references, [401] and [540]. Nagumo's equation can be regarded as preliminary to the Hodgkin–Huxley system. In some ranges of the potential the Hodgkin-Huxley equation can be approximated by the cable equation:

$$\frac{\partial V}{\partial t}(t, \xi) = \frac{\partial^2 V}{\partial \xi^2}(t, \xi) - V(t, \xi) + \dot{W}(t, \xi), \quad t \geq 0, \ \xi \in [0, L], \tag{23}$$

where $\dot{W}(t, \xi)$ is the current arriving at ξ at moment t. It is reasonable to conjecture that impulses arrive according to a Poisson process, in both time and space variables. If the intensity of the impulses is high one can assume that $\dot{W}$ is of white noise type. We refer to [702] for more details. In this way one arrives again at a stochastic equation of type (1). More sophisticated arguments suggest that a nonlinear term $p(V)$, where p is a cubic polynomial, should be added to the right hand side of (23); see the paper by McKean [540].

0.9 Equation of population genetics

Stochastic semilinear equations have also been used in population genetics to model changes in the structure of a population in time and in space. In particular, Dawson [225] proposed the following equation

$$dp(t, \xi) = a\Delta p(t, \xi)dt + b\sqrt{p^+(t, \xi)}dW_t, \quad \xi \in \mathbb{R}^d, \tag{24}$$

for the mass distribution $p(t,\cdot)$ of the population at time $t \geq 0$. A slightly different equation

$$dp(t,\xi) = (\Delta p(t,\xi) + ap(t,\xi) - b)dt + b\left(\frac{1}{2}\, p^+(t,\xi)(1-p(t,\xi)^+)\right)^{1/2} dW_t, \quad \xi \in \mathbb{R}^d, \tag{25}$$

was proposed by Fleming [311]. In equations (24) and (25), a, b are positive constants and W is a U-valued Wiener process with a nuclear covariance operator Q. Both the equations are of type (1). Here $E = L^2(\mathbb{R}^d)$, $A = a\Delta$ or $A = \Delta + aI - c$, $D(A) = H^2(\mathbb{R}^d)$ and

$$(Bx)u(\xi) = b\left(x^+(\xi)\right)^{1/2} u(\xi)$$

or

$$(Bx)u(\xi) = b\left((1/2)x^+(\xi)(1-x(\xi))^+\right)^{1/2} u(\xi).$$

0.10 Musiela's equation of the bond market

Let $P(t,T),\ 0 \leq t \leq T$, denote the price, at moment t, of a bond which matures at $T \geq t$. It is often represented in the form

$$P(t,T) = e^{-\int_t^T f(t,s)ds},$$

where $f(t,s),\ 0 \leq t \leq s$, is a random field called the "forward rate." In the approach proposed by Heath, Jarrow and Morton [400], one assumes that for every T, the process $f(t,T),\ t \leq T$, has a representation of the form:

$$\begin{cases} df(t,x) = \alpha(t,T)dt + \sigma(t,T)dW(t), \\ f(0,T) = f_0(T), \end{cases} \tag{26}$$

where W is a Wiener process, for simplicity assume that it is one dimensional. Musiela [555] noticed that analysis of the forward rates is simplified if they are considered in the "moving frame" resulting in the new parameterization

$$f(t,t+\xi) = r(t,\xi), \quad t \geq 0,\ \xi \geq 0$$

with ξ being the "time to maturity." Then r satisfies the equation

$$dr(t,\xi) = \left(\frac{\partial r}{\partial \xi}(t,\xi) + \widetilde{\alpha}(t,\xi)\right) dt + \widetilde{\sigma}(t,\xi)dW_t$$

where

$$\widetilde{\alpha}(t,\xi) = \alpha(t,t+\xi), \quad \widetilde{\sigma}(t,\xi) = \sigma(t,t+\xi).$$

Assuming, for modeling purposes, that the "volatility" $\widetilde{\sigma}$ is a function of forward rates and requiring that the model is arbitrage free, one arrives at the equation

$$dr(t,\xi) = \left[\frac{\partial r}{\partial \xi}(t,\xi) + \left(\int_0^{\xi} \psi(r(t,\eta))d\eta\right) \psi(r(t,\xi))\right] dt + \psi(r(t,\xi))dW(t).$$

For more information see for example Filipović [293] and Barski and Zabczyk [49].

Part I

Foundations

1
Random variables

In this chapter we recall basic definitions and results from the measure theoretic foundations of probability theory to make the content of the book self contained. We also add less standard material on operator and Banach space valued random variables needed later.

1.1 Random variables and their integrals

A measurable space is a pair $(\Omega, \mathscr{F})$ where Ω is a nonempty set and $\mathscr{F}$ is a σ-field, also called a σ-algebra, of subsets of Ω. This means that the family $\mathscr{F}$ contains the set Ω and is closed under the operation of taking complements and countable unions of its elements. If $(\Omega, \mathscr{F})$ and $(E, \mathscr{G})$ are two measurable spaces, then a mapping X from Ω into E such that the set $\{\omega \in \Omega : X(\omega) \in A\} = \{X \in A\}$ belongs to $\mathscr{F}$ for arbitrary $A \in \mathscr{G}$ is called a *measurable mapping* or a *random variable* from $(\Omega, \mathscr{F})$ into $(E, \mathscr{G})$ or an E*-valued random variable*. A random variable is called *simple* if it takes on only a finite number of values.

Assume that E is a metric space, then the *Borel σ-field* of E is the smallest σ-field containing all closed (or open) subsets of E; it will be denoted as $\mathscr{B}(E)$.

If E is a Banach space we shall denote its norm by $\|\cdot\|$ and its topological dual by E^*. Given $x \in E$ and $r > 0$ we set

$$B(x,r) = \{a \in E : \|x - a\| < r\}, \quad \overline{B}(x,r) = \{a \in E : \|x - a\| \le r\}.$$

Proposition 1.1 *Let E be a separable Banach space. Then $\mathscr{B}(E)$ is the smallest σ-field of subsets of E containing all sets of the form*

$$\{x \in E : \varphi(x) \le \alpha\}, \quad \varphi \in E^*, \ \alpha \in \mathbb{R}^1. \tag{1.1}$$

Proof Since E is separable, there exists a sequence $\{\varphi_n\}_{n\in\mathbb{N}} \subset E^*$ such that [(1)]

$$\|x\| = \sup_{n\in\mathbb{N}} |\varphi_n(x)|, \quad x \in E. \tag{1.2}$$

[(1)] $\mathbb{N}$ denotes the set of all natural numbers 1,2,3, ...

Consequently, for arbitrary $a \in E, r \geq 0$

$$B(a,r) = \bigcup_{m=1}^{\infty} \overline{B}\,(a, r(1-1/m))$$
$$= \bigcup_{m=1}^{\infty} \bigcap_{n=1}^{\infty} \{x \in E : |\varphi_n(x-a)| \leq r\,(1-1/m)\}\,.$$

This implies that the smallest σ-field containing all sets of form (1.1) contains open balls of E, and thus it coincides with $\mathscr{B}(E)$. □

Proposition 1.2 *Let E be a separable Banach space and let X and Y be E-valued random variables defined on $(\Omega, \mathscr{F})$. Then*

(i) $\alpha X + \beta Y$ is an E-valued random variable for any $\alpha, \beta \in \mathbb{R}$,
(ii) $\|X(\cdot)\|$ is a real valued random variable.

Proof (i) We only show that $X + Y$ is a random variable. For this it is enough to prove that for an arbitrary open set $A \subset E$ we have

$$\{\omega : \; X(\omega) + Y(\omega) \in A\} \in \mathscr{F}.$$

Let $\{x_n\} \subset E$ be a dense countable subset of E and

$$I = \{(n,m,r) : \; n, m \in \mathbb{N}, \; r \in \mathbb{Q}^+ : \; B(x_n, r) + B(x_m, r) \subset A\}.$$

Then

$$\{\omega : \; X(\omega) + Y(\omega) \in A\}$$
$$= \bigcup_{(n,m,r)\in I} \Big[\{\omega : \; X(\omega) \in B(x_n, r)\} \cup \{\omega : \; Y(\omega) \in B(x_m, r)\}\Big],$$

and the result follows.

(ii) Since E is separable, there exists a sequence $\{\varphi_n\} \subset E^*$ such that (1.2) holds. Consequently

$$\|X(\omega)\| = \sup_{n \in N} |\varphi_n(X(\omega))|\,, \qquad \forall\, \omega \in \Omega,$$

so $\|X(\cdot)\|$ is a real valued random variable. □

We shall need the following result.

Lemma 1.3 *Let E be a separable metric space with metric ρ and let X be an E-valued random variable. Then there exists a sequence $\{X_m\}$ of simple E-valued random variables such that, for arbitrary $\omega \in \Omega$, the sequence $\{\rho(X(\omega), X_m(\omega))\}$ is monotonically decreasing to 0.*

Proof Let $E_0 = \{e_k\}_{k\in\mathbb{N}}$ be a countable dense subset of E. For $m \in \mathbb{N}$ define for $\omega \in \Omega$,

$$\begin{aligned} \rho_m(\omega) &= \min\{\rho(X(\omega), e_k),\ k = 1, \dots, m\} \\ k_m(\omega) &= \min\{k \le m :\ \rho_m(\omega) = \rho(X(\omega), e_k)\} \\ X_m(\omega) &= e_{k_m(\omega)}. \end{aligned}$$

Obviously X_m are simple random variables since

$$X_m(\Omega) \subset \{e_1, e_2, \dots, e_m\}.$$

Moreover, by the density of E_0, the sequence $\{\rho_m(\omega)\}$ is monotonically decreasing to 0 for arbitrary $\omega \in \Omega$. Since $\rho_m(\omega) = \rho(X(\omega), X_m(\omega))$, the conclusion follows. □

Let $\mathscr{K}$ be a collection of subsets of Ω. The smallest σ-field on Ω which contains $\mathscr{K}$ is denoted by $\sigma(\mathscr{K})$ and is called the *σ-field generated by* $\mathscr{K}$. It is also the intersection of all σ-fields which contain $\mathscr{K}$. Analogously, let $\{X_i\}_{i\in I}$ be a family of mappings from Ω into E. Then the smallest σ-field $\sigma(X_i : i \in I)$ on Ω such that all functions X_i are measurable from $(\Omega, \sigma(X_i : i \in I))$ into $(E, \mathscr{G})$ is called the σ-field *generated by* $\{X_i\}_{i\in I}$.

If $\mathscr{G} = \sigma(\mathscr{K})$ and for arbitrary $A \in \mathscr{K}$

$$\{\omega \in \Omega :\ X(\omega) \in A\} \in \mathscr{F}, \tag{1.3}$$

then X is a measurable mapping from $(\Omega, \mathscr{F})$ into $(E, \mathscr{G})$. This is because the family of all sets $A \in \mathscr{G}$ for which (1.3) holds is a σ-field.

A collection $\mathscr{K}$ of subsets of Ω is said to be a *π-system* if $\varnothing \in \mathscr{K}$ and if $A, B \in \mathscr{K}$ then $A \cap B \in \mathscr{K}$.

The following proposition, due to Dynkin [274], will often be used for proving that a given mapping or a given set is measurable.

Proposition 1.4 *Assume that $\mathscr{K}$ is a π-system and let $\mathscr{G}$ be the smallest family of subsets of Ω such that*

(i) $\mathscr{K} \subset \mathscr{G}$,
(ii) if $A \in \mathscr{G}$ then $A^c \in \mathscr{G}$, [2]
(iii) if $A_i \in \mathscr{G}$ for all $i \in \mathbb{N}$ and $A_n \cap A_m = \varnothing$ for $n \neq m$, then $\bigcup_{n=1}^{\infty} A_n \in \mathscr{G}$.

Then $\mathscr{G} = \sigma(\mathscr{K})$.

Proof Since $\sigma(\mathscr{K})$ satisfies (i), (ii) and (iii), we have $\mathscr{G} \subset \sigma(\mathscr{K})$. To prove the opposite inclusion it suffices to show that $\mathscr{G}$ is a π-system, because it can be easily proved that if a π-system $\mathscr{G}$ satisfies (ii) and (iii), then it is a σ-field. Let $B \in \mathscr{G}$ and define

$$\mathscr{G}_B = \{A \in \mathscr{G} :\ A \cap B \in \mathscr{G}\}.$$

[2] For any set $A \subset \Omega$, A^c denotes its complement.

$\mathscr{G}_B$ satisfies (ii) since if $B \in \mathscr{G}$ and $A \cap B \in \mathscr{G}$ then $A \cap B^c = A \cap (A \cap B)^c \in \mathscr{G}$. Moreover $\mathscr{G}_A$ obviously satisfies (iii) and if $A \in \mathscr{K}$ then the condition (i) is also satisfied. Thus, for $A \in \mathscr{K}$, $\mathscr{G}_A = \mathscr{G}$ and we have proved that if $A \in \mathscr{K}$ and if $B \in \mathscr{G}$ then $A \cap B \in \mathscr{G}$. But this implies $\mathscr{G}_B \supset \mathscr{K}$ and consequently $\mathscr{G}_B = \mathscr{G}$ for any $B \in \mathscr{G}$. □

By Proposition 1.1 it follows that if E is a separable Banach space, then a mapping $X : \Omega \to E$ is an E-valued random variable if and only if for arbitrary $\varphi \in E^*$, $\varphi(X) : \Omega \to \mathbb{R}^1$ is an $\mathbb{R}^1$-valued random variable.

A *probability measure* on a measurable space $(\Omega, \mathscr{F})$ is a σ-additive function $\mathbb{P}$ from $\mathscr{F}$ into $[0, 1]$ such that $\mathbb{P}(\Omega) = 1$. The triplet $(\Omega, \mathscr{F}, \mathbb{P})$ is called a *probability space*. If $(\Omega, \mathscr{F}, \mathbb{P})$ is a probability space, we set

$$\overline{\mathscr{F}} = \{A \subset \Omega : \exists\, B, C \in \mathscr{F}\,;\, B \subset A \subset C,\ \mathbb{P}(B) = \mathbb{P}(C)\}.$$

$\overline{\mathscr{F}}$ is a σ-field, called the *completion* of $\mathscr{F}$. If $\mathscr{F} = \overline{\mathscr{F}}$, then the probability space $(\Omega, \mathscr{F}, \mathbb{P})$ is said to be *complete*.

A measure on $(\Omega, \mathscr{F})$ is determined by its values on an arbitrary π-system $\mathscr{K}$ which generates $\mathscr{F}$. We have in fact the following result.

Proposition 1.5 *Let $\mathbb{P}_1$ and $\mathbb{P}_2$ be probability measures on $(\Omega, \mathscr{F})$ and let $\mathscr{K}$ be a π-system such that $\sigma(K) = \mathscr{F}$. If $\mathbb{P}_1 = \mathbb{P}_2$ on $\mathscr{K}$ then $\mathbb{P}_1 = \mathbb{P}_2$ on $\mathscr{F}$.*

Proof It suffices to check that, if we set

$$\mathscr{G} = \{F \in \mathscr{F} :\ \mathbb{P}_1(F) = \mathbb{P}_2(F)\},$$

then $\mathscr{G}$ satisfies properties (i), (ii) and (iii) of Proposition 1.4. □

If X is a random variable from $(\Omega, \mathscr{F})$ into $(E, \mathscr{E})$ and $\mathbb{P}$ a probability measure on Ω then by $\mathscr{L}(X)$ we will denote the image of $\mathbb{P}$ by the mapping X

$$\mathscr{L}(X)(A) = \mathbb{P}(\omega \in \Omega :\ X(\omega) \in A), \quad \forall\, A \in \mathscr{E}. \tag{1.4}$$

The measure $\mu = \mathscr{L}(X)$ is called the *distribution* or the *law* of X.

We take for granted the definition of the Lebesgue integral of real valued random variables. If X is a real valued random variable then its integral

$$\int_\Omega X(\omega)\mathbb{P}(d\omega)$$

will also be denoted by $\mathbb{E}(X)$ or $\mathbb{E}X$.

Assume now that E is a separable Banach space; Lemma 1.3 allows us to define the integral for an E-valued random variable on $(\Omega, \mathscr{F})$. For a simple random variable X

$$X = \sum_{i=1}^{N} x_i 1\!\!1_{A_i}, \quad A_i \in \mathscr{F},\ x_i \in E,\ N \in \mathbb{N},$$

we set

$$\int_B X(\omega)\mathbb{P}(d\omega) = \int_B X d\mathbb{P} = \sum_{i=1}^{N} x_i \, \mathbb{P}(A_i \cap B), \tag{1.5}$$

for all $B \in \mathscr{F}$. Here $1\!\!1_{A_i}$ denotes the indicator function of A_i.[3] One can check easily that the above definition does not depend on the choice of the particular representation of X. Moreover the usual properties of additivity and linearity of the integral hold true and

$$\left\| \int_B X(\omega)\mathbb{P}(d\omega) \right\| \le \int_B \|X(\omega)\| \, \mathbb{P}(d\omega). \tag{1.6}$$

We want now to define the integral for general random variables. The random variable X is said to be *Bochner integrable* or *shortly integrable* if

$$\int_\Omega \|X(\omega)\| \, \mathbb{P}(d\omega) < \infty. \tag{1.7}$$

Note that, in view of Proposition 1.2, we know that $\|X\|$ is a random variable, so the Lebesgue integral in (1.7) is well defined. Let X be Bochner integrable; by Lemma 1.3 there exists a sequence $\{X_m\}$ of simple random variables such that the sequence $\{\|X(\omega) - X_m(\omega)\|\}$ decreases to 0 for all $\omega \in \Omega$. It follows that

$$\left\| \int_\Omega X_m(\omega)\mathbb{P}(d\omega) - \int_\Omega X_n(\omega)\mathbb{P}(d\omega) \right\|$$
$$\le \int_\Omega \|X(\omega) - X_m(\omega)\| \, \mathbb{P}(d\omega) + \int_\Omega \|X(\omega) - X_n(\omega)\| \, \mathbb{P}(d\omega) \downarrow 0$$

as $m, n \to \infty$. Therefore the integral of X can be defined by

$$\int_\Omega X(\omega)\mathbb{P}(d\omega) = \lim_{n\to\infty} \int_\Omega X_n(\omega)\mathbb{P}(d\omega).$$

We will get the same limit if we approximate X by an arbitrary sequence $\{X_n\}$ of simple random variables satisfying

$$\int_\Omega \|X - X_n\| \, d\mathbb{P} \to 0$$

as $n \to \infty$. The integral $\int_\Omega X d\mathbb{P}$ will often be denoted by $\mathbb{E}(X)$. The integral defined this way is called *Bochner's integral* and has many properties of the Lebesgue integral. In particular the estimate (1.6) is valid for all random variables satisfying (1.7). We use the same notation for both integrals. Here are some other useful properties of the integral.

Let X be a random variable in $(\Omega, \mathscr{F})$ with values in a measurable space $(G, \mathscr{G})$ and let $\mu = \mathscr{L}(X)$ be its law. If ψ is a measurable mapping from $(G, \mathscr{G})$ into $(E, \mathscr{B}(E))$

[3] For any set $I \in \Omega$ we define $1\!\!1_I(\omega)$ to be equal to 1 if $\omega \in I$ and equal to 0 if $\omega \notin I$.

integrable with respect to μ then, by a standard limit argument, one obtains

$$\mathbb{E}(\psi(X)) = \int_E \psi(x)\mu(dx). \tag{1.8}$$

If $(\Omega_1, \mathscr{F}_1)$ and $(\Omega_2, \mathscr{F}_2)$ are two measurable spaces then we denote by $\mathscr{F}_1 \times \mathscr{F}_2$ the smallest σ-field of subsets of $\Omega_1 \times \Omega_2$ containing all sets of the form $A_1 \times A_2$ with $A_1 \in \mathscr{F}_1, A_2 \in \mathscr{F}_2$.

If $\mathbb{P}_1$ and $\mathbb{P}_2$ are nonnegative measures on $(\Omega_1, \mathscr{F}_1)$ and $(\Omega_2, \mathscr{F}_2)$ respectively, then $\mathbb{P}_1 \times \mathbb{P}_2$ is a measure on $(\Omega_1 \times \Omega_2, \mathscr{F}_1 \times \mathscr{F}_2)$ such that

$$(\mathbb{P}_1 \times \mathbb{P}_2)(A_1 \times A_2) = \mathbb{P}_1(A_1)\mathbb{P}_2(A_2), \quad A_1 \in \mathscr{F}_1, A_2 \in \mathscr{F}_2.$$

If E is a separable Banach space and $X \colon \Omega_1 \times \Omega_2 \to E$ is a random variable from $(\Omega_1 \times \Omega_2, \mathscr{F}_1 \times \mathscr{F}_2)$ into $(E, \mathscr{B}(E))$ then, by the classical Fubini theorem (see e.g. [559]), for all $\varphi \in E^*$, $\omega_1 \in \Omega_1$, $\omega_2 \in \Omega_2$ the functions $\varphi(X(\omega_1, \cdot))$(respectively $\varphi(X(\cdot, \omega_2)))$ are measurable mappings from $(\Omega_2, \mathscr{F}_2)$ into $(\mathbb{R}^1, \mathscr{B}(\mathbb{R}^1))$ (respectively from $(\Omega_1, \mathscr{F}_1)$ into $(\mathbb{R}^1, \mathscr{B}(\mathbb{R}^1)))$. Consequently also the transformations $X(\omega_1, \cdot) \colon \Omega_2 \to E$ and $X(\omega_2, \cdot) \colon \Omega_1 \to E$ are measurable mappings with respect to the appropriate σ-fields. By a similar argument, if $X \colon \Omega_1 \times \Omega_2 \to E$ is, in addition, Bochner integrable:

$$\int_{\Omega_1 \times \Omega_2} \|X(\omega_1, \omega_2)\| (\mathbb{P}_1 \times \mathbb{P}_2)(d\omega_1 \times \omega_2) < +\infty,$$

then

$$\begin{aligned}
&\int_{\Omega_1 \times \Omega_2} X(\omega_1, \omega_2)(\mathbb{P}_1 \times \mathbb{P}_2)(d\omega_1 \times \omega_2) \\
&= \int_{\Omega_2} \int_{\Omega_1} X(\omega_1, \omega_2)\mathbb{P}_1(d\omega_1)\mathbb{P}_2(d\omega_2) \\
&= \int_{\Omega_1} \int_{\Omega_2} X(\omega_1, \omega_2)\mathbb{P}_2(d\omega_2)\mathbb{P}_1(d\omega_1).
\end{aligned} \tag{1.9}$$

We will need the following.

Proposition 1.6 *Assume that E and F are separable Banach spaces and $A \colon D(A) \subset E \to F$ is a closed operator with the domain $D(A)$ a Borel subset of E. If $X \colon \Omega \to E$ is a random variable such that $X(\omega) \in D(A)$, $\mathbb{P}$-almost surely (a.s.), then AX is an F-valued random variable, and X is a $D(A)$-valued random variable, where $D(A)$ is endowed with the graph norm of A.* [(4)] *If moreover*

$$\int_\Omega \|AX(\omega)\| \mathbb{P}(d\omega) < +\infty, \tag{1.10}$$

(4) If A is closed, the graph norm on $D(A)$ is defined as: for any $x \in D(A)$, $\|x\|_{D(A)} := \|x\|_E + \|Ax\|_F$. Then $(D(A), \|\cdot\|_{D(A)})$ is a Banach space.

then

$$A \int_\Omega X(\omega)\mathbb{P}(d\omega) = \int_\Omega AX(\omega)\mathbb{P}(d\omega). \tag{1.11}$$

Proof We only prove the last statement. By Lemma 1.3 there exists a sequence $\{X_m\}$ of simple $D(A)$-valued random variables, such that

$$\begin{aligned}
&\|X_m(\omega) - X(\omega)\|_{D(A)} \\
&\quad = \|X_m(\omega) - X(\omega)\| + \|AX_m(\omega) - AX(\omega))\| \downarrow 0 \quad \text{as } m \to \infty.
\end{aligned}$$

Consequently

$$\int_\Omega \|X_m(\omega) - X(\omega)\|\mathbb{P}(d\omega) + \int_\Omega \|AX_m(\omega) - AX(\omega)\|\mathbb{P}(d\omega) \downarrow 0$$

as well. Then

$$\int_\Omega X_m(\omega)\mathbb{P}(d\omega) \to \int_\Omega X(\omega)\mathbb{P}(d\omega),$$

$$\int_\Omega AX_m(\omega)\mathbb{P}(d\omega) \to \int_\Omega AX(\omega)\mathbb{P}(d\omega).$$

But

$$\int_\Omega AX_m(\omega)\mathbb{P}(d\omega) = A \int_\Omega X_m(\omega)\mathbb{P}(d\omega)$$

from the very definition of the integral, and therefore, by the closedness of A,

$$\int_\Omega AX(\omega)\mathbb{P}(d\omega) = A \int_\Omega X(\omega)\mathbb{P}(d\omega).$$

□

Finally we introduce some basic function spaces. We denote by $L^1(\Omega, \mathscr{F}, \mathbb{P}; E)$ the set of all equivalence classes of E-valued random variables (with respect to the equivalence relation $X \sim Y \Leftrightarrow X = Y$ a.s.). In the same way as for real random variables, one can check that $L^1(\Omega, \mathscr{F}, \mathbb{P}; E)$, equipped with the norm

$$\|X\|_1 = \mathbb{E}(\|X\|), \tag{1.12}$$

is a Banach space. In a similar way one can define $L^p(\Omega, \mathscr{F}, \mathbb{P}; E)$, for arbitrary $p > 1$ with norms

$$\|X\|_p = (\mathbb{E}\|X\|^p)^{1/p}, \quad p \in [1, +\infty),$$

and

$$\|X\|_\infty = \operatorname*{ess.\,sup}_{\omega\in\Omega} \|X(\omega)\|.$$

If Ω is an interval (a, b), $\mathscr{F} = \mathscr{B}((a, b))$ and $\mathbb{P}$ is the Lebesgue measure on (a, b), we write $L^p(a, b; E)$.

We will often implicitly use the following fact.

Proposition 1.7 *Assume that E is a separable Banach space, then for arbitrary $T > 0$ and $r \geq p \geq 1$ the spaces $L^r(0, T; E)$ and $C([0, T]; E)$ are Borel subsets of $L^p(0, T; E)$.*

Proof (i) Assume that $r \geq p \geq 1$. Then for arbitrary $f \in L^p(0, T; E)$ we have $\|f\|_p \leq T^{\frac{r-p}{pr}} \|f\|_r$, where $\|f\|_r = +\infty$ if $f \notin L^p(0, T; E)$. It is easy to see that for arbitrary $n \in \mathbb{N}$ the set

$$\Gamma_n = \{f \in L^p(0, T; E) : \ \|f\|_r \leq n\}$$

is closed in $L^p(0, T; E)$. Since we have

$$L^r(0, T; E) = \bigcup_{n \in \mathbb{N}} \Gamma_n,$$

we see that $L^r(0, T; E)$ is a countable sum of closed sets of $L^p(0, T; E)$ and so $L^r(0, T; E)$ is a Borel subset of $L^p(0, T; E)$.

(ii) We show first that the space

$$C_0([0, T]; E) = \{f \in C([0, T]; E) : \ f(0) = 0\}$$

is a Borel subset of $L^p(0, T; E)$.

For arbitrary $f \in L^p(0, T; E)$ or $f \in C([0, T]; E)$ we extend f to the whole $\mathbb{R}^1$ by setting $f(t) = 0$ for all $t \notin [0, T]$. Since for arbitrary $\alpha \geq 0$ the set

$$\{f \in L^p(0, T; E) : \ \|f\|_\infty \leq \alpha\}$$

is closed, we see that the function

$$L^p(0, T; E) \to \mathbb{R}, \quad f \to \|f\|_\infty$$

is Borel. Now for arbitrary $n \in \mathbb{N}$ and $f \in L^p$ define

$$\Phi_n f(t) = n \int_{t-\frac{1}{n}}^{t} f(s) ds, \quad t \in [0, T].$$

Then $\Phi_n f \in C_0([0, T]; E)$ and we can show easily that

$$\Phi_n f \to f \quad \text{as } n \to \infty \text{ in } L^p(0, T; E), \quad \text{for } f \in L^p(0, T; E),$$

$$\Phi_n f \to f \quad \text{as } n \to \infty \text{ in } C_0([0, T]; E), \quad \text{for } f \in C_0([0, T]; E).$$

Since

$$C_0([0, T]; E) = \left\{f \in L^p(0, T; E) : \ \lim_{n \to \infty} \|\Phi_n f - f\|_\infty = 0\right\},$$

and since for each $n \in \mathbb{N}$, $f \to \|\Phi_n f - f\|_\infty$ is a measurable function on $L^p(0, T; E)$, the proof that $C_0([0, T]; E)$ is a Borel subset of $L^p(0, T; E)$ follows.

To complete the proof define a sequence $\{\varphi_n\}$ of continuous linear functionals on $L^p(0, T; E)$ setting

$$\varphi_n(f) = \frac{1}{n}\int_0^{1/n} f(s)ds.$$

The mapping φ_∞ defined as

$$\varphi_\infty(f) = \begin{cases} \lim_{n\to\infty} \varphi_n(f) & \text{if the limit exists and is finite} \\ 0 & \text{otherwise,} \end{cases}$$

is measurable. Moreover, for all $f \in C([0, T]; E)$, we have $\varphi_\infty(0) = f(0)$. Now it is easy to see that the mapping

$$L^p(0, T; E) \to L^p(0, T; E), \;\; f \to \varphi_\infty(f),$$

is measurable. Since we have

$$C([0, T]; E) = \{f \in L^p(0, T; E) : \;\; f - \varphi_\infty(f) \in C_0([0, T]; E)\},$$

the proof is complete. □

1.2 Operator valued random variables

Of great interest to us will be operator valued random variables and their integrals. Let U and H be two separable Hilbert spaces and denote by $L = L(U, H)$ the set of all linear bounded operators from U into H. The set L is a linear space and, equipped with the operator norm, becomes a Banach space. However, if both spaces are infinite dimensional, then L is not a separable space. To see this we can assume that $U = H = L^2(\mathbb{R}^1)$. Define for arbitrary $t \in \mathbb{R}^1$ the isometry $S(t)$ from H onto H

$$S(t)x(z) = x(z + t), \quad z \in \mathbb{R}^1, \; x \in H. \tag{1.13}$$

Assume that $t > s, \;\; x \in H$, then

$$|(S(t) - S(s))x| = |S(s)(S(t - s)x - x)| = |S(t - s)x - x|.$$

If $x \in H$ has support in the interval $((s - t)/2, (t - s)/2)$ then the functions x and $S(t - s)x$ have disjoint supports and therefore $|(S(t) - S(s))x|^2 = 2|x|^2$. Consequently $\|S(t) - S(s)\| \;\geq\!\sqrt{2}$ and $L = L(H, H)$ is not separable.

The nonseparability of L has several consequences. First of all the corresponding Borel σ-field $\mathscr{B}(L)$ is very rich, to the extent that very simple L-valued functions turn out to be nonmeasurable. In particular, the function $S(\cdot)$ defined by (1.13), considered as a mapping from $(\mathbb{R}^1, \mathscr{B}(\mathbb{R}^1))$ into $(L, \mathscr{B}(L))$, is not measurable. To see this fix a non-Borel subset Γ of $\mathbb{R}^1$ and define an open subset D of L by the formula

$$D = \{G \in L : \|G - S(t)\| < \sqrt{2}/2, \;\text{ for some } t \in \Gamma\}.$$

Since $\{t \in \mathbb{R}^1 : \;\; S(t) \in D\} = \Gamma$, therefore $S(\cdot)$ cannot be measurable.

The lack of separability of L implies also that Bochner's definition cannot be applied directly to the L-valued functions. To overcome these difficulties it is convenient to introduce a weaker concept of measurability.

A function $\Phi(\cdot)$ from Ω into L is said to be *strongly measurable* if for arbitrary $u \in U$ the function $\Phi(\cdot)u$ is measurable as a mapping from $(\Omega, \mathscr{F})$ into $(H, \mathscr{B}(H))$. Let $\mathscr{L}$ be the smallest σ-field of subsets of L containing all sets of the form

$$\{\Phi \in L : \ \Phi u \in A\}, \quad u \in U, \ A \in \mathscr{B}(H),$$

then $\Phi\colon \Omega \to L$ is a strongly measurable mapping from $(\Omega, \mathscr{F})$ into $(L, \mathscr{L})$. Elements of $\mathscr{L}$ are called *strongly measurable*. If $\mathbb{P}$ is a nonnegative (and not necessarily a normalized) measure on $\mathscr{F}$ then Φ is said to be *Bochner integrable* if for arbitrary u the function $\Phi(\cdot)u$ is Bochner integrable and there exists a bounded linear operator $\Psi \in L(U, H)$ such that

$$\int_{\Omega} \Phi(\omega)u \ \mathbb{P}(d\omega) = \Psi u, \quad \forall\, u \in U.$$

The operator Ψ is then denoted as

$$\Psi = \int_{\Omega} \Phi(\omega)\mathbb{P}(d\omega)$$

and is called the *strong Bochner integral* of Φ. It is an easy exercise to show that if U and H are separable then $\|\Phi(\cdot)\|$ is a measurable function and

$$\|\Psi\| \le \int_{\Omega} \|\Phi(\omega)\|\mathbb{P}(d\omega).$$

It is obvious that the function $S(\cdot)$ defined by (1.13) is Bochner integrable over arbitrary finite interval $[0, T]$ and that

$$\left\| \int_0^T S(t)dt \right\| \le \int_0^T \|S(t)\|dt.$$

If we restrict our investigation to smaller spaces – the space $L_1(U, H)$ of all nuclear operators from U into H or the space $L_2(U, H)$ of all Hilbert–Schmidt operators from U into H – then the nonmeasurability problem mentioned above does not appear. See Appendix C for basic definitions and properties of nuclear and Hilbert–Schmidt operators. This is because the spaces $L_1(U, H)$ and $L_2(U, H)$ are separable Banach spaces ($L_2(U, H)$ is a Hilbert space). *It is useful to note that $L_1(U, H)$ and $L_2(U, H)$ are strongly measurable subsets of $L(U, H)$.*

We will need the following result on measurable decomposition of a $L_1(U, U) = L_1(U)$ valued random variable.

Proposition 1.8 *Let U be a separable Hilbert space and assume that Φ is an $(L_1(U), \mathscr{B}(L_1(U)))$ random variable on $(\Omega, \mathscr{F})$ such that $\Phi(\omega)$ is a nonnegative symmetric operator for all $\omega \in \Omega$. Then there exists a decreasing sequence $\{\lambda_n\}$ of* *nonnegative random variables and a sequence* $\{g_n\}$ *of* U*-valued random variables such that* [5]

$$\Phi(\omega) = \sum_{n=1}^{\infty} \lambda_n(\omega) g_n(\omega) \otimes g_n(\omega), \tag{1.14}$$

for $\omega \in \Omega$. *Moreover the sequences* $\{\lambda_n\}$ *and* $\{g_n\}$ *can be chosen in such a way that*

$$|g_n(\omega)| = \begin{cases} 1 & \text{if } \lambda_n(\omega) > 0 \\ 0 & \text{if } \lambda_n(\omega) = 0 \end{cases} \tag{1.15}$$

and

$$\langle g_n(\omega), g_m(\omega)\rangle = 0, \quad \forall\, n \neq m \text{ and } \forall\, \omega \in \Omega. \tag{1.16}$$

Note that, since for each ω the operator $\Phi(\omega)$ is compact and nonnegative, there exists a decreasing sequence $\{\lambda_n(\omega)\}$ of real numbers and a sequence $\{g_n(\omega)\}$ in U such that (1.15)–(1.16) hold. It remains to show the measurability of λ_n and g_n. This is provided by the following classical result of Kuratowski and Ryll-Nardzewski [480] on measurable selectors.

Lemma 1.9 *Let* E *be a compact metric space and let* $\psi : E \times \Omega \to \mathbb{R}^1$ *be a mapping such that* $\psi(x, \cdot)$ *is measurable for arbitrary* $x \in E$ *and* $\psi(\cdot, \omega)$ *is a continuous mapping for arbitrary* $\omega \in \Omega$. *Then there exists an* E*-valued random variable* $X : \Omega \to E$ *such that*

$$\psi(X(\omega), \omega) = \sup_{x \in E} \psi(x, \omega), \quad \omega \in \Omega. \tag{1.17}$$

Proof of Proposition 1.8 We apply Lemma 1.9 to the set $E = \{x \in U : |x|_U \leq 1\}$ endowed with the weak topology and the function

$$\psi(x, \omega) = \langle \Phi(\omega)x, x\rangle, \quad x \in E,\ \omega \in \Omega.$$

All conditions of the lemma are satisfied and therefore there exists an E-valued random variable which we denote by g_1 such that

$$\langle \Phi(\omega)g_1(\omega), g_1(\omega)\rangle = \sup_{|x|\leq 1} \langle \Phi(\omega)x, x\rangle = \lambda_1(\omega), \quad \omega \in \Omega.$$

With the random variables g_1 and λ_1 defined in this way, we repeat the same procedure for the new operator valued random variable $\Phi_1(\omega) = \Phi(\omega) - \lambda_1(\omega)g_1(\omega) \otimes g_1(\omega)$, $\omega \in \Omega$, to define g_2 and λ_2, and by induction to define the sequences $\{g_n\}$, $\{\lambda_n\}$. By standard results on nonnegative compact operators, the sequences $\{g_n\}$ and $\{\lambda_n\}$ have the required properties. □

If $X, Y \in L^2(\Omega, \mathscr{F}, \mathbb{P}; H)$ and H is a Hilbert space, with inner product $\langle \cdot, \cdot \rangle$, we define the *covariance operator* of X and the *correlation operator* of (X, Y) by the

[5] For arbitrary $a, b \in H$ we denote by $a \otimes b$ the linear operator defined by $(a \otimes b)h = a\langle b, h\rangle$, $h \in H$.

formulae

$$\mathrm{Cov}(X) = \mathbb{E}\left(X - \mathbb{E}(X)) \otimes (X - \mathbb{E}(X))\right),$$

and

$$\mathrm{Cor}(X, Y) = \mathbb{E}[(X - \mathbb{E}(X)) \otimes (Y - \mathbb{E}(Y))].$$

$\mathrm{Cov}(X)$ is a symmetric positive and nuclear operator (see Appendix C) and

$$\mathrm{Tr}[\mathrm{Cov}(X)] = \mathbb{E}\left(|X - \mathbb{E}(X)|^2\right).$$

In fact if $\{e_k\}$ is a complete orthonormal basis in H and, for simplicity, $\mathbb{E}(X) = 0$, we have

$$\begin{aligned} \mathrm{Tr}[\mathrm{Cov}(X)] &= \sum_{h=1}^{\infty} \langle \mathrm{Cov}(X) e_h, e_h \rangle \\ &= \sum_{h=1}^{\infty} \int_{\Omega} |\langle X(\omega), e_h \rangle|^2 \, \mathbb{P}(d\omega) = \mathbb{E}|X|^2. \end{aligned}$$

1.3 Conditional expectation and independence

We have the following important result.

Proposition 1.10 *Assume that E is a separable Banach space. Let X be a Bochner integrable E-valued random variable defined on $(\Omega, \mathscr{F}, \mathbb{P})$ and let $\mathscr{G}$ be a σ-field contained in $\mathscr{F}$. There exists a unique, up to a set of $\mathbb{P}$-probability zero, integrable E-valued random variable Z, measurable with respect to $\mathscr{G}$ such that*

$$\int_A X d\mathbb{P} = \int_A Z d\mathbb{P}, \quad \forall\, A \in \mathscr{G}. \tag{1.18}$$

The random variable Z will be denoted as $\mathbb{E}(X|\mathscr{G})$ and called the conditional expectation of X given $\mathscr{G}$.

Proof We will first show uniqueness. Assume that there exist two random variables Z and $\widetilde{Z}$ having the required properties and such that $\mathbb{P}(Z - \widetilde{Z} \neq 0) > 0$. Separability of E implies that

$$\mathbb{P}\left(\|Z - \widetilde{Z} - a\| < \frac{1}{3}\|a\|\right) > 0,$$

for some $a \in E$. Let φ be a continuous linear functional such that $\varphi(a) = \|a\|$, $\|\varphi\| = 1$. Then

$$\mathbb{P}\left(\|Z - \widetilde{Z} - a\| < \frac{1}{3}\|a\|\right) > 0.$$

So $\mathbb{P}(\varphi(Z - \widetilde{Z}) > \frac{2}{3}\|a\|) > 0$. However, $\varphi(Z) = \mathbb{E}(\varphi(X)|\mathscr{G}) = \varphi(\widetilde{Z})$, $\mathbb{P}$-a.s. because the result is true for real valued random variables. The obtained contradiction implies uniqueness.

We now prove existence. If X is a simple random variable, say $X = \sum_{j=1}^{k} 1\!\!1_{A_j} x_j$, then one defines

$$Z = \sum_{j=1}^{k} x_j \mathbb{P}(A_j|\mathscr{G}),$$

where $\mathbb{P}(A_j|\mathscr{G})$ represents the classical notion of the conditional expectation of A_j given $\mathscr{G}$ (see e.g. [559]). It is clear that Z fulfills (1.18) and moreover

$$\mathbb{E}(\|Z\|) \le \sum_{j=1}^{k} \mathbb{E}\left(\|x_j\| |\mathbb{P}(A_j|\mathscr{G})|\right) \le \sum_{j=1}^{k} \|x_j\| \mathbb{P}(A_j) \le \mathbb{E}(\|X\|).$$

For general X let $\{X_n\}$ be the sequence defined in Lemma 1.3 and $Z_n = \mathbb{E}(X_n|\mathscr{G}),\ n \in \mathbb{N}$. Then

$$\mathbb{E}\|Z_n - Z_m\| \le \mathbb{E}\|X_n - X_m\| \to 0 \quad \text{as } n, m \to \infty.$$

Therefore there exists a subsequence $\{Z_{n_k}\}$ convergent $\mathbb{P}$-a.s. to a $\mathscr{G}$ measurable random variable Z. Since

$$\mathbb{E}\|Z_{n_k} - Z_{n_l}\| \le \mathbb{E}\|X_{n_k} - X_{n_l}\|, \quad k, l \in \mathbb{N}$$

so, letting l tend to ∞,

$$\mathbb{E}\|Z_{n_k} - Z\| \le \mathbb{E}\|X_{n_k} - X\| \downarrow 0 \quad \text{as } k \to \infty. \tag{1.19}$$

Moreover, for arbitrary $A \in \mathscr{G}$,

$$\int_A Z_{n_k} d\mathbb{P} = \int_A X_{n_k} d\mathbb{P}, \quad k \in \mathbb{N}$$

and by (1.19), (1.18) follows and the proof is complete. □

We remark that from the proof it follows that

$$\|\mathbb{E}(X|\mathscr{G})\| \le \mathbb{E}(\|X\| \,|\mathscr{G}). \tag{1.20}$$

Finally we give the definition of independence. Let $\{\mathscr{F}_i\}_{i\in I}$ be a family of sub σ-fields of $\mathscr{F}$. These σ-fields are said to be *independent* if, for every finite subset $J \subset I$ and every family $\{A_i\}_{i\in J}$ such that $A_i \in \mathscr{F}_i,\ i \in J$,

$$\mathbb{P}\left(\bigcap_{i\in J} A_i\right) = \prod_{i\in J} \mathbb{P}(A_i). \tag{1.21}$$

Random variables $\{X_i\}_{i\in I}$ are *independent* if the σ-fields $\{\sigma(X_i)\}_{i\in I}$ are independent. The following fact is a consequence of Proposition 1.4.

Proposition 1.11 *Let $\mathscr{K}_i$ be π-systems on Ω and let $\mathscr{F}_i = \sigma(\mathscr{K}_i),\ i \in I$. The σ-fields $\{\mathscr{F}_i\}_{i\in I}$ are independent if for every finite set $J \subset I$ and for sets $A_i \in \mathscr{K}_i,\ i \in J$,* (1.21) *holds.*

Proof Let $i_1, i_2 \in I$ and let $A_2 \in \mathscr{K}_{i_2}$. Set

$$\mathscr{G} = \{A_1 \in \sigma(\mathscr{K}_{i_1}) : \mathbb{P}(A_1 \cap A_2) = \mathbb{P}(A_1)\mathbb{P}(A_2))\}.$$

Obviously $\mathscr{G}$ satisfies the assumptions of Proposition 1.4, so that $\mathscr{G} = \sigma(\mathscr{K}_{i_1})$. Analogously we find

$$\mathbb{P}(A_1 \cap A_2) = \mathbb{P}(A_1)\mathbb{P}(A_2)), \quad \forall\, A_1 \in \mathscr{K}_{i_1}, \quad \forall\, A_2 \in \mathscr{K}_{i_2}.$$

Proceeding by induction, we arrive at the conclusion. □

We end this section with the following result.

Proposition 1.12 *Let $(E_1, \mathscr{E}_1)$ and $(E_2, \mathscr{E}_2)$ be two measurable spaces and $\psi : E_1 \times E_2 \to \mathbb{R}^1$ a bounded measurable function. Let ξ_1, ξ_2 be two random variables in $(\Omega, \mathscr{F}, \mathbb{P})$ with values in $(E_1, \mathscr{E}_1)$ and $(E_2, \mathscr{E}_2)$ respectively, and let $\mathscr{G} \subset \mathscr{F}$ be a fixed σ-field.*

Assume that ξ_1 is $\mathscr{G}$-measurable, then there is a bounded $\mathscr{E}_1 \times \mathscr{G}$-measurable function $\widehat{\psi}(x_1, \omega)$, $x_1 \in E_1$, $\omega \in \Omega$ such that

$$\mathbb{E}(\psi(\xi_1, \xi_2)|\mathscr{G})(\omega) = \widehat{\psi}(\xi_1(\omega), \omega), \quad \omega \in \Omega. \tag{1.22}$$

If in addition ξ_2 is independent of $\mathscr{G}$, then

$$\widehat{\psi}(x_1, \omega) = \widehat{\psi}(x_1) = \mathbb{E}(\psi(x_1, \xi_2)), \quad x_1 \in E_1. \tag{1.23}$$

Proof Assume first that

$$\psi(x_1, x_2) = \psi_1(x_1)\psi_2(x_2), \quad x_1 \in E_1,\ x_2 \in E_2,$$

with $\psi_1 : E_1 \to \mathbb{R}^1$ and $\psi_2 : E_2 \to \mathbb{R}^1$ bounded and measurable functions. Then

$$\mathbb{E}(\psi(\xi_1, \xi_2)|\mathscr{G}) = \mathbb{E}(\psi_1(\xi_1)\psi_2(\xi_2)|\mathscr{G}) = \psi_1(\xi_1)\mathbb{E}(\psi_2(\xi_2)|\mathscr{G})$$

and it is enough to set

$$\widehat{\psi}(x_1, \omega) = \psi_1(x_1)\mathbb{E}(\psi_2(x_2)|\mathscr{G})(\omega).$$

So the result is true in this case.

Denote by $\mathscr{G}_1$ the family of all sets $\Gamma \in \mathscr{E}_1 \times \mathscr{E}_2$ such that the representation (1.22) holds for $\psi = \mathbb{1}_\Gamma$, and by $\mathscr{K}$ the family of all sets $\Gamma = \Gamma_1 \times \Gamma_2$ where $\Gamma_1 \in \mathscr{E}_1$, $\Gamma_2 \in \mathscr{E}_2$. Then $\mathscr{K}$ is a π-system and $\mathscr{K} \subset \mathscr{G}_1$. One can check that all conditions of Proposition 1.4 hold true and therefore (1.22) holds for all $\Gamma \in \mathscr{E}_1 \times \mathscr{E}_1$. Consequently the result is true for all measurable simple functions ψ and, by a monotonic passage to the limit in (1.22), we get in general the representation (1.22). The proof of the representations (1.22)–(1.23) when ξ_2 is independent of $\mathscr{G}$ is similar. □

2
Probability measures

This chapter is devoted to basic facts about probability measures in Banach spaces. We review concepts such as regularity and tightness. Special attention is paid to *Gaussian measures* in Banach and Hilbert spaces and their reproducing kernels. In particular, we prove *Fernique's theorem* on the integrability of Gaussian vectors, an infinite dimensional extension of *Bochner's theorem* and the *Feldman–Hajek theorem* on absolute continuity. We also give basic results on *white noise* expansions.

2.1 General properties

Let E be a separable, Banach space with the norm $\|\cdot\|$ and let $\mathscr{B}(E)$ be its Borel σ-field. Given $x \in E$ and $r > 0$ we set, as in Chapter 1,

$$B(x,r) = \{y \in E : \ \|x - y\| < r\}, \quad \overline{B}(x,r) = \{y \in E : \ \|x - y\| \leq r\},$$

and

$$\partial B(x,r) = \{y \in E : \ \|x - y\| = r\}.$$

We denote by $C_b(E)$ the Banach space of all bounded, continuous real valued functions defined on E equipped with the norm

$$\|\varphi\| = \sup_{x \in E} |\varphi(x)|$$

and by $UC_b(E)$ the closed subspace of $C_b(E)$ of all functions which are uniformly continuous. By a *probability measure on E* we will usually mean a probability measure defined on the Borel σ-field $\mathscr{B}(E)$. Such measures on $(E, \mathscr{B}(E))$ are completely determined by their integrals with respect to all $\varphi \in C_b(E)$. This means that if μ, ν are probability measures on E and for all $\varphi \in C_b(E)$,

$$\int_E \varphi(x)\mu(dx) = \int_E \varphi(x)\nu(dx), \tag{2.1}$$

then $\mu = \nu$. Note in fact that if (2.1) holds for all $\varphi \in C_b(E)$ then it holds for $\varphi = 1\!\!1_K$ where K is a closed set. Since the family

$$\mathscr{G} := \{F \in \mathscr{B}(E) : \ \mu(F) = \nu(F)\}$$

satisfies the conditions of Proposition 1.4, with $\mathscr{K}$ being the family of all closed sets, therefore $\mathscr{G} = \sigma(\mathscr{K})$.

It is easy to check, using Proposition 1.4, that any probability measure μ on $(E, \mathscr{B}(E))$ is *regular* in the sense that for every $\Gamma \in \mathscr{B}(E)$ we have

$$\begin{aligned} \mu(\Gamma) &= \sup\{\mu(F);\ F \subset \Gamma, F \ \text{ closed}\} \\ &= \inf\{\mu(G) :\ G \supset \Gamma, G \ \text{ open}\}. \end{aligned} \tag{2.2}$$

In fact this result holds for arbitrary complete metric spaces E, see for example Kallenberg [433, page 18].

Proposition 2.1 *Let E be a separable Banach space and let μ be a probability measure on $(E, \mathscr{B}(E))$. Then for arbitrary $\varepsilon > 0$ there exists a compact set $K_\varepsilon \subset E$ such that*

$$\mu(K_\varepsilon) \geq 1 - \varepsilon.$$

Proof Fix $\varepsilon > 0$ and let $\{a_k\}$ be a dense sequence in E. Then for any $k \in \mathbb{N}$ there exists $m_k > 0$ such that

$$\mu(F_k) \geq 1 - 2^{-k}\varepsilon,$$

where

$$F_k = \bigcup_{i=1}^{m_k} \overline{B}(a_i, 1/k).$$

Define $F = \bigcap_{k=1}^{\infty} F_k$. We recall that a set on E is relatively compact if and only if it is totally bounded, that is for arbitrary $\varepsilon > 0$ it can be covered with a finite number of balls of radius less than or equal to ε (see e.g. [272, page 22]). It is clear that F is compact. Finally we have

$$\mu(F) = 1 - \mu(F^c) = 1 - \mu\left(\bigcup_{k=1}^{\infty} F_k^c\right) \geq 1 - \varepsilon,$$

which ends the proof. □

The property of measures expressed by Proposition 2.1 generalizes to a family Λ of probability measures on E and has important applications to weak convergence of measures. A family Λ of probability measures on E is said to be *tight* if for arbitrary $\varepsilon > 0$ there exists a compact set $K_\varepsilon \subset E$ such that

$$\mu(K_\varepsilon) \geq 1 - \varepsilon, \quad \text{for all } \mu \in \Lambda.$$

The following proposition is a useful reformulation of the original definition; its proof is very similar to that of Proposition 2.1.

Proposition 2.2 *A family Λ of probability measures on E is tight if and only if for arbitrary $\varepsilon > 0$ and $r > 0$ there exists a finite family of balls $\overline{B}(a_i, r)$, $i = 1, \dots, m$, such that*

$$\mu\left(\bigcup_{i=1}^{m} B(a_i, r)\right) \geq 1 - \varepsilon, \quad \text{for all } \mu \in \Lambda.$$

A sequence of measures $\{\mu_n\}$ on $(E, \mathscr{B}(E))$ is said to be *weakly convergent* to a measure μ in E if one of the following equivalent conditions holds.

(i) For every $\varphi \in C_b(E)$ we have

$$\lim_{n\to\infty} \int_E \varphi(x)\mu_n(dx) = \int_E \varphi(x)\mu(dx).$$

(ii) For every $\varphi \in UC_b(E)$ we have

$$\lim_{n\to\infty} \int_E \varphi(x)\mu_n(dx) = \int_E \varphi(x)\mu(dx).$$

(iii) For every closed set $F \subset E$ we have

$$\limsup_{n\to\infty} \mu_n(F) \leq \mu(F).$$

(iv) For every open set $G \subset E$ we have

$$\liminf_{n\to\infty} \mu_n(G) \geq \mu(G).$$

(v) For every Borel set A such that $\mu(\partial A) = 0$, [(1)] we have

$$\lim_{n\to\infty} \mu_n(A) = \mu(A).$$

Clearly any sequence $\{\mu_n\}$ can have only one weak limit.

A family Λ is said to be *compact* (respectively *relatively compact*) if an arbitrary sequence $\{\mu_n\}$ of elements from Λ contains a subsequence $\{\mu_{n_k}\}$ weakly convergent to a measure $\mu \in \Lambda$ (respectively to a measure μ on $(E, \mathscr{B}(E))$).

The following theorem is due to Prokhorov.

Theorem 2.3 *Let E be a complete, separable, metric space. A set Λ of probability measures on $(E, \mathscr{B}(E))$ is relatively compact if and only if it is tight.*

Proof Assume first that the space E is compact. In this case any family Λ of probability measures is tight and therefore we have only to show that Λ is relatively compact. We recall that $C(E)$ is separable.

(1) $\partial A = \overline{A} \backslash \mathring{A}$, where $\overline{A}$ and $\mathring{A}$ are the closure and the interior of A respectively.

Let $\{f_n\}$ be a dense subset of $C(E)$ and $\{\mu_n\}$ an arbitrary sequence of probability measures on E. By a diagonal procedure we can extract a subsequence such that the sequences $\{\mu_{n_k}(f_m)\}_{m\in\mathbb{N}}$ are convergent as $k\to\infty$. Here $\mu_{n_k}(f_m)=\int_E f_m d\mu_{n_k}$. We deduce easily that the limit $\lim_{k\to\infty}\mu_{n_k}(f)$ exists for every $f\in C(E)$ and defines a linear nonnegative functional. By the Riesz representation theorem this functional is a probability measure.

In the general case, consider a sequence of compact sets $\{K_m\}$ such that

$$\mu(K_m)\geq 1-\frac{1}{m},\quad \mu\in\Lambda,\ m\in\mathbb{N}.$$

By the diagonalization procedure and by the preceding arguments, we can construct a sequence of measures $\{\mu_n\}\subset\Lambda$ such that, for each $m\in\mathbb{N}$, their restrictions to $\widehat{K}_m=\bigcup_{j=1}^m K_j$ converge weakly to a measure $\widehat{\mu}_m$ concentrated on K_m such that

$$\widehat{\mu}_m(\widehat{K}_m)\geq 1-\frac{1}{m}.$$

If $l>m$ then $\widehat{\mu}_m$ is the restriction of $\widehat{\mu}_l$ to $\widehat{K}_m$. It is now easy to check that the function $\widehat{\mu}$

$$\widehat{\mu}(\Gamma)=\lim_{m\to\infty}\widehat{\mu}_m(\Gamma\cap\widehat{K}_m),\quad \Gamma\in\mathscr{B}(E),$$

is a probability measure on $\mathscr{B}(E)$ and is the weak limit of the sequence $\{\mu_n\}$. □

The following theorem, due to Skorohod, links the concept of weak convergence of a probability measure with that of almost sure convergence of random variables.

Theorem 2.4 *Let E be a complete, separable, metric space. For an arbitrary sequence of probability measures $\{\mu_n\}$ on $\mathscr{B}(E)$ weakly convergent to a probability measure μ, there exists a probability space $(\Omega,\mathscr{F},\mathbb{P})$ and a sequence of random variables, X, $\{X_n\}$ such that their laws are μ and $\{\mu_n\}$ respectively and*

$$\lim_{n\to\infty}X_n=X,\quad \mathbb{P}\text{-a.s.}$$

Proof We define $\Omega=[0,1)$, $\mathscr{F}=\mathscr{B}([0,1))$, $\mathbb{P}=\lambda$, the Lebesgue measure, and divide the proof into several steps.

Step 1 Construction of a special family of coverings of E.

Let us fix a countable dense set $\{a_k\}$ on E. For any $C\in\mathscr{B}(E)$ and any $r>0$, we set

$$\begin{aligned}C_1^r&=C\cap B(a_1,r),\\ C_{k+1}^r&=\Big(C\cap\textstyle\bigcup_{j=1}^{k+1}B(a_j,r)\Big)\setminus\bigcup_{j=1}^k C_j^r,\quad k\in\mathbb{N}.\end{aligned}$$

The sets $C_1^r,C_2^r,\ldots,$ are disjoint, cover the set C and are of diameter less than or equal to $2r$.

Let $\{r_m\}$ be a sequence in $\mathbb{R}^1$ decreasing to 0. For arbitrary natural numbers $i_1, \ldots, i_n$ we define sets $A_{i_1,\ldots,i_n}$ inductively as follows. We set $A_k = E_k^{r_1}$, $k \in \mathbb{N}$, and given $m \in \mathbb{N}$ and $(i_1, \ldots, i_m) \in \mathbb{N}^m$ we define

$$A_{i_1,\ldots,i_m,k} = (A_{i_1,\ldots,i_m})_k^{r_{m+1}}, \quad k \in \mathbb{N}.$$

Step 2 Construction of a family of coverings of $[0, 1)$ related to a probability measure ν.

We construct inductively a family $I^\nu_{i_1,\ldots,i_m}$ of subintervals of $[0, 1)$, by setting

$$I_1^\nu = [0, \nu(A_1)), \ldots, I_j^\nu = \left[\sum_{k=1}^{j-1} \nu(A_k), \sum_{k=1}^{j} \nu(A_k)\right), \quad j \in \mathbb{N}$$

and if

$$I^\nu_{i_1,\ldots,i_m} = [\alpha, \beta), \quad \text{with } \beta - \alpha = \nu(A_{i_1,\ldots,i_m}),$$

then

$$I^\nu_{i_1,\ldots,i_m,j} = \left[\alpha + \sum_{k=1}^{j-1} \nu(A_{i_1,\ldots,i_m,k}), \alpha + \sum_{k=1}^{j} \nu(A_{i_1,\ldots,i_m,k})\right).$$

It is easy to see that for each $m \in \mathbb{N}$ the sets $\{I^\nu_{i_1,\ldots,i_m} : (i_1, \ldots, i_m) \in \mathbb{N}^m\}$ are disjoint and cover $[0, 1)$.

Step 3 Construction of random variables $X, X_1, \ldots$ from $[0, 1)$ into E, such that $\mathscr{L}(X) = \mu$, $\mathscr{L}(X_i) = \mu_i$, $i \in \mathbb{N}$.

Set

$$F = \bigcup_{m=1}^{\infty} \bigcup_{k=1}^{\infty} \partial B(a_k, r_m).$$

We choose the sequence $\{r_m\}$ in such a way that

$$\mu(F) = \mu_n(F) = 0, \quad n \in \mathbb{N}.$$

This is possible because for arbitrary probability measure ν and arbitrary $a \in E$, the set of all positive numbers r such that $\nu(\partial B(a, r)) > 0$ is at most countable. This is because for different positive r, sets $\partial B(a, r)$ are disjoint.

With such a choice of the sequence $\{r_m\}$, we have

$$\mu_n(\partial A_{i_1,\ldots,i_m}) = 0, \quad \mu(\partial A_{i_1,\ldots,i_m}) = 0, \quad m \in \mathbb{N},\ i_1, \ldots, i_m \in \mathbb{N}^m,$$

where $\partial A = \overline{A} \backslash \mathring{A}$.

If ν is a probability measure on $(E, \mathscr{B}(E))$ such that

$$\nu(\partial A_{i_1,\ldots,i_m}) = 0, \quad m \in \mathbb{N},\ i_1, \ldots, i_m \in \mathbb{N}^m,$$

we define

$$Z_m^\nu(\omega) = x_{i_1,\ldots,i_m}, \quad \text{for all } m \in \mathbb{N} \text{ and } \omega \in I^\nu_{i_1,\ldots,i_m},$$

where $x_{i_1,\ldots,i_m}$ is a point selected in $\mathring{A}_{i_1,\ldots,i_m}$.

It follows from the construction that

$$\rho(Z_m^\nu(\omega), Z_n^\nu(\omega)) \leq r_m, \quad \text{for all } n \geq m, \ \omega \in [0, 1),$$

and therefore the limit

$$\lim_{n\to\infty} Z_m^\nu(\omega) := Z_\infty^\nu(\omega)$$

exists. Moreover, let $f \in UC_b(E)$ and $\nu_m = \mathscr{L}(Z_m^\nu)$. Then

$$\left|\int_E f(x)\nu(dx) - \int_E f(x)\nu_m(dx)\right| \leq \sum_{i_1,\dots,i_m\in\mathbb{N}} \left|\int_{A_{i_1,\dots,i_m}} (f(x) - f(x_{i_1,\dots,i_m}))\nu(dx)\right|$$
$$\leq \omega_f(r_m) \to 0 \quad \text{as } m \to \infty$$

and therefore $\mathscr{L}(Z_\infty^\nu) = \nu$. Here ω_f denotes the modulus of continuity of f.

We finally set

$$X = Z_\infty^\nu, \quad X_m = Z_m^\nu, \quad m \in \mathbb{N}.$$

We have just shown that

$$\mathscr{L}(X) = \mu, \quad \mathscr{L}(X) = \mu_m, \quad m \in \mathbb{N}.$$

Assume that ω belongs to the interior of $I^\mu_{i_1,\dots,i_k}$. From the weak convergence $\mu_m \rightharpoonup \mu$, we have for arbitrary $j_1, \dots, j_k$

$$\lambda(I^{\mu_m}_{j_1,\dots,j_k}) = \mu_m(A_{j_1,\dots,j_k}) \to \mu(A_{j_1,\dots,j_k}) = \lambda(I^\mu_{j_1,\dots,j_k}).$$

Therefore, we have for $\omega \in I^{\mu_m}_{i_1,\dots,i_k}$, $m \geq m_k$ and $Z_k^{\mu_m}(\omega) = Z_k^\mu(\omega)$.

Consequently, for $m \geq m_k$ we have

$$\begin{aligned}\rho(X_m(\omega), X(\omega)) &= \rho(Z_\infty^{\mu_m}(\omega), Z_\infty^\mu(\omega)) \\ &\leq \rho(Z_\infty^{\mu_m}(\omega), Z_k^\mu(\omega)) + \rho(Z_k^{\mu_m}(\omega), Z_k^\mu(\omega)) + \rho(Z_k^\mu(\omega), Z_\infty^\mu(\omega)) \\ &\leq 2r_k.\end{aligned}$$

In this way we have shown that $X_n(\omega) \to X_n(\omega)$ for all

$$\omega \in \bigcap_{k\in\mathbb{N}} \bigcup_{i_1,\dots,i_k\in\mathbb{N}} \text{int.}(I^\mu_{i_1,\dots,i_k}).$$

This ends the proof since

$$\lambda\left(\bigcap_{k\in\mathbb{N}} \bigcup_{i_1,\dots,i_k\in\mathbb{N}} \text{int.}(I^\mu_{i_1,\dots,i_k})\right) = 1.$$

□

If μ is a probability measure on $(E, \mathscr{B}(E))$, then the following function φ_μ on E^*

$$\varphi_\mu(y^*) = \int_E e^{iy^*(x)}\mu(dx), \quad y^* \in E^*,$$

is called the *characteristic function* of μ. Instead of φ_μ we often write $\widehat{\mu}$. If $E = \mathbb{R}^n$ or, more generally, if E is a Hilbert space H with inner product $\langle\cdot,\cdot\rangle$, then φ_μ is regarded as a function on $\mathbb{R}^n$ or on H and

$$\varphi_\mu(\lambda) = \int_H e^{i\langle\lambda,x\rangle}\mu(dx), \quad \lambda \in H.$$

An arbitrary subset of E of the form

$$\{x \in E : (y_1^*(x), \dots, y_n^*(x)) \in \Gamma\}, \quad y_1^*, \dots, y_n^* \in E^*, \ \Gamma \in \mathscr{B}(\mathbb{R}^n), \tag{2.3}$$

is called *cylindrical*, and an arbitrary function $\Psi(y_1^*, \dots, y_n^*)$, where Ψ is a Borel function on $\mathbb{R}^n$, is called *cylindrical*. Cylindrical sets form a π-system, so if two measures are identical on cylindrical sets they are equal on $\mathscr{B}(E)$, by Proposition 1.4.

Proposition 2.5 *Assume that M is a linear subspace of E^* which generates $\mathscr{B}(E)$.*

(i) If μ and ν are two probability mesures on $\mathscr{B}(E)$ such that

$$\varphi_\mu(y^*) = \varphi_\nu(y^*), \quad \forall\, y^* \in M,$$

then $\mu = \nu$.

(ii) If X and Y are two E-valued random variables such that

$$\mathscr{L}(y^*(X)) = \mathscr{L}(y^*(Y)), \quad \forall\, y^* \in M,$$

then $\mathscr{L}(X) = \mathscr{L}(Y)$.

Proof It is enough to show (i). Fix $y_1^*, \dots, y_n^* \in M, \lambda_1, \dots, \lambda_n \in \mathbb{R}$. By the hypothesis we have

$$\int_E e^{i\lambda_1 y_1^*(x)+\cdots+i\lambda_n y_n^*(x)}\mu(dx) = \int_E e^{i\lambda_1 y_1^*(x)+\cdots+i\lambda_n y_n^*(x)}\nu(dx). \tag{2.4}$$

Identity (2.4) implies that the $\mathbb{R}^n$-valued mapping

$$x \to (y_1^*(x), \dots, y_n^*(x))$$

maps measures μ and ν onto measures $\widetilde{\mu}$ and $\widetilde{\nu}$ on $(\mathbb{R}^n, \mathscr{B}(\mathbb{R}^n))$ with identical characteristic functions. So the measures $\widetilde{\mu}$ and $\widetilde{\nu}$ are identical. But this implies that the measures μ and ν are equal on the π-system of all cylindrical sets generated by M, so $\mu = \nu$ on $\mathscr{B}(E)$. □

2.2 Gaussian measures in Banach spaces

2.2.1 Fernique theorem

A Gaussian measure μ on $\mathbb{R}^1$ is either concentrated at one point $\mu = \delta_m$ or has a density

$$\frac{1}{\sqrt{2\pi q}} e^{-\frac{1}{2q}(x-m)^2}, \quad x \in \mathbb{R}^1,$$

where $q > 0$ and $m \in \mathbb{R}^1$. Such a measure is denoted by $\mathcal{N}(m, q)$. One can check directly that

$$\int_{\mathbb{R}^1} x \mathcal{N}(m,q)(dx) = m, \qquad \int_{\mathbb{R}^1} (x-m)^2 \mathcal{N}(m,q)(dx) = q,$$

$$\int_{\mathbb{R}^1} e^{i\lambda x} \mathcal{N}(m,q)(dx) = e^{im\lambda - \frac{1}{2} q\lambda^2}, \qquad \int_{\mathbb{R}^1} e^{\lambda x} \mathcal{N}(m,q)(dx) = e^{m\lambda + \frac{1}{2} q\lambda^2}.$$

Moreover

$$\int_{\mathbb{R}^1} e^{\lambda |x|^2} \mathcal{N}(m,q)(dx) = \begin{cases} +\infty & \text{if } \lambda \geq 1/2q, \\ (1-2\lambda q)^{-1/2} & \text{if } \lambda < 1/2q. \end{cases}$$

The following useful characterization of Gaussian measures $\mathcal{N}(0, q)$ can be easily proved by considering characteristic functions.

Proposition 2.6 *A measure μ on $\mathbb{R}^1$ is of the form $\mathcal{N}(0, q)$ for some $q \geq 0$ if and only if, for arbitrary independent real valued random variables ξ and η such that $\mathcal{L}(\xi) = \mathcal{L}(\eta) = \mu$ and numbers α, β such that $\alpha^2 + \beta^2 = 1$, we have $\mathcal{L}(\alpha\xi + \beta\eta) = \mu$.*

Now let E be a separable Banach space. A probability measure μ on $(E, \mathcal{B}(E))$ is said to be a *Gaussian measure* if and only if the law of an arbitrary linear functional $h \in E^*$, considered as a random variable on $(E, \mathcal{B}(E), \mu)$, is a Gaussian measure on $(\mathbb{R}^1, \mathcal{B}(\mathbb{R}^1))$. If the law of each $h \in E^*$ is in addition a symmetric (zero mean) Gaussian law on $\mathbb{R}^1$ then μ is called a *symmetric Gaussian measure*.

The characteristic function $\hat{\mu}$ of a Gaussian measure μ in $\mathbb{R}^n$ is of the form

$$\hat{\mu}(\lambda) = e^{i\langle \lambda, m\rangle - \frac{1}{2}\langle Q\lambda, \lambda\rangle}, \quad \lambda \in \mathbb{R}^n, \tag{2.5}$$

where $m \in \mathbb{R}^n$ and Q is a nonnegative symmetric $n \times n$ matrix. The Gaussian measure μ on $\mathbb{R}^n$ with the characteristic function (2.5) will be denoted by $\mathcal{N}(m, Q)$. Similar to the case $n = 1$, for $m = (m_1, \ldots, m_n)$, $Q = (q_{jk})$,

$$\int_{\mathbb{R}^n} x_j \mathcal{N}(m, Q)(dx) = m_j, \quad j = 1, \ldots, n$$

$$\int_{\mathbb{R}^n} (x_j - m_j)(x_k - m_k) \mathcal{N}(m, Q)(dx) = q_{jk}, \quad j, k = 1, \ldots, n.$$

Thus, if $\mathcal{L}(X) = \mathcal{N}(m, Q)$, we have $\mathbb{E}X = m$ and $\mathrm{Cov}X = Q$.

If Q is positive definite $\mathcal{N}(m, Q)$ has density $g_{m,Q}$ given by

$$g_{m,Q}(x) = \frac{1}{\sqrt{(2\pi)^n \det Q}}\, e^{-\frac{1}{2}\langle Q^{-1}(x-m), x-m\rangle}, \quad x \in \mathbb{R}^n.$$

One can check easily that if $X = (X_1, \dots, X_n)$ and $\mathcal{L}(X) = \mathcal{N}(m, Q)$, then the random variables $X_1, \dots, X_n$ are independent if and only if the matrix Q is diagonal.

We have the following fundamental result due to Fernique [291].

Theorem 2.7 *Let μ be an arbitrary symmetric Gaussian measure on E. Let $\lambda > 0$ and $r > 0$ be such that* (2)

$$\log\left(\frac{1-\mu(\overline{B}(0,r))}{\mu(\overline{B}(0,r))}\right) + 32\lambda r^2 \le -1. \tag{2.6}$$

Then

$$\int_E e^{\lambda\|x\|^2}\mu(dx) \le e^{16\lambda r^2} + \frac{e^2}{e^2-1}. \tag{2.7}$$

Proof Let X and Y be two independent E-valued symmetric Gaussian random variables defined on a probability space $(\Omega, \mathcal{F}, \mathbb{P})$ such that $\mathcal{L}(X) = \mathcal{L}(Y) = \mu$. We show first that the random variables

$$\widetilde{X} = \frac{1}{\sqrt{2}}\,(X+Y), \quad \widetilde{Y} = \frac{1}{\sqrt{2}}\,(X-Y),$$

are independent with law also equal to μ. To prove independence it is enough to show that arbitrary random variables $h(\widetilde{X})$, $g(\widetilde{X})$, h, $g \in E^*$ are independent. Since X and Y are independent and $\mathcal{L}(X) = \mathcal{L}(Y)$ therefore

$$\begin{aligned}\mathbb{E}(h(\widetilde{X})g(\widetilde{Y})) &= \frac{1}{2}\,\mathbb{E}(h(X+Y)g(X-Y))\\ &= \frac{1}{2}\,[\mathbb{E}(h(X)g(Y)) - \mathbb{E}(h(Y)g(X))] = 0.\end{aligned}$$

So, the independence follows. Note also that for arbitrary $\lambda \in \mathbb{R}^1$ and $h \in E^*$

$$\begin{aligned}\mathbb{E}\left(\exp\left\{i\tfrac{\lambda}{\sqrt{2}}\,h(X-Y)\right\}\right) &= \mathbb{E}\left(\exp\left\{i\tfrac{\lambda}{\sqrt{2}}\,h(X+Y)\right\}\right)\\ &= \exp\left\{-\tfrac{1}{2}\,\mathbb{E}\left(h(X)^2\right)\right\}.\end{aligned}$$

Consequently, for arbitrary $h \in E^*$, $h(\widetilde{X})$ and $h(\widetilde{Y})$ have the same law as $h(X)$. Thus $\mathcal{L}(\widetilde{X})=\mathcal{L}(\widetilde{Y}) = \mathcal{L}(X)$. Assume that $t \ge s \ge 0$. Then

$$\begin{aligned}\mathbb{P}(\|X\| \le s)\,\mathbb{P}(\|Y\| > t) &= \mathbb{P}\left(\tfrac{\|X-Y\|}{\sqrt{2}} \le s\right)\mathbb{P}\left(\tfrac{\|X+Y\|}{\sqrt{2}} \ge t\right)\\ &= \mathbb{P}\left(\tfrac{\|X-Y\|}{\sqrt{2}} \le s \text{ and } \tfrac{\|X+Y\|}{\sqrt{2}} > t\right)\\ &\le \mathbb{P}\left(|\|X\| - \|Y\|| \le \sqrt{2}\,s \text{ and } \|X\| + \|Y\| > \sqrt{2}\,t\right).\end{aligned}$$

(2) $\overline{B}(0,r) = \{x \in E : \|x\| \le r\}$.

Since the set

$$\{(\xi,\eta)\in\mathbb{R}^2_+ : |\xi-\eta|\le\sqrt{2}\,s,\ \xi+\eta>\sqrt{2}t\},$$

is contained in

$$\left\{(\xi,\eta)\in\mathbb{R}^2_+ : \xi>\frac{t-s}{\sqrt{2}}\,,\ \eta>\frac{t-s}{\sqrt{2}}\right\},$$

therefore

$$\begin{aligned}&\mathbb{P}(\|X\|\le s)\,\mathbb{P}(\|Y\|>t)\\&\quad\le\mathbb{P}\left(\|X\|>\frac{t-s}{\sqrt{2}}\ \text{and}\ \|Y\|>\frac{t-s}{\sqrt{2}}\right)\le\left[\mathbb{P}\left(\|X\|>\frac{t-s}{\sqrt{2}}\right)\right]^2.\end{aligned}$$

So, since X and Y have the same law, we find

$$\mathbb{P}(\|X\|>t)\le\frac{\left[\mathbb{P}(\|X\|>\frac{t-s}{\sqrt{2}})\right]^2}{\mathbb{P}(\|X\|\le s)}. \tag{2.8}$$

Now fix $r>0$ and define $t_0=r>0$, $t_{n+1}=r+\sqrt{2}\,t_n$ and

$$\alpha_n(r)=\frac{\mathbb{P}(\|X\|>t_n)}{\mathbb{P}(\|X\|\le r)},\quad n\in\mathbb{N}\cup\{0\}.$$

By (2.8) it follows that

$$\begin{aligned}\alpha_{n+1}(r)&=\frac{\mathbb{P}(\|X\|>t_{n+1})}{\mathbb{P}(\|X\|\le r)}=\frac{\mathbb{P}(\|X\|>r+\sqrt{2}\,t_n)}{\mathbb{P}(\|X\|\le r)}\\&\le\left(\frac{\mathbb{P}(\|X\|>t_n)}{\mathbb{P}(\|X\|\le r)}\right)^2=\alpha_n(r)^2,\quad n\in\mathbb{N}\cup\{0\}.\end{aligned}$$

Therefore $\alpha_n(r)\le e^{2^n\log\alpha_0(r)}$, $n\in\mathbb{N}\cup\{0\}$. Moreover, since $(\sqrt{2})^{n+4}r>t_n$,

$$\begin{aligned}\mathbb{P}\left(\|X\|>(\sqrt{2})^{n+4}r\right)&\le\mathbb{P}(\|X\|>t_n)\\&=\alpha_n(r)\,\mathbb{P}(\|X\|\le r)\le e^{2^n\log\alpha_0(r)},\quad n\in\mathbb{N}\cap\{0\}.\end{aligned} \tag{2.9}$$

Consequently, for $\lambda>0$, setting

$$\Sigma_n=\{x\in E:(\sqrt{2})^{n+4}r<\|x\|\le(\sqrt{2})^{n+5}r\},$$

we have

$$\begin{aligned}\int_{\|x\|>4r}e^{\lambda\|x\|^2}\mu(dx)&=\sum_{n=0}^\infty\int_{\Sigma_n}e^{\lambda\|x\|^2}\mu(dx)\\&\le\sum_{n=0}^\infty\mathbb{P}(\|X\|>(\sqrt{2})^{n+4}r)\,e^{\lambda r^2 2^{n+5}}\\&\le\sum_{n=0}^\infty e^{2^n(\log\,\alpha_0(r)+32\lambda r^2)}.\end{aligned}$$

Choosing r, λ as in (2.6), we find estimate (2.7). [3] □

[3] We have used the rough estimate $\sum_{n=0}^\infty e^{-2^n}<\sum_{n=0}^\infty e^{-2n}=\frac{e^2}{e^2-1}$.

Remark 2.8 It follows from the theorem that for a sufficiently small positive λ we have

$$\int_E e^{\lambda|x|^2}\mu(dx) < +\infty.$$

In fact, choosing $r > 0$ such that

$$\log\left(\frac{1-\mu(\overline{B}(0,r))}{\mu(\overline{B}(0,r))}\right) \le -2,$$

we can take $\lambda < 1/32r^2$.

2.2.2 Reproducing kernels

Let μ be a symmetric Gaussian measure on a Banach space E. A linear subspace $H \subset E$ equipped with a Hilbert norm $|\cdot|_H$ is said to be a *reproducing kernel space* for μ if H is complete, continuously embedded in E and such that for arbitrary $\varphi \in E^*$

$$\mathscr{L}(\varphi) = \mathscr{N}(0, |\varphi|_H^2), \tag{2.10}$$

where

$$|\varphi|_H = \sup_{|h|_H \le 1} |\varphi(h)|.$$

Theorem 2.9 *For an arbitrary symmetric Gaussian measure μ on a separable Banach space, there exists a unique reproducing kernel space $(H, |\cdot|_H)$.*

Proof By the Fernique theorem (Theorem 2.7) the measure μ has all moments finite. In particular, an arbitrary $\varphi \in E^*$ can be identified with an element of the Hilbert space $L^2 = L^2(E, \mathscr{B}(E), \mu)$. We shall denote by $\overline{E^*}$ the closure of E^* in L^2. Since arbitrary $\varphi \in \overline{E^*}$ is an almost sure limit of functions $\psi \in E^*$ which have Gaussian distributions, then for arbitrary $\varphi \in \overline{E^*}$ we have

$$\mathscr{L}(\varphi) = \mathscr{N}(0, |\varphi|_{L^2}^2).$$

Define a mapping $J : \overline{E^*} \to E$ by setting

$$J(\varphi) = \int_E x\varphi(x)\mu(dx), \quad \forall\, \varphi \in \overline{E^*}. \tag{2.11}$$

J is one-to-one, since if $J(\varphi) = 0$ then

$$\varphi(J(\varphi)) = \int_E |\varphi(x)|^2\mu(dx) = 0$$

and so $\varphi = 0$, a.s. J is continuous, in fact

$$\|J(\varphi)\|^2 = \left|\int_E x\varphi(x)\mu(dx)\right|^2 \le \int_E \|x\|^2|\mu(dx)\, |\varphi|_{L^2}^2.$$

Thus the image H of J in E, $H = J(E)$, is a Hilbert space with the scalar product

$$\langle J(\varphi), J(\psi)\rangle_H = \int_E \varphi(x)\psi(x)\mu(dx).$$

We shall denote by $|\cdot|_H$ the norm in H. The space H is a reproducing kernel space of μ since, for any $\psi \in E^*$,

$$\begin{aligned}|\psi|_H &= \sup\{|\psi(x)| : x \in H, |x|_H = 1\}\\ &= \sup\{|\psi(J(\varphi))| : |\varphi|_{L^2} = 1\}\\ &\le \sup\left\{\left|\int_E \varphi(x)\psi(x)\mu(dx)\right| : |\varphi|_{L^2} = 1\right\} = |\psi|_{L^2}.\end{aligned}$$

Taking $\varphi = \psi/|\psi|_{L^2}$ we get $|\psi|_H = |\psi|_{L^2}$.

To prove uniqueness assume that a Hilbert space $\widetilde{H} \subset E$ with a scalar product $\langle\cdot,\cdot\rangle_{\widetilde{H}}$ is also a reproducing kernel for μ. Then, by definition, for arbitrary $\varphi \in E^*$ there exists a unique element $\widetilde{J}\varphi \in \widetilde{H}$ such that

$$\varphi(h) = \langle \widetilde{J}\varphi, h\rangle_{\widetilde{H}}, \quad |\widetilde{J}\varphi|_{\widetilde{H}} = |\varphi|_{L^2}, \quad h \in \widetilde{H}.$$

The transformation $\widetilde{J}$ can be extended by continuity to all $\varphi \in \overline{E^*}$ and, for any $\varphi, \psi \in \overline{E^*}$,

$$\langle \widetilde{J}\varphi, \widetilde{J}\psi\rangle_{\widetilde{H}} = \int_E \varphi(x)\psi(x)\mu(dx).$$

Thus for $\varphi \in E^*$, $\psi \in \overline{E^*}$ we have

$$\varphi(\widetilde{J}\psi) = \langle \widetilde{J}\varphi, \widetilde{J}\psi\rangle_{\widetilde{H}} = \int_E \varphi(x)\psi(x)\mu(dx) = \varphi(J(\psi)).$$

Consequently $\widetilde{J}(\psi) = J(\psi)$ and this implies $H \subset \widetilde{H}$ with isometric embedding.

If $\widetilde{H} \neq H$ there exists $\widetilde{h} \in \widetilde{H}$, $\widetilde{h} \neq 0$ and $\widetilde{h}$ orthogonal to H in the $\widetilde{H}$ sense. Consequently, for all $\varphi \in E^*$ we have

$$0 = \langle \widetilde{h}, J\varphi\rangle_{\widetilde{H}} = \varphi(\widetilde{h}).$$

So $\widetilde{h} = 0$ and $\widetilde{H} = H$. □

The reproducing kernel space of μ will be denoted by H_μ. In a sense it is independent of the Banach space E.

Proposition 2.10 *Assume that a Banach space E_1 is continuously and as a Borel set imbedded in E. If the measure μ is symmetric and Gaussian on E and E_1, then the reproducing kernel space calculated with respect to E or E_1 is the same.*

Proof Note that we have the following obvious inclusion $H_1 \subset E_1 \subset E$. Since restriction of any $\varphi \in E^*$ to E_1 belongs to E_1^*, the space H_1 satisfies all the conditions required by the reproducing kernel space H. By uniqueness $H_1 = H$. □

Note that if $h = J\varphi,\ \varphi \in E^*$, then we have

$$\langle h, x\rangle_H = \varphi(x), \quad x \in H. \tag{2.12}$$

Therefore the functional $\langle h, \cdot\rangle_H$ can be naturally extended to the whole space E. If h is an arbitrary element on H, then there exists a sequence $\{h_n\} = \{J\varphi_n\},\ \varphi_n \in E^*,\ n \in \mathbb{N}$, such that

$$\lim_{n\to\infty} |h - h_n| = 0.$$

Moreover, for some $\varphi \in L^2$, $\varphi_n \to \varphi$. Functional φ does not belong in general to E^* but is defined in a unique way as an element of $L^2(E, \mathscr{B}(E), \mu)$. In this way definition (2.12) can be extended to all remaining $h \in H$ and μ-almost all $x \in E$.

We have the following *reproducing kernel formula*:

$$\int_E \langle h, x\rangle_H \langle g, x\rangle_H \mu(dx) = \langle h, g\rangle_H, \quad h, g \in H.$$

The following result [651] will be used to check that a given measure μ on a separable Banach space is Gaussian and to identify its reproducing kernel.

Proposition 2.11 *Let μ be a measure in a separable Banach space E and M a linear subspace of E^* generating the Borel σ-field $\mathscr{B}(E)$.*

(i) If arbitrary $\varphi \in M$ has a symmetric Gaussian law then μ is symmetric Gaussian.
(ii) If in addition H_0 is a Hilbert space continuously embedded into E and such that $\mathscr{L}(\varphi) = \mathscr{N}(0, |\varphi|_0^2)$ for arbitrary $\varphi \in M$, then H_0 is the reproducing kernel of μ.

Proof We show first that μ is Gaussian. Let X and Y be independent E-valued random variables with the law μ and α, β real numbers satisfying $\alpha^2 + \beta^2 = 1$. For arbitrary $\varphi \in M$

$$\begin{aligned}\mathscr{L}(\varphi(\alpha X + \beta Y)) &= \mathscr{L}(\varphi(\alpha X) + \varphi(\beta Y)) = \mathscr{L}(\alpha\varphi(X) + \beta\varphi(Y)) \\ &= \mathscr{N}(0, \alpha^2|\varphi|_0^2 + \beta^2|\varphi|_0^2) = \mathscr{L}(\varphi(X)).\end{aligned}$$

By an application of Proposition 2.5(ii), the random variables $\alpha X + \beta Y$ and X have the same distribution. If φ is now an arbitrary element of E^* then

$$\mathscr{L}(\alpha\varphi(X) + \beta\varphi(Y)) = \mathscr{L}(\varphi(\alpha X + \beta Y)) = \mathscr{L}(\varphi(X)).$$

Proposition 2.6 implies that $\varphi(X)$ has a Gaussian symmetric distribution. Consequently μ is a symmetric Gaussian measure.

To show that H_0 is the reproducing kernel of μ we consider M and E^* as subspaces of $L^2(E, \mathscr{B}(E), \mu)$ and show first that $\overline{M} = \overline{E^*}$. Suppose that $\varphi \in E^*\backslash\overline{M}$ then there exists $\varphi_0 \in \overline{M}$ such that $\psi = \varphi - \varphi_0$ is orthogonal to $\overline{M}$. Since $\overline{E^*}$ is a linear set of symmetric Gaussian random variables, ψ is independent of the σ-field generated by

M. It follows that $\psi = 0$ and $\varphi = \varphi_0$, μ-a.s. To complete the proof of the proposition, denote by $J\varphi,\ \varphi \in E^*$, the unique element in H such that

$$\langle h, J\varphi\rangle_0 = \varphi(h), \quad \forall\, h \in H_0.$$

Arguing as in the proof of Theorem 2.9 one obtains that

$$J\varphi = \int_E x\varphi(x)\mu(dx)$$

and

$$|J\varphi|_0^2 = \int_E \varphi^2(x)\mu(dx).$$

This implies the result. □

2.2.3 White noise expansions

The aim of this section is to show that an arbitrary symmetric Gaussian measure μ in a separable Banach space E can be regarded as a distribution of the Gaussian *white noise*. Namely, following [651], we prove the following result.

Theorem 2.12 *Let μ be a symmetric Gaussian measure on a separable Banach space E and let H_μ be its reproducing kernel space. Let $\{e_n\}$ be an orthonormal and complete basis in H_μ and $\{\xi_n\}$ a sequence of independent real valued random variables such that $\mathscr{L}(\xi_n) = \mathscr{N}(0, 1),\ n \in \mathbb{N}$. Then the series $\sum_{k=1}^{\infty} \xi_k e_k$ converges $\mathbb{P}$-a.s. in E and*

$$\mathscr{L}\left(\sum_{k=1}^{\infty} \xi_k e_k\right) = \mu.$$

For the proof we will need a result on a series of independent random variables in a separable Banach space E, $\sum_{j=1}^{\infty} \xi_j$, where the laws $\mathscr{L}(\xi_i)$ are symmetric measures on E. We prove first a basic inequality.

Proposition 2.13 *Let E be a separable Banach space. Assume that $\xi_1, \ldots, \xi_N$ are E-valued independent random variables with symmetric distributions. Let $S_k = \xi_1 + \cdots + \xi_k,\ k = 1, 2, \ldots, N$; then for arbitrary $r \geq 0$,*

$$\mathbb{P}\left(\sup_{k\leq N} \|S_k\| \geq r\right) \leq 2\mathbb{P}(\|S_N\| \geq r). \tag{2.13}$$

Proof We first notice that

$$\begin{aligned}\mathbb{P}\left(\sup_{k\leq N} \|S_k\| \geq r\right) &= \mathbb{P}\left(\sup_{k\leq N} \|S_k\| \geq r \text{ and } \|S_N\| \geq r\right)\\ &\quad + \mathbb{P}\left(\sup_{k\leq N} \|S_k\| \geq r \text{ and } \|S_N\| < r\right)\\ &= I_1 + I_2.\end{aligned} \tag{2.14}$$

Let

$$A_1 = \{\|S_1\| \geq r\}\,,$$
$$\dots\dots\dots\dots\dots\dots$$
$$A_k = \{\|S_1\| < r, \dots, \|S_{k-1}\| < r,\ \|S_k\| \geq r\}, \quad k = 2, \dots, N,$$

then

$$I_1 \leq \mathbb{P}\left(\|S_N\| \geq r\right) \tag{2.15}$$

and, since

$$\bigcup_{i=1}^{N} A_i = \{\sup\{\|S_k\| \geq r :\ k \leq N\}\}\,,$$

we have

$$I_2 = \sum_{k=1}^{N} \mathbb{P}\left(A_k \cap \{\|S_N\| < r\}\right)$$

and

$$\mathbb{P}(\|S_1\| \geq r \text{ and } \|S_N\| < r) = \mathbb{P}(\|S_1\| \geq r \text{ and } \|\xi_1 + \xi_2 + \cdots + \xi_N\| < r).$$

By symmetry

$$\begin{aligned}\mathbb{P}(A_1 \cap \{\|S_N\| < r\}) &= \mathbb{P}(\|S_1\| \geq r \text{ and } \|S_N\| < r)\\ &= \mathbb{P}(\|S_1\| \geq r \text{ and } \|S_1 - \xi_2 - \cdots - \xi_N\| < r)\\ &= \mathbb{P}(\|S_1\| \geq r \text{ and } \|2S_1 - S_N\| < r)\\ &\leq \mathbb{P}(\|S_1\| \geq r \text{ and } \|S_N\| \geq r) = \mathbb{P}(A_1 \cap \{\|S_N\| \geq r\}).\end{aligned}$$

In a similar way

$$\mathbb{P}(A_k \cap \{\|S_N\| < r\}) \leq \mathbb{P}(A_k \cap \{\|S_N\| \geq r\}),$$

so that

$$I_2 \leq \mathbb{P}\left(\bigcup_{i=1}^{N} A_i \cap \{\|S_N\| \geq r\}\right) \leq \mathbb{P}\left(\|S_N\| \geq r\right). \tag{2.16}$$

Now the conclusion follows from (2.14), (2.15) and (2.16). □

Proposition 2.14 *Let E be a separable Banach space. Assume that $\{\xi_i\}_{i\in N}$ is a sequence of E-valued independent random variables with symmetric distributions. Assume that there exists a random variable S such that, for arbitrary $r > 0$,*

$$\lim_{N\to\infty} \mathbb{P}(\|S - S_N\| \geq r) = 0$$

where $S_k = \xi_1 + \cdots + \xi_k,\ k \in \mathbb{N}$. *Then*

$$\lim_{N\to\infty} S_N = S = \sum_{k=1}^{\infty} \xi_k, \quad \mathbb{P}\text{-a.s.}$$

Proof Using the Borel–Cantelli lemma one can extract a subsequence $\{S_{n_k}\}$ converging $\mathbb{P}$-a.s. to S. Fix k and let $m > n_k$, by Proposition 2.13

$$\begin{aligned}\mathbb{P}\left(\sup_{n_k<l\le m} \|S_l - S_{n_k}\| \ge r\right) &\le 2\mathbb{P}(\|S_m - S_{n_k}\| \ge r)\\ &\le 2\mathbb{P}(\|S_m - S\| \ge \tfrac{p}{2}) + 2\mathbb{P}(\|S - S_{n_k}\| \ge \tfrac{p}{2}).\end{aligned}$$

Consequently, letting $m \to \infty$,

$$\mathbb{P}\left(\sup_{n_k<l} \|S_l - S_{n_k}\| \ge r\right) \le 2\mathbb{P}\left(\|S - S_{n_k}\| \ge \frac{p}{2}\right)$$

and this implies the conclusion. □

Proof of Theorem 2.12 As in the proof of Theorem 2.9 we denote by $\overline{E^*}$ the closure of E^* in $L^2 = L^2(E, \mathscr{B}(E), \mu)$ and by $\varphi_n \in \overline{E^*}$ elements such that $\varphi_n = e_n,\ n \in \mathbb{N}$. We can assume that the sequence $\{\varphi_n\}$ is linearly dense in $\overline{E^*}$. Note that the real random variables $\varphi_n,\ n \in \mathbb{N}$, are independent and $\mathscr{N}(0, 1)$ distributed. Our aim is to show that

$$x = \sum_{k=1}^{\infty} \varphi_n(x)e_n \quad \text{for } \mu\text{-almost all } x \in E.$$

Let

$$\mu_n = \mathscr{L}\left(\sum_{k=1}^{n} \xi_k e_k\right), \qquad n \in \mathbb{N}.$$

We prove first that the sequence $\{\mu_n\}$ is tight. Denote $S(x) = x$ and

$$S_N(x) = \sum_{n=1}^{N} \varphi_n(x)e_n = \sum_{n=1}^{N} \varphi_n(x)J\varphi_n, \qquad x \in E.$$

For arbitrary $\varphi \in E^*$ the sequence

$$\varphi(S_N(x)) = \sum_{n=1}^{N} \varphi_n(x)\varphi(e_n)$$

is μ-almost surely convergent because

$$\sum_{n=1}^{\infty} |\varphi(e_n)|^2 = \sum_{n=1}^{\infty} \langle J\varphi, e_n\rangle_{H_\mu}^2 = |J\varphi|_{H_\mu}^2 < +\infty.$$

Moreover

$$\mathscr{L}\left(\sum_{n=1}^{\infty}\varphi_n(x)e_n\right) = \mathscr{N}(0, |J\varphi|^2_{H_\mu}).$$

Since $\{\varphi_n\}$ is a complete orthonormal basis in $\overline{E^*} \subset L^2$, for $\varphi \in E^*$ we have in particular,

$$\varphi(x) = \sum_{n=1}^{\infty}\varphi(y)\int_E \varphi_n(y)\varphi_n(x)\mu(dy),$$

in L^2 and μ-a.e. Consequently

$$\varphi(Sx) = \varphi(x) = \sum_{n=1}^{\infty}\varphi_n\langle J\varphi, J\varphi_n\rangle_{H_\mu} = \lim_{N\to\infty}\varphi(S_N(x)), \quad \mu\text{-a.s.}$$

Therefore, for all $N \in \mathbb{N}$

$$\lim_{M\to\infty}\varphi(S_M(x) - S_N(x)) = \varphi(S(x) - S_N(x)),$$

and independence of $S_M - S_N$ and S_N also implies that $S - S_N$ and S_N are independent.

Let $\mu_M^{\perp}$ denote the distribution of $S - S_M$, $M \in \mathbb{N}$. Then [(4)]

$$\mu_M * \mu_M^{\perp} = \mu, \quad M \in \mathbb{N}.$$

For a given ε, let K be a compact set such that $\mu(K) > 1 - \varepsilon$. Since

$$\mu_M * \mu_M^{\perp}(K) = \int_E \mu_M(K - x)\mu_M^{\perp}(dx),$$

there exists $x_0 \in E$ such that

$$\mu_M(K - x_0) \geq 1 - \varepsilon.$$

But μ is symmetric,

$$\mu_M(K - x_0) = \mu_M(-K + x_0).$$

Consequently

$$\mu_M\left(2^{-1}(K - K)\right) \geq \mu_M\left((K - x_0) \cap (-K + x_0)\right) \geq 1 - 2\varepsilon, \quad M \in \mathbb{N}.$$

(4) If μ and ν are two probability measures on$(E, \mathscr{B}(E))$ then $\mu * \nu$ denotes their convolution $\mu * \nu(\Gamma) = \int_E \mu(\Gamma - x)\nu(dx)$, $\Gamma \in \mathscr{B}(E)$.

The set $\frac{K-K}{2}$ is compact, so the tightness of $\{\mu_n\}$ follows. In the same way one can show tightness of $\mu_M^{\perp}$. By a direct calculation we have

$$\widehat{\mu}(\varphi) = \int_E e^{i\varphi(x)}\mu(dx) = e^{-\frac{1}{2}|J\varphi|_H^2},$$

$$\widehat{\mu}_M(\varphi) = \int_E e^{i\varphi(S_M x)}\mu(dx) = e^{-\frac{1}{2}\sum_{n=1}^{M}\langle J\varphi, e_n\rangle_H},$$

$$\widehat{\mu}_M^{\perp}(\varphi) = \int_E e^{i\varphi(x-S_M x)}\mu(dx) = e^{-\frac{1}{2}\sum_{n=M+1}^{\infty}\langle J\varphi, e_n\rangle_H}, \quad \varphi \in E^*.$$

Consequently

$$\lim_{M\to\infty} \widehat{\mu}_M(\varphi) = \widehat{\mu}(\varphi), \quad \lim_{M\to\infty} \widehat{\mu}_M^{\perp}(\varphi) = 1$$

and therefore any weakly convergent subsequence of $\{\mu_M\}$ has the same limit. Therefore μ_N converges to μ weakly. In the same way $\{\widehat{\mu}_M^{\perp}\}$ converges to $\delta_{\{0\}}$ weakly and therefore $S - S_M \to 0$ in probability. It is enough to apply Proposition 2.14. □

2.3 Probability measures on Hilbert spaces

2.3.1 Gaussian measures on Hilbert spaces

In this subsection we show that for Gaussian measures on Hilbert spaces more precise information can be given. Our presentation is direct and independent of the Fernique theorem.

According to the general definition (see Section 2.2.1) a probability measure μ on $(H, \mathscr{B}(H))$ is called *Gaussian* if for arbitrary $h \in H$ there exist $m \in \mathbb{R}^1$, $q \geq 0$ such that,

$$\mu(\{x \in H; \langle h, x\rangle \in A\}) = \mathscr{N}(m, q)(A), \quad \forall\, A \in \mathscr{B}(\mathbb{R}^1).$$

In particular, if μ is Gaussian, the following functionals,

$$H \to \mathbb{R}^1, \quad h \to \int_H \langle h, x\rangle \mu(dx), \tag{2.17}$$

$$H \times H \to \mathbb{R}^1, \quad (h_1, h_2) \to \int_H \langle h_1, x\rangle\langle h_2, x\rangle \mu(dx), \tag{2.18}$$

are well defined. We show now that they are continuous. We need a lemma on general probability measures.

Lemma 2.15 *Let ν be a probability measure on $(H, \mathscr{B}(H))$. Assume that for some $k \in \mathbb{N}$*

$$\int_H |\langle z, x\rangle|^k \nu(dx) < +\infty, \quad \forall\, z \in H,$$

then there exists a constant $c > 0$ such that

$$\left| \int_H \langle h_1, x\rangle \cdots \langle h_k, x\rangle \nu(dx) \right| \le c|h_1|\cdots|h_k|, \quad h_1 \cdots h_k \in H.$$

Proof Define for any $n \in \mathbb{N}$ the set,

$$U_n = \left\{ z \in H : \int_H |\langle z, x\rangle|^k \nu(dx) \le n \right\}.$$

By the hypothesis $H = \bigcup_{n=1}^\infty U_n$; since H is a complete metric space and U_n are closed sets, by the Baire category argument, there exist $n_0 \in \mathbb{N}$, $z_0 \in U_{n_0}$ and $r_0 > 0$ such that $B(z_0, r_0) \subset U_{n_0}$, where $B(z_0, r_0)$ is the ball in H with center z_0 and radius r_0. Hence

$$\int_H |\langle z_0 + y, x\rangle|^k \, \nu(dx) \le n_0, \quad \forall\, y \in B(0, r_0).$$

But, for any $y \in B(0, r_0)$,

$$\int_H |\langle y, x\rangle|^k \nu(dx) \le 2^k \int_H |\langle z_0 + y, x\rangle|^k \nu(dx) + 2^k \int_H |\langle z_0, x\rangle|^k \nu(dx) \le 2^{k+1} n_0. \tag{2.19}$$

So, for all z in H different from 0 we can apply (2.19) to $y = r_0 z/|z|$ and obtain

$$\int_H |\langle z, x\rangle|^k \, \nu(dx) \le 2^{k+1} n_0 |z|^k r_0^{-k}.$$

By the obvious inequality

$$|\xi_1 \xi_2 \cdots \xi_k| \le |\xi_1|^k + |\xi_2|^k + \cdots + |\xi_k|^k, \quad \forall\, (\xi_1, \xi_2, \ldots, \xi_k) \in \mathbb{R}^k$$

it follows that the transformation

$$H^k \to \mathbb{R}^1, \quad (h_1, \ldots h_k) \to \int_H \langle h_1, x\rangle \cdots \langle h_k, x\rangle \nu(dx)$$

is continuous. □

It follows from the lemma that if μ is Gaussian, then there exist an element $m \in H$ and a linear operator Q, such that

$$\int_H \langle h, x\rangle \mu(dx) = \langle m, h\rangle, \quad \forall\, h \in H, \tag{2.20}$$

$$\int_H \langle h_1, x - m\rangle \langle h_2, x - m\rangle \mu(dx) = \langle Q h_1, h_2\rangle, \quad \forall\, h_1, h_2 \in H. \tag{2.21}$$

The vector m is called the *mean* and Q is called the *covariance operator* of μ. It is clear that the operator Q is symmetric. Moreover, since

$$\langle Qh, h\rangle = \int_H \langle h, x - m\rangle^2 \mu(dx) \ge 0, \quad h \in H,$$

it is also nonnegative. It follows from (2.20)–(2.21) that a Gaussian measure μ on H with mean m and covariance Q has the following characteristic functional

$$\widehat{\mu}(\lambda) = \int_H e^{i\langle\lambda,x\rangle}\mu(dx) = e^{i\langle\lambda,m\rangle - \frac{1}{2}\langle Q\lambda,\lambda\rangle}, \quad \lambda \in H.$$

It is therefore uniquely determined by m and Q. It is denoted by $\mathcal{N}(m, Q)$.

It turns out that the covariance operator has to be nuclear (see Appendix C for the definition and basic properties).

Proposition 2.16 *Let μ be a Gaussian probability measure with mean* 0 *and covariance Q. Then Q has finite trace.*

Proof Consider the characteristic function of the measure μ

$$\varphi(h) = \int_H e^{\iota\langle h,x\rangle}\mu(dx) = e^{-\frac{1}{2}\langle Qh,h\rangle}, \quad h \in H.$$

For arbitrary $c > 0$, we have:

$$\begin{aligned}1 - \varphi(h) &= \int_H [1 - \cos\langle h, x\rangle]\mu(dx)\\ &\le \frac{1}{2}\int_{|x|\le c} \langle h, x\rangle^2\mu(dx) + 2\mu(\{x \in H : |x| \ge c\})\\ &\le \frac{1}{2}\langle Q_c h, h\rangle + 2\mu(\{x \in H : |x| \ge c\}), \quad h \in H,\end{aligned}$$

where Q_c is the linear bounded operator defined by

$$\langle Q_c h, h\rangle = \int_{|x|\le c} \langle h, x\rangle^2\mu(dx).$$

Since for any complete orthonormal basis $\{e_n\}$ in H we have

$$\begin{aligned}\mathrm{Tr}\,(Q_c) &= \sum_{n=1}^{\infty}\langle Q_c e_n, e_n\rangle = \int_{|x|\le c}\sum_{n=1}^{\infty}\langle x, e_n\rangle^2\mu(dx)\\ &= \int_{|x|\le c} |x|^2\mu(dx) \le c^2,\end{aligned}$$

the operator Q_c has finite trace. To show that Q has finite trace, it suffices to prove that there exist $\beta > 0$ and $c > 0$ such that

$$h \in H,\ \langle Q_c h, h\rangle \le 1 \Rightarrow \langle Qh, h\rangle \le \beta, \tag{2.22}$$

as this implies that $\langle Qh, h\rangle \le \beta\langle Q_c h, h\rangle,\ h \in H$. We have in fact

$$1 - e^{-\frac{1}{2}\langle Qh,h\rangle} \le \frac{1}{2}\langle Q_c h, h\rangle + 2\mu(\{x \in H : |x| \ge c\}), \quad \forall\, h \in H,\ c > 0. \tag{2.23}$$

Now let $h \in H$ be such that $\langle Q_c h, h\rangle \le 1$; then by (2.23)

$$e^{\frac{1}{2}\langle Qh,h\rangle} \le \left[\tfrac{1}{2} - 2\mu(\{x \in H : |x| \le c\})\right]^{-1}$$

thus (2.22) holds, provided $\mu(\{x \in H : |x| \ge c\}) < \frac{1}{4}$. The proof is complete. □

It follows from the proof that $\mathrm{Tr}(Q) = \mathbb{E}|X|^2 < +\infty$. So, the definition of the covariance operator given above is a special case of the generalí definition given in Section 2.1.

In the following considerations we denote by $\{e_k\}$ a complete orthonormal basis on H which diagonalizes Q, and by $\{\lambda_k\}$ the corresponding set of eigenvalues of Q. Moreover, for any $x \in H$ we set $x_k = \langle x, e_k \rangle$, $k \in \mathbb{N}$. Note that the random variables $x_1, x_2, \ldots, x_n$ are independent, because the covariance matrix of the $\mathbb{R}^n$-valued random variable $(x_1, x_2, \ldots, x_n)$ is precisely $(\lambda_i \delta_{ij})$.

Proposition 2.17 *Let μ be a Gaussian probability measure with mean* 0 *and covariance Q. If $s < (2 \,\mathrm{Tr}\, Q)^{-1}$ then*

$$\int_H e^{s|x|^2} \mu(dx) = \exp\left\{-\frac{1}{2} \,\mathrm{Tr}\,[\log(1 - 2sQ)]\right\}$$
$$\leq \frac{1}{\sqrt{1 - 2s \,\mathrm{Tr}\, Q}}. \tag{2.24}$$

Proof Fix $s < (2 \,\mathrm{Tr}\, Q)^{-1}$ and set

$$I_n(s) = \int_H e^{s \sum_{i=1}^n x_i^2} \mu(dx).$$

Since the random variables $x_1, x_2, \ldots, x_n$ are independent, we have

$$I_n(s) = \prod_{i=1}^n \int_H \exp\{s x_i^2\} \mu(dx).$$

Since x_i has the density $\mathscr{N}(0, \lambda_i)$, it follows that

$$I_n(s) = \prod_{i=1}^n \frac{1}{\sqrt{2\pi\lambda_i}} \int_{-\infty}^{\infty} e^{s\xi^2 - \frac{\xi^2}{2\lambda_i}} \, d\xi$$
$$= \prod_{i=1}^n \frac{1}{\sqrt{1 - 2\lambda_i s}} = \frac{1}{\sqrt{\prod_{i=1}^n (1 - 2\lambda_i s)}}.$$

However, if $2\lambda_i s \leq 1, \; i = 1, \ldots, n$, then

$$\prod_{i=1}^n (1 - 2\lambda_i s) \geq 1 - 2s \sum_{i=1}^n \lambda_i. \tag{2.25}$$

The inequality (2.25) is certainly true for $s \leq 0$ and, if $0 < 2\lambda_i s \leq 1$, it follows from an induction argument. By letting n tend to infinity we obtain formula (2.24). □

Proposition 2.18 *Let Q be a positive, symmetric, trace class operator in H and let $m \in H$. Then there exists a Gaussian measure in H with mean m and covariance Q.*

Proof Let $\{\xi_n\}$ be a sequence of independent $\mathcal{N}(0, 1)$ real variables, on a probability space $(\Omega, \mathcal{F}, \mathbb{P})$. Set

$$\xi = m + \sum_{j=1}^{\infty} \sqrt{\lambda_j}\, \xi_j e_j. \tag{2.26}$$

Notice that the series in (2.26) is convergent in $L^2(\Omega, \mathcal{F}, \mathbb{P}; H)$ since

$$\mathbb{E} \sum_{j=1}^{\infty} \left(\sqrt{\lambda_j}\, \xi_j\right)^2 = \sum_{j=1}^{\infty} \lambda_j = \mathrm{Tr}(Q).$$

Fix $h \in H$ then

$$\begin{aligned} \mathbb{E}(e^{i\langle h,\xi\rangle}) &= e^{i\langle h,m\rangle} \lim_{n\to\infty} \mathbb{E}\left(e^{\sum_{j=1}^{n}\sqrt{\lambda_j}\,\xi_j\langle h,e_j\rangle}\right) \\ &= e^{i\langle h,m\rangle - \frac{1}{2}\sum_{j=1}^{n}\lambda_j\langle h,e_j\rangle^2} = e^{i\langle h,m\rangle-\frac{1}{2}\langle Qh,h\rangle}. \end{aligned}$$

Thus ξ is Gaussian, $\mathbb{E}\xi = m$ and

$$\langle \mathrm{Cov}(\xi)x, y\rangle = \mathbb{E} \sum_{j=1}^{n} \sqrt{\lambda_j}\, \xi_j \langle e_j, x\rangle \sum_{k=1}^{n} \sqrt{\lambda_k}\, \xi_k \langle e_k, y\rangle = \langle Qx, y\rangle,$$

for all $x, y \in H$. Thus ξ has a Gaussian distribution as required. □

Proposition 2.19 *For arbitrary $m \in \mathbb{N}$ there exists a constant C_m such that*

$$\int_H |x|^{2m} \mu(dx) \le C_m [\mathrm{Tr}(Q)]^m \tag{2.27}$$

for arbitrary $\mu = \mathcal{N}(0, Q)$.

Proof Note that, using the notation from Proposition 2.18,

$$\begin{aligned} \int_H |x|^{2m} \mu(dx) &= \mathbb{E}\left(\left|\sum_{j=1}^{\infty}\sqrt{\lambda_j}\,\xi_j e_j\right|^2\right)^m = \mathbb{E}\left(\sum_{j=1}^{\infty}\lambda_j\,\xi_j^2\right)^m \\ &= \sum_{j_1,\dots j_m=1}^{\infty} \lambda_{j_1}\cdots\lambda_{j_m}\mathbb{E}\left(\xi_{j_1}^2\cdots\xi_{j_m}^2\right) \le c\left(\sum_{j=1}^{\infty}\lambda_j\right)^m, \end{aligned}$$

where $c = \sup_{j_1,\dots,j_m=1} \mathbb{E}\left(\xi_{j_1}^2 \cdots \xi_{j_m}^2\right)$. The constant c is finite by a consequence of the Schwartz inequality. □

2.3.2 Feldman–Hajek theorem

In this section, following [691] and [348], we give necessary and sufficient conditions under which two Gaussian measures, $\mu = \mathcal{N}(m_1, Q_1)$, $\nu = \mathcal{N}(m_2, Q_2)$ on the

Hilbert space H, are absolutely continuous. We will need some preliminary facts concerned with the so called *Hellinger integral*.

Let μ and ν be probability measures on a measurable space $(E, \mathscr{E})$. Measure μ is *absolutely continuous* with respect to ν (we write $\mu << \nu$), if $\mu(\Gamma) = 0$ for all $\Gamma \in \mathscr{E}$ such that $\nu(\Gamma) = 0$. If $\mu << \nu$ and $\nu << \mu$ then μ and ν are said to be *equivalent* and we write $\mu \sim \nu$. If μ and ν are concentrated on disjoint sets then they are called *singular*; in this case we write $\mu \perp \nu$.

The Hellinger integral $H(\mu, \nu)$ is defined by the formula

$$H(\mu, \nu) = \int_E \left(\frac{d\mu}{d\lambda} \frac{d\nu}{d\lambda} \right)^{1/2} d\lambda, \tag{2.28}$$

where λ stands for any probability measure on $(E, \mathscr{E})$ with respect to which μ and ν are absolutely continuous. Note that a probability measure λ such that $\mu << \lambda$, $\nu << \lambda$ always exists; it suffices to take $\lambda = (\mu + \nu)/2$. The definition (2.28) is independent of the choice of the measure λ. To see this, take two measures $\lambda, \widetilde{\lambda}$ such that

$$\mu << \lambda, \ \nu << \lambda \quad \text{and} \quad \mu << \widetilde{\lambda}, \ \nu << \widetilde{\lambda}.$$

Then, setting $\zeta = (\lambda + \widetilde{\lambda})/2$ and noticing that

$$\frac{d\mu}{d\zeta} = \frac{d\mu}{d\lambda} \frac{d\lambda}{d\zeta},$$

it is easy to see that the right hand side of (2.28) is equal to

$$\int_E \left(\frac{d\mu}{d\zeta} \frac{d\nu}{d\zeta} \right)^{1/2} d\zeta.$$

Basic properties of the Hellinger integral are given in the following proposition.

Proposition 2.20 *The following properties hold for an arbitrary pair μ, ν of probability measures on $(E, \mathscr{E})$.*

(i) $0 \leq H(\mu, \nu) \leq 1$.
(ii) $H(\mu, \nu) = 0$ *if and only if* $\mu \perp \nu$.
(iii) Let $\mathscr{G} \subset \mathscr{E}$ be a σ-field and $H_{\mathscr{G}}(\mu, \nu)$ the Hellinger integral calculated for the restrictions of μ and ν to $\mathscr{G}$, then $H_{\mathscr{G}}(\mu, \nu) \geq H(\mu, \nu)$.

Proof Property (i) follows from the Schwartz inequality

$$\int_E \left(\frac{d\mu}{d\lambda} \frac{d\nu}{d\lambda} \right)^{1/2} d\lambda \leq \left(\int_E \frac{d\nu}{d\lambda} d\lambda \right)^{1/2} \left(\int_E \frac{d\mu}{d\lambda} d\lambda \right)^{1/2} \leq 1.$$

Property (ii) is obvious. To prove (iii) consider the probability space $(E, \mathscr{E}, \lambda)$, with the expectation denoted by $\mathbb{E}_\lambda$. The densities of the restrictions of μ and ν to $\mathscr{G}$ with respect to the restrictions of λ to $\mathscr{G}$ are then equal respectively to $\mathbb{E}_\lambda(\frac{d\mu}{d\lambda} | \mathscr{G})$ and

$\mathbb{E}_\lambda(\frac{d\nu}{d\lambda}|\mathscr{G})$ and consequently

$$H_{\mathscr{G}}(\mu,\nu) = \int_E \left[\mathbb{E}_\lambda\left(\frac{d\mu}{d\lambda}\Big|\mathscr{G}\right)\mathbb{E}_\lambda\left(\frac{d\nu}{d\lambda}\Big|\mathscr{G}\right)\right]^{1/2} d\lambda.$$

Since λ-a.s.

$$\frac{\left(\frac{d\mu}{d\lambda}\right)^{1/2}}{\left(\mathbb{E}_\lambda\left(\frac{d\mu}{d\lambda}|\mathscr{G}\right)\right)^{1/2}} \cdot \frac{\left(\frac{d\nu}{d\lambda}\right)^{1/2}}{\left(\mathbb{E}_\lambda\left(\frac{d\nu}{d\lambda}|\mathscr{G}\right)\right)^{1/2}} \le \frac{1}{2}\left(\frac{\frac{d\mu}{d\lambda}}{\mathbb{E}_\lambda\left(\frac{d\mu}{d\lambda}|\mathscr{G}\right)} + \frac{\frac{d\nu}{d\lambda}}{\mathbb{E}_\lambda\left(\frac{d\nu}{d\lambda}|\mathscr{G}\right)}\right), \quad (2.29)$$

therefore calculating conditional expectations of both sides of (2.29) one gets λ-a.e.

$$\mathbb{E}_\lambda\left(\frac{d\mu}{d\lambda}\Big|\mathscr{G}\right)^{1/2}\mathbb{E}_\lambda\left(\frac{d\nu}{d\lambda}\Big|\mathscr{G}\right)^{1/2} \ge \mathbb{E}_\lambda\left(\left(\frac{d\mu}{d\lambda}\right)^{1/2}\left(\frac{d\nu}{d\lambda}\right)^{1/2}\Big|\mathscr{G}\right). \quad (2.30)$$

Integrating (2.30) with respect to λ the required result follows. □

It follows from Proposition 2.20 that the condition $H(\mu,\nu) > 0$ is necessary for the equivalence of μ and ν. It turns out however that for product measures on $(\mathbb{R}^\infty, \mathscr{B}(\mathbb{R}^\infty))$ it is also a sufficient condition.

Let $\mu_k, \nu_k,\ k \in \mathbb{N}$, be probability measures on $(\mathbb{R}^1, \mathscr{B}(\mathbb{R}^1))$, assume that $\mu_k < \nu_k,\ k \in \mathbb{N}$ and let $\mu = \bigtimes_{k=1}^\infty \mu_k, \quad \nu = \bigtimes_{k=1}^\infty \nu_k$. Then

$$\prod_{k=1}^\infty H(\mu_k,\nu_k) = \prod_{k=1}^\infty \int_{\mathbb{R}^1}\left(\frac{d\mu_k}{d\nu_k}\right)^{1/2} d\nu_k.$$

Proposition 2.21 *If* $\prod_{k=1}^\infty H(\mu_k,\nu_k) > 0$ *then the measure* μ *is absolutely continuous with respect to* ν*. Moreover the sequence*

$$f_N(x) = \prod_{k=1}^N \frac{d\mu_k}{d\nu_k}(x_k), \quad x = \{x_k\} \in \mathbb{R}^\infty,\ N \in \mathbb{N}$$

converges in $L^1(\mathbb{R}^\infty, \mathscr{B}(\mathbb{R}^\infty), \nu)$ *to* $d\mu/d\nu$.

Proof Note that for $N \ge 1, m \ge 0$

$$\int_{\mathbb{R}^\infty}\left|f_{N+m}^{1/2}(x) - f_N^{1/2}(x)\right|^2 \nu(dx)$$

$$= \int_{\mathbb{R}^\infty}\prod_{k=1}^N \frac{d\mu_k}{d\nu_k}(x_k)\left|\left(\prod_{k=N+1}^{N+m}\frac{d\mu_k}{d\nu_k}(x_k)\right)^{1/2} - 1\right|^2 \nu(dx)$$

$$= \int_{\mathbb{R}^\infty}\left|\left(\prod_{k=N+1}^{N+m}\frac{d\mu_k}{d\nu_k}(x_k)\right)^{1/2} - 1\right|^2 \nu(dx)$$

$$= 2 - 2\prod_{k=N+1}^{N+m}\int_{\mathbb{R}^1}\left(\frac{d\mu_k}{d\nu_k}\right)^{1/2} d\nu_k = 2 - 2\prod_{k=N+1}^{N+m} H(\mu_k,\nu_k) \to 0$$

as $N, m \to \infty$. Consequently $\{f_N^{1/2}\}$ converges in $L^2(\mathbb{R}^\infty, \mathscr{B}(\mathbb{R}^\infty), \nu)$.

Since

$$\begin{aligned}
&\int_{\mathbb{R}^\infty} |f_{N+m}(x) - f_N(x)|\, \nu(dx) \\
&= \int_{\mathbb{R}^\infty} \left| f_{N+m}^{1/2}(x) - f_N^{1/2}(x) \right| \left| f_{N+m}^{1/2}(x) + f_N^{1/2}(x) \right| \nu(dx) \\
&\le \left(\int_{\mathbb{R}^\infty} \left| f_{N+m}^{1/2}(x) - f_N^{1/2}(x) \right|^2 \nu(dx) \right)^{1/2} \\
&\quad \times \left(2 \int_{\mathbb{R}^\infty} \left| f_{N+m}^{1/2}(x) + f_N^{1/2}(x) \right|^2 \nu(dx) \right)^{1/2} \\
&\le \sqrt{2} \left(\int_{\mathbb{R}^\infty} \left| f_{N+m}^{1/2}(x) - f_N^{1/2}(x) \right|^2 \nu(dx) \right)^{1/2} \to 0,
\end{aligned}$$

as $N, m \to \infty$, therefore

$$f_N(x) \to f(x) = \prod_{k=1}^{\infty} \frac{d\mu_k}{d\nu_k}(x_k),$$

in $L^1(\mathbb{R}^\infty, \mathscr{B}(\mathbb{R}^\infty), \nu)$. If $\varphi = \varphi(x_1, \ldots, x_k)$ is a bounded Borel function on $\mathbb{R}^k$ then for $N \ge k$

$$\int_{\mathbb{R}^\infty} \varphi(x)\mu(dx) = \int_{\mathbb{R}^\infty} \varphi(x) f_N(x)\nu(dx) \to \int_{\mathbb{R}^\infty} \varphi(x) f(x)\mu(dx).$$

The identity

$$\int_{\mathbb{R}^\infty} \varphi(x)\mu(dx) = \int_{\mathbb{R}^\infty} \varphi(x) f(x)\nu(dx),$$

can be easily extended from cylindrical bounded Borel functions φ to all bounded Borel functions on $\mathbb{R}^\infty$, using Proposition 1.4. □

Remark 2.22 If measure μ is absolutely continuous with respect to measure ν, we can see easily that

$$H(\mu, \nu) = \prod_{k=1}^{\infty} H(\mu_k, \nu_k).$$

Before giving a complete characterization of equivalent pairs of measures $\mu = \mathscr{N}(m_1, Q_1)$, $\nu = \mathscr{N}(m_2, Q_2)$ we deduce from Proposition 2.21 an important special case.

Theorem 2.23 *The following statements hold.*

(i) Gaussian measures $\mu = \mathscr{N}(m_1, Q)$, $\nu = \mathscr{N}(m_2, Q)$ are either singular or equivalent.

(ii) They are equivalent if and only if $m_1 - m_2 \in Q^{1/2}(H)$. Moreover

$$\frac{d\mu}{d\nu}(x) = \exp\Big\{\langle Q^{-1/2}(m_1 - m_2), Q^{-1/2}(x - m_1)\rangle - \frac{1}{2}|Q^{-1/2}(m_1 - m_2)|^2\Big\}, \quad \nu\text{-a.s.}, x \in H.$$

Remark 2.24 By $Q^{-1/2}$ we denote the pseudo-inverse of $Q^{1/2}$, see Appendix B.2. The random variable $\langle Q^{-1/2}(m_1 - m_2), Q^{-1/2}(x - m_2)\rangle$, $x \in H$, is defined as the sum of the series

$$\sum_{j=1}^{\infty} \langle x - m_2, e_j\rangle\langle m_1 - m_2, e_j\rangle\lambda_j^{-1}, \quad x \in H,$$

where $\{e_j, \lambda_j\}$ is the eigensequence associated with Q. Since

$$\begin{aligned}\sum_{j=1}^{\infty} \langle m_1 - m_2, e_j\rangle^2\lambda_j^{-2} \int_H \langle x - m_2, e_j\rangle^2 \nu(dx) &= \sum_{j=1}^{\infty} \langle m_1 - m_2, e_j\rangle^2\lambda_j^{-2}\langle Qe_i, e_j\rangle = \sum_{j=1}^{\infty} \langle m_1 - m_2, e_j\rangle^2\lambda_j^{-1} \\ &= |Q^{-1/2}(m_1 - m_2)|^2 < +\infty,\end{aligned}$$

the series converges in $L^2(H, \mathscr{B}(H), \nu)$.

Proof of Theorem 2.23 We first remark that the space H can be identified with ℓ^2 [(5)] and as a Borel subset of $\mathbb{R}^\infty$ and the measures μ and ν as product measures

$$\mu = \bigtimes_{k=1}^{\infty} \mu_k, \quad \nu = \bigtimes_{k=1}^{\infty} \nu_k,$$

where

$$\mu_k = \mathscr{N}(m_{1k}, \lambda_k), \quad \nu_k = \mathscr{N}(m_{2k}, \lambda_k),$$

and

$$m_{1k} = \langle m_1, e_k\rangle, \quad m_{2k} = \langle m_2, e_k\rangle, \quad k \in \mathbb{N}.$$

Note that

$$\frac{d\mu_k}{d\nu_k}(x_k) = \exp\left\{-\tfrac{1}{2\lambda_k}\ \left[(m_{1k} - m_{2k})^2 - 2(m_{1k} - m_{2k})(x_k - m_{2k})\right]\right\}$$

and

$$H(\mu_k, \nu_k) = \exp\left\{-\frac{(m_{1k} - m_{2k})^2}{8\lambda_k}\right\}, \quad k \in \mathbb{N}.$$

(5) ℓ^2 is the linear space of all sequences $\{\xi_k\}$ of real numbers such that $\sum_{k=1}^{\infty} \xi_k^2 < +\infty$ endowed with the inner product $\langle \xi, \eta\rangle = \sum_{k=1}^{\infty} \xi_k\eta_k$.

So $\prod_{k=1}^{\infty} H(\mu_k, \nu_k) > 0$ if and only if

$$\sum_{k=1}^{\infty} \frac{(m_{1k} - m_{2k})^2}{\lambda_k} < +\infty,$$

which is the condition $m_1 - m_2 \in Q^{1/2}(H)$. It is therefore enough to apply Proposition 2.21. □

Now we pass to the general case.

Theorem 2.25 (Feldman–Hajek theorem) *The following statements hold.*

(1) Gaussian measures $\mu = \mathcal{N}(m_1, Q_1)$, $\nu = \mathcal{N}(m_2, Q_2)$ are either singular or equivalent.

(2) They are equivalent if and only if the following conditions hold.

(i) $Q_1^{1/2}(H) = Q_2^{1/2}(H) = H_0$.

(ii) $m_1 - m_2 \in H_0$.

(iii) The operator $(Q_1^{-1/2}Q_2^{1/2})(Q_1^{-1/2}Q_2^{1/2})^* - I$ is a Hilbert–Schmidt operator on $\overline{H_0}$.

Note that if (i) holds then by the closed graph theorem the operator $Q_1^{-1/2}Q_2^{1/2}$ is an isomorphism of $\overline{H_0}$ onto $\overline{H_0}$. So the operator

$$R = (Q_1^{-1/2}Q_2^{1/2})(Q_1^{-1/2}Q_2^{1/2})^*, \tag{2.31}$$

is an invertible positive operator on H.

Proof Step 1 We show first that if $\mathcal{N}(m_1, Q_1)$ and $\mathcal{N}(m_2, Q_2)$ are nonsingular for some $m_1, m_2 \in H$, then the operators Q_1 and Q_2 have the properties formulated in (i) and (iii). In fact, let $m_1, m_2 \in H$ and assume that $\mathcal{N}(m_1, Q_1)$ and $\mathcal{N}(m_2, Q_2)$ are nonsingular, then their shifts $\mathcal{N}(0, Q_1)$ and $\mathcal{N}(m_2 - m_1, Q_2)$ are nonsingular too, as well as their symmetric images $\mathcal{N}(0, Q_1)$ and $\mathcal{N}(m_1 - m_2, Q_2)$. But then also

$$\mathcal{N}(0, 2Q_1) = \mathcal{N}(0, Q_1) * \mathcal{N}(0, Q_1)$$

and

$$\mathcal{N}(0, 2Q_2) = \mathcal{N}(m_2 - m_1, Q_2) * \mathcal{N}(m_1 - m_2, Q_2)$$

are nonsingular. So, without any loss of generality, one can assume that $m_1 = m_2 = 0$. It follows from Appendix B.2 that $Q_1^{1/2}(H) = Q_2^{1/2}(H)$ if and only if, for some constants $c \le C$, we have

$$c|Q_2 x| \le |Q_1 x| \le C|Q_2 x|, \quad x \in H.$$

Assume that $Q_1^{1/2}(H) \neq Q_2^{1/2}(H)$, then there exists a sequence $\{a_n\}$ with $|a_n| \le 1$ such that

$$\text{either } \lim_{n\to\infty} \frac{|Q_2^{1/2}a_n|}{|Q_1^{1/2}a_n|} = +\infty, \quad \text{or } \lim_{n\to\infty} \frac{|Q_1^{1/2}a_n|}{|Q_2^{1/2}a_n|} = +\infty.$$

Without any loss of generality we can assume that

$$\lim_{n\to\infty} \frac{|Q_2^{1/2} a_n|}{|Q_1^{1/2} a_n|} = +\infty.$$

Define on the measurable space $(H, \mathscr{B}(H))$ random variables

$$\xi_n(x) = \frac{\langle x, a_n\rangle}{|Q_1^{1/2} a_n|}, \qquad n \in \mathbb{N}.$$

Then $\int_H \xi_n^2 \mu(dx) = 1$ and

$$\lim_{n\to\infty} \int_H \xi_n^2 \nu(dx) = \lim_{n\to\infty} \left(\frac{|Q_1^{1/2} a_n|}{|Q_2^{1/2} a_n|}\right)^2 = 0.$$

One can therefore find sequences $\gamma_k \uparrow \infty$ and $n_k \uparrow +\infty$ such that

$$\mu\left(\left\{x : \lim_{k\to\infty} |\gamma_k \xi_{n_k}(x)| = +\infty\right\}\right) = 1,$$
$$\nu\left(\left\{x : \lim_{k\to\infty} |\gamma_k \xi_{n_k}(x)| = 0\right\}\right) = 1.$$

So the measures μ and ν are singular. In this way the necessity of condition (i) has been proved. To show (iii) we can assume, without any loss of generality, that $Q_1^{1/2}(H) = Q_2^{1/2}(H) = H_0$ and $\overline{H}_0 = H$. Let R be the operator defined by (2.31) and $\{e_j, \lambda_j\}$ the eigensequence determined by Q_1. One can check directly that the operator R is represented in the basis $\{e_j\}$ by the matrix $\{r_{ij}\}$ where

$$r_{ij} = \frac{\langle Q_2 e_i, e_j\rangle}{\sqrt{\lambda_i \lambda_j}}, \qquad i, j \in \mathbb{N}.$$

Our aim is to show that

$$\sum_{i,j=1}^{\infty} \left(r_{ij} - \delta_{ij}\right)^2 < +\infty,$$

where $\delta_{ii} = 1$ and $\delta_{ij} = 0$ if $i \neq j$. Let

$$\xi_j = \frac{\langle x, e_j\rangle}{\sqrt{\lambda_j}}, \qquad j \in \mathbb{N},\ x \in H.$$

Then

$$\int_H \xi_j\, \mu(dx) = 0, \qquad \int_H \xi_i\, \xi_j\, \mu(dx) = \delta_{ij}, \quad i, j \in \mathbb{N},$$
$$\int_H \xi_j\, \nu(dx) = 0, \qquad \int_H \xi_i \xi_j\, \nu(dx) = r_{ij}, \quad i, j \in \mathbb{N}.$$

If $\mathscr{G}_n = \sigma(\xi_1, \ldots, \xi_n)$, $n \in \mathbb{N}$, then

$$H_{\mathscr{G}_n}(\mu, \nu) \geq H(\mu, \nu) > 0, \quad n \in \mathbb{N},$$

by Proposition 2.20(i). If $R_n = \{r_{ij}\}_{i,j=1,\ldots,n}$ and $I_n = \{\delta_{ij}\}_{i,j=1,\ldots,n}$, one gets by direct calculation that for $n \in \mathbb{N}$,

$$H_{\mathscr{G}_n}(\mu, \nu) = H(\mathscr{N}(0, I_n), \mathscr{N}(0, R_n)) = \frac{(\det R_n)^{1/4}}{\left(\det\left(\frac{I_n + R_n}{2}\right)\right)^{1/2}} \geq H(\mu, \nu).$$

If $\lambda_{nj}, = 1, \ldots, n$ are eigenvalues of R_n, then

$$\begin{aligned} -\log H_{\mathscr{G}_n}(\mu, \nu) &= \sum_{j=1}^{n} \log\left[\frac{(1+\lambda_{nj})^2}{4\lambda_{nj}}\right] \\ &\leq -\log H(\mu, \nu) < +\infty. \end{aligned} \tag{2.32}$$

Note that $\log \frac{(1+\lambda)^2}{4\lambda} \geq 0$ for all $\lambda > 0$ with the equality taking place only for $\lambda = 1$. Consequently for some constants $C_1, C_2 > 0$

$$0 < C_1 \leq \lambda_{nj} < C_2, \quad j = 1, \ldots, n, \ n \in \mathbb{N}.$$

There exists also a constant $C_3 > 0$ such that

$$(1-\lambda)^2 \leq C_3 \log \frac{(1+\lambda)^2}{4\lambda}, \quad \forall\, \lambda \in (C_1, C_2). \tag{2.33}$$

Taking into account (2.32) and (2.33) one obtains that

$$\sup_{n \in \mathbb{N}} \sum_{j=1}^{n} (1-\lambda_{nj})^2 \leq -C_3 \log H(\mu, \nu) < +\infty.$$

But $\sum_{j=1}^{n}(1-\lambda_{nj})^2$ is the Hilbert–Schmidt norm of the matrix $I_n - R_n$. Consequently the Hilbert–Schmidt norm of $I - R$ is finite. Hence the proof of property (iii) is complete as well.

Step 2 If the covariance operators Q_1 and Q_2 satisfy (i) and (iii) then $\mathscr{N}(0, Q_1) \sim \mathscr{N}(0, Q_2)$.

Let $\{f_j, \tau_j\}$ be the eigensequence corresponding to the operator R:

$$R f_j = \tau_j f_j, \quad j \in \mathbb{N}.$$

Then $C_1 \geq \tau_j \geq C_2 > 0$, for all $j \in \mathbb{N}$ and some $C_1, C_2 > 0$. Let $\{\xi_j\}$ be a sequence of $\mathscr{N}(0, 1)$ independent real random variables. We claim that

$$\mathscr{L}\left(\sum_{j=1}^{\infty} \xi_j Q_1^{1/2} f_j\right) = \mathscr{N}(0, Q_1), \tag{2.34}$$

$$\mathscr{L}\left(\sum_{j=1}^{\infty} \sqrt{\tau_j}\xi_j Q_1^{1/2} f_j\right) = \mathscr{N}(0, Q_2). \tag{2.35}$$

Since $\sum_{j=1}^{\infty} |Q_1^{1/2} f_j|^2 < +\infty$ therefore the series in (2.34) and (2.35) do converge in $L^2(\Omega, \mathscr{F}, \mathbb{P}; H)$ and their distributions have symmetric Gaussian laws on H.

We check for instance (2.35). Note that

$$\begin{aligned}
&\mathbb{E}\left(\left\langle \sum_{j=1}^{\infty} \sqrt{\tau_j}\xi_j Q_1^{1/2} f_j, a\right\rangle \left\langle \sum_{l=1}^{\infty} \sqrt{\tau_l}\xi_l Q_1^{1/2} f_l, b\right\rangle\right) \\
&= \sum_{j=1}^{\infty} \tau_j \langle Q_1^{1/2} f_j, a\rangle\langle Q_1^{1/2} f_j, b\rangle \\
&= \sum_{j=1}^{\infty} \left\langle Q_1^{1/2}\left(Q_1^{-1/2} Q_2^{1/2}\right)\left(Q_1^{-1/2} Q_2^{1/2}\right)^* f_j, a\right\rangle \left\langle f_j, Q_1^{1/2} b\right\rangle \\
&= \sum_{j=1}^{\infty} \left\langle f_j, \left(Q_1^{-1/2} Q_2^{1/2}\right) Q_2^{1/2} a\right\rangle \left\langle f_j, Q_1^{1/2} b\right\rangle \\
&= \left\langle \left(Q_1^{-1/2} Q_2^{1/2}\right) Q_2^{1/2} a, Q_1^{1/2} b\right\rangle \\
&= \langle Q_2 a, b\rangle, \quad a, b \in H.
\end{aligned}$$

Let $\widetilde{\mu}$ and $\widetilde{\nu}$ be measures generated on $\mathbb{R}^\infty$ by sequences $\{\xi_j\}$ and $\{\sqrt{\tau_j}\xi_j\}$. Since [(6)]

$$H(\mathscr{N}(0, 1), \mathscr{N}(0, \tau_j)) = \sqrt{\frac{2\sqrt{\tau_j}}{1 + \tau_j}}$$

and $\sum_{j=1}^{\infty}(1 - \tau_j)^2 < \infty$, it follows easily from Proposition 2.21 that $\widetilde{\mu} \sim \widetilde{\nu}$. But μ and ν are the images of the measures $\widetilde{\mu}$ and $\widetilde{\nu}$ respectively by the measurable mapping $V : \mathbb{R}^\infty \to H$, $\{v_k\} \to V(v_k)$,

$$V(v_k) = \begin{cases} \lim\limits_{N\to\infty} \sum\limits_{k=1}^{\infty} v_k Q^{1/2} f_k & \text{if the limit exists,} \\ 0 \quad \text{otherwise.} \end{cases}$$

So the measures μ and ν are equivalent as well.

Step 3 If $\mathscr{N}(m_1, Q_1)$ and $\mathscr{N}(m_2, Q_2)$ are not singular, then (i), (ii) and (iii) hold and $\mathscr{N}(m_1, Q_1) \sim \mathscr{N}(m_2, Q_2)$.

To see this, note that $\mathscr{N}(0, Q_1)$ and $\mathscr{N}(m_2 - m_1, Q_2)$ are not singular. By Steps 1 and 2, (i) and (iii) hold and $\mathscr{N}(0, Q_1) \sim \mathscr{N}(0, Q_2)$. Consequently $\mathscr{N}(m_2 - m_1, Q_2)$

[(6)] Here H denotes the Hellinger integral.

and $\mathcal{N}(0, Q_2)$ are not singular. By Theorem 2.23 the condition (ii) holds as well and

$$\mathcal{N}(m_2 - m_1, Q_2) \sim \mathcal{N}(0, Q_2) \sim \mathcal{N}(0, Q_1),$$

so also

$$\mathcal{N}(m_2, Q_2) \sim \mathcal{N}(m_1, Q_2) \sim \mathcal{N}(m_1, Q_1).$$

Step 4 If (i), (ii) and (iii) hold then $\mathcal{N}(m_1, Q_1) \sim \mathcal{N}(m_2, Q_2)$.

This is a direct consequence of Theorem 2.23 and Step 2. □

2.3.3 An application to a general Cameron–Martin formula

We will deduce now from the results of Section 2.3.2, the so called *Cameron–Martin* formula for Gaussian measures on a Banach space.

Proposition 2.26 *Let μ be a symmetric Gaussian measure on a separable Banach space E and let H_μ, with the norm $|\cdot|_\mu$, be its reproducing kernel. For arbitrary $h \in H_\mu$, the measure*

$$\mu^h(\Gamma) = \mu(\Gamma - h), \quad \Gamma \in \mathscr{B}(E),$$

is absolutely continuous with respect to μ and

$$\frac{d\mu^h}{d\mu}(x) = \exp\left\{ \langle x, h\rangle_\mu - \frac{1}{2}|h|_\mu^2 \right\}, \quad x \in E, \ \mu\text{-a.s.} \tag{2.36}$$

Proof Let $\{e_i\}$ be a complete orthonormal basis in H_μ and $\{\xi_j\}$ a sequence of independent $\mathcal{N}(0, 1)$ random variables such that

$$\mathscr{L}\left(\sum_{j=1}^{\infty} \xi_j e_j\right) = \mu.$$

Then

$$\mathscr{L}\left(h + \sum_{j=1}^{\infty} \xi_j e_j\right) = \mu^h.$$

Note that

$$h = \sum_{j=1}^{\infty} \langle h, e_j\rangle_\mu e_j$$

and

$$|h|_\mu^2 = \sum_{j=1}^{\infty} [\langle h, e_j \rangle_\mu]^2 < +\infty.$$

If $\widetilde{\mu}$ and $\widetilde{\mu}^h$ are the laws of $\mathbb{R}^\infty$-valued random variables $\{\xi_j,\ j \in \mathbb{N}\}$ and $\{\xi_j + \langle h, e_j \rangle_\mu,\ j \in \mathbb{N}\}$, then μ and μ^h are the images of $\widetilde{\mu}$ and $\widetilde{\mu}^h$ respectively by the transformation

$$\mathscr{T} : \mathbb{R}^\infty \to H_\mu, \alpha = \{\alpha_j\} \to \mathscr{T}(\alpha)$$

defined by

$$\mathscr{T}(\alpha) = \begin{cases} \lim_{N\to\infty} \sum_{j=1}^{N} \alpha_j e_j & \text{if the limit exists,} \\ 0 \quad \text{otherwise.} \end{cases}$$

By Theorem 2.23 measures $\widetilde{\mu}$ and $\widetilde{\mu}^h$ are equivalent and

$$\frac{d\widetilde{\mu}^h}{d\widetilde{\mu}}(\alpha) = e^{\left\langle h, \sum_{j=1}^\infty \alpha_j e_j \right\rangle_\mu - \frac{1}{2}|h|_\mu^2}, \qquad \alpha \in \mathbb{R}^\infty,\ \widetilde{\mu}\text{-a.s.},$$

which implies (2.36). □

2.3.4 The Bochner theorem

Let H be a separable Hilbert space with inner product $\langle \cdot, \cdot \rangle$. Let us recall that for arbitrary probability measures μ the characteristic function φ_μ (or $\widehat{\mu}$) of μ is defined as

$$\widehat{\mu}(\lambda) = \varphi_\mu(\lambda) = \int_H e^{i\langle \lambda, x \rangle} \mu(dx), \quad \lambda \in H. \tag{2.37}$$

We are going now to give a complete description of characteristic functions of all probability measures.

First notice that it follows from Lebesgue's dominated convergence theorem that the characteristic function φ_μ of μ is continuous; moreover for arbitrary vectors $\lambda_1, \lambda_2, \ldots, \lambda_N \in H$ and complex numbers $z_1, z_2, \ldots, z_N \in \mathbb{C}$, $N \in \mathbb{N}$, one has

$$\sum_{k,j=1}^{N} \varphi_\mu(\lambda_k - \lambda_j) z_k \overline{z}_j = \int_H \left| \sum_{k=1}^{N} e^{i\langle \lambda_k, x \rangle} z_k \right|^2 \mu(dx)$$

and therefore

$$\sum_{k,j=1}^{N} \varphi_\mu(\lambda_k - \lambda_j) z_k \overline{z}_j \geq 0, \quad \forall\, N \in \mathbb{N},\ z_1, z_2, \ldots, z_N \in \mathbb{C}. \tag{2.38}$$

A function φ_μ that satisfies (2.38) is said to be *positive definite*.

If the space H is finite dimensional then, by the classical Bochner theorem the property (2.38), together with continuity of φ and the obvious relation $\varphi(0) = 1$, characterizes characteristic functions of probability measures completely.

The extension of the Bochner theorem to arbitrary separable Hilbert spaces is given by the following theorem.

Theorem 2.27 *If $\varphi : H \to \mathbb{C}$ is a characteristic function of a probability measure μ on $(H, \mathscr{B}(H))$ then*

(i) φ is a continuous, positive definite function such that $\varphi(0) = 1$,
(ii) for arbitrary $\varepsilon > 0$ there exists a nonnegative nuclear operator S_ε such that

$$1 - \mathfrak{Re}\,\varphi(\lambda) \le \varepsilon \quad \text{for all } \lambda \text{ satisfying } \langle S_\varepsilon \lambda, \lambda\rangle \le 1. \tag{2.39}$$

Conversely, an arbitrary function $\varphi : H \to \mathbb{C}$ satisfying (i) and (ii) is the characteristic function of a probability measure μ on H.

Proof Assume that $\varphi\colon H \to \mathbb{C}$ is the characteristic function of a probability measure μ on H. Since (i) is clear, it is enough to show (ii). Define the linear operator $S_R,\ R > 0$, by the relation

$$\langle S_R \lambda, \lambda\rangle = \int_{B(0,R)} |\langle x, \lambda\rangle|^2 \mu(dx), \quad \forall\, \lambda \in H.$$

It is obvious that S_R is a positive definite operator with the finite trace

$$\mathrm{Tr}\, S_R = \int_{B(0,R)} |x|^2 \mu(dx) < +\infty.$$

However

$$\begin{aligned}
1 - \mathfrak{Re}\,\varphi(\lambda) &= \int_H (1 - \cos\langle x, \lambda\rangle)\mu(dx) \\
&= 2\int_H \left[\sin\frac{\langle x, \lambda\rangle}{2}\right]^2 \mu(dx) \\
&\le 2\int_{B(0,R)} \left[\sin\frac{\langle x, \lambda\rangle}{2}\right]^2 \mu(dx) + 2\mu(B(0, R)^c) \\
&\le \frac{1}{2}\langle S_R \lambda, \lambda\rangle + 2\mu(B(0, R))^c).
\end{aligned}$$

Now it suffices to choose R such that $2\mu(B(0, R)^c) < \varepsilon/2$ and set $S_\varepsilon = S_R/\varepsilon$.

Assume conversely that φ verifies (i) and (ii). Let $\{e_k\}$ be a fixed orthonormal basis in H and let $\Pi_n = \sum_{i=1}^n e_i \otimes e_i$ be the orthogonal projector of H onto

$H_n = \text{lin}\{e_1, \dots, e_n\}$ [7] and $\Pi_{n.m}$ the orthogonal projector on the linear span of $\{e_{n+1}, \dots, e_m\}$. Define

$$\varphi_n(\lambda) = \varphi(\Pi_n \lambda), \quad \lambda \in H.$$

Then φ_n is the characteristic function of a measure μ_n concentrated on H_n by the Bochner theorem in finite dimensions. Our aim is to show that the sequence $\{\mu_n\}$ is tight. This will imply, by Theorem 2.3, that there exists a subsequence $\{\mu_{n_j}\}$ weakly convergent to a measure μ and it is clear that φ is the characteristic function of μ.

To prove tightness of $\{\mu_n\}$ we will show that for arbitrary $r \in (0, 1)$, $\varepsilon \in (0, 1/2)$, there exist a natural number M and a positive number k, such that

$$\begin{aligned} \mu_n\left(\{x \in H : |\Pi_M x| \le k\}\right) = \mu_n(A_{M,k}) \ge 1 - \varepsilon \\ \mu_n\left(\{x \in H : |x - \Pi_M x| \le r\}\right) = \mu_n(C_{M,r}) \ge 1 - \varepsilon. \end{aligned} \tag{2.40}$$

From this and Proposition 2.1 tightness of $\{\mu_n\}$ easily follows. To see this let $a_1, \dots, a_l$ be points in $B(0, k) \cap H_M$ such that

$$\bigcup_{i=1}^{l} B(a_i, r) \supset B(0, k) \cap H_M.$$

Since

$$\bigcup_{i=1}^{l} B(a_i, 2r) \supset A_{M,k} \cap C_{M,r},$$

we have

$$\mu\left(\bigcup_{i=1}^{l} B(a_i, 2r)\right) \ge 1 - 2\varepsilon,$$

and the condition of Proposition 2.1 holds.

It remains to prove (2.40). Let $r \in (0, 1)$, $\varepsilon \in (0, 1/2)$, and let $S \ge 0$ be nuclear operator such that $1 - \mathfrak{Re}\, \varphi(\lambda) < \varepsilon/L$ if $\langle S\lambda, \lambda\rangle \le 1$, where $L > 0$ is a positive number to be chosen later. Since, by the Fubini theorem

$$\begin{aligned} \int_H e^{-\frac{1}{2}|\Pi_{M,N} x|^2} \mu_n(dx) &= \int_H \left[\int_H e^{i\langle \lambda, x\rangle} \mathcal{N}(0, \Pi_{M,N})(d\lambda)\right] \mu_n(dx) \\ &= \int_H \varphi_n(\lambda) \mathcal{N}(0, \Pi_{M,N})(d\lambda), \end{aligned}$$

[7] $\text{lin}\{e_1, \dots, e_n\}$ is the subspace in H generated by $\{e_1, \dots, e_n\}$.

we have

$$\begin{aligned}\int_H \left(1 - e^{-\frac{1}{2}|\Pi_{M,N}x|^2}\right) \mu_n(dx) &= \int_H (1 - \mathfrak{Re}\ \varphi(\lambda))\mathcal{N}(0, \Pi_{M,N}\Pi_n)(d\lambda) \\ &= \int_{\langle S\lambda,\lambda\rangle \le 1} (1 - \mathfrak{Re}\ \varphi(\lambda))\mathcal{N}(0, \Pi_{M,N}\Pi_n)(d\lambda) \\ &\quad + \int_{\langle S\lambda,\lambda\rangle \ge 1} (1 - \mathfrak{Re}\ \varphi(\lambda))\mathcal{N}(0, \Pi_{M,N}\Pi_n)(d\lambda) \\ &\le \frac{\varepsilon}{L} + \int_H \langle S\lambda, \lambda\rangle \mathcal{N}(0, \Pi_{M,N}\Pi_n)(d\lambda) \\ &\le \frac{\varepsilon}{L} + \mathrm{Tr}\,[S\Pi_{M,N}\Pi_n].\end{aligned}$$

Let M now be such that

$$\mathrm{Tr}\,[S\Pi_{M,N}\Pi_n] \le \frac{\varepsilon}{L}, \quad n \in \mathbb{N},\ N > M.$$

Consequently

$$\int_H \left(1 - e^{-\frac{1}{2}\,|\Pi_{M,N}x|^2}\right) \mu_n(dx) \le \frac{2\varepsilon}{L}, \quad n \in \mathbb{N},\ N > M.$$

For arbitrary $r > 0$ we then have the estimate

$$\begin{aligned}\frac{2\varepsilon}{L} &\ge \int_{\{x\in H:\ |\Pi_{M,N}x|>r\}} \left(1 - e^{-\frac{1}{2}|\Pi_{M,m}x|^2}\right) \mu_n(dx) \\ &\ge \left(1 - e^{-\frac{1}{2}\,r^2}\right) \mu_n\left(\{x \in H:\ |\Pi_{M,N}x| > r\}\right).\end{aligned}$$

It follows that

$$\mu_n\left(\left\{x \in H:\ |\Pi_{M,N}x| \le r\right\}\right) \ge 1 - \frac{2\varepsilon}{L\left(1 - e^{-r^2/2}\right)}.$$

Letting $N \to \infty$ and choosing L such that $L > 2(1 - e^{-r^2/2})^{-1}$ we see that

$$\mu_n\left(\left\{x \in H:\ |x - \Pi_M x|^2 \le r\right\}\right) \ge 1 - \varepsilon.$$

Thus the second inequality in (2.40) is fulfilled.

To prove the first inequality denote by μ_n^j the image of the measure μ_n by the transformation $x \to \langle x, e_j\rangle$. Then the characteristic function φ_n^j of the measure μ_n^j is

$$\varphi_n^j(t) = \varphi(t\Pi_n e_j), \quad t \in \mathbb{R}^1.$$

Since $\varphi_n^j = \varphi$ for $n \ge j$, we have $\mu_n^j = \mu_j^j$ for $n = j+1, \ldots$ and the sequence $\{\mu_n^j\}$ is tight for any choice of j. In particular, for arbitrary $\varepsilon > 0$ and $M \in \mathbb{N}$ there exists $k > 0$ such that

$$\mu_n^j([-kM^{-1/2}, kM^{-1/2}]) \ge 1 - \varepsilon, \quad n \in \mathbb{N},\ j = 1, \ldots, M.$$

Consequently

$$\mu_n(\{x \in H : |\langle x, e_j\rangle| \le kM^{-1/2}\}) \ge 1 - \varepsilon, \quad n \in \mathbb{N},\ j = 1, \ldots, M,$$

which implies

$$\mu_n\left(\left\{x \in H : \sum_{j=1}^{M} |\langle x, e_j\rangle|^2 \le k^2\right\}\right) \ge 1 - \varepsilon, \quad n \in \mathbb{N},\ j = 1, \dots, M.$$

In this way the proof of (2.40) is complete. □

3

Stochastic processes

In this chapter we summarize basic definitions and facts about real valued stochastic processes and generalize some of these results to processes with values in Banach or Hilbert spaces.

3.1 General concepts

Assume that E is a separable Banach space and let $\mathscr{B}(E)$ be the σ-field of its Borel subsets. Let $(\Omega, \mathscr{F}, \mathbb{P})$ be a probability space and let I be an interval of $\mathbb{R}^1$. An arbitrary family $X = \{X(t)\}_{t\in I}$, of E-valued random variables $X(t)$, $t \in I$, defined on Ω is called a *stochastic process*. We say also that $X(t)$ is a stochastic process on I. We set $X(t, \omega) = X(t)(\omega)$ for all $t \in I$ and $\omega \in \Omega$. Functions $X(\cdot, \omega)$ are called the *trajectories* of $X(t)$.

A stochastic process Y is called a *modification* or a *version* of X if

$$\mathbb{P}(\omega \in \Omega : \; X(t, \omega) \neq Y(t, \omega)) = 0 \quad \text{for all } t \in I.$$

We now give several definitions of regularity for a process X on $I = [0, T]$.

- X is *measurable* if the mapping $X(\cdot, \cdot) : I \times \Omega \to E$ is $\mathscr{B}(I) \times \mathscr{F}$-measurable.
- X is *stochastically continuous* at $t_0 \in I$ if for all $\varepsilon > 0$ and all $\delta > 0$ there exists $\rho > 0$ such that

$$\mathbb{P}(\|X(t) - X(t_0)\| \geq \varepsilon) \leq \delta \quad \text{for all } t \in [t_0 - \rho, t_0 + \rho] \cap [0, T].$$

- X is *stochastically continuous* in I if it is stochastically continuous at any point of I.
- X is *stochastically uniformly continuous* in I if for all $\varepsilon > 0$ and all $\delta > 0$ there exists $\rho > 0$ such that

$$\mathbb{P}(\|X(t) - X(s)\| \geq \varepsilon) \leq \delta \quad \text{for all } t, s \in I \text{ such that } |t - s| < \rho.$$

- X is *mean square continuous* at $t_0 \in I$ if
$$\lim_{t \to t_0} \mathbb{E}\left(\|X(t) - X(t_0)\|^2\right) = 0.$$
- X is *mean square continuous* in I if it is mean square continuous at any point of I.
- X is *continuous with probability* 1 (or *continuous*) if its trajectories $X(\cdot, \omega)$ are continuous $\mathbb{P}$-almost surely.
- X is α*-Hölder continuous with probability* 1 (or α*-Hölder continuous*) if its trajectories $X(\cdot, \omega)$ are α-Hölder continuous $\mathbb{P}$-almost surely.

Lemma 3.1 *A stochastically continuous process on* $[0, T]$ *is uniformly stochastically continuous on* $[0, T]$.

Proof Let X be a stochastically continuous process on $[0, T]$ and $\varepsilon > 0$, $\delta > 0$ arbitrary numbers. For any $r \in [0, T]$ there exists an open interval $I(r) \neq \varnothing$ with center at r such that
$$\mathbb{P}\left(\|X(s) - X(r)\| \geq \frac{\varepsilon}{2}\right) \leq \frac{\delta}{2} \quad \text{for all } s \in [0, T] \cap I(r).$$
Consequently for $s, t \in [0, T] \cap I(r)$
$$\mathbb{P}(\|X(s) - X(t)\| \geq \varepsilon) \leq \mathbb{P}\left(\|X(s) - X(r)\| \geq \tfrac{\varepsilon}{2} \text{ or } \|X(t) - X(r)\| \geq \tfrac{\varepsilon}{2}\right) \leq \delta.$$
Since the interval $[0, T]$ is compact, there exists a finite family of intervals, $I(r_1), I(r_2), \ldots, I(r_k)$ which covers $[0, T]$. This implies easily the result. □

We remark that if $X(t)$ is a stochastic process on $[0, T]$ the function $X(\cdot, \cdot)$ need not to be measurable in the product space $\mathscr{B}([0, T]) \times \mathscr{F}$. However, we have the following result.

Proposition 3.2 *Let* $X(t)$, $t \in [0, T]$, *be a stochastically continuous process with values in a separable Banach space* E. *Then* X *has a measurable modification.*

Proof By Lemma 3.1, X is stochastically uniformly continuous. Thus, for any positive integer m there exists a partition $0 = t_{m,0} < t_{m,1} < \cdots < t_{m,n(m)} = T$, such that for $t \in (t_{m,k}, t_{m,k+1}]$,
$$\mathbb{P}(\|X(t_{m,k}) - X(t))\| \geq 2^{-m}) \leq 2^{-m}, \quad k = 0, 1, \ldots, n(m) - 1.$$
Define
$$X_m(t, \omega) = \begin{cases} X(0, \omega) & \text{if } t = 0 \\ X(t_{m,k}, \omega) & \text{if } t \in (t_{m,k}, t_{m,k+1}] \text{ and } k \leq n(m) - 1 \end{cases}$$
for $\omega \in \Omega$ and $t \in [0, T]$. It is clear that X_m is measurable with respect to the σ-algebra $\mathscr{B}([0, T]) \times \mathscr{F}$. Denote by A the set of all those $(t, \omega) \in [0, T] \times \Omega$ for which the sequence $\{X_m(t, \omega)\}$ is convergent. Then $A \in \mathscr{B}([0, T]) \times \mathscr{F}$ and the

process Y defined as

$$Y(t,\omega) = \begin{cases} \lim_{m\to\infty} X_m(t,\omega) & \text{if } (t,\omega) \in A \\ 0 \quad \text{if } (t,\omega) \notin A \end{cases}$$

is $\mathscr{B}([0,T]) \times \mathscr{F}$ measurable. The Borel–Cantelli lemma implies that for arbitrary fixed $t \in [0,T]$ and $\mathbb{P}$-almost all $\omega \in \Omega$, the inequality $\|X_m(t,\omega) - X(t,\omega)\| \le 2^{-m}$ is satisfied for all sufficiently large $m \in \mathbb{N}$. Therefore, for all $t \in [0,T]$, $X(t,\cdot) = Y(t,\cdot)$ $\mathbb{P}$-a.s. and the process Y is the required modification. □

3.2 Kolmogorov test

An important result on existence of regular modifications is provided by the following theorem of Kolmogorov.

Theorem 3.3 *Let $X(t)$, $t \in [0,T]$, be a stochastic process with values in a complete metric space (E,ρ) such that for some constants $C > 0$, $\varepsilon > 0$, $\delta > 1$ and all $t, s \in [0,T]$,*

$$\mathbb{E}\left((\rho(X(t),X(s)))^\delta\right) \le C|t-s|^{1+\varepsilon}. \tag{3.1}$$

Then there exists a version of X with $\mathbb{P}$-almost all trajectories being Hölder continuous functions with an arbitrary exponent smaller than ε/δ. In particular X has a continuous version.

Proof We can assume that $T = 1$. We first remark that the process X is stochastically uniformly continuous, as for any $\beta > 0$

$$\mathbb{P}(\rho(X(t),X(s)) \ge \beta) \le \beta^{-\delta}\mathbb{E}\left((\rho(X(t),X(s)))^\delta\right) \le C\beta^{-\delta}\,|t-s|^{1+\varepsilon}. \tag{3.2}$$

Let $0 < \gamma < \varepsilon/\delta$. By (3.2), for $k = 1, 2, \ldots, 2^n$, $n \in \mathbb{N}$,

$$\mathbb{P}\left(\rho(X(k2^{-n})), X((k-1)2^{-n})) \ge 2^{-\gamma n}\right) \le C2^{-n(1+\varepsilon-\gamma\delta)}$$

and therefore

$$\mathbb{P}\left(\max_{1\le k\le 2^n} \rho(X(k2^{-n}), X((k-1)2^{-n}) \ge 2^{-\gamma n}\right)$$
$$\le \sum_{k=1}^{2^n} \mathbb{P}\left(\rho(X(k2^{-n}), X((k-1)2^{-n}) \ge C2^{-\gamma n}\right) \le C2^{-n(\varepsilon-\gamma\delta)}.$$

Since

$$\sum_{n=1}^{\infty} 2^{-(\varepsilon-\gamma\delta)n} < +\infty,$$

by the Borel–Cantelli lemma, there exists a set $\tilde{\Omega} \in \mathscr{F}$ and a random variable $\tilde{N}(\omega)$, $\omega \in \Omega$, taking values in $\mathbb{N}$ such that for $\omega \in \tilde{\Omega}$ and $n \geq \tilde{N}(\omega)$,

$$\max_{1 \leq k \leq 2^n} \rho(X(k2^{-n}), X((k-1)2^{-n}) < 2^{-\gamma n}. \tag{3.3}$$

Let

$$D_n = \left\{0, \tfrac{1}{2^n}, \cdots, \tfrac{2^n-1}{2^n}\right\}, \quad D = \bigcup_{n=1}^{\infty} D_n.$$

Note that the arbitrary number $x = k2^{-n}$ has a unique representation in the form,

$$x = \sum_{j=1}^{m} \varepsilon_j 2^{-j}, \quad \text{where } \varepsilon_j = 0 \text{ or } 1. \tag{3.4}$$

Now let $m > n \geq \tilde{N}(\omega)$ and $t = k2^{-m}$, $s = l2^{-m}$, $0 \leq l \leq k < 2^m$ be such that $t - s < 2^{-n}$. By (3.4) it follows that

$$t - s = k2^{-m} - l2^{-m} = \sum_{j=n+1}^{m} \varepsilon_j 2^{-j}, \quad \varepsilon_j = 0 \text{ or } 1, \; j = n+1, \ldots, m. \tag{3.5}$$

Consequently, by (3.3) and (3.5)

$$\begin{aligned}
\rho(X(t), X(s)) &\leq \rho\left(X\left(\frac{l}{2^m}\right), X\left(\frac{l}{2^m} + \frac{\varepsilon_1}{2^{n+1}}\right)\right) \\
&\quad + \rho\left(X\left(\frac{l}{2^m} + \frac{\varepsilon_1}{2^{n+1}}\right), X\left(\frac{l}{2^m} + \frac{\varepsilon_1}{2^{n+1}} + \frac{\varepsilon_2}{2^{n+2}}\right)\right) \\
&\quad + \cdots + \rho\left(X\left(\frac{l}{2^m} + \frac{\varepsilon_1}{2^{n+1}} \cdots + \frac{\varepsilon_{m-1}}{2^{m-1}}\right),\right. \\
&\qquad \left. X\left(\frac{l}{2^m} + \frac{\varepsilon_1}{2^{n+1}} + \cdots + \frac{\varepsilon_m}{2^m}\right)\right) \\
&\leq 2^{-\gamma(n+1)} + \cdots + 2^{-\gamma m} \leq 2^{-\gamma(n+1)}(1 - 2^{-\gamma})^{-1}.
\end{aligned}$$

Selecting n such that $2^{-n-1} \leq t - s \leq 2^{-n}$ one gets,

$$\rho(X(t), X(s)) \leq \frac{(t-s)^\gamma}{1 - 2^{-\gamma}}. \tag{3.6}$$

Therefore $X(t, \omega)$, $t \in D$, is a uniformly continuous function on D and has unique extension to a continuous function $\tilde{X}(t, \omega)$, $t \in [0, T]$. Set $\tilde{X}(t, \omega) = 0$ for $t \in [0, T]$ and $\omega \notin \tilde{\Omega}$. Stochastic continuity implies that the process $\tilde{X}$ is a modification of X and inequality (3.6) implies that trajectories of $\tilde{X}$ are γ-Hölder continuous. □

The Kolmogorov theorem has been extended to $\mathbb{R}^m$-valued families of random variables $X(\xi)$, $\xi \in \mathscr{O} \subset \mathbb{R}^d$, indexed from the open bounded set $\mathscr{O}$. As in [702] we derive it from a version of the Garsia, Rademich and Rumsay lemma, see [335]. In our presentation we weaken assumptions assuming that the set $\mathscr{O}$ is *regular* in the following sense. There exists a sequence $\{\mathcal{O}_n\}$ of open sets included in $\mathcal{O}$ such that $\bigcup_{x\in\mathcal{O}_n} B(x,\frac{1}{n}) \subset \mathcal{O}$ and a number $\kappa > 0$ such that

$$n^d \lambda\left(\mathcal{O}_n \cap B(x,\tfrac{1}{n})\right) \ge \kappa \quad \text{for all } x \in \mathcal{O}_n,\ n \in \mathbb{N}.$$

Here λ denotes the Lebesgue measure in $\mathbb{R}^d$. Parallelepipeds are trivially regular.

Let us first prove an approximation result.

Proposition 3.4 *Assume that $r > 0$, $\mathcal{O} \subset \mathbb{R}^d$, $\mathcal{O}' \subset \mathcal{O}$ are such that $B(x,r) \subset \mathcal{O}$ for all $x \in \mathcal{O}'$. For $x \in \mathcal{O}'$ and for a locally integrable function f on $\mathcal{O}$ we define*

$$f_r(x) = \int_{\mathbb{R}^d} f(x-z)\varphi_r(z)dz = \int_{B(0,r)} f(x-z)\varphi_r(z)dz, \tag{3.7}$$

where

$$\varphi_r(z) = \begin{cases} c_d^{-1} r^{-d} & \text{if } z \in B(0,r), \\ 0 & \text{otherwise}, \end{cases}$$

and c_d is the volume of $B(0,1)$. Then for arbitrary $\delta > 1$ and $\beta > 0$

$$\int_{\mathcal{O}'\times\mathcal{O}'} \frac{|f_r(x)-f_r(y)|^\delta}{|x-y|^\beta}\,dx\,dy \le \int_{\mathcal{O}\times\mathcal{O}} \frac{|f(x)-f(y)|^\delta}{|x-y|^\beta}\,dx\,dy. \tag{3.8}$$

Proof If $x, y \in \mathcal{O}'$, then by the Schwartz inequality

$$\begin{aligned} |f_r(x)-f_r(y)|^\delta &\le \left|\int_{B(0,r)} |f(x-z)-f(y-z)|\varphi_r(z)dz\right|^\delta \\ &\le \left(\int_{B(0,r)} |\varphi_r(z)|^{\delta^*}\right)^{\frac{\delta}{\delta^*}} \int_{B(0,r)} |f(x-z)-f(y-z)|^\delta dz, \end{aligned}$$

where

$$\frac{1}{\delta} + \frac{1}{\delta^*} = 1. \tag{3.9}$$

Consequently

$$\begin{aligned} I &= \int_{\mathcal{O}'\times\mathcal{O}'} \frac{|f_r(x)-f_r(y)|^\delta}{|x-y|^\beta}\,dx\,dy \\ &\le \left(\int_{B(0,r)} |\varphi_r(z)|^{\delta^*}\right)^{\delta/\delta^*} \int_{B(0,r)} \left[\int_{\mathcal{O}'\times\mathcal{O}'} \frac{|f(x-z)-f(y-z)|^\delta}{|(x-z)-(y-z)|^\beta}\,dx\,dy\right] dz \\ &= \left(\int_{B(0,r)} |\varphi_r(z)|^{\delta^*}\right)^{\delta/\delta^*} \int_{B(0,r)} \left[\int_{\mathcal{O}_z\times\mathcal{O}_z} \frac{|f(x')-f(y')|^\delta}{|x'-y'|^\beta}\,dx'\,dy'\right] dz, \end{aligned}$$

where $\mathcal{O}_z = \mathcal{O}' - z$. Since $\mathcal{O}_z \subset \mathcal{O}$ we have that

$$I \le \left(\int_{B(0,r)} |\varphi_r(z)|^{\delta^*}\right)^{\delta/\delta^*} c_d r^d \int_{B(0,r)} \int_{\mathcal{O}\times\mathcal{O}} \frac{|f(x)-f(y)|^\delta}{|x-y|^\beta}\,dx\,dy.$$

Taking into account (3.9) we see that

$$\left(\int_{B(0,r)} |\varphi_r(z)|^{\delta^*}\right)^{\delta/\delta^*} c_d r^d = 1,$$

and the result follows. □

Now we are ready to prove the following theorem.

Theorem 3.5 *Assume that $\mathscr{O} \in \mathbb{R}^d$ is regular and that for some constants $C > 0$, $\varepsilon > 0$, $\delta > 1$ and all $\xi, \eta \in \mathscr{O}$*

$$\mathbb{E}\left(|X(\xi) - X(\eta)|^\delta\right) \le C|\xi - \eta|^{d+\varepsilon}.$$

Then, there exists a modification of X which has Hölder continuous realizations with an arbitrary exponent smaller than ε/δ.

Proof We will apply a version of Garsia, Rademich and Rumsay, from [220]. By [220, Theorem B.1.5] it follows that there exists a universal constant $\bar{c} > 0$ such that for continuous functions g on $\mathscr{O}_n$

$$|g(x) - g(y)| \le \bar{c}|x - y|^{\frac{\beta - 2d}{\delta}} \left(\int_{\mathscr{O}_n \times \mathscr{O}_n} \frac{|g(x) - g(y)|^\delta}{|x - y|^\beta}\, dx\, dy\right)^{1/\delta}.$$

In particular the function f_n

$$f_n(x) = \int_{\mathbb{R}^d} f(z)\varphi_{r_n}(x - z)dz, \quad x \in \mathscr{O}_n,\ r_n = \frac{1}{n},$$

is continuous on $\mathscr{O}_n$ and by Proposition 3.4

$$|f_n(x) - f_n(y)| \le \bar{c}|x - y|^{\frac{\beta - 2d}{\delta}} \left(\int_{\mathscr{O} \times \mathscr{O}} \frac{|f(\xi) - f(\eta)|^\delta}{|\xi - \eta|^\beta}\, d\xi\, d\eta\right)^{1/\delta}. \tag{3.10}$$

Since the random field X is stochastically continuous one can assume that $X(\xi, \omega)$, $\xi \in \mathscr{O}$, $\omega \in \Omega$, is a measurable function. Setting

$$X_n(x) = \int_{\mathbb{R}^d} X(\xi)\varphi_{r_n}(x - \xi)d\xi, \quad \forall\, x \in \mathscr{O}_n,$$

one has by (3.10)

$$|X_n(x) - X_n(y)|^\delta \le \bar{c}|x - y|^{\beta - 2d} \int_{\mathscr{O} \times \mathscr{O}} \frac{|X(\xi) - X(\eta)|^\delta}{|\xi - \eta|^\beta}\, d\xi\, d\eta. \tag{3.11}$$

However

$$\mathbb{E}\int_{\mathscr{O} \times \mathscr{O}} \frac{|X(\xi) - X(\eta)|^\delta}{|\xi - \eta|^\beta}\, d\xi\, d\eta \le c\int_{\mathscr{O} \times \mathscr{O}} \frac{|\xi - \eta|^{d+\varepsilon}}{|\xi - \eta|^\beta}\, d\xi\, d\eta < +\infty,$$

provided $2d - \beta + \varepsilon > 0$. Thus, if $\beta < 2d + \varepsilon$, the integral on the right hand side of (3.11) is finite with probability one. Consequently if $\gamma < \varepsilon/\delta$ there exists a finite random variable Y such that

$$|X_n(x) - X_n(y)| \leq Y|x - y|^{\gamma}, \quad \forall\, x, y \in \mathscr{O}_n. \tag{3.12}$$

However, for λ-almost all $x \in \mathscr{O}$,

$$\lim_{n\to\infty} X_n(x) = X(x).$$

Define

$$\widetilde{X}(x) = \begin{cases} \lim\limits_{n\to\infty} X_n(x) & \text{if the limit exists} \\ 0 \quad \text{otherwise.} \end{cases}$$

Then $\mathbb{P}$-almost surely for λ-almost all $x, y \in \mathscr{O}$,

$$|\widetilde{X}(x) - \widetilde{X}(y)| \leq Y|x - y|^{\gamma}. \tag{3.13}$$

Thus $\widetilde{X}$ has a continuous extension satisfying (3.13) for all $x, y \in \mathscr{O}$. Since X is stochastically continuous, the extended family $\widetilde{X}$ is the required modification. □

3.3 Processes with filtration

Let us assume that $I = [0, T]$ or $[0, +\infty)$ and that the probability space $(\Omega, \mathscr{F}, \mathbb{P})$ is equipped with an increasing family of σ-fields $\{\mathscr{F}_t\}$, $t \in I$, called a *filtration*. Denote by $\mathscr{F}_{t+}$ the intersection of all $\mathscr{F}_s$ where $s > t$. The filtration is said to be *normal* if

(i) $\mathscr{F}_0$ contains all $A \in \mathscr{F}$ such that $\mathbb{P}(A) = 0$,
(ii) $\mathscr{F}_t = \mathscr{F}_{t+}$ for all $t \in T$.

If for arbitrary $t \in I$ the random variable $X(t)$ is $\mathscr{F}_t$-measurable then the process X is said to be *adapted* (to the family $\mathscr{F}_t$).

X is *progressively measurable* if for every $t \in [0, T]$ the mapping

$$[0, t] \times \Omega \to E, \quad (s, \omega) \to X(s, \omega)$$

is $\mathscr{B}([0, t]) \times \mathscr{F}_t$-measurable.

The following result can be proved as Proposition 3.2.

Proposition 3.6 *Let $X(t)$, $t \in [0, T]$, be a stochastically continuous and adapted process with values in a separable Banach space E. Then X has a progressively measurable modification.*

The following σ-field $\mathscr{P}_\infty$ of subsets of $[0, +\infty) \times \Omega$ will play an important role in what follows. $\mathscr{P}_\infty$ is the σ-field generated by sets of the form,

$$(s, t] \times F, \quad 0 \leq s < t < \infty, \; F \in \mathscr{F}_s \quad \text{and} \quad \{0\} \times F, \; F \in \mathscr{F}_0. \tag{3.14}$$

This σ-field is called a *predictable* σ-field and its elements are *predictable sets*. The restriction of the σ-field $\mathscr{P}_\infty$ to $[0, T] \times \Omega$ will be denoted by $\mathscr{P}_T$.

An arbitrary measurable mapping from $([0, +\infty) \times \Omega, \mathscr{F}_\infty)$ or $([0, T)\times \Omega, \mathscr{F}_T)$ into $(E, \mathscr{B}(E))$ is called a *predictable process*. A predictable process is necessarily an adapted one. Predictable processes form a large class of processes, as can be seen from the next proposition.

Proposition 3.7

(i) An adapted process Φ with values in $L(U, H)$ such that for arbitrary $u \in U$ and $h \in H$ the process $\langle \Phi(t)u, h\rangle$, $t \geq 0$, has left continuous trajectories is predictable.

(ii) Assume that Φ is an adapted and stochastically continuous process on an interval $[0, T]$. Then the process Φ has a predictable version on $[0, T]$.

Proof (i) We can assume, without any loss of generality, that $U = H = \mathbb{R}^1$. Define processes Φ_m

$$\Phi_m(t, \omega) = \sum_{k=1}^{\infty} \Phi_{m,k}(t, \omega), \quad t \in [0, T],\ \omega \in \Omega,$$

where, for any $k \in \mathbb{N}$,

$$\Phi_{m,k}(t, \omega) = \Phi((k-1)2^{-m}, \omega), \quad \text{for } t \in ((k-1)2^{-m}, k2^{-m}],\ \omega \in \Omega.$$

Then they are predictable and, by the left continuity of the process Φ, they do converge to Φ. So the process Φ is predictable as well.

(ii) We argue as in the proof of Proposition 3.2. Since Φ is stochastically continuous there exists a partition $0 = t_{m,0} < t_{m,1} \cdots < t_{m,n(m)} = T$, such that for $t \in (t_{m,k}, t_{m,k+1}]$,

$$\mathbb{P}(\|\Phi(t_{m,k}) - \Phi(t)\| \geq 2^{-m}) \leq 2^{-m}, \quad k = 0, 1, \ldots, n(m) - 1. \tag{3.15}$$

Define

$$\Phi_m(t, \omega) = \begin{cases} \Phi(0, \omega) & \text{if } t = 0 \\ \Phi(t_{m,k}, \omega) & \text{if } t \in (t_{m,k}, t_{m,k+1}], \end{cases}$$

for $\omega \in \Omega$ and $k \leq n(m) - 1$. It is clear that Φ_m is a predictable process. Denote by A the set of all those $(t, \omega) \in [0, T] \times \Omega$ for which the sequence $\{\Phi_m(t, \omega)\}$ is convergent. Then A is a predictable set and the process Ψ defined as

$$\Psi(t, \omega) = \begin{cases} \lim\limits_{n\to\infty} \Phi_m(t, \omega) & \text{if } (t, \omega) \in A, \\ 0 \quad \text{if } (t, \omega) \notin A, \end{cases}$$

is predictable. The Borel–Cantelli lemma and (3.15) imply that for arbitrary fixed $t \in [0, T]$ and $\mathbb{P}$-almost all $\omega \in \Omega$ the inequality $\|\Phi_m(t, \omega) - \Phi(t, \omega)\| \leq 2^{-m}$ is satisfied for all sufficiently large $m \in \mathbb{N}$. Therefore, for all $t \in [0, T]$, $\Phi(t, \cdot) = \Psi(t, \cdot)$ $\mathbb{P}$-a.s. and the process Ψ is the required modification. □

3.4 Martingales

If $\mathbb{E}\|X(t)\| < +\infty$ for all $t \in I$ then the process is called *integrable*. An integrable and adapted E-valued process $X(t)$, $t \in I$, is said to be a *martingale* if

$$\mathbb{E}(X(t)|\mathscr{F}_s) = X(s), \quad \mathbb{P}\text{-a.s.} \tag{3.16}$$

for arbitrary $t, s \in I$, $t \geq s$. According to the definition of the conditional expectation, the identity (3.16) is equivalent to the statement

$$\int_F X(t)d\mathbb{P} = \int_F X(s)d\mathbb{P}, \quad \forall\, F \in \mathscr{F}_s,\ s \leq t,\ s, t \in I.$$

We recall that a real valued integrable and adapted process $X(t)$, $t \in I$, is said to be a *submartingale* (respectively a *supermartingale*) if

$$\mathbb{E}(X(t)|\mathscr{F}_s) \geq X(s) \text{ (respectively } \mathbb{E}(X(t)|\mathscr{F}_s) \leq X(s)), \ \mathbb{P}\text{-a.s.}$$

We will need the following result.

Proposition 3.8 *The following statements hold.*

(i) If $M(t)$ is a martingale then $\|M(t)\|$, $t \in [0, T]$, is a submartingale.
(ii) If g is an increasing convex function from $[0, +\infty)$ into $[0, +\infty)$ and $\mathbb{E}(g(\|M(t)\|)) < \infty$ for $t \in [0, T]$, then $g(\|M(t)\|)$, $t \in [0, T]$, is a submartingale.

Proof (i) Let $t > s$, $t, s \in [0, T]$, then

$$\begin{aligned}\|M(s)\| &\leq \|\mathbb{E}(M(t) - M(s)|\mathscr{F}_s)\| + \|\mathbb{E}(M(t)|\mathscr{F}_s)\| \\ &\leq \mathbb{E}(\|M(t)\||\mathscr{F}_s), \quad \mathbb{P}\text{-a.s.},\end{aligned}$$

where the last estimate follows from the standard properties of the conditional expectation.

(ii) If $\|M(s)\| \leq \mathbb{E}(\|M(t)\||\mathscr{F}_s)$, $\mathbb{P}$-a.s., $s < t$, then monotonicity and convexity of g together with Jensen's inequality imply

$$g(\|M(s)\|) \leq g(\mathbb{E}(\|M(t)\||\mathscr{F}_s)) \leq \mathbb{E}(g(\|M(t)\||\mathscr{F}_s)), \quad \mathbb{P}\text{-a.s.}$$ □

As an immediate consequence of the proposition and the maximal inequalities for real valued submartingales, see for example [559], we have the following theorem.

Theorem 3.9 *The following statements hold.*

(i) If $M(t)$, $t \in I$, is an E-valued martingale, I a countable set and $p \geq 1$, then for arbitrary $\lambda > 0$,

$$\mathbb{P}\left(\sup_{t\in I}\|M(t)\| \geq \lambda\right) \leq \lambda^{-p}\sup_{t\in I}\mathbb{E}(\|M(t)\|^p). \tag{3.17}$$

(ii) If in addition $p > 1$ then,

$$\mathbb{E}\left(\sup_{t\in I}\|M(t)\|^p\right) \leq (p/(p-1))^p\sup_{t\in I}\mathbb{E}(\|M(t)\|^p). \tag{3.18}$$

(iii) The above estimates remain true if the set I is uncountable and the martingale M is continuous.

Let us fix a number $T > 0$ and denote by $\mathscr{M}_T^2(E)$ the space of all E-valued continuous, square integrable martingales M, such that $M(0) = 0$. Since $\|M(t)\|^2$, $t \in [0, T]$, is a submartingale we have

$$\mathbb{E}\|M(t)\|^2 \leq \mathbb{E}\|M(T)\|^2, \quad t \in [0, T].$$

We will need the following.

Proposition 3.10 *The space $\mathscr{M}_T^2(E)$, equipped with the norm*

$$\|M\|_{\mathscr{M}_T^2(E)} = \left(\mathbb{E}\sup_{t\in[0,T]}\|M(t)\|^2\right)^{1/2}, \tag{3.19}$$

is a Banach space.

Proof Since $\|M(t)\|$ is a submartingale, by Theorem 3.9, identity (3.19) defines a norm. To prove completeness assume that $\{M_n\}$ is a Cauchy sequence, i.e.

$$\mathbb{E}\left(\sup_{t\in[0,T]}\|M_n(t) - M_m(t)\|^2\right) \to 0 \quad \text{as } n, m \to \infty.$$

Then one can find a subsequence $\{M_{n_k}\}$ such that

$$\mathbb{P}\left(\sup_{t\in[0,T]}\|M_{n_{k+1}} - M_{n_k}\| \geq 2^{-k}\right) \leq 2^{-k}.$$

The Borel–Cantelli lemma implies that $\{M_{n_k}\}$ converges $\mathbb{P}$-a.s. to a process $M(t)$, $t \in [0, T]$, uniformly on $[0, T]$. So M is a continuous process. It is clear that, for arbitrary $t \in [0, T]$, the sequence $\{M_{n_k}(t)\}$ converges to $M(t)$ in the mean square. If $0 \leq s \leq t \leq T$ and $k \in \mathbb{N}$ then

$$\mathbb{E}(M_{n_k}(t)|\mathscr{F}_s) = M_{n_k}(s), \quad \mathbb{P}\text{-a.s.} \tag{3.20}$$

and one can let k tend to infinity in (3.8) to get

$$\mathbb{E}(M(t)|\mathscr{F}_s) = M(s), \quad \mathbb{P}\text{-a.s.}$$

So $M \in \mathscr{M}_T^2(E)$ and obviously $M_n \to M$ in $\mathscr{M}_T^2(E)$. □

If $M \in \mathscr{M}_T^2(\mathbb{R}^1)$ then there exists a unique increasing predictable process $\langle\langle M(\cdot)\rangle\rangle$, starting from 0, such that the process

$$M^2(t) - \langle\langle M(\cdot)\rangle\rangle, \quad t \in [0, T],$$

is a continuous martingale. The process $\langle\langle M(\cdot)\rangle\rangle$ is called the *quadratic variation* of M. If $M_1, M_2 \in \mathscr{M}_T^2(\mathbb{R}^1)$ then the process

$$\langle\langle M_1(t), M_2(t)\rangle\rangle = \frac{1}{4}\left[\, \langle\langle (M_1 + M_2)(t)\rangle\rangle - \langle\langle (M_1 - M_2)(t)\rangle\rangle\right]$$

is called the *cross quadratic variation* of M_1, M_2. It is the unique, predictable process with trajectories of bounded variation, starting from 0 such that

$$M_1(t)M_2(t) - \langle\langle M_1(t), M_2(t)\rangle\rangle, \quad t \in [0, T]$$

is a continuous martingale.

The following propositions will be frequently used (see [418]).

Proposition 3.11 (Lévy's theorem) *If $M \in \mathscr{M}_T^2(\mathbb{R}^1)$, $M(0) = 0$ and $\langle\langle M(t)\rangle\rangle = t$, $t \in [0, T]$, then $M(\cdot)$ is a standard Wiener process adapted to $\mathscr{F}_t$ and with increments $M(s) - M(t)$, $s > t$, independent of $\mathscr{F}_t$, $t \in [0, T]$.*

Proposition 3.12

(i) If $M \in \mathscr{M}_T^2(\mathbb{R}^1)$, and $\langle\langle M(t)\rangle\rangle = 0$ for all $t \in [0, T]$, then $M(t) = M(0)$, $t \in [0, T]$, $\mathbb{P}$-a.s.
(ii) If the martingales $M_1, M_2 \in \mathscr{M}_T^2(\mathbb{R}^1)$ are independent then

$$\langle\langle M_1(t), M_2(t)\rangle\rangle = 0, \quad t \in [0, T].$$

We are now going to define the quadratic variation process for $M \in \mathscr{M}_T^2(H)$ where H is a separable Hilbert space. Denote by $L_1(H)$ the space of all nuclear operators on H equipped with the nuclear norm, see Appendix C. Then $L_1 = L_1(H)$ is a separable Banach space and for each $a, b \in H$ the mapping $T \to \langle Ta, b\rangle$ is a continuous functional on L_1. An L_1-valued process $V(\cdot)$ is said to be *increasing* if the operators $V(t)$, $t \in [0, T]$, are nonnegative and $V(t) \le V(s)$ if $0 \le t \le s \le T$. An L_1-valued continuous, adapted and increasing process V such that $V(0) = 0$ is said to be a *quadratic variation* process of the martingale $M(\cdot)$ if and only if for arbitrary $a, b \in H$ the process

$$\langle M(t), a\rangle\langle M(t), b\rangle - \langle V(t)a, b\rangle, \quad t \in [0, T],$$

is an $\{\mathscr{F}_t\}$-martingale, [(1)] or equivalently if and only if the process

$$M(t) \otimes M(t) - V(t), \quad t \in [0, T],$$

is an $\{\mathscr{F}_t\}$-martingale. By the proposition below the process $V(\cdot)$ is uniquely determined and it is also denoted by $\langle\langle M(t) \rangle\rangle$, $t \in [0, T]$.

Proposition 3.13 *An arbitrary $M \in \mathscr{M}_T^2(H)$ has exactly one quadratic variation process.*

Proof (sketch) Let $\{e_i\}$ be an orthonormal complete basis in H. Then processes $M_i(t) = \langle M(t), e_i \rangle$, $i \in \mathbb{N}$, are continuous real valued square integrable martingales. Note that

$$\mathbb{E} \sum_{i=1}^{\infty} \langle\langle M_i(t) \rangle\rangle = \mathbb{E} \sum_{i=1}^{\infty} |\langle M(t), e_i \rangle|^2 = \mathbb{E}|M(t)|^2 < +\infty.$$

Consequently the sum is convergent $\mathbb{P}$-a.s. and the formula,

$$\langle\langle M(t) \rangle\rangle = \sum_{i,j=1}^{\infty} \langle\langle M_i(t), M_j(t) \rangle\rangle e_i \otimes e_j, \quad t \in [0, T],$$

defines an $L_1(H)$-valued adapted process. It is easy to see that

$$\langle M(t), a \rangle \langle M(t), b \rangle - \Big\langle \langle\langle M(t) \rangle\rangle a, b \Big\rangle$$

is a continuous martingale.

Moreover $\langle\langle M(t) \rangle\rangle$ is $\mathbb{P}$-a.s. a nonnegative operator. One can also show that the constructed process is $L_1(H)$-continuous. □

In a similar way one can define a *cross quadratic variation* for $M^1 \in \mathscr{M}_T^2(H_1)$, $M^2 \in \mathscr{M}_T^2(H_2)$ where H_1 and H_2 are two Hilbert spaces. Namely we define

$$\langle\langle M^1(t), M^2(t) \rangle\rangle = \sum_{i,j=1}^{\infty} \langle\langle M_i^1(t), M_j^2(t) \rangle\rangle e_i^1 \otimes e_j^2, \quad t \in [0, T],$$

where $\{e_i^1\}$ and $\{e_j^2\}$ are complete orthonormal bases in H_1 and H_2 respectively. From the real valued case, we get the following generalization of Proposition 3.11.

Proposition 3.14

(i) If $M \in \mathscr{M}_T^2(H)$, and $\langle\langle M(t) \rangle\rangle = 0$ for all $t \in [0, T]$, then $M(t) = M(0)$, $t \in [0, T]$, $\mathbb{P}$-a.s.

(ii) If the martingales $M_1 \in \mathscr{M}_T^2(H_1)$, $M_2 \in \mathscr{M}_T^2(H_2)$ are independent then

$$\langle\langle M_1(t), M_2(t) \rangle\rangle = 0, \quad t \in [0, T].$$

We will need the following generalization of an inequality due to B. Davis.

(1) If $a \in H_1, b \in H_2$ then $a \otimes b$ denotes a linear operator from H_2 into H_1 given by the formula $(a \otimes b)x = a\langle b, x \rangle_{H_2}$, $x \in H_1$.

Theorem 3.15 *If $M \in \mathcal{M}_T^2(H)$, then*

$$\mathbb{E}\left(\sup_{t\in[0,T]} |M(t)|\right) \leq 3\mathbb{E}\left(\text{Tr}\,\langle\langle M(T)\rangle\rangle^{1/2}\right). \tag{3.21}$$

The proof can be found in [576, pages 17–18].

3.5 Stopping times and Markov processes

A nonnegative random variable τ defined on $(\Omega, \mathscr{F})$ is said to be an $\mathscr{F}_t$-*stopping time* if for arbitrary $t \geq 0$, $\{\omega \in \Omega : \tau(\omega) \leq t\} \in \mathscr{F}_t$. If a stopping time takes on only a finite number of values it is called *simple*. For any stopping time τ the following decreasing sequence of stopping times $\{\tau_n\}$ converges to τ,

$$\tau_n = (k+1)2^{-n} \quad \text{if } k\,2^{-n} \leq \tau < (k+1)2^{-n}, \quad k \in \mathbb{N}.$$

We finish this section with the definition of a Markov process. An $\mathscr{F}_t$-adapted process $X(t)$, $t \in [0, T]$, with values in a measurable space $(E, \mathscr{E})$ is said to be *Markov* if, for arbitrary $A \in \mathscr{E}$ and $0 \leq s < t < T$,

$$\mathbb{P}(X(t) \in A|\mathscr{F}_s) = \mathbb{P}(X(t) \in A|\sigma(X(s))), \quad \mathbb{P}\text{-a.s.}$$

where $\sigma(X(s))$ denotes the σ-field of subsets of Ω generated by $X(s)$.

A family $P_{s,t}$, $0 \leq s \leq t \leq T$, of linear operators acting on the space of bounded measurable functions defined on E is said to be *Markovian* if

(i) $P_{s,t}h \geq 0$ for $h \geq 0$, bounded and measurable, $t \geq s$,
(ii) $P_{s,t}1 = 1$ for $t \geq s$,
(iii) $P_{s,t}P_{t,u} = P_{s,u}$ for $s \leq t \leq u$.

An $\mathscr{F}_t$-adapted process $X(t)$, $t \in [0, T]$, is said to be *Markov* with transition operators $\{P_{s,t}\}_{0\leq s\leq t\leq T}$, if for arbitrary bounded measurable function $h : E \to \mathbb{R}^1$ and $0 \leq s \leq t \leq T$

$$\mathbb{E}(h(X(t))|\mathscr{F}_s) = P_{s,t}h(X(s)), \quad \mathbb{P}\text{-a.s.}$$

It is obvious that if a process is Markov with respect to transition operators $\{P_{s,t}\}_{0\leq s\leq t\leq T}$ then it is Markov.

3.6 Gaussian processes in Hilbert spaces

Let $(\Omega, \mathscr{F}, \mathbb{P})$ be a probability space. An H-valued stochastic process X in $[0, \infty)$ is said to be *Gaussian* if for any $n \in \mathbb{N}$ and for arbitrary positive numbers $t_1, t_2, \ldots, t_n$, the H^n-valued random variable $(X(t_1), \ldots, X(t_n))$ is Gaussian.

Proposition 3.16 *Let X be a Gaussian process on H. Assume that $\mathbb{E}(X(t)) = 0,\ t \geq 0$, and that there exist $M > 0$ and $\gamma \in (0, 1]$ such that*

$$\mathbb{E}(\|X(t) - X(s)\|^2) \leq M(t-s)^\gamma, \quad \forall\, t, s \geq 0. \tag{3.22}$$

Then X has an α-Hölder continuous version, for any $\alpha \in (0, \gamma/2)$.

Proof From (3.22) it follows that

$$\mathbb{E}(\| X(t) - X(s)\|^{2m}) \leq C_m(t-s)^{m\gamma}, \quad \forall\, t, s \geq 0$$

so, by the Kolmogorov test (see Theorem 3.3), X is α_m-Hölder continuous with $\alpha_m = (m\gamma - 1)/(2m)$. Since m is arbitrary, the conclusion follows. □

Let X be a Gaussian process in a Hilbert space H. Let

$$m(t) = \mathbb{E}(X(t)), \quad Q(t) = \mathbb{E}(X(t) - m(t)) \otimes (X(t) - m(t)), \quad t \geq 0$$

and

$$B(t, s) = \mathbb{E}(X(t) - m(t)) \otimes (X(s) - m(s)), \quad t,\ s \geq 0.$$

The process X is said to be *stationary* if

$$\mathbb{E}\left(e^{i\sum_{k=1}^n \langle X(t_k+r), h_k\rangle}\right) = \mathbb{E}\left(e^{i\sum_{k=1}^n \langle X(t_k), h_k\rangle}\right)$$

for all $n \in \mathbb{N},\ t_1, \ldots, t_n \in [0, \infty),\ h_1, \ldots, h_n \in H,$ and $r \in [0, \infty)$.

Proposition 3.17 *A Gaussian process X is stationary if and only if*

(i) $m(t + r) = m(t)$, for all $t, r \geq 0$.
(ii) $B(t + r, s + r) = B(t, s)$, for all $t, s, r > 0$.

Proof Let $n \in \mathbb{N},\ t_1, \ldots, t_n \in [0, \infty),\ \xi_1, \ldots, \xi_n \in H$ and $r \in [0, \infty)$. Then $(\langle X(t_1), \xi_1\rangle, \ldots, \langle X(t_n), \xi_n\rangle)$ is an $\mathbb{R}^n$-valued Gaussian variable, so that

$$\mathbb{E}(e^{i\sum_{k=1}^n \langle X(t_k), \xi_k\rangle}) = e^{-\frac{1}{2}\sum_{h,k=1}^n \langle B(t_h, t_k)\xi_h, \xi_k\rangle + i\sum_{k=1}^n \langle m(t_k), \xi_k\rangle}$$

and the conclusion follows immediately. □

3.7 Stochastic processes as random variables

Let $X(\cdot)$ be a measurable, E-valued process on $(\Omega, \mathscr{F}, \mathbb{P})$. It will be convenient to regard X as a mapping from Ω into a Banach space of functions (equivalent classes of functions) like $C = C([0, T]; E)$ or $L^p = L^p(0, T; E)$, $1 \leq p < +\infty$, by associating $\omega \in \Omega$ with the trajectory $X(\cdot, \omega)$. In almost all cases of interest the mapping $\omega \to X(\cdot, \omega)$ will be a random variable.

Proposition 3.18 *Let $X(t)$, $t \in [0, T]$, be a continuous process with values in a separable Banach space E. Then the mapping $\widetilde{X} : \Omega \to C$,*

$$\widetilde{X}(\omega) = \begin{cases} X(t, \omega) & \text{if } X(\cdot, \omega) \text{ is continuous} \\ 0 & \text{otherwise} \end{cases}$$

is a C-valued random variable.

Proof One can assume that $X(\cdot, \omega)$ is a continuous function for all ω. Let $\{t_i\}$ be a dense sequence in $[0, T]$ and x an arbitrary element of C. Then

$$\|\widetilde{X}(\omega) - x\|_C = \sup\{\|X(t_i, \omega) - x(t_i)\| : \ t_i \in [0, T]\}, \qquad \omega \in \Omega$$

is a real valued random variable. Consequently, for arbitrary $r > 0$,

$$\left\{\omega \in \Omega : \ \|\widetilde{X}(\omega) - x\|_C < r\right\} \in \mathscr{F}.$$

From the separability of C one therefore obtains that for an arbitrary open set $\mathscr{U} \subset C$ the set $\{\omega \in \Omega : \ \widetilde{X}(\omega) \in \mathscr{U}\} \in \mathscr{F}$. This implies the required measurability. □

For any p-integrable f we denote by $[f]$ its equivalence class in L^p.

Proposition 3.19 *Let $X(t)$, $t \in [0, T]$, be a measurable process with values in a separable Banach space E. Then the mapping $\widetilde{X} : \Omega \to L^p$,*

$$\widetilde{X}(\omega) = \begin{cases} [X(\cdot, \omega)] & \text{if } \displaystyle\int_0^T \|X(t, \omega)\|_E^p \, dt < +\infty, \\ 0 & \text{otherwise} \end{cases}$$

is an L^p-valued random variable.

Proof For arbitrary Borel p-integrable function $x(t)$, $t \in [0, T]$, the process $X(\cdot) - x(\cdot)$ is also E-valued and measurable. By the Fubini theorem the mapping

$$\omega \to \left(\int_0^T \|X(t, \omega) - x(t)\|^p dt\right)^{1/p},$$

is a random variable. Consequently, for arbitrary $r > 0$ the set

$$\left\{\omega \in \Omega : \|\widetilde{X}(\omega) - x\|_{L^p} < r\right\}$$

is measurable. So, as in the proof of Proposition 3.18, $\widetilde{X}$ is a measurable mapping. □

4
The stochastic integral

Let H and U be two separable Hilbert spaccs. This chapter is devoted to the construction of the stochastic Itô integral

$$\int_0^t \Phi(s)dW(s), \quad t \in [0, T],$$

where $W(\cdot)$ is a Wiener process on a Hilbert space U and Φ is a process with values that are linear but not necessarily bounded operators from U into a Hilbert space H.

We will start by collecting basic facts on Hilbert space valued Wiener processes, including cylindrical Wiener processes. Then we define the stochastic integral in steps starting from elementary processes and ending up with the most general. We also establish basic properties of the stochastic integral, including the Itô formula and the stochastic Fubini theorem.

4.1 Wiener processes

A central role in the present book is played by a Wiener process and some of its generalizations.

Definition 4.1 *A real valued stochastic process $W(t)$, $t \geq 0$, is called a Wiener process if*

(i) W has continuous trajectories and $W(0) = 0$,
(ii) W has independent increments and

$$\mathscr{L}(W(t) - W(s)) = \mathscr{L}(W(t-s)), \quad t \geq s \geq 0,$$

(iii) $\mathscr{L}(W(t)) = \mathscr{L}(-W(t))$, $t \geq 0$.

Equivalently, see for example [348], a real valued stochastic process $W(t)$, $t \geq 0$, with continuous trajectories is called a Wiener process if it is Gaussian and there exists $\sigma \geq 0$ such that,

$$\mathbb{E}(W(t)) = 0, \quad \mathbb{E}(W(t)W(s)) = \sigma\, t \wedge s.$$

The avantage of Definition 4.1 is that it can be naturally generalized to processes taking values in general linear spaces E. Let E be a linear topological space and let $\mathscr{B}(E)$ the σ-field generated by all open subsets of E. Then an E-valued stochastic process $W(t),\ t \geq 0$, satisfying (i)–(iii) is called a *Wiener process* in E.

The most important cases are when E is either a Hilbert or a Banach space of functions or distributions defined on a subset $\mathscr{O} \subset \mathbb{R}^d$. Let us examine some of the classes with some details.

4.1.1 Hilbert space valued Wiener processes

Assume that E is a separable Hilbert space U, with the inner product $\langle \cdot, \cdot \rangle$ and W is a U-valued Wiener process. Then, for each $u \in U$, the process

$$\langle W(t), u \rangle, \quad t \geq 0,$$

is a real valued Wiener process. This implies in particular that $\mathscr{L}(W(t))$ is a Gaussian measure with mean vector 0. Note also that for arbitrary $u, v \in U, t, s \geq 0$,

$$\mathbb{E}[\langle W(t), u \rangle \langle W(s), u \rangle] = t \wedge s\, \mathbb{E}[\langle W(1), u \rangle^2]$$

and

$$\mathbb{E}[\langle W(t), u \rangle \langle W(s), v \rangle] = t \wedge s\ \mathbb{E}[\langle W(1), u \rangle \langle W(1), v \rangle] = t \wedge s\ \langle Qu, v \rangle,$$

where Q is the covariance operator of the Gaussian measure $\mathscr{L}(W(1))$, see Section 2.3.1. The operator Q is of trace class and it completely characterizes distributions of W.

Let Q be a trace class nonnegative operator on a Hilbert space U.

Definition 4.2 *A U-valued stochastic process $W(t),\ t \geq 0$, is called a Q-*Wiener process *if*

(i) $W(0) = 0$,
(ii) W *has continuous trajectories,*
(iii) W *has independent increments,*
(iv) $\mathscr{L}(W(t) - W(s)) = \mathscr{N}(0, (t-s)Q), \quad t \geq s \geq 0.$

Note that there exists a complete orthonormal system $\{e_k\}$ in U and a bounded sequence of nonnegative real numbers $\{\lambda_k\}$ such that

$$Qe_k = \lambda_k e_k, \quad k \in \mathbb{N}.$$

Proposition 4.3 *Assume that $W(t)$ is a Q-Wiener process. Then the following statements hold.*

(i) W is a Gaussian process on U and

$$\mathbb{E}(W(t)) = 0, \quad \mathrm{Cov}(W(t)) = tQ, \quad t \geq 0. \tag{4.1}$$

(ii) For arbitrary $t \geq 0$, W has the expansion

$$W(t) = \sum_{j=1}^{\infty} \sqrt{\lambda_j}\, \beta_j(t) e_j \tag{4.2}$$

where

$$\beta_j(t) = \frac{1}{\sqrt{\lambda_j}} \langle W(t), e_j \rangle, \quad j \in \mathbb{N}, \tag{4.3}$$

are real valued Brownian motions mutually independent on $(\Omega, \mathscr{F}, \mathbb{P})$ and the series in (4.2) is convergent in $L^2(\Omega, \mathscr{F}, \mathbb{P})$.

Proof Let $0 < t_1 < \cdots < t_n$ and let $u_1, \ldots, u_n \in U$. Let us consider the random variable Z defined by

$$Z = \sum_{j=1}^{n} \langle W(t_j), u_j \rangle = \sum_{k=1}^{n} \langle W(t_1), u_k \rangle + \sum_{k=2}^{n} \langle W(t_2) - W(t_1), u_k \rangle + \cdots + \langle W(t_n) - W(t_{n-1}), u_n \rangle.$$

Since W has independent increments, Z is Gaussian for any choice of $u_1, \ldots, u_n$ and (i) follows.

We now prove (ii). Let $t > s > 0$, then by (4.3) it follows that

$$\begin{aligned}
\mathbb{E}(\beta_i(t)\beta_j(s)) &= \frac{1}{\sqrt{\lambda_i \lambda_j}} \mathbb{E}(\langle W(t), e_i \rangle \langle W(s), e_j \rangle) \\
&= \frac{1}{\sqrt{\lambda_i \lambda_j}} [\langle \mathbb{E}(\langle W(t) - W(s), e_i \rangle \langle W(s), e_j \rangle) \\
&\quad + \mathbb{E}(\langle W(s), e_i \rangle \langle W(s), e_j \rangle)\rangle] \\
&= \frac{1}{\sqrt{\lambda_i \lambda_j}} s \langle Q e_i, e_j \rangle = s \delta_{ij}
\end{aligned}$$

then independence of $\beta_i,\ i \in \mathbb{N}$ follows. To prove representation (4.2) it is enough to notice that, for $m \geq n \geq 1$,

$$\mathbb{E} \left\| \sum_{j=n}^{m} \sqrt{\lambda_j} \beta_j(t) e_j \right\|^2 = t \sum_{j=n}^{m} \lambda_j, \tag{4.4}$$

and recall that $\sum_{j=1}^{\infty} \lambda_j < \infty$. □

Proposition 4.4 *For any trace class symmetric nonnegative operator Q on a separable Hilbert space U there exists a Q-Wiener process $W(t),\ t \geq 0$.*

Proof The proof of existence of a process W satisfying conditions (i), (iii) and (iv) of Definition 4.2 is a straightforward consequence of the Kolmogorov extension theorem. To find a version which also satisfies condition (ii) it is enough to use the Kolmogorov test, Theorem 3.3. □

If it will not lead to confusion we will say simply Wiener process instead of Q-Wiener process. We will now considerably strengthen part (ii) of Proposition 4.3 and show the following.

Theorem 4.5 *Let W be a Q-Wiener process such that (4.1) holds. Then the series (4.2) is $\mathbb{P}$-a.s. uniformly convergent on $[0, T]$ for arbitrary $T > 0$.*

Proof To prove the required convergence of (4.2) consider the $C = C([0, T]; U)$-valued random variables $\xi_j,\ j \in \mathbb{N}$ defined by

$$\xi_j(t) = \sqrt{\lambda_j}\ \beta_j(t) e_j, \quad t \in [0, T].$$

By (4.4) and Theorem 3.9(ii)

$$\mathbb{P}\left(\sup_{t\in[0,T]} \|\xi_n(t) + \cdots + \xi_m(t)\| > r\right)$$
$$\leq \frac{1}{r^2}\,\mathbb{E}\left(\|\xi_n(T) + \cdots + \xi_m(T)\|^2\right) \leq \frac{T}{r^2}\sum_{j=n}^{m}\lambda_j.$$

So, the sequence $S_N = \sum_{j=1}^N \xi_j,\ N \in \mathbb{N}$, of $C([0, T]; U)$-valued random variables converges in probability to its sum S which can be easily identified with the Wiener process W. Proposition 2.14 finally implies that $S_N \to W$ in $C([0, T]; U)$, $\mathbb{P}$–a.s. □

Note that the quadratic variation of a Q-Wiener process in U, with Tr $Q < +\infty$, is given by the formula $\langle\langle W(t)\rangle\rangle = tQ,\ t \geq 0$. We have in fact the following direct generalization of Lévy's one dimensional result, see Proposition 3.11.

Theorem 4.6 *A martingale $M \in \mathcal{M}_T^2(U)$, $M(0) = 0$, is a Q-Wiener process on $[0, T]$ adapted to the filtration $\mathscr{F}_t,\ t \geq 0$, and with increments $M(t) - M(s), t \geq s$, independent of $\mathscr{F}_s,\ s \geq 0$, if and only if $\langle\langle M(t)\rangle\rangle = tQ,\ t \geq 0$.*

Proof We have only to show that if $\langle\langle M(t)\rangle\rangle = tQ,\ t \geq 0$, then M is a Q-Wiener process. Note that for arbitrary $j \in \mathbb{N}$ the process $M_j(\cdot) = \langle M(\cdot), e_j\rangle$ belongs to $\mathcal{M}_T^2(\mathbb{R}^1)$ and moreover

$$\langle\langle M_j(t)\rangle\rangle = \lambda_j t, \quad t \geq 0.$$

So by Proposition 3.11 M_j is a λ_j-Wiener process with increments $M_j(t) - M_j(s),\ t > s$, independent from $\mathscr{F}_s$. By the same argument, finite dimensional processes $(M_1(\cdot), \ldots, M_N(\cdot))$ are Wiener processes with the diagonal quadratic variation process $\operatorname{diag}(\lambda_1, \dots, \lambda_N)$. Consequently

$$M(t) = \sum_{j=1}^{\infty} \sqrt{\lambda_j} \beta_j(t)_j, \quad t \geq 0$$

where the processes $\beta_j(\cdot)$,

$$\beta_j(t) = \frac{M_j(t)}{\sqrt{\lambda_j}}, \quad t \geq 0, \ j \in \mathbb{N}$$

are normalized independent Wiener processes. □

4.1.2 Generalized Wiener processes on a Hilbert space

Let $W(t)$, $t \geq 0$, be a Wiener process on a Hilbert space U and let Q be its covariance operator. For each $a \in U$ define a real valued Wiener process $W_a(t)$, $t \geq 0$, by the formula

$$W_a(t) = \langle a, W(t) \rangle, \quad t \geq 0. \tag{4.5}$$

The transformation $a \to W_a$ is linear from U to the space of stochastic processes. Moreover it is continuous in the following sense:

$$t \geq 0, \ \{a_n\} \subset U, \ \lim_{n\to\infty} a_n = a \Rightarrow \lim_{n\to\infty} \mathbb{E}|W_a(t) - W_{a_n}(t)|^2 = 0. \tag{4.6}$$

Any linear transformation $a \to W_a$ whose values are real valued Wiener processes on $[0, +\infty)$ satisfying (4.6) is called a *generalized Wiener process*.

From this definition it follows that there exists a bilinear form $K(a, b)$, $a, b \in U$, such that

$$\mathbb{E}[W_a(t)W_b(s)] = t \wedge s \ K(a, b), \quad t, s \geq 0, \ a, b \in U. \tag{4.7}$$

Condition (4.6) easily implies that K is a continuous bilinear form in U and therefore there exists $Q \in L(U)$ such that

$$\mathbb{E}[W_a(t)W_b(s)] = t \wedge s \ \langle Qa, b \rangle, \quad t, s \geq 0, \ a, b \in U. \tag{4.8}$$

The operator Q is self-adjoint and positive definite; we call it the *covariance* of the generalized Wiener process $a \to W_a$. If the covariance Q is the identity operator I then the generalized Wiener process is called a *cylindrical Wiener process* in U.

Denote by U_0 the image $Q^{1/2}(U)$ with the induced norm. We call $Q^{1/2}(U)$ the *reproducing kernel* of the generalized Wiener process $a \to W_a$.

It is easy to construct, for an arbitrary self-adjoint and positive definite operator Q, a generalized Wiener process $a \to W_a$ satsfying (4.8). Let in fact $\{e_j\}$ be a complete and orthonormal basis in U and $\{\beta_j\}$ a sequence of independent real valued standard

Wiener processes. Define

$$W_a(t) = \sum_{j=1}^{\infty} \langle Q^{1/2} e_j, a \rangle \beta_j(t), \quad t \geq 0, \ a \in U. \tag{4.9}$$

Since

$$\sum_{j=1}^{\infty} |\langle Q^{1/2} e_j, a \rangle|^2 = |Q^{1/2} a|^2 < +\infty,$$

for each $a \in U$ there exists a version of W_a which is a Wiener process. Since

$$\mathbb{E}[W_a(t) W_b(s)] = (t \wedge s) \sum_{j=1}^{\infty} \langle Q^{1/2} e_j, a \rangle \ \langle Q^{1/2} e_j, b \rangle = (t \wedge s) \ \langle Qa, b \rangle,$$

the result follows.

Proposition 4.7 *Let U_1 be a Hilbert space such that $U_0 = Q^{1/2}(U)$ is embedded into U_1 with a Hillbert–Schmidt embedding J. Then the formula*

$$W(t) = \sum_{j=1}^{\infty} Q^{1/2} e_j \beta_j(t), \quad t \geq 0, \tag{4.10}$$

defines a U_1-valued Wiener process. Moreover, if Q_1 is the covariance of W then the spaces $Q_1^{1/2}(U_1)$ and $Q^{1/2}(U)$ are identical.

Proof Note that the elements $g_j = Q^{1/2} e_j, \ j \in \mathbb{N}$, form an orthonormal and complete basis in U_0 and therefore

$$\sum_{j=1}^{\infty} |J g_j|_{U_1}^2 < +\infty.$$

Consequently the series in (4.10) defines a Wiener process in U_1. For $a, b \in U_1$ we have

$$\begin{aligned} \langle Qa, b \rangle_{U_1} &= \mathbb{E}[\langle a, W(1) \rangle_{U_1} \ \langle b, W(1) \rangle_{U_1}] = \sum_{j=1}^{\infty} \langle a, J g_j \rangle_{U_1} \ \langle b, J g_j \rangle_{U_1} \\ &= \sum_{j=1}^{\infty} \langle J^* a, g_j \rangle_{U_0} \ \langle J^* b, g_j \rangle_{U_0} = \langle J^* a, J^* b \rangle_{U_0} = \langle J J^* a, b \rangle_{U_1}. \end{aligned}$$

Consequently $J J^* = Q_1$. In particular

$$|Q_1^{1/2} a|_{U_1}^2 = \langle J^* a, J^* a \rangle_{U_1} = |J^* a|_{U_0}^2, \quad a \in U_1.$$

Thus by Proposition B.1 applied to operators $Q^{1/2} : U_1 \to U_1$ and $J : U_0 \to U_1$ we have $Q_1^{1/2}(U_1) = J(U_0) = U_0$ and $|Q_1^{-1/2} u|_{U_1} = |u|_{U_0}$, as required. □

Thus with some abuse of language we can say that an arbitrary generalized Wiener process on U is a classical Wiener process in some larger Hilbert space U_1. Reproducing kernels related to all these extensions are the same.

To complete the picture we will show that an arbitrary generalized Wiener process W is of the form (4.10). To do so let $\{a_j\}$ be a sequence of linearly independent elements, linearly dense in U. Set

$$V = Q^{1/2}(\mathrm{lin}\{a_j\}).$$

Without any loss of generality we assume that $Q^{1/2}(U)$ is dense in U. Then V is also dense in U. By the Hilbert–Schmidt orthogonalization of $\{Q^{1/2}a_j\}$ we create an orthonormal complete basis $\{e_i\}$ in U. Note that $\mathrm{lin}\{e_i\} = V$ and

$$U = \overline{\mathrm{lin}\{Q^{1/2}a_j\}}. \tag{4.11}$$

Proposition 4.8 *For every $a \in U$ we have*

$$W_a(t) = \sum_{j=1}^{\infty} \langle Q^{1/2}e_j, a\rangle W_{Q^{-1/2}e_j}(t), \quad t \geq 0. \tag{4.12}$$

Proof It is easily checked that the right hand side of (4.12), denoted by $\widetilde{W_a}(t)$, defines a generalized Wiener process with covariance Q. Now fix $a = Q^{-1/2}e_k$, for some $k \in \mathbb{N}$. Then

$$\widetilde{W}_{Q^{-1/2}e_k}(t) = \sum_{j=1}^{\infty} \langle Q^{1/2}e_j, Q^{-1/2}e_k\rangle W_{Q^{-1/2}e_j}(t) = W_{Q^{-1/2}e_k}(t), \quad t \geq 0.$$

Thus $\widetilde{W_a}(t) = W_a(t)$. However, the set of all elements $a = Q^{-1/2}e_j$ is dense in U so the identity holds for all $a \in U$. □

4.1.3 Wiener processes in $U = L^2(\mathscr{O})$

If $U = L^2(\mathscr{O})$ and W is a U-valued Q-Wiener process, then the operator $Q^{1/2}$ is Hilbert–Schmidt on U and so it is an integral operator,

$$Q^{1/2}\varphi(\xi) = \int_{\mathscr{O}} r(\xi,\eta)\varphi(\eta)d\eta, \quad \xi \in \mathscr{O},\ \varphi \in L^2(\mathscr{O}),$$

where the kernel r is square integrable on $\mathscr{O} \times \mathscr{O}$, see for example [271]. Consequently, the operator $Q = Q^{1/2}Q^{1/2}$ is also integral

$$Q\varphi(\xi) = \int_{\mathscr{O}} g(\xi,\eta)\varphi(\eta)d\eta, \quad \xi \in \mathscr{O},\ \varphi \in L^2(\mathscr{O}),$$

where

$$g(\xi,\eta) = \int_{\mathscr{O}} r(\xi,\zeta)r(\zeta,\eta)d\zeta, \quad \xi,\eta \in \mathscr{O}.$$

The kernel g is a symmetric function which can be interpreted as a spatial correlation of the Q-Wiener process W in $L^2(\mathscr{O})$. In fact, for each $t > 0$, $W(t, \cdot)$ is an $L^2(\mathscr{O})$-valued random variable, thus $W(\cdot, \xi)$ is a real valued random variable for almost all $\xi \in \mathscr{O}$.

Without any loss of generality one can assume that the random field $W(t, \xi),\ t \geq 0,\ \xi \in \mathscr{O}$, is measurable. For $\varphi, \psi \in L^2(\mathscr{O})$ we have

$$\mathbb{E}\left[\langle W(t), \varphi\rangle_H \langle W(t), \psi\rangle_H\right] = \mathbb{E}\left[\int_{\mathscr{O}} W(t, \xi)\varphi(\xi)d\xi \int_{\mathscr{O}} W(t, \eta)\psi(\eta)d\eta\right].$$

By the Schwartz inequality

$$\begin{aligned}
&\mathbb{E}\left[\int_{\mathscr{O}}\int_{\mathscr{O}} |W(t, \xi)|\, |W(t, \eta)|\, |\varphi(\xi)|\, |\psi(\eta)|\, d\xi\, d\eta\right] \\
&\leq |\varphi|_H\, |\psi|_H \mathbb{E}\left[\left(\int_{\mathscr{O}} |W(t, \xi)|^2 d\xi\right)^{1/2} \left(\int_{\mathscr{O}} |W(t, \eta)|^2 d\eta\right)^{1/2}\right] \\
&\leq |\varphi|_H\, |\psi|_H\, \mathbb{E}|W(t)|_H^2 < +\infty.
\end{aligned}$$

Thus, one can change the order of expectation on the Lebesgue integral to find that

$$\begin{aligned}
t\langle Q\varphi, \psi\rangle &= \int_{\mathscr{O}}\int_{\mathscr{O}} \varphi(\xi)\psi(\eta)\, \mathbb{E}[W(t, \xi)W(t, \eta)]d\xi\, d\eta \\
&= t\int_{\mathscr{O}}\int_{\mathscr{O}} \varphi(\xi)\psi(\eta)g(\xi, \eta)d\xi\, d\eta.
\end{aligned}$$

Consequently, for almost all $\xi, \eta \in \mathscr{O}$

$$g(\xi, \eta) = \frac{1}{t}\, \mathbb{E}[W(t, \xi)W(t, \eta)], \tag{4.13}$$

as claimed.

The kernel g, defined by (4.13), is *positive definite* in the sense that for arbitrary $n \in \mathbb{N}, \lambda_1, \ldots, \lambda_n \in \mathbb{R}, \xi_1, \ldots, \xi_n \in \mathscr{O}$, for which $g(\xi_i, \xi_j),\ i, j = 1, \ldots, n$, is well defined one has

$$\sum_{i,j=1}^{n} \lambda_i\lambda_j g(\xi_i, \xi_j) \geq 0. \tag{4.14}$$

The inequality follows directly from (4.13) as

$$\begin{aligned}
\sum_{i,j=1}^{n} \lambda_i\lambda_j g(\xi_i, \xi_j) &= \mathbb{E}\sum_{i,j=1}^{n} \lambda_i\lambda_j W(1, \xi_i)W(1, \xi_j) \\
&= \mathbb{E}\left|\sum_{i}^{n} \lambda_i W(1, \xi_i)\right|^2.
\end{aligned}$$

One can also show that if $\mathcal{O} \subset \mathbb{R}^d$ is an open (closed) bounded set and g is a bounded continuous function on $\mathcal{O}$, then the operator Q with kernel g is a covariance operator on H of trace class.

Example 4.9 Take $\mathcal{O} = [0, \pi]$ and

$$g(\xi, \eta) = \xi \wedge \eta, \quad \xi, \eta \in [0, \pi]. \tag{4.15}$$

The function g is obviously continuous. Since for a one dimensional Wiener process β

$$\xi \wedge \eta = \mathbb{E}[\beta(\xi)\beta(\eta)],$$

we immediately see that g, given by (4.15), is positive definite. The eigenvalues and eigenfunctions of the corresponding operator Q are of the form

$$\lambda_n = \frac{1}{\left(n + \frac{1}{2}\right)^2}, \quad e_n = \sqrt{\frac{2}{\pi}} \sin\left(n + \frac{1}{2}\right)\xi, \quad \xi \in [0, \pi], \ n \in \mathbb{N}.$$

Thus we have the following expansion of W

$$W(t, \xi) = \sum_{n=0}^{\infty} \frac{\beta_n(t)}{n + \frac{1}{2}} \sqrt{\frac{2}{\pi}} \sin\left(n + \frac{1}{2}\right)\xi, \quad t \geq 0, \ \xi \in [0, \pi],$$

where $\{\beta_n\}_{n \in \mathbb{N}}$ are independent real standard Wiener processes. It is obvious that

$$\mathbb{E}[W(t, \xi)W(s, \eta)] = (t \wedge s)(\xi \wedge \eta), \quad t, s \geq 0, \ \xi, \eta \in [0, \pi]. \tag{4.16}$$

Fields W with covariance given by (4.16) are called *Brownian sheets* on $[0, \infty] \times [0, \pi]$. □

The following result on positive definite functions is instructive.

Proposition 4.10 *A function $g(\xi, \eta)$, $\xi, \eta \in \mathcal{O}$, where $\mathcal{O}$ is an arbitrary set, is positive definite if and only if there exists a family $X(\xi)$, $\xi \in \mathcal{O}$, such that for arbitrary distinct elements $\xi_1, \ldots, \xi_j \in \mathcal{O}$ the random vector $(X(\xi_1), \ldots, X(\xi_j))$ is Gaussian with mean 0 and*

$$g(\xi, \eta) = \mathbb{E}[X(\xi)X(\eta)], \quad \xi, \eta \in \mathcal{O}. \tag{4.17}$$

Proof If g is given by (4.17), then for arbitrary $\lambda_1, \ldots, \lambda_j \in \mathbb{R}$ and $\xi_1, \ldots, \xi_j \in \mathcal{O}$,

$$\sum_{k,l=1}^{j} g(\xi_k, \xi_l)\lambda_k\lambda_l = \mathbb{E}\left(\sum_{k=1}^{j} X(\xi_k)\lambda_k\right)^2 \geq 0.$$

So, g is positive definite. To prove the converse we apply Kolmogorov's existence theorem, see for example [433, page 115].

To check consistency denote by μ_j, μ_{j+1} Gaussian measures on $\mathbb{R}^j, \mathbb{R}^{j+1}$ with 0 mean vectors and covariances

$$Q_1 = (g(\xi_k, \xi_l) : k, l = 1, \dots, j), \quad Q_2 = (g(\xi_k, \xi_l) : k, l = 1, \dots, j, j+1).$$

Then

$$\int_{\mathbb{R}^j} e^{i \sum_{k=1}^{j} \alpha_i \beta_j} \mu_j(d\beta_1, \dots, d\beta_j) = e^{-\frac{1}{2} \sum_{k,l=1}^{j} g(\xi_k, \xi_l) \alpha_k \alpha_l}$$

$$\int_{\mathbb{R}^{j+1}} e^{i \sum_{k=1}^{j+1} \alpha_i \beta_j} \mu_{j+1}(d\beta_1, \dots, d\beta_j) = e^{-\frac{1}{2} \sum_{k,l=1}^{j+1} g(\xi_k, \xi_l) \alpha_k \alpha_l}.$$

Let $\pi : \mathbb{R}^{j+1} \to \mathbb{R}^j$ be the projection

$$\pi(\beta_1, \dots, \beta_{j+1}) = (\beta_1, \dots, \beta_j), \quad (\beta_1, \dots, \beta_{j+1}) \in \mathbb{R}^{j+1}$$

and let $\pi \circ \mu_{j+1}$ be the image of μ_{j+1} by π. Then

$$\begin{aligned}
\int_{\mathbb{R}^j} e^{i\langle \alpha, \zeta \rangle} \pi \circ \mu_{j+1}(d\zeta_1, \dots, d\zeta_j) &= \int_{\mathbb{R}^{j+1}} e^{i\langle \alpha, \pi(\rho) \rangle} \mu_{j+1}(d\rho_1, \dots, d\rho_{j+1}) \\
&= \int_{\mathbb{R}^{j+1}} e^{i \sum_{k=1}^{j} \alpha_j \rho_j} \mu_{j+1}(d\rho_1, \dots, d\rho_{j+1}) \\
&= e^{-\frac{1}{2} \sum_{k,l=1}^{j} g(\xi_k \xi_l) \alpha_k \alpha_l} = \int_{\mathbb{R}^j} e^{i\langle \alpha, \rho \rangle} \mu_j(d\rho_1, \dots, d\rho_j)
\end{aligned}$$

and thus $\mu_j = \pi \circ \mu_{j+1}$. So, Kolmogorov's existence theorem is applicable and the result follows. □

As an easy corollary we have the following proposition.

Proposition 4.11 *If g_1 and g_2 are positive definite functions on $\mathscr{O}_1$ and $\mathscr{O}_2$ then*

$$g((\xi_1, \eta_1), (\xi_2, \eta_2)) := g_1(\xi_1, \xi_2) g_2(\eta_1, \eta_2), \quad (\xi_1, \xi_2) \in \mathscr{O}_1, \quad (\eta_1, \eta_2) \in \mathscr{O}_2,$$

is a positive definite function on $\mathscr{O}_1 \times \mathscr{O}_2$.

Proof With obvious notations one has from the previous proposition

$$\begin{aligned}
g((\xi_1, \eta_1), (\xi_2, \eta_2)) &= \mathbb{E}_1[X^1(\xi_1) X^1(\xi_2)] \mathbb{E}_2[X^2(\eta_1) X^2(\eta_2)] \\
&= \mathbb{E}[X^1(\xi_1) X^1(\xi_2) X^2(\eta_1) X^2(\eta_2)],
\end{aligned}$$

so the result follows easily. □

Example 4.12 As a generalization of Example 4.9 consider $\mathscr{O} \subset \mathbb{R}^d$ and a function g positive definite on $\mathscr{O}$. Then by Proposition 4.11 the function

$$t \wedge s \; g(\xi, \eta), \quad (t, \xi), \; (s, \eta) \in [0, +\infty) \times \mathscr{O},$$

is also positive definite and there exists a Gaussian family $W(t,\xi)$, $t \geq 0$, $\xi \in \mathscr{O}$, such that

$$\mathbb{E}[W(t,\xi)W(s,\eta)] = t \wedge s\ g(\xi,\eta), \quad (t,\xi),\ (s,\eta) \in [0,+\infty) \times \mathscr{O}.$$

For each ξ, $W(\cdot,\xi)$ has all the distributional properties of the real Wiener process with the exception of the regularity of trajectories. Existence of a proper modification requires additional properties of g. □

4.1.4 Spatially homogeneous Wiener processes

In applications in physics and in the theory of particle systems one has to deal with Wiener processes on $\mathbb{R}^d$. In general they are not even function valued but take values in some spaces of Schwartz distributions. As we shall see, however, they can always be regarded as processes with values in appropriately chosen Hilbert spaces.

A typical example of a positive definite function (introduced in Section 4.1.2) is provided by Bochner's theorem. In fact if μ is a finite, nonnegative measure on $\mathbb{R}^d$ and

$$\Gamma(\lambda) := \int_{\mathbb{R}^d} e^{i\langle\lambda,\xi\rangle}\mu(d\xi), \quad \lambda \in \mathbb{R}^d, \tag{4.18}$$

then $g(\xi,\eta) = \Gamma(\xi-\eta)$, $\xi,\ \eta \in \mathbb{R}^d$, is a continuous, bounded, positive definite function. Under mild conditions on the measure μ one can show that there exists a continuous random field $W(t,\xi)$, $t \geq 0$, $\xi \in \mathbb{R}^d$, such that the process $W(t,\cdot)$ is a Wiener process in $L^2_\rho(\mathbb{R}^d)$, for arbitrary integrable weight ρ, such that

$$\mathbb{E}[W(t,\xi)W(s,\eta)] = t \wedge s\ \Gamma(\xi-\eta).$$

We note only that $W(t,\cdot)$ is certainly $L^2_\rho(\mathbb{R}^d)$ valued because

$$\begin{aligned}\mathbb{E}|W(t,\cdot)|^2_{L^2_\rho(\mathbb{R}^d)} &= \mathbb{E}\int_{\mathbb{R}^d}|W(t,x)|^2\rho(x)dx\\ &= \int_{\mathbb{R}^d}\mathbb{E}|W(t,x)|^2\rho(x)dx\\ &= t\Gamma(0)\int_{\mathbb{R}^d}\rho(x)dx.\end{aligned}$$

We now generalize this setting, first extending considerably Bochner's theorem. Let $\mathscr{S}_c(\mathbb{R}^d)$ denote the space of infinitely differentiable functions ψ on $\mathbb{R}^d$ taking complex values, for which all the seminorms

$$|\psi|_{\alpha,\beta} = \sup_{\xi\in\mathbb{R}^d}|\xi^\alpha D^\beta\psi(\xi)|,$$

are finite. The dual space $\mathscr{S}'_c(\mathbb{R}^d)$ is the space of *tempered distributions*. By $\mathscr{S}(\mathbb{R}^d)$ and $\mathscr{S}'(\mathbb{R}^d)$ we denote the spaces of real functions from $\mathscr{S}_c(\mathbb{R}^d)$ and of real functionals from $\mathscr{S}'_c(\mathbb{R}^d)$. The value of the distribution u on the test function ϕ will be denoted by (u, ϕ).

For $\psi \in \mathscr{S}_c(\mathbb{R}^d)$ define

$$\psi_{(s)}(x) = \overline{\psi(-x)}, \quad x \in \mathbb{R}^d.$$

By $\mathscr{S}_{(s)}(\mathbb{R}^d)$ and $\mathscr{S}'_{(s)}(\mathbb{R}^d)$ we denote the spaces of $\psi \in S_c(\mathbb{R}^d)$ such that $\psi = \psi_{(s)}$ and the space of all $u \in \mathscr{S}'_{(s)}(\mathbb{R}^d)$ such that $(u, \psi) = (u, \psi_{(s)})$ for all $\psi \in \mathscr{S}_c(\mathbb{R}^d)$. Let us denote by $\mathscr{F}\varphi$ the Fourier transform of $\varphi \in S_c(\mathbb{R}^d)$:

$$\mathscr{F}\varphi(\lambda) = (2\pi)^{-d/2} \int_{\mathbb{R}^d} e^{-2\pi i \langle \lambda, \xi \rangle} \varphi(\xi) d\xi.$$

For every $u \in \mathscr{S}'_c(\mathbb{R}^d)$ we define $\mathscr{F}u$ by the formula

$$(\mathscr{F}u, \varphi) = (u, \mathscr{F}^{-1}\varphi), \quad \varphi \in \mathscr{S}_c(\mathbb{R}^d),$$

where

$$\mathscr{F}^{-1}\varphi(\lambda) = (2\pi)^{-d/2} \int_{\mathbb{R}^d} e^{2\pi i \langle \lambda, \xi \rangle} \varphi(\xi) d\xi.$$

Since $\mathscr{F}^{-1}$ is a continuous mapping on $\mathscr{S}_c(\mathbb{R}^d)$ the above definition of $\mathscr{F}u$ is meaningful. Note that $\mathscr{F}$ and $\mathscr{F}^{-1}$ transform $\mathscr{S}'(\mathbb{R}^d)$ onto $\mathscr{S}'_{(s)}(\mathbb{R}^d)$ and $\mathscr{S}'_{(s)}(\mathbb{R}^d)$ onto $\mathscr{S}'(\mathbb{R}^d)$.

A distribution $\Gamma \in \mathscr{S}'(\mathbb{R}^d)$ is said to be *positive definite* if for every $\varphi \in \mathscr{S}(\mathbb{R}^d)$ we have

$$(\Gamma, \varphi * \varphi_{(s)}) \geq 0.$$

We have the following generalization of Bochner's theorem, see [345].

Theorem 4.13 *A distribution $\Gamma \in \mathscr{S}'(\mathbb{R}^d)$ is positive definite if there exists a measure $\mu \in \mathscr{S}'(\mathbb{R}^d)$ symmetric with respect to 0 such that $\Gamma = \mathscr{F}\mu$.*

The measure μ is called the *spectral measure* of Γ.

An $\mathscr{S}'(\mathbb{R}^d)$-valued, continuous, stochastic process $W(t),\ t \geq 0$, is called a *Wiener process with covariance kernel* Γ if

(i) for each $\varphi \in \mathscr{S}(\mathbb{R}^d)$, $(W(t), \varphi),\ t \geq 0$, is a real valued Wiener process,
(ii) for all $t, s \geq 0$, $\varphi, \psi \in \mathscr{S}(\mathbb{R}^d)$ we have,

$$\mathbb{E}[(W(t), \varphi)(W(s), \psi)] = t \wedge s\ (\Gamma, \varphi * \psi_{(s)}). \tag{4.19}$$

Theorem 4.14 *For arbitrary positive definite distribution Γ there exists a Wiener process W with covariance kernel Γ. Moreover, there is a Hilbert space U_Γ, continuously embedded in $\mathscr{S}'(\mathbb{R}^d)$ such that*

$$\mathscr{L}((W(t), \varphi)) = \mathscr{N}(0, t|\varphi|^2_{U_\Gamma}), \quad \forall\, \varphi \in \mathscr{S}(\mathbb{R}^d). \tag{4.20}$$

Proof Let μ be the measure corresponding to Γ by Theorem 4.13. Consider the complex Hilbert space $L^2(\mathbb{R}^d, \mu)$ endowed with the usual inner product

$$\langle \varphi, \psi \rangle_{L^2(\mathbb{R}^d,\mu)} = \int_{\mathbb{R}^d} \varphi(\xi)\overline{\psi}(\xi)\mu(d\xi)$$

and denote by $L^2_{(s)}$ the closed subspace of $L^2(\mathbb{R}^d, \mu)$ of all symmetric functions. We consider now the space

$$V := \{\psi\mu : \psi \in L^2_{(s)}\}.$$

Equipped with the inner product

$$\langle \varphi\mu, \psi\mu \rangle_V = \int_{\mathbb{R}^d} \varphi(\xi)\overline{\psi}(\xi)\mu(d\xi),$$

V is a Hilbert space continuously embedded in $\mathscr{S}'_{(s)}(\mathbb{R}^d)$. Let $\{e_j\}$ be a complete orthonormal basis in $L^2_{(s)}$ and $\widetilde{V}$ a Hilbert subspace of $\mathscr{S}'_{(s)}(\mathbb{R}^d)$ such that the embedding $V \subset \widetilde{V}$ is Hilbert–Schmidt. Let $\{\beta_j\}$ be a sequence of real valued standard Wiener processes which are mutually independent. Define

$$W(t) = \sum_{j=1}^{\infty} \beta_j(t)\mathscr{F}(e_j\mu), \quad t \geq 0$$

and $U_\Gamma = \mathscr{F}(V)$ with the induced norm. From the very definition, $\{\mathscr{F}(e_j\mu)\}$ is a complete orthonormal basis in U_Γ. Moreover the process

$$\widetilde{W}(t) = \sum_{j=1}^{\infty} \beta_j(t)e_j\mu, \quad t \geq 0,$$

is a well defined Wiener process with values in $\widetilde{V}$ and V is the reproducing kernel of $\mathscr{L}(\widetilde{W}(1))$, see Section 2.2.2. Since $\mathscr{F}$ is a continuous transformation from $\mathscr{S}'_{(s)}(\mathbb{R}^d)$ into $\mathscr{S}'(\mathbb{R}^d)$, we deduce that $W(t)$, $t \geq 0$, is a continuous $\mathscr{S}'(\mathbb{R}^d)$ valued process. Moreover, for each $\varphi \in \mathscr{S}(\mathbb{R}^d)$ the process

$$(W(t), \varphi) = \sum_{j=1}^{\infty} \beta_j(t)(\mathscr{F}(e_j\mu), \varphi) \quad t \geq 0,$$

is a real valued Wiener process. Moreover

$$\begin{aligned}
(W(t), \varphi) &= \sum_{j=1}^{\infty} \beta_j(t)(e_j\mu, \mathscr{F}^{-1}\varphi) \\
&= \sum_{j=1}^{\infty} \beta_j(t) \int_{\mathbb{R}^d} e_j(x)\mathscr{F}^{-1}\varphi(x)\mu(dx) \\
&= \sum_{j=1}^{\infty} \beta_j(t)\ \langle e_j, \overline{\mathscr{F}^{-1}\varphi} \rangle_{L^2(\mu)}.
\end{aligned}$$

Since $\{e_j\}$ is a complete orthonormal basis in $L^2_{(s)}$ we have

$$\begin{aligned}\mathbb{E}[(W(t),\varphi)(W(s),\psi)] &= t\wedge s\sum_{j=1}^{\infty}\langle e_j,\overline{\mathscr{F}^{-1}\varphi}\rangle_{L^2(\mu)}\langle e_j,\overline{\mathscr{F}^{-1}\psi}\rangle_{L^2(\mu)}\\ &= t\wedge s\,\langle\overline{\mathscr{F}^{-1}\psi},\overline{\mathscr{F}^{-1}\varphi}\rangle_{L^2(\mu)} = t\wedge s\int_{\mathbb{R}^d}\mathscr{F}^{-1}(\varphi*\psi_{(s)})d\mu\\ &= t\wedge s\,(\mathscr{F}(\mu),\varphi*\psi_{(s)}) = t\wedge s\,(\Gamma,\varphi*\psi_{(s)}),\quad t,s\geq 0,\end{aligned}$$

thus equality (4.19) holds. It is therefore clear that (4.20) holds as well. □

Remark 4.15 If the function $\varphi\in\mathscr{S}(\mathbb{R}^d)$ is treated as a functional on $\mathscr{S}'(\mathbb{R}^d)$ then we will write (φ,u) instead of (u,φ), for $u\in\mathscr{S}'(\mathbb{R}^d)$. Identity (4.20) implies that

$$\mathbb{E}|(\varphi,W(t))|^2 = t|\varphi|^2_{U_\Gamma}. \tag{4.21}$$

Thus U_Γ has all properties required for a reproducing kernel of $\mathscr{L}(W(1))$, see Section 2.2.2.

Example 4.16 Assume that $\mu=\delta_0$, then

$$\Gamma=\mathscr{F}(\delta_0)=(2\pi)^{-d/2}\lambda_d,$$

where λ_d is the Lebesgue measure in $\mathbb{R}^d$. The space U_Γ is one dimensional and consists of constant functions.

The process W is of the form

$$W(t)=(2\pi)^{-d/2}\beta(t)\lambda_d,$$

where β is a standard Wiener process. □

Example 4.17 Assume that $\mu=\lambda_d$, then

$$\Gamma=\mathscr{F}(\lambda_d)=(2\pi)^{d/2}\delta_0.$$

For any $\varphi,\psi\in\mathscr{S}(\mathbb{R}^d)$ we have

$$\begin{aligned}(\Gamma,\varphi*\psi_{(s)}) &= (2\pi)^{d/2}\int_{\mathbb{R}^d}\varphi(-\eta)\psi_{(s)}(\eta)d\eta\\ &= (2\pi)^{d/2}\int_{\mathbb{R}^d}\varphi(-\eta)\psi(-\eta)d\eta\\ &= (2\pi)^{d/2}\int_{\mathbb{R}^d}\varphi(\eta)\psi(\eta)d\eta.\end{aligned}$$

Consequently $(2\pi)^{d/2}W(t),\ t\geq 0$, is a cylindrical Wiener process on $L^2(\mathbb{R}^d)$. □

Remark 4.18 The case of finite probability measure μ was discussed earlier. Many examples of positive definite kernels Γ can be found in the monographs

[345, 484, 550]. In particular the functions

$$\Gamma(\lambda) = |\lambda|^{-\alpha}, \quad \alpha \in (0, d),$$

are positive definite kernels and the corresponding measures μ are given by

$$\mu(d\xi) = c_\alpha |\xi|^{d-\alpha} d\xi,$$

for some constant c_α. □

4.1.5 Complements on a Brownian sheet

We extend now the definition of Brownian sheet B to $\mathbb{R}_+ \times \mathbb{R}^d$ and show that the distributional derivative

$$\frac{\partial^d}{\partial \xi_1, \ldots, \partial \xi_j} B(t, \xi_1, \ldots, \xi_d) = W(t, \xi_1, \ldots, \xi_d), \tag{4.22}$$

can be identified up to a constant with the cylindrical Wiener process of Example 4.17.

A Brownian sheet B in $\mathbb{R}_+ \times \mathbb{R}_+^d$ is a Gaussian random field with correlation function

$$t \wedge s \prod_{j=1}^{d} \xi_j \wedge \eta_j, \quad t, s, \xi_j, \eta_j \geq 0, \; j = 1, \ldots, d.$$

Note that the function

$$g(\xi, \eta) := \prod_{j=1}^{d} \xi_j \wedge \eta_j$$

is positive definite in view of Proposition 4.11.

Let $B_{\epsilon_1, \ldots, \epsilon_d}$ be a family of independent Brownian sheets on $\mathbb{R}_+ \times \mathbb{R}_+^d$ indexed by the family of all sequences $\epsilon_1, \ldots, \epsilon_d$ of numbers 0 or 1. For arbitrary $t \geq 0$ and $(\xi_1, \ldots, \xi_d) \in \mathbb{R}^d$ we set

$$B(t, \xi_1, \ldots, \xi_d) = B_{\epsilon_1, \ldots, \epsilon_d}(t, (-1)^{\epsilon_1}\xi_1, \ldots, (-1)^{\epsilon_d}\xi_d), \text{ if } (-1)^{\epsilon_j}\xi_j \geq 0, \quad j = 1, \ldots, d$$

and call B the *Brownian sheet* on $\mathbb{R}_+ \times \mathbb{R}^d$. By the Kolmogorov continuity test one may assume that the realizations of B are continuous. We could also regard B as a Wiener process in $L^2_\rho(\mathbb{R}^d)$ where the weight ρ is such that

$$\int_{\mathbb{R}^d} |x_1| \cdots |x_d| \rho(x_1, \ldots, x_d) dx_1 \cdots dx_d < +\infty.$$

Proposition 4.19 *If B is a Brownian sheet on $\mathbb{R}_+ \times \mathbb{R}^d$ then for arbitrary $\varphi, \psi \in \mathscr{S}(\mathbb{R}^d)$ we have*

$$\mathbb{E}\left(\frac{\partial^d}{\partial\xi_1,\ldots,\partial\xi_d}\,B(t,\cdot),\varphi\right)\left(\frac{\partial^d}{\partial\xi_1,\ldots,\partial\xi_d}\,B(s,\cdot),\psi\right) = t\wedge s\int_{\mathbb{R}^d}\varphi(\xi)\psi(\xi)d\xi.$$

Proof We have to show that

$$\mathbb{E}\left(B(t,\cdot),\frac{\partial^d}{\partial\xi_1,\ldots,\partial\xi_d}\,\varphi\right)\left(B(s,\cdot),\frac{\partial^d}{\partial\xi_1,\ldots,\partial\xi_d}\,\psi\right) = t\wedge s\int_{\mathbb{R}^d}\varphi(\xi)\psi(\xi)d\xi.$$

To do this we can restrict integration only to positive arguments. Then we have

$$\begin{aligned}
&\mathbb{E}\left(\int_{\mathbb{R}^d_+}B(t,\xi)\frac{\partial^d\varphi(\xi)}{\partial\xi_1,\ldots,\partial\xi_d}\,d\xi\right)\left(\int_{\mathbb{R}^d_+}B(s,\eta)\frac{\partial^d\psi(\eta)}{\partial\eta_1,\ldots,\partial\eta_d}\,d\eta\right)\\
&\quad= t\wedge s\int_{(\mathbb{R}^d_+)^3}\frac{\partial^d\varphi(\xi)}{\partial\xi_1,\ldots,\partial\xi_d}\,\frac{\partial^d\psi(\eta)}{\partial\eta_1,\ldots,\partial\eta_d}\prod_{j=1}^d \mathbb{1}_{[0,\xi_j]}(\zeta_j)\mathbb{1}_{[0,\eta_j]}(\zeta_j)\,d\xi d\eta\, d\zeta\\
&\quad= t\wedge s\int_{\mathbb{R}^d_+}\left[\int_{\{\xi\geq\zeta\}}\frac{\partial^d\varphi(\xi)}{\partial\xi_1,\ldots,\partial\xi_d}\,d\xi\int_{\{\eta\geq\zeta\}}\frac{\partial^d\psi(\eta)}{\partial\eta_1,\ldots,\partial\eta_d}\,d\eta\right]d\zeta\\
&\quad= t\wedge s\int_{\mathbb{R}^d_+}\varphi(\zeta)\psi(\zeta)d\zeta,
\end{aligned}$$

as required. All the changes of order of integration and expectation can be easily justified by taking into account that $\varphi, \psi \in \mathscr{S}(\mathbb{R}^d)$. □

4.2 Definition of the stochastic integral

We are given here a Q-Wiener process in $(\Omega, \mathscr{F}, \mathbb{P})$ having values in U. By Proposition 4.3, $W(t)$ is given by (4.2). For the sake of simplicity of notation, we require that $\lambda_k > 0$ for all $k \in \mathbb{N}$. We are also given a normal filtration $\{\mathscr{F}_t\}_{t\geq 0}$ in $\mathscr{F}$ and we assume that

(i) $W(t)$ is $\mathscr{F}_t$-measurable,
(ii) $W(t+h) - W(t)$ is independent of $\mathscr{F}_t$, $\quad\forall\, h \geq 0,\ \forall\, t \geq 0$.

If a Q-Wiener process W satisfies (i) and (ii) we say that W is a *Q-Wiener process with respect to* $\{\mathscr{F}_t\}_{t\geq 0}$. However, to shorten the formulation we usually avoid stressing the dependence on the filtration.

Let us fix $T < \infty$. An $L = L(U,H)$-valued process $\Phi(t)$, $t \in [0,T]$ taking only a finite number of values is said to be *elementary* if there exists a sequence $0 = t_0 < t_1 < \cdots < t_k = T$ and a sequence $\Phi_0, \Phi_1, \ldots, \Phi_{k-1}$ of L-valued random variables taking on only a finite number of values such that Φ_m are $\mathscr{F}_{t_m}$-measurable and

$$\Phi(t) = \Phi_m, \quad \text{for } t \in (t_m, t_{m+1}],\ m = 0, 1, \ldots, k-1.$$

For elementary processes Φ one defines the stochastic integral by the formula

$$\int_0^t \Phi(s)dW(s) = \sum_{m=0}^{k-1} \Phi_m(W_{t_{m+1}\wedge t} - W_{t_m \wedge t})$$

and denote it by $\Phi \cdot W(t),\ t \in [0, T]$.

It is useful, at this moment, to introduce the subspace $U_0 = Q^{1/2}(U)$ of U which, endowed with the inner product

$$\langle u, v\rangle_0 = \sum_{k=1}^{\infty} \frac{1}{\lambda_k}\langle u, e_k\rangle\langle v, e_k\rangle = \langle Q^{-1/2}u, Q^{-1/2}v\rangle, \quad u, v \in U_0,$$

is a Hilbert space.

In the construction of the stochastic integral for more general processes, an important role will be played by the space of all Hilbert–Schmidt operators $L_2^0 = L_2(U_0, H)$ from U_0 into H. The space L_2^0 is also a separable Hilbert space, equipped with the norm

$$\begin{aligned}\|\Psi\|^2_{L_2^0} &= \sum_{h,k=1}^{\infty} |\langle \Psi g_h, f_k\rangle|^2 = \sum_{h,k=1}^{\infty} \lambda_h |\langle \Psi e_h, f_k\rangle|^2 \\ &= \| \Psi Q^{\frac{1}{2}} \|^2_{L_2(U;H)} = \mathrm{Tr}\ [(\Psi Q^{1/2})(\Psi Q^{1/2})^*],\end{aligned}$$

where $\{g_j\}$, with $g_j = \sqrt{\lambda_j}e_j$, $\{e_j\}$ and $\{f_j\}$ $j \in \mathbb{N}$, are complete orthonormal bases in U_0, U and H respectively. Clearly, $L \subset L_2^0$, but not all operators from L_2^0 can be regarded as restrictions of operators from L. The space L_2^0 contains genuinely unbounded operators on U.

Let $\Phi(t),\ t \in [0, T]$, be a measurable L_2^0-valued process; we define the norms

$$\begin{aligned}|||\Phi|||_t &= \left[\mathbb{E}\int_0^t \|\Phi(s)\|^2_{L_2^0}\, ds\right]^{1/2} \\ &= \left[\mathbb{E}\int_0^t \mathrm{Tr}\ [(\Phi(s)Q^{1/2})(\Phi(s)Q^{1/2})^*]ds\right]^{1/2}, \quad t \in [0, T].\end{aligned}$$

Proposition 4.20 *If a process Φ is elementary and $|||\Phi|||_T < \infty$ then the process $\Phi \cdot W$ is a continuous, square integrable H-valued martingale on* $[0, T]$ *and*

$$\mathbb{E}|\Phi \cdot W(t)|^2 = |||\Phi|||_t^2, \quad 0 \le t \le T. \tag{4.23}$$

Proof The proof is straightforward. We will check for instance that (4.23) holds for $t = t_m \le T$. Define $\zeta_j = W(t_{j+1}) - W(t_j)$, $j = 1, \dots, m-1$. Then

$$\begin{aligned}\mathbb{E}|\Phi \cdot W(t_m)|^2 &= \mathbb{E}\left|\sum_{j=0}^{m-1} \Phi(t_j)\zeta_j\right|^2 \\ &= \mathbb{E}\sum_{j=0}^{m-1} |\Phi(t_j)\zeta_j|^2 + 2\mathbb{E}\sum_{i<j=1}^{n} \langle \Phi(t_i)\zeta_i, \Phi(t_j)\zeta_j\rangle.\end{aligned}$$

We will show first that

$$\mathbb{E}\sum_{j=0}^{m-1}|\Phi(t_j)\zeta_j|^2 = \sum_{j=0}^{m-1}(t_{j+1}-t_j)\mathbb{E}\|\Phi(t_j)\|^2_{L_2^0}, \quad j=0,1,\dots,m-1. \tag{4.24}$$

To this purpose note that the random variable $\Phi^*(t_j)f_l$ is $\mathscr{F}_{t_j}$-measurable, and ζ_j is a random variable independent of $\mathscr{F}_{t_j}$. Consequently (see Proposition 1.12)

$$\begin{aligned}
\mathbb{E}|\Phi(t_j)\zeta_j|^2 &= \sum_{l=1}^{\infty}\mathbb{E}(|\langle\Phi(t_j)\zeta_j, f_l\rangle|^2)\\
&= \sum_{l=1}^{\infty}\mathbb{E}(\mathbb{E}[\langle\zeta_j, \Phi^*(t_j)f_l\rangle]^2|\mathscr{F}_{t_j}))\\
&= (t_{j+1}-t_j)\sum_{l=1}^{\infty}\mathbb{E}(\langle Q\Phi^*(t_j)f_l, \Phi^*(t_j)f_l\rangle)\\
&= (t_{j+1}-t_j)\sum_{l=1}^{\infty}|Q^{1/2}\Phi^*(t_j)f_l|^2 = (t_{j+1}-t_j)\mathbb{E}\|\Phi(t_j)\|^2_{L_2^0}.
\end{aligned}$$

This shows (4.24). Similarly one has

$$\mathbb{E}\langle\Phi(t_i)\zeta_i, \Phi(t_j)\zeta_j\rangle = 0 \quad \text{if } i \neq j$$

and the conclusion follows. □

Remark 4.21 *Note that the stochastic integral is a linear transformation from the space of all elementary processes equipped with the norm* $|||\cdot|||_T$ *into the space* $\mathscr{M}_T^2(H)$ *of H-valued martingales.*

To extend the definition of the stochastic integral to more general processes it is convenient to regard integrands as random variables defined on the product space $\Omega_\infty = [0,+\infty)\times\Omega$ (respectively $\Omega_T = [0,T]\times\Omega$), equipped with the product σ-field: $\mathscr{B}([0,+\infty))\times\mathscr{F}$ (respectively $\mathscr{B}([0,T])\times\mathscr{F}$). The product of the Lebesgue measure on $[0,+\infty)$ (respectively on [0,T]) and the probability measure $\mathbb{P}$ is denoted by $\mathbb{P}_\infty$ (respectively $\mathbb{P}_T$).

The σ-field just introduced does not take into account adaptability of the considered processes and therefore is not proper for our purposes. The natural and the right choice is provided by the σ-field generated by the adapted simple processes. It is easy to see that these are exactly the predictable σ-field $\mathscr{P}_\infty$ (respectively $\mathscr{P}_T$) introduced in Section 3.3. It turns out that the proper class of integrands are predictable processes with values in L_2^0, more precisely, measurable mappings from $(\Omega_\infty, \mathscr{P}_\infty)$ (respectively $(\Omega_T, \mathscr{P}_T)$) into $(L_2^0, \mathscr{B}(L_2^0))$.

Proposition 4.22 *The following statements hold.*

(i) *If a mapping Φ from Ω_T, into L is L-predictable then Φ is also L_2^0-predictable. In particular, elementary processes are L_2^0-predictable.*
(ii) *If Φ is an L_2^0-predictable process such that $|||\Phi|||_T < \infty$ then there exists a sequence $\{\Phi_n\}$ of elementary processes such that $|||\Phi - \Phi_n|||_T \to 0$ as $n \to \infty$.*

Proof We will use the following elementary fact, the proof of which is left as an exercise.

Lemma 4.23 *Assume that K is a separable Hilbert space and K_1 is a linearly dense subset of K. If X is a mapping from Ω into K such that for arbitrary $k \in K_1$, $\langle k, X\rangle$ is $\mathscr{F}$-measurable then X is a random variable from $(\Omega, \mathscr{F})$ into $(K, \mathscr{B}(K))$.*

Proof of Proposition 4.22 Since the operators

$$f_k \otimes e_j \cdot u = f_k\langle e_j, u\rangle, \quad u \in U, \ k, j \subset \mathbb{N}$$

are linearly dense in L_2^0 and for arbitrary $T \in L_2^0$

$$\langle f_k \otimes e_j, T\rangle_{L_2^0} = \lambda_j\langle Te_j, f_k\rangle_H,$$

part (i) follows.

Let us prove (ii). Since the space L is densely embedded into L_2^0, by Lemma 1.3 there exists a sequence $\{\Phi_n\}$ of L-valued predictable processes on $[0, T]$ taking on only a finite numbers of values such that

$$\|\Phi(t, \omega) - \Phi_n(t, \omega)\|_{L_2^0} \downarrow 0$$

for all $(t, \omega) \in \Omega_T$. Consequently $|||\Phi - \Phi_n|||_T \downarrow 0$. It is therefore sufficient to prove that for arbitrary $A \in \mathscr{P}_T$ and arbitrary $\varepsilon > 0$ there exists a finite sum Γ of disjoint sets of the form (3.14) with $s, t \le T$ such that

$$\mathbb{P}_T\{(A\backslash\Gamma) \cup (\Gamma\backslash A)\} < \varepsilon.$$

To show this let us denote by $\mathscr{K}$ the family of all finite sums of sets of the form (3.14) with $s \le t \le T$. It is easy to check that $\mathscr{K}$ is a π-system (see Section 1.1). Let $\mathscr{G}$ be the family of all $A \in \mathscr{P}_T$ which can be approximated in the above sense by elements from $\mathscr{K}$. One can check easily that $\mathscr{K} \subset \mathscr{G}$ and that the conditions of Proposition 1.4 are satisfied. Therefore $\sigma(\mathscr{K}) = \mathscr{P}_T = \mathscr{G}$ as required. □

We are able now to extend the definition of stochastic integral to all L_2^0-predictable processes Φ such that $|||\Phi|||_T < +\infty$. Note that they form a Hilbert space denoted by $\mathscr{N}_W^2(0, T; L_2^0)$, more simply $\mathscr{N}_W^2(0, T)$ or $\mathscr{N}_W^2$, and, by the previous proposition, elementary processes form a dense set in $\mathscr{N}_W^2(0, T)$. By Proposition 4.20 the stochastic integral $\Phi \cdot W$ is an isometric transformation from that dense set into $\mathscr{M}_T^2(H)$, therefore the definition of the integral can be immediately extended to all elements of $\mathscr{N}_W^2(0, T)$. Moreover (4.23) holds and $\Phi \cdot W$ is a continuous square integrable martingale.

As a final step we extend the definition of the stochastic integral to L_2^0-predictable processes satisfying the even weaker condition

$$\mathbb{P}\left(\int_0^T \|\Phi(s)\|_{L_2^0}^2 ds < \infty\right) = 1. \tag{4.25}$$

All such processes are called *stochastically integrable* on $[0, T]$. They form a linear space denoted by $\mathcal{N}_W(0, T; L_2^0)$, more simply $\mathcal{N}_W(0, T)$ or even $\mathcal{N}_W$. The extension can be accomplished by the so called *localization procedure*. To do so we need the following.

Lemma 4.24 *Assume that $\Phi \in \mathcal{N}_W^2(0, T; L_2^0)$ and that τ is an $\mathcal{F}_\tau$-stopping time such that $\mathbb{P}(\tau \le T) = 1$. Then*

$$\int_0^t I_{[0,\tau]}(s)\Phi(s)dW(s) = \Phi \cdot W(\tau \wedge t), \quad t \in (0, T], \ \mathbb{P}\text{-a.s.} \tag{4.26}$$

Proof Assume that Φ is elementary and that τ is a simple stopping time (see Section 3.5), then (4.26) holds by inspection. If Φ is elementary and τ general, then there exists a sequence of simple stopping times $\{\tau_n\}$ such that $\tau_n \downarrow \tau$, and $\mathbb{P}$-a.s. $\Phi \cdot W(\tau_n \wedge t) \to \Phi \cdot W(\tau \wedge t)$. On the other hand

$$|||I_{[0,\tau]}\Phi - I_{[0,\tau_n]}\Phi|||_T^2 = \mathbb{E}\int_0^t I_{[0,\tau_n]}(s)\|\Phi(s)\|_{L_2^0}^2 ds \downarrow 0$$

and therefore, for a subsequence, still denoted by $\{\tau_n\}$,

$$I_{[0,\tau_n]}\Phi \cdot W \to I_{[0,\tau]}\Phi \cdot W, \quad \mathbb{P}\text{-a.s. and uniformly in } [0, T].$$

If Φ is general and $|||\Phi - \Phi_m|||_T \to 0$ for a sequence of elementary processes, we have $\Phi_m \cdot W \to \Phi \cdot W$ and, for an appropriate subsequence, $I_{[0,\tau]}\Phi_{m_k} \cdot W \to I_{[0,\tau]}$ $\Phi \cdot W$. □

Assume that condition (4.25) holds and define

$$\tau_n = \inf\left\{t \in [0, T] : \int_0^t \|\Phi(s)\|_{L_2^0}^2 ds \ge n\right\}$$

with the convention that the infimum of an empty set is T. Then $\{\tau_n\}$ is a sequence such that

$$\mathbb{E}\int_0^t \|I_{[0,\tau_n)}(s)\Phi(s)\|_{L_2^0}^2 ds < \infty. \tag{4.27}$$

Consequently stochastic integrals $I_{[0,\tau_n]}\Phi \cdot W(t)$, $t \in [0, T]$, are well defined for all $n \in \mathbb{N}$. Moreover if $n < m$ then $\mathbb{P}$-a.s.

$$I_{[0,\tau_n]}\Phi \cdot W(t) = \big(I_{[0,\tau_n]}(I_{[0,\tau_m]}\Phi) \cdot W(t)\big) = (I_{[0,\tau_m]}\Phi) \cdot (\tau_n \wedge t), \quad t \in [0, T].$$

Therefore one can assume that (4.27) holds for all $\omega \in \Omega,\ t \in [0, T],\ n < m$. For arbitrary $t \in [0, T]$ define

$$\Phi \cdot W(t) = I_{[0,\tau_n]}\Phi \cdot W(t) \tag{4.28}$$

where n is an arbitrary natural number such that $\tau_n \geq t$. Note that if also $\tau_m \geq t$ and $m > n$ then

$$(I_{[0,\tau_m]}\Phi) \cdot W(t) = (I_{[0,\tau_m]}\Phi) \cdot W(\tau_n \wedge t) = I_{[0,\tau_n]}\Phi \cdot W(t)$$

and therefore the definition (4.28) is consistent. By an analogous reasoning if $\{\tau_n'\} \uparrow \tau$ is another sequence satisfying (4.27) then the definition (4.28) leads to a stochastic process identical $\mathbb{P}$-a.s. for all $t \in [0, T]$. Note that for arbitrary $n \in \mathbb{N},\ \omega \in \Omega,\ t \in [0, T]$

$$\Phi \cdot W(\tau_n \wedge t) = I_{[0,\tau_n]}\Phi \cdot W(\tau_n \wedge t) = M_n(\tau_n \wedge t), \quad t \in [0, T],$$

where M_n is a square integrable continuous H-valued martingale. This property will be referred as the *local martingale property* of stochastic integral.

Remark 4.25 *It follows from the above construction that Lemma 4.24 is valid for all* $\Phi \in \mathcal{N}_W(0, T; L_2^0)$. □

4.2.1 Stochastic integral for generalized Wiener processes

The construction of the stochastic integral required the assumption that Q was a nuclear operator; only then does the Q-Wiener process have values in U. We can, however, easily extend the definition of the integral to the case of generalized Wiener processes with a covariance operator Q not necessarily of trace class. One can perform this in several equivalent ways. The following simple proposition plays an important role in these extensions. As before we denote by $U_0 = Q^{1/2}(U)$ (with the induced norm $\|u\|_0 = \|Q^{-1/2}(u)\|,\ u \in U_0$) the reproducing kernel of W. We shall use again the notation,

$$L_2^0 = L_2(U_0, H).$$

Proposition 4.26 *Assume that Z is a U-valued random variable with mean* 0 *and covariance Q and that R is a Hilbert–Schmidt operator from U_0 into H. If $\{R_n\} \subset L_2(U_0, H)$ is such that*

$$\lim_{n\to\infty} \|R - R_n\|_{L_2(U_0,H)} = 0,$$

there exists a random variable RZ such that

$$\lim_{n\to\infty} \mathbb{E}\|RZ - R_n Z\|^2_{L_2(U_0,H)} = 0.$$

RZ is independent of the sequence $\{R_n\}$.

Proof The proof is a straightforward consequence of the identity

$$\mathbb{E}|SZ|^2 = \|SQ^{1/2}\|^2_{L_2(U,H)},$$

valid for arbitrary linear bounded operators $S : U \to H$. Note that

$$\mathbb{E}|Z_n - Z_m|^2 = \|R_n - R_m\|^2_{L_2(U_0,H)} \to 0 \quad \text{as } n, m \to \infty.$$

If $\{R'_n\}$ is another sequence with all the required properties then

$$\mathbb{E}|Z_n - Z'_n|^2 = \mathbb{E}\|R_n Z - R'_n Z\|^2_{L_2(U_0,H)} = \|R_n - R'_n\|^2_{L_2(U_0,H)} \to 0 \quad \text{as } n \to \infty,$$

so the result follows. □

If now W_a, $a \in U$, is a generalized Wiener process with covariance Q, then by Proposition 4.7 there exists a sequence $\{\beta_j\}$ of independent Wiener processes and an orthonormal basis $\{e_j\}$ in U such that

$$W_a(t) = \sum_{j=1}^{\infty} \langle a, Q^{1/2} e_j \rangle \beta_j(t), \quad a \in U,\ t \geq 0.$$

Moreover the formula

$$W(t) = \sum_{j=1}^{\infty} Q^{1/2} e_j \beta_j(t), \quad t \geq 0,$$

defines a Wiener process on any Hilbert space $U_1 \supset U_0$ with Hilbert–Schmidt embedding. If $\Phi \in L_2^0$ then the random variables $\Phi W(t)$, $t \geq 0$, described in Proposition 4.26, are given by the formula,

$$\Phi W(t) = \sum_{j=1}^{\infty} \Phi Q^{1/2} e_j \beta_j(t), \quad t \geq 0, \tag{4.29}$$

and in particular we have

$$W_a(t) = \langle a, W(t) \rangle, \quad t \geq 0.$$

Thus the construction of the stochastic integral

$$\int_0^t \Phi(s) dW(s), \quad t \geq 0,$$

can be done as in the case when $\mathrm{Tr}Q < +\infty$. It is enough to take into account that random variables of the form

$$\Phi_{t_j}(W_{t_{j+1}} - W_{t_j}),$$

are defined in a unique way provided $\Phi_{t_j} \in L_2^0$. The basic formula

$$\mathbb{E}\left|\int_0^t \Phi(s) dW(s)\right|^2 = \mathbb{E}\left(\int_0^t \|\Phi(s)\|^2_{L_2^0} ds\right), \quad t \geq 0, \tag{4.30}$$

remains the same.

Equivalently one can repeat the definition of the stochastic integral for a U_1-valued Wiener process W determined by $W_a,\ a \in U$. Again, the space of integrands and formula (4.30) remain the same.

Let us finally consider stochastic integration with respect to a Wiener process W_Γ with covariance kernel Γ introduced in Section 4.1.4. If $\Psi(s),\ s \geq 0$, is a predictable process with values in $L^2(U_\Gamma, H)$, where U_Γ is the subspace of $\mathscr{S}'(\mathbb{R}^d)$ introduced in Theorem 4.14, then the stochastic integral

$$\int_0^t \Psi(s) dW_\Gamma(s), \quad t \geq 0,$$

should be understood in the classic Hilbertian meaning. Assume, in particular, that $H = \mathbb{R}$, then for any $s \geq 0$ the operator $\Psi(s)$ can be identified with a function $\psi(s) \in \mathscr{S}(\mathbb{R}^d)$. Then we have

$$\mathbb{E}\,|(\psi(s), dW_\Gamma(s))|^2 = \mathbb{E}\left(\int_0^t |\psi(s)|^2_{U_\Gamma}\right). \tag{4.31}$$

Therefore formula (4.31) is a generalization of (4.19).

Let us now give some comments about Walsh's approach to stochastic integration. In his seminal paper [702] on stochastic PDEs, Walsh introduced the stochastic integral with respect to the Brownian sheet rather than with respect to a Hilbert valued Wiener process as we do in the present book. In fact both approaches lead to equivalent theories. Let B be the Brownian sheet discussed in Section 4.1.4 and let $W_\Gamma(t),\ t \geq 0$, be given by consecutive derivations of B as in Proposition 4.19. Then $\Gamma = \delta_0$ and we have

$$\int_0^t (\psi(s), dW_\Gamma(s)) = \int_0^t \int_{\mathbb{R}^d} \psi(s,\xi) B(ds, d\xi), \tag{4.32}$$

where $\psi(s,\cdot),\ s \geq 0$, is an $L^2(\mathbb{R}^d)$-valued predictable process such that

$$\int_0^t |\psi(s)|^2_{L^2(\mathbb{R}^d)} ds < +\infty, \quad t \geq 0.$$

To check (4.32) assume that $\psi(s) \equiv \psi \in \mathscr{S}_c(\mathbb{R}^d)$. Then the left hand side of (4.32) reduces to $(W_\Gamma(t), \psi)$ and the right hand side, see [702, page 287], becomes

$$\int_{\mathbb{R}^d} \psi(\xi) \frac{\partial^d}{\partial \xi_1 \cdots \partial \xi_d}(t,\xi)\, d\xi,$$

so that they coincide.

4.2.2 Approximations of stochastic integrals

We now describe a way of approximating stochastic integrals which could also serve as a different way of defining the stochastic integral with respect to an infinite dimensional Q-Wiener process (Tr $Q \leq +\infty$). Let

$$W_N(t) = \sum_{j=1}^{N} \sqrt{\lambda_j}\, \beta_j(t) e_j, \quad t \in [0, T],$$

where $\{\lambda_j, e_j\}$ is an eigensequence defined by Q, and let Φ be stochastically integrable on $[0, T]$, with respect to the Q-Wiener process W. Notice that W_N and $W^N = W - W_N$ are respectively $Q_N = \sum_{j=1}^{N} \lambda_j e_j \otimes e_j$ and $Q^N(t) = \sum_{j=N+1}^{\infty} \lambda_j e_j \otimes e_j$ Wiener processes. It is easy to see that $\Phi \cdot W = \Phi \cdot W_N + \Phi \cdot W^N$. Thus

$$\mathbb{E}\|\Phi \cdot W(T) - \Phi \cdot W_N(T)\|^2 = \mathbb{E} \int_0^T \|\Phi(s)(Q^N)^{1/2}\|^2_{L_2^0}\, ds.$$

If $|||\Phi|||_T < \infty$ then

$$\mathbb{E} \int_0^T \|\Phi(s)(Q^N)^{1/2}\|^2_{L_2^0}\, ds \to 0 \quad \text{as } N \to \infty.$$

Then by the martingale property of the stochastic integral

$$\mathbb{E}\left(\sup_{0 \leq t \leq T} \|\Phi \cdot W(t) - \Phi \cdot W_N(t)\|^2 \right) \to 0 \quad \text{as } N \to \infty$$

and consequently one can consider a subsequence $\{\Phi \cdot W_{N_k}\}$ converging $\mathbb{P}$-a.s. and uniformly in $[0, T]$. Thus the stochastic integral with respect to an infinite dimensional Wiener process, also cylindrical, can be obtained, in the above sense, as a limit of stochastic integrals with respect to finite dimensional Wiener processes. The sequence $\Phi \cdot W_N(\cdot)$ contains a subsequence convergent $\mathbb{P}$-a.s. uniformly with respect to $t \in [0, T]$. The limit is independent of the subsequence chosen and perhaps gives a more intuitive definition of the stochastic integral of a predictable process Φ such that $\mathbb{E} \int_0^T \|\Phi(s)\|^2_{L_2^0}\, ds < +\infty$. The case of general $\Phi \in \mathcal{N}_W(0, T; L_2^0)$ can be obtained by localization.

4.3 Properties of the stochastic integral

The following theorem summarizes the results from Sections 4.1 and 4.2.

Theorem 4.27 *Assume that $\Phi \in \mathcal{N}_W^2(0, T, L_2^0)$, then the stochastic integral $\Phi \cdot W$ is a continuous square integrable martingale, and its quadratic variation is of the form*

$$\langle\!\langle \Phi \cdot W(t) \rangle\!\rangle = \int_0^t Q_\Phi(s) ds,$$

where

$$Q_\Phi(s) = (\Phi(s) Q^{1/2})(\Phi(s) Q^{1/2})^*, \quad s, t \in [0, T].$$

If $\Phi \in \mathcal{N}_W(0, T, L_2^0)$, then $\Phi \cdot W$ is a local martingale.

In calculations, we will frequently use the following.

Proposition 4.28 *Assume that* $\Phi_1, \Phi_2 \in \mathcal{N}_W^2(0, T; L_0^2)$. *Then*

$$\mathbb{E}(\Phi_i \cdot W(t)) = 0, \quad \mathbb{E}(\|\Phi_i \cdot W(t)\|^2) < \infty, \quad t \in [0, T],\ i = 1, 2.$$

Moreover, the correlation operators

$$V(t, s) = \mathrm{Cor}(\Phi_1 \cdot W(t), \Phi_2 \cdot W(s)), \quad t, s \in [0, T]$$

are given by the formulae

$$V(t, s) = \mathbb{E} \int_0^{t\wedge s} (\Phi_2(r)Q^{1/2})(\Phi_1(r)Q^{1/2})^* dr. \tag{4.33}$$

Proof Note that $\Phi_2(r)Q^{1/2}$ and $(\Phi_1(r)Q^{1/2})^*$, $r \in [0, T]$, are respectively $L_2(U, H)$ and $L_2(H, U)$-valued processes. Therefore the process

$$\Phi_2(r)Q^{1/2}(\Phi_1(r)Q^{1/2})^*, \quad r \in [0, T],$$

is an $L_1(H, H)$-valued process and (see Appendix C)

$$\begin{aligned}&\|(\Phi_2(r)Q^{1/2})(\Phi_1(r)Q^{1/2})^*\|_1 \\ &\quad \le \|(\Phi_2(r)Q^{1/2})\|_{L_2(U,H)}\|(\Phi_1(r)Q^{1/2})\|_{L_2(U,H)}, \quad r \in [0, T].\end{aligned}$$

Consequently

$$\begin{aligned}&\mathbb{E}\int_0^T \|(\Phi_2(r)Q^{1/2})(\Phi_1(r)Q^{1/2})^*\|_1 dr \\ &\le \mathbb{E}\int_0^T \|\Phi_2(r)Q^{1/2}\|_{L_2(U,H)}\|\Phi_1(r)Q^{1/2}\|_{L_2(U,H)} dr \\ &\le \mathbb{E}\left[\int_0^T \|\Phi_1(r)Q^{1/2}\|^2_{L_2(U,H)} dr\right]^{1/2}\left[\int_0^T \|\Phi_2(r)Q^{1/2}\|^2_{L_2(U,H)} dr\right]^{1/2} \\ &\le |||\Phi_1|||_T \cdot |||\Phi_2|||_T < \infty,\end{aligned} \tag{4.34}$$

and therefore the integral (4.33) exists as an $L_1(H, H)$-valued Bochner integral. According to the definition (see Section 1.2), the operator $V(t, s)$ is defined by

$$\mathbb{E}\langle \Phi_1 \cdot W(t), a\rangle\langle \Phi_2 \cdot W(s), b\rangle = \langle V(t, s)a, b\rangle, \quad a, b \in H.$$

One can easily see that if, in addition, Φ_1 and Φ_2 are simple processes then

$$\begin{aligned}&\mathbb{E}\langle \Phi_1 \cdot W(t), a\rangle\langle \Phi_2 \cdot W(s), b\rangle \\ &= \mathbb{E}\int_0^{t\wedge s} \langle \Phi_1(r)dW(r), a\rangle \int_0^{t\wedge s} \langle \Phi_2(r)dW(r), b\rangle \\ &= \mathbb{E}\int_0^{t\wedge s} \langle Q^{1/2}\Phi_1^*(r)a, Q^{1/2}\Phi_2^*(r)b\rangle dr.\end{aligned}$$

So the result is true for simple processes. The estimate (4.34) and the approximation Proposition 4.22 imply that the result holds in general. □

From the definition of the correlation operator we have the following.

Corollary 4.29 *Under the hypothesis of Proposition 4.28, we have*

$$\mathbb{E}\langle \Phi_1 \cdot W(t), \Phi_2 \cdot W(s)\rangle = \mathbb{E}\int_0^{t\wedge s} \mathrm{Tr}[(\Phi_2(r)Q^{1/2})(\Phi_1(r)Q^{1/2})^*]dr. \tag{4.35}$$

We remark that if the processes Φ_1 and Φ_2 are $L = L(U, H)$-valued then we can write formula (4.35) in a shorter way

$$\mathbb{E}\langle \Phi_1 \cdot W(t), \Phi_2 \cdot W(s)\rangle = \mathbb{E}\int_0^{t\wedge s} \mathrm{Tr}[(\Phi_2(r)Q\Phi_1(r)^*]dr.$$

Several results valid for the Bochner integral have their counterparts for stochastic integrals. In particular an analog to Proposition 1.6 holds. Assume that $\Phi(t)$, $t \in [0, T]$, is an $L_2^0(H) = L_2^0(U_0, H)$-predictable process and let $A : D(A) \subset H \to E$ be a closed operator with the domain $D(A)$ a Borel subset of H.

Proposition 4.30 *If $\Phi(t)(L_2^0(H)) \subset D(A)$, $\mathbb{P}$-a.s. for all $t \in [0, T]$ and*

$$\mathbb{P}\left(\int_0^T \|\Phi(s)\|^2_{L_0^2(D(A))} ds < \infty\right) = 1,$$

$$\mathbb{P}\left(\int_0^T \|A\Phi(s)\|^2_{L_0^2(D(A))} ds < \infty\right) = 1,$$

then $\mathbb{P}\big(\int_0^T \Phi(s)dW(s) \in D(A)\big) = 1$ and

$$A\int_0^T \Phi(s)dW(s) = \int_0^T A\Phi(s)dW(s), \quad \mathbb{P}\text{-a.s.}$$

Proof One can check immediately that the result is true for elementary processes. The general case can be obtained in a similar way to the proof of Proposition 1.6 using Proposition 4.22(ii) and the following fact. If $D(A)$ is endowed with its graph norm and S is a linear operator from U_0 into $D(A)$ then $S \in L_2^0(U_0, D(A))$ if and only if $S \in L_2^0(U_0, H)$, $AS \in L_2^0(U_0, H)$. Moreover

$$\|S\|^2_{L_2^0(H)} + \|AS\|^2_{L_2^0(H)} = \|S\|^2_{L_2(U_0, D(A))}.$$

□

We will finish this section with a useful estimate valid for a general integrand Φ.

Proposition 4.31 *Assume that $\Phi \in \mathcal{N}_W(0, T; L_2^0)$. Then for arbitrary $a > 0$, $b > 0$*

$$\mathbb{P}\left(\sup_{t\in[0,T]} |\Phi \cdot W(t)| > a\right) \le \frac{b}{a^2} + \mathbb{P}\left(\int_0^T \|\Phi(t)\|^2_{L_2^0} dt > b\right).$$

Proof Define

$$\tau_b = \inf \left\{ t \in [0, T] : \int_0^T \|\Phi(t)\|_{L_2^0}^2 \, dt > b \right\}.$$

Then

$$\mathbb{P}\left(\sup_{t \in [0,T]} |\Phi \cdot W(t)| > a \right) = I_1 + I_2,$$

where

$$I_1 = \mathbb{P}\left(\sup_{t \in [0,T]} |\Phi \cdot W(t)| > a \ \text{ and } \ \int_0^T \|\Phi(s)\|_{L_2^0}^2 ds > b \right)$$

$$I_2 = \mathbb{P}\left(\sup_{t \in [0,T]} |\Phi \cdot W(t)| > a \ \text{ and } \ \int_0^T \|\Phi(s)\|_{L_2^0}^2 ds \le b \right).$$

But

$$I_2 \le \mathbb{P}\left(\sup_{t \in [0,T]} \left| \int_0^t I_{[0,\tau_b]}(s)\Phi(s) dW(s) \right| > a \right)$$

and from Theorem 3.9 and the definition of τ_b

$$I_2 \le \frac{1}{a^2} \mathbb{E} \int_0^t \left\| I_{[0,\tau_b]}(s)\Phi(s) \right\|_{L_2^0}^2 ds \le \frac{b}{a^2}.$$

Since $I_1 \le \mathbb{P}\big(\int_0^t \|\Phi(s)\|_{L_2^0}^2 ds \| > b \big)$ therefore the result follows. □

4.4 The Itô formula

Assume that Φ is an L_2^0-valued process stochastically integrable in $[0, T]$, φ a H-valued predictable process Bochner integrable on $[0, T]$, $\mathbb{P}$-a.s., and $X(0)$ a $\mathscr{F}_0$-measurable H-valued random variable. Then the following process

$$X(t) = X(0) + \int_0^t \varphi(s) ds + \int_0^t \Phi(s) dW(s), \quad t \in [0, T],$$

is well defined. Assume that a function $F\colon [0, T] \times H \to \mathbb{R}^1$ and its partial derivatives F_t, F_x, F_{xx}, are uniformly continuous on bounded subsets of $[0, T] \times H$.

Theorem 4.32 *Under the above conditions, $\mathbb{P}$-a.s., for all $t \in [0, T]$*

$$\begin{aligned} F(t, X(t)) = F(0, X(0)) &+ \int_0^t \langle F_x(s, X(s)), \Phi(s) dW(s) \rangle \\ &+ \int_0^t \Big\{ F_t(s, X(s)) + \langle F_x(s, X(s)), \varphi(s) \rangle \\ &+ \frac{1}{2} \operatorname{Tr} \left[F_{xx}(s, X(s))(\Phi(s)Q^{1/2})(\Phi(s)Q^{1/2})^* \right] \Big\} ds. \end{aligned} \tag{4.36}$$

Proof We will sketch first how we can reduce the proof to the case of constant processes $\varphi(s) = \varphi_0$ and $\Phi(s) = \Phi_0,\ s \in [0, T]$, and so to the case when

$$X(t) = X(0) + t\varphi_0 + \Phi_0 W(t), \quad t \in [0, T]. \tag{4.37}$$

We can assume that the process $X(t),\ t \in [0, T]$, and the integrals $\int_0^T |\varphi(s)|ds$ and $\int_0^T \|\Phi(s)\|^2_{L_2^0} ds$ are bounded. This can be shown by localization. Namely for arbitrary constant $C > 0$ define a stopping time τ_C:

$$\tau_C = \inf\left\{t \in [0, T] :\ |X(t)| \geq C \text{ or } \int_0^t |\varphi(s)|ds \geq C \text{ or } \int_0^t \|\Phi(s)\|^2_{L_2^0} ds \geq C\right\}$$

with the convention that the infimum of the empty set is T. If one defines

$$\varphi_C(t) = I_{[0,\tau_C]}(t)\varphi(t), \quad \Phi_C(t) = I_{[0,\tau_C]}(t)\Phi(t), \quad t \in [0, T]$$

and $X_C(t) = X(t \wedge \tau_C)$, then

$$X_C(t) = X_C(0) + \int_0^t \varphi_C(s)ds + \int_0^t \Phi_C(s)dW(s), \quad t \in [0, T],$$

compare Lemma 4.24. Moreover the processes φ_C, Φ_C, and X_C have the required properties. If the formula (4.36) is true for φ_C, Φ_C and X_C for arbitrary $C > 0$ then, using again Lemma 4.24, the formula (4.36) is true in the general case. Thus in particular it can be assumed that

$$\mathbb{E}\int_0^T |\varphi(s)|ds < +\infty, \quad \mathbb{E}\int_0^T \|\Phi(s)\|^2_{L_2^0} ds < +\infty.$$

Consequently using Lemma 1.3 and the basic approximation Proposition 4.22, one can limit considerations to elementary processes $\varphi(\cdot)$ and $\Phi(\cdot)$ and consequently to processes $\varphi(\cdot)$ and $\Phi(\cdot)$ constant on intervals and this finally shows that the required reduction is possible.

Let the points $t_0 = 0 < t_1 < \cdots < t_k < t$ define a partition of a fixed time interval $[0, t] \subset [0, T]$. Then

$$\begin{aligned} F(t, X(t)) - F(0, X(0)) &= \sum_{j=0}^{k-1} [F(t_{j+1}, X(t_{j+1})) - F(t_j, X(t_{j+1}))] \\ &\quad + \sum_{j=0}^{k-1} [F(t_j, X(t_{j+1})) - F(t_j, X(t_j))]. \end{aligned}$$

Applying Taylor's formula one gets for some (random) $\theta_{00}, \theta_{01}, \ldots, \theta_{0(k-1)}$, $\theta_{10}, \theta_{11}, \ldots, \theta_{1(k-1)} \in [0, 1]$,

$$\begin{aligned}
F(t, X(t)) - F(0, X(0)) &= \sum_{j=0}^{k-1} F_t(t_{j+1}, X(t_{j+1}))\Delta t_j + \sum_{j=0}^{k-1} \langle F_x(t_j, X(t_j)), \Delta X_j\rangle \\
&\quad + \frac{1}{2}\sum_{j=0}^{k-1} \langle F_{xx}(t_j, X(t_j)) \cdot \Delta X_j, \Delta X_j\rangle \\
&\quad + \sum_{j=0}^{k-1} [F_t(\widetilde{t}_j, X_{j+1}) - F_t(t_{j+1}, X(t_{j+1}))]\Delta_{t_j} \\
&\quad + \frac{1}{2}\sum_{j=0}^{k-1} \langle [F_{xx}(t_j, \widetilde{X}_j) - F_{xx}(t_j, X(t_j))] \cdot \Delta X_j, \Delta X_j\rangle \\
&= I_1 + I_2 + I_3 + I_4 + I_5,
\end{aligned}$$

where

$$\begin{aligned}
&t_{j+1} - t_j = \Delta t_j, \quad X(t_{j+1}) - X(t_j) = \Delta X_j \\
&t_j + \theta_{0j}(t_{j+1} - t_j) = \widetilde{t}_j, \quad X(t_j) + \theta_{1j}(X(t_{j+1}) - X(t_j)) = \widetilde{X}_j.
\end{aligned}$$

Taking into account (4.37) and assuming that the partition $t_0 = 0 < t_1 < \cdots < t_k < t$ becomes finer and finer, one sees that

$$I_1 \to \int_0^t F_t(s, X(s))ds, \quad \mathbb{P}\text{-a.e.}$$

$$I_2 \to \int_0^t \langle F_x(s, X(s)), \varphi(s)\rangle ds + \int_0^t \langle F_x(s, X(s)), \Phi(s)dW(s)\rangle, \quad \mathbb{P}\text{-a.e.}$$

To find the limit of I_3 let us remark that

$$\begin{aligned}
I_3 &= \frac{1}{2}\sum_{j=0}^{k-1} \langle \Phi_0^* F_{xx}(t_j, X(t_j))\Phi_0 \Delta W_j, \Delta W_j\rangle + \frac{1}{2}\sum_{j=0}^{k-1} \langle F_{xx}(t_j, X(t_j))\varphi_0, \varphi_0\rangle (\Delta t_j)^2 \\
&\quad + \sum_{j=0}^{k-1} \langle F_{xx}(t_j, X(t_j))\Phi_0 \Delta W_j, \varphi_0\rangle \Delta t_j \\
&= I_{3,1} + I_{3,2} + I_{3,3}
\end{aligned}$$

where $W(t_{j+1}) - X(t_j) = \Delta W_j$.

We will show first that for a subsequence

$$I_{3,1} \to \frac{1}{2}\int_0^t \text{Tr}\,[\Phi_0^* F_{xx}(s, X(s))\Phi_0 Q]ds. \tag{4.38}$$

Denote $\xi_j = \Phi_0^* F_{xx}(t_j, X(t_j))\Phi_0$, then

$$J = \mathbb{E}\left[\sum_{j=0}^{k-1}\langle \Phi_0^* F_{xx}(t_j, X(t_j))\Phi_0 \Delta W_j, \Delta W_j\rangle \right.$$
$$\left. - \sum_{j=0}^{k-1} \mathrm{Tr}\left[\Phi_0^* F_{xx}(t_j, X(t_j))\Phi_0 Q \Delta t_j\right]\right]^2$$
$$= \mathbb{E}\left[\sum_{j=0}^{k-1} \mathbb{E}(\langle \xi_j \Delta W_j, \Delta W_j\rangle^2 | \mathscr{F}_j) - (\mathrm{Tr}\,[\xi_j Q])^2 (\Delta t_j)^2\right].$$

This follows from the elementary fact that if $\eta_0, \ldots, \eta_{k-1}$ are random variables with finite second moments and $\mathscr{G}_0, \ldots, \mathscr{G}_{k-1}$ an increasing sequence of σ-fields such that η_i are measurable with respect to $\mathscr{G}_j$, $0 \le j \le k-1$, then

$$\mathbb{E}\left(\sum_{j=0}^{k-1}\eta_j - \sum_{j=0}^{k-1}\mathbb{E}(\eta_j|\mathscr{G}_j)\right)^2 = \sum_{j=0}^{k-1}\left(\mathbb{E}(\eta_j)^2 - \mathbb{E}(\mathbb{E}(\eta_j|\mathscr{G}_j)^2)\right).$$

Let M be a constant such that $|\xi_j| \le M$, $j = 0, 1, \ldots, k-1$. Then

$$J \le M^2\left(\sum_{j=0}^{k-1}\mathbb{E}|W(t_{j+1}) - W(t_j)|^4 + \mathrm{Tr}(Q)(t_{j+1} - t_j)^2\right)$$

and we see that $J \to 0$. Consequently, taking a subsequence one gets (4.38). From the paths continuity of the Wiener process and boundedness of $F_{xx}(s, X(s))$, $s \in [0, T]$, one deduces that $I_{3,2} \to 0$ and $I_{3,3} \to 0$. It remains to show that there exist subsequences of I_4 and I_5, $\mathbb{P}$-a.s. converging to 0. The convergence of I_4 to 0 is a consequence of uniform continuity of F_t. By the uniform continuity of F_{xx} and taking into account that the sequence $\sum_{j=0}^{k-1}|X(t_{j+1}) - X(t_j)|^2$ contains a $\mathbb{P}$-a.s. bounded subsequence, a subsequence of I_5 tends to 0. The proof is complete. □

Let us apply the Itô formula to

$$X(t) = \int_0^t \Phi(s)dW(s), \quad t \ge 0$$

and $F(x) = |x|^2$, $x \in H$. Then

$$|X(t)|^2 = 2\int_0^t \langle X(s), \Phi(s)\rangle dW(s) + \int_0^t \mathrm{Tr}[(\Phi(s)Q^{1/2})(\Phi(s)Q^{1/2})^*]ds$$
$$= 2\int_0^t \langle X(s), \Phi(s)\rangle dW(s) + \int_0^t \|\Phi(s)\|^2_{L_2^0}\, ds.$$

Consequently the process

$$|X(t)|^2 - \int_0^t \|\Phi(s)\|^2_{L_2^0}\, ds, \quad t \geq 0,$$

is a local martingale. We will often use the notation

$$\langle X \rangle_t = \int_0^t \|\Phi(s)\|^2_{L_2^0}\, ds, \quad t \geq 0,$$

in analogy with real valued martingales.

4.5 Stochastic Fubini theorem

Let $(E, \mathscr{E})$ be a measurable space and let $\Phi(t, \omega, x) \to \varphi(t, \omega, x)$ be a measurable mapping from $(\Omega_T \times E, \mathscr{P}_T \times \mathscr{B}(E))$ into $(L_2^0, \mathscr{B}(L_2^0))$. Thus, in particular, for arbitrary $x \in E$, $\Phi(\cdot, \cdot, x)$ is a predictable L_2^0-valued process. Let in addition μ be a finite positive measure on $(E, \mathscr{E})$.

The following stochastic version of the Fubini theorem will be frequently used. It generalizes a similar result due to [163], see [428] for the finite dimensional case. (Sometimes to simplify notation we will not indicate the dependence of Φ on ω.)

Theorem 4.33 *Assume that $(E, \mathscr{E})$ is a measurable space and let*

$$\Phi : (t, \omega, x) \to \Phi(t, \omega, x)$$

be a measurable mapping from $(\Omega_T \times E, \mathscr{P}_T \times \mathscr{B}(E))$ into $L_2^0, \mathscr{B}(L_2^0))$. Assume moreover that

$$\int_E |||\Phi(\cdot, \cdot, x)|||_T \mu(dx) < +\infty \tag{4.39}$$

then $\mathbb{P}$-a.s.

$$\int_E \left[\int_0^T \Phi(t, x) dW(t)\right] \mu(dx) = \int_0^T \left[\int_E \Phi(t, x)\mu(dx)\right] dW(t). \tag{4.40}$$

Proof It follows from (4.39) that for μ-almost any $x \in E$ the predictable process $\Phi(\cdot, \cdot, x)$ is stochastically integrable. We have to show that there exists an $\mathscr{F}_T \times \mathscr{B}(E)$-measurable version of the integral

$$\xi(\omega, x) = \int_0^T \Phi(t, \omega, x) dW(t, \omega), \quad (\omega, x) \in \Omega \times E$$

which is μ-Bochner integrable for $\mathbb{P}$-almost all $\omega \in \Omega$ and such that

$$\int_E \xi(\omega, x)\mu(dx) = \int_0^T \eta(t, \omega) dW(t, \omega), \quad \text{for } \mathbb{P}\text{-a.s. } \omega \in \Omega,$$

where $\eta(\cdot,\cdot)$ is a predictable version of the strong Bochner integral

$$\eta(t,\omega) = \int_E \Phi(t,\omega,x)\mu(dx), \quad (t,\omega) \in \Omega_T.$$

We first generalize the basic approximation result from Proposition 4.22.

Proposition 4.34 *Assume that $(E,\mathscr{E})$ is a measurable space and let $\Phi : (t,\omega,x) \to \varphi(t,\omega,x)$ be a measurable mapping from $(\Omega_T \times E, \mathscr{P}_T \times \mathscr{B}(E))$ into $L_2^0, \mathscr{B}(L_2^0))$. Assume moreover that* (4.39) *holds. Then there exists a sequence $\{\Phi_n\}$ of measurable mappings from $(\Omega_T \times E, \mathscr{P}_T \times \mathscr{B}(E))$ into $(L_2^0, \mathscr{B}(L_2^0))$ of the form*

$$\Phi_n(t,\omega,x) = \sum_{j=1}^{J_n}\sum_{k=1}^{K_n}\sum_{l=1}^{L_n} \alpha_{k,l}^n H_j \varphi_{j,k}^n(x) 1\!\!1_{(s_{k,l}^n, t_{k,l}^n]} 1\!\!1_{F_{k,l}^n} \tag{4.41}$$

where $\alpha_{k,l}^n$ are constants, H_j Hilbert–Schmidt operators from U into H, $\varphi_{j,k}^n$ bounded real valued $\mathscr{B}(E)$-measurable functions, $(s_{k,l}^n, t_{k,l}^n]$ subinterval of $[0,T]$ and $F_{k,l}^n \in \mathscr{F}_{s_{k,l}^n}$ such that

$$\lim_{n\to\infty} \int_E \left(\int_{\Omega_T} \|\Phi(t,\omega,x) - \Phi_n(t,\omega,x)\|_{L_2^0}^2 \, \mathbb{P}_T(dt,d\omega)\right)^{1/2} \mu(dx) = 0. \tag{4.42}$$

Proof Since there exists a sequence $\{\Phi_n\}$ of uniformly bounded mappings for which (4.42) holds, we can assume, without any loss of generality, that $\Phi(\cdot,\cdot,\cdot)$ is a bounded mapping. Let $\{H_j\}$ be a complete orthonormal system on L_2^0 and $\{\varepsilon_k(\cdot,\cdot)\}$ a complete orthonormal system on $\mathscr{L}_T^2 = L^2(\Omega_T, \mathscr{P}_T, \mathbb{P}_T)$ composed of bounded processes. Taking into account expansions with respect to bases $\{\varepsilon_n\}$ and $\{H_j\}$ we define approximating processes

$$\Phi_J^1(t,\omega,x) = \sum_{j=1}^{J} H_j \varphi_j(t,\omega,x)$$

where

$$\varphi_j(t,\omega,x) = \langle \Phi(t,\omega,x), H_j \rangle_{L_2^0},$$

$$\Phi_{J,K}^2(t,\omega,x) = \sum_{j=1}^{J} H_j \left(\sum_{k=1}^{K} \varepsilon_k(t,\omega)\varphi_{j,k}(x)\right)$$

and

$$\varphi_{j,k}(x) = \langle \varphi_j(\cdot,\cdot,x), \varepsilon_k(\cdot,\cdot)\rangle_{\mathscr{L}_T^2}.$$

The third approximating process replaces predictable processes ε_k by simple processes

$$\Phi_{J,K,L}^3(t,\omega,x) = \sum_{j=1}^{J} H_j \left[\sum_{k=1}^{K} \varphi_{j,k}(x) \sum_{l=1}^{L} \alpha_{k,l} 1\!\!1_{]s_{k,l},t_{k,l}]}(t) 1\!\!1_{F_{k,l}^n}\right],$$

for $(t, \omega, x) \in \Omega_T \times E$. Denote the integral in (4.39) by $|||\Phi|||$. By the Lebesque monotone convergence theorem we have

$$\lim_{J\to\infty} |||\Phi - \Phi^1_J||| = 0, \quad \lim_{K\to\infty} |||\Phi^1_J - \Phi^2_{J,K}||| = 0. \tag{4.43}$$

Finally, proceeding as in the proof of Proposition 4.22(ii), for fixed J and K one can find a sequence of approximations of the third type such that

$$\lim_{L\to\infty} |||\Phi^2_{J,K} - \Phi^3_{J,K,L}||| = 0. \tag{4.44}$$

It follows from the boundedness of Φ and the construction of the approximations that $\varphi_{j,k}(\cdot)$ are bounded functions and the approximations are predictable processes. Taking into account (4.43) and (4.44) we arrive at (4.42).

Proof of Theorem 4.33 It is obvious that the theorem is true for the approximating processes Φ_n, $n \in \mathbb{N}$,

$$\int_E \left[\int_0^T \Phi_n(t, x) dW(t)\right] \mu(dx) = \int_0^T \left[\int_E \Phi_n(t, x)\mu(dx)\right] dW(t).$$

Moreover, by the Fubini theorem (see Section 1.9) and the preceding discussion, the process

$$\eta(t, \omega) = \int_E \Phi(t, \omega, x)\mu(dx), \quad (t, \omega) \in \Omega_T,$$

is predictable. We now show that there exists a version of $\xi(\cdot, \cdot)$ which is $\mathscr{F}_T \times \mathscr{B}(E)$ measurable. The processes

$$\xi_n(\omega, x) = \int_0^T \Phi_n(t, \omega, x) dW(t, \omega), \quad \omega \in \Omega, x \in E$$

are $\mathscr{F}_T \times \mathscr{B}(E)$-measurable and, by the standard use of Chebyshev's inequality and the Borel–Cantelli lemma, we get from (4.42) that there exists a set $E_0 \in \mathscr{B}(E)$, $\mu(E \setminus E_0) = 0$ and a subsequence $\{\Phi_{n_m}\}$ such that for all $x \in E_0$ and $m \geq m(x)$

$$|||\Phi(\cdot, \cdot, x) - \Phi_{n_m}(\cdot, \cdot, x)||| \leq \frac{1}{2^m}.$$

Consequently, if we define,

$$\xi(\omega, x) = \begin{cases} \lim\limits_{m\to\infty} \int_0^T \Phi_{n_m}(t, \omega, x) dW(t, \omega) & \text{if the limit exists} \\ 0 \quad \text{otherwise,} \end{cases}$$

then $\xi(\cdot, \cdot)$ is a $\mathscr{F}_T \times \mathscr{B}(E)$-measurable random variable and, by the elementary properties of the stochastic integral,

$$\xi(\omega, x) = \int_0^T \Phi(t, \omega, x) dW(t, \omega)$$

for all $x \in E_0$ and $\mathbb{P}$-a.s. $\omega \in \Omega$. Note that

$$\begin{aligned}
&\mathbb{E}\left|\int_E \left(\int_0^T \Phi(t,x)dW(t)\right)\mu(dx) - \int_E \left(\int_0^T \Phi_{n_m}(t,x)dW(t)\right)\mu(dx)\right| \\
&\quad \le \int_E \left(\mathbb{E}\left|\int_0^{T^{(}} (\Phi(t,x) - \Phi_{n_m}(t,x))dW(t)\right|\right)\mu(dx) \\
&\quad \le \int_E |||\Phi(t,x) - \Phi_{n_m}(t,x)|||\mu(dx) \to 0 \quad \text{as } m \to \infty. \qquad (4.45)
\end{aligned}$$

Moreover

$$\begin{aligned}
&\mathbb{E}\left|\int_0^T \left(\int_E \Phi(t,x)\mu(dx)\right)dW(t) - \int_E \left(\int_0^T \Phi_{n_m}(t,x)dW(t)\right)\mu(dx)\right|^2 \\
&= \mathbb{E}\left|\int_0^T \left(\int_E \Phi(t,x)\mu(dx)\right)dW(t) - \int_0^T \left(\int_E (\Phi_{n_m}(t,x)\mu(dx)\right)dW(t)\right|^2 \\
&= \mathbb{E}\left|\int_0^T \left(\int_E (\Phi(t,x) - \Phi_{n_m}(t,x))\mu(dx)\right)dW(t)\right|^2.
\end{aligned}$$

Therefore

$$\begin{aligned}
&\mathbb{E}\left|\int_0^T \left(\int_E \Phi(t,x)\mu(dx)\right)dW(t) - \int_E \left(\int_0^T \Phi_{n_m}(t,x)dW(t)\right)\mu(dx)\right|^2 \\
&\le |||\int_E (\Phi(t,x) - \Phi_{n_m}(t,x))\mu(dx)|||^2 \\
&\le \left(\int_E |||\Phi(t,x) - \Phi_{n_m}(t,x)|||\mu(dx)\right)^2 \to 0 \text{ as } m \to \infty. \qquad (4.46)
\end{aligned}$$

The final estimate follows by regarding $\Phi(\cdot,\cdot,\cdot) - \Phi_{n_m}(\cdot,\cdot,\cdot)$ as a measurable mapping from $(E, \mathscr{E}, \mu)$ into $L^2(\Omega_T, \mathbb{P}_T, H)$ (equipped with the norm $|||\cdot|||$) and by applying (1.6) (in fact this is also the generalized Minkowski inequality). From the estimates (4.45)–(4.46) the result follows. □

Remark 4.35 The condition (4.39), written in an expanded version, has the form

$$\int_E \left[\int_\Omega \int_0^T \|\Phi(t,\omega,x)\|^2_{L_2^0})dt)\mathbb{P}(d\omega)\right]^{1/2} \mu(dx) < \infty.$$

In the case when Φ is a nonstochastic mapping, one has simply

$$\int_E \left[\int_0^T \|\Phi(t,x)\|^2_{L_2^0})dt\right]^{1/2} \mu(dx) < \infty.$$

□

4.6 Basic estimates

Our aim in this section is to prove the following two theorems.

Theorem 4.36 *For every $p > 0$ there exists $c_p > 0$ such that for every $t \geq 0$,*

$$\mathbb{E} \sup_{s\in[0,t]} \left| \int_0^s \Phi(\tau) dW(\tau) \right|^p \leq c_p \left[\mathbb{E} \int_0^t \|\Phi(s)\|^2_{L^0_2} ds \right]^{p/2}. \tag{4.47}$$

Theorem 4.37 *For every $p \geq 2$ there exists $c'_p > 0$ such that for every $t \geq 0$,*

$$\mathbb{E} \sup_{s\in[0,t]} \left| \int_0^s \Phi(\tau) dW(\tau) \right|^p \leq c'_p \left[\int_0^t \left(\mathbb{E}\|\Phi(s)\|^p_{L^0_2} \right)^{2/p} ds \right]^{p/2}. \tag{4.48}$$

In the proof of both theorems we may assume that the process

$$Z(t) := \int_0^t \Phi(s) dW(s), \quad t \geq 0,$$

is bounded by considering stopping times

$$\tau := \inf\{t \geq 0 : \ |Z(t)| \geq k\}$$

and proving the estimates first for integrands

$$\widetilde{\Phi} = \mathbb{1}_{[0,\tau]}(s)\Phi(s), \quad s \geq 0$$

and the stochastic integral

$$\widetilde{Z}(t) := \int_0^t \widetilde{\Phi}(s) dW(s), \quad t \geq 0.$$

Proof of Theorem 4.36
Step 1 $p \geq 2$. By the martingale inequality (3.18) and Corollary 4.29 the result is true for $p = 2$. Assume now that $p > 2$, set

$$Z(t) = \int_0^t \Phi(s) dW(s), \quad t \geq 0,$$

and apply Itô's formula to $f(Z(\cdot))$ where $f(x) = |x|^p$, $x \in H$. Since

$$f_{xx}(x) = p(p-2)|x|^{p-4} x \otimes x + p|x|^{p-2} I, \quad x \in H,$$

we have

$$\|f_{xx}(x)\| \leq p(p-1)|x|^{p-2},$$

therefore

$$|\mathrm{Tr}\ \Phi^*(t) f_{xx}(Z(t))\Phi(t)Q| \leq p(p-1)|Z(t)|^{p-2} \|\Phi(t)\|^2_{L^0_2}.$$

By taking expectation in the identity

$$|Z(t)|^p = p \int_0^t |Z(s)|^{p-2} \langle \Phi(s), d\mathbb{Z}(s) \rangle + \frac{1}{2} \int_0^t \mathrm{Tr}\ [\Phi^*(s) f_{xx}(Z(s))\Phi(s)Q] ds,$$

we obtain

$$\mathbb{E}|Z(t)|^p \le \frac{p(p-1)}{2}\,\mathbb{E}\left(\int_0^t |Z(s)|^{p-2}\|\Phi(s)\|^2_{L_2^0}\,ds\right)$$
$$\le \frac{p(p-1)}{2}\,\mathbb{E}\left(\sup_{s\in[0,t]}|Z(s)|^{p-2}\int_0^t \|\Phi(s)\|^2_{L_2^0}ds\right). \qquad (4.49)$$

By Hölder's inequality with exponents $\frac{p}{p-2}, \frac{p}{2}$

$$\mathbb{E}|Z(t)|^p \le \frac{p(p-1)}{2}\left[\mathbb{E}\sup_{s\in[0,t]}|Z(s)|^p\right]^{(p-2)/p}\left[\left(\mathbb{E}\int_0^t \|\Phi(s)\|^2_{L_2^0}\,ds\right)^{p/2}\right]^{2/p}.$$

Using the martingale inequality (3.18) one arrives at

$$\mathbb{E}|Z(t)|^p \le \frac{p(p-1)}{2}\left[\left(\frac{p}{p-1}\right)^p\mathbb{E}|Z(t)|^p\right]^{1-2/p}\left[\left(\mathbb{E}\int_0^t \|\Phi(s)\|^2_{L_2^0}\,ds\right)^{p/2}\right]^{2/p}.$$

Dividing both sides by $(\mathbb{E}|Z(t)|^p)^{1-2/p}$ one gets

$$(\mathbb{E}|Z(t)|^p)^{2/p} \le \frac{p(p-1)}{2}\left(\frac{p}{p-1}\right)^{p-2}\left[\left(\mathbb{E}\int_0^t \|\Phi(s)\|^2_{L_2^0}\,ds\right)^{p/2}\right]^{2/p}.$$

Taking both sides the power $p/2$ yields (4.47).

Step 2 $p \in (0,2)$.

We need two lemmas.

Lemma 4.38 *For arbitrary nonnegative random variable ξ and $p \in (0,2)$, we have*

$$\mathbb{E}(\xi^p) = \frac{(2-p)p}{2}\int_0^\infty \lambda^{p-3}\mathbb{E}(\xi^2 \wedge \lambda^2)d\lambda. \qquad (4.50)$$

Proof Note that

$$p\int_0^\infty \lambda^{p-3}\mathbb{E}(\xi^2\wedge\lambda^2)d\lambda = \mathbb{E}\left(\int_0^\xi p\lambda^{p-1}d\lambda\right) + \mathbb{E}\left(\xi^2\int_\xi^{+\infty} p\lambda^{p-3}d\lambda\right)$$
$$= \mathbb{E}(\xi^p) + \frac{p}{2-p}\,\mathbb{E}(\xi^p) = \frac{2}{2-p}\,\mathbb{E}(\xi^p).$$

□

Recall that we introduced the notation

$$\langle Z\rangle_t = \int_0^t |\Phi(s)|^2_{L_2^0}\,ds, \quad t \ge 0.$$

Lemma 4.39 *For arbitrary $\lambda > 0$ and $T > 0$ we have*

$$\mathbb{P}\left(\sup_{t\in[0,T]}|Z(t)| > \lambda\right) \le \mathbb{P}\left(\langle Z\rangle_T > \lambda^2\right) + \frac{1}{\lambda^2}\,\mathbb{E}\left(\lambda^2 \wedge \langle Z\rangle_T\right). \qquad (4.51)$$

Proof Define

$$\tau_\lambda = \inf\{t \in [0, T] : \langle Z \rangle_t > \lambda^2\}$$

and note that

$$\begin{cases} \tau_\lambda = T & \text{if and only if } \langle Z \rangle_t \le \lambda^2 \quad \text{for all } t \in [0, T], \\ \tau_\lambda < T & \text{if and only if } \langle Z \rangle_T > \lambda^2. \end{cases}$$

Consequently

$$\begin{aligned} \mathbb{P}\left(\sup_{t\in[0,T]} |Z(t)| > \lambda\right) &= \mathbb{P}\left(\sup_{t\in[0,T]} |Z(t)| > \lambda \text{ and } \tau_\lambda = T\right) \\ &\quad + \mathbb{P}\left(\sup_{t\in[0,T]} |Z(t)| > \lambda \text{ and } \tau_\lambda < T\right) \\ &\le \mathbb{P}\left(\sup_{t\in[0,\tau_\lambda\wedge T]} |Z(t)| > \lambda \text{ and } \tau_\lambda = T\right) + \mathbb{P}(\langle Z \rangle_T > \lambda^2) \\ &\le \mathbb{P}\left(\sup_{t\in[0,\tau_\lambda\wedge T]} |Z(t)| > \lambda\right) + \mathbb{P}(\langle Z \rangle_T > \lambda^2) \\ &\le \mathbb{P}\left(\sup_{t\in[0,T]} |Z(\tau_\lambda \wedge t)| > \lambda\right) + \mathbb{P}(\langle Z \rangle_T > \lambda^2). \end{aligned}$$

By the martingale inequality (3.17) with $p = 2$ we get

$$\mathbb{P}\left(\sup_{t\in[0,T]} |Z(\tau_\lambda \wedge t)| > \lambda\right) \le \frac{1}{\lambda^2}\mathbb{E}(\langle Z \rangle_{\tau_\lambda\wedge T}) \le \frac{1}{\lambda^2}\mathbb{E}(\lambda^2 \wedge \langle Z \rangle_T).$$

Therefore (4.51) follows. □

Proof of step 2 of Theorem 4.36 Since

$$\mathbb{E}\left(\sup_{s\in[0,T]} |Z(s)|^p\right) = p\int_0^{+\infty} \lambda^{p-1}\mathbb{P}\left(\sup_{s\in[0,T]} |Z(s)| > \lambda\right) d\lambda,$$

therefore, by Lemma 4.39,

$$\begin{aligned} \mathbb{E}\left(\sup_{s\in[0,T]} |Z(s)|^p\right) &\le p\int_0^{+\infty} \lambda^{p-1}\mathbb{P}\left(\langle Z \rangle_T > \lambda^2\right) d\lambda \\ &\quad + p\int_0^{+\infty} \lambda^{p-3}\mathbb{E}\left(\lambda^2 \wedge \langle Z \rangle_T\right) d\lambda. \end{aligned}$$

But

$$p\int_0^{+\infty} \lambda^{p-1}\mathbb{P}\left(\langle Z \rangle_T > \lambda^2\right) d\lambda = \mathbb{E}(\langle Z \rangle_T)^{p/2}$$

and by Lemma 4.38

$$p\int_0^{+\infty} \lambda^{p-3}\mathbb{E}\left(\lambda^2 \wedge \langle Z\rangle_T\right) d\lambda = \frac{2}{2-p}\,\mathbb{E}(\langle Z\rangle_T)^{p/2}.$$

Thus the result is true if $p \in (0, 2)$. □

Proof of Theorem 4.37 By (3.18) we have

$$\mathbb{E}\left(\sup_{s\in[0,T]} |Z(s)|^p\right) \le \left(\frac{p}{p-1}\right)^p \sup_{s\in[0,T]} \mathbb{E}|Z(s)|^p.$$

We will show that

$$\sup_{s\in[0,t]} \mathbb{E}|Z(s)|^p \le \frac{p(p-1)}{2}\left(\int_0^t \left(\mathbb{E}\|\Phi(s)\|^p_{L_2^0}\right)^{2/p} ds\right)^{p/2}. \tag{4.52}$$

The inequality (4.52) is obviously true for $p = 2$. So we can assume that $p > 2$. By (4.49) we have

$$\mathbb{E}|Z(t)|^p \le \frac{p(p-1)}{2}\int_0^t \mathbb{E}\left(|Z(s)|^{p-2}\|\Phi(s)\|^2_{L_2^0}\right) ds.$$

By Hölder's inequality with exponents $\frac{p}{p-2}$, $\frac{p}{2}$ we have

$$\mathbb{E}(|Z(s)|^{p-2}\|\Phi(s)\|^2_{L_2^0}) \le (\mathbb{E}|Z(s)|^p)^{(p-2)/p}\left(\mathbb{E}\|\Phi(s)\|^p_{L_2^0}\right)^{2/p}.$$

Consequently

$$\mathbb{E}|Z(t)|^p \le \frac{p(p-1)}{2}\int_0^t (\mathbb{E}|Z(s)|^p)^{(p-2)/p}\left(\mathbb{E}\|\Phi(s)\|^p_{L_2^0}\right)^{2/p} ds$$

and

$$\mathbb{E}|Z(t)|^p \le \frac{p(p-1)}{2}\int_0^t \left(\sup_{u\in[0,s]} \mathbb{E}|Z(u)|^p\right)^{(p-2)/p}\left(\mathbb{E}\|\Phi(s)\|^p_{L_2^0}\right)^{2/p} ds.$$

Since the right hand side is increasing in t, it follows that

$$\sup_{u\in[0,t]} \mathbb{E}|Z(u)|^p \le \frac{p(p-1)}{2}\left(\sup_{u\in[0,t]} \mathbb{E}|Z(u)|^p\right)^{(p-2)/p}\int_0^t \left(\mathbb{E}\|\Phi(s)\|^p_{L_2^0}\right)^{2/p} ds.$$

Now the result follows with $c_p' = (\frac{p(p-1)}{2})^{p/2}$. □

4.7 Remarks on generalization of the integral

Stochastic integration theory with respect to martingales $M \in \mathscr{M}_T^2(H)$, completely analogous to the one with respect to a Wiener process described in the preceeding sections, can be developed, see [543]. The role of the process tQ is played by the quadratic variation $\langle\langle M(t)\rangle\rangle$, $t \in [0, T]$. We will need this extension in the case when the martingale M is itself a stochastic integral, say $M = \Phi \cdot W$ with $\Phi \in \mathcal{N}_W^2(0, T; L_2^0)$. Then the extension is straightforward, since we can define the stochastic integral $\Phi \cdot M$ simply by

$$\Psi \cdot M(t) = \int_0^t \Psi(s) dM(s) = \int_0^t \Psi(s)\Phi(s) dW(s), \quad t \in [0, T]. \tag{4.53}$$

Note that

$$\langle\langle \Psi \cdot M(t)\rangle\rangle = \int_0^T \Psi(s) Q_\Phi(s) \Psi^*(s) ds, \quad t \in [0, T], \tag{4.54}$$

where

$$Q_\Phi(s) = (\Phi(s)Q^{1/2})(\Phi(s)Q^{1/2})^*.$$

Since in the present case

$$\langle\langle M(t)\rangle\rangle = \int_0^t Q_\Phi(s) ds, \quad s \in [0, T],$$

so (4.54) can be intrinsically written as

$$\langle\langle \Psi \cdot M(t)\rangle\rangle = \int_0^t \Psi(s) \frac{d}{ds} \langle\langle M(s)\rangle\rangle \, \Psi^*(s) ds, \quad t \in [0, T]. \tag{4.55}$$

Finally (4.55) can be extended to general martingales and general integrands.

Part II

Existence and uniqueness

5

Linear equations with additive noise

We first introduce various concepts of solutions and discuss their elementary properties. Then we prove existence of weak solutions and their continuity by the *factorization method*. Temporal and spatial regularity results are examined next in the case of the drift operator generating an analytic semigroup on a Hilbert and some Banach spaces. At the end we give sufficient conditions for existence of strong solutions.

5.1 Basic concepts

5.1.1 Concept of solutions

We are given a probability space $(\Omega, \mathscr{F}, \mathbb{P})$ together with a normal filtration $\mathscr{F}_t,\ t \geq 0$. We consider two Hilbert spaces H and U, and a Q-Wiener process $W(t)$ on $(\Omega, \mathscr{F}, \mathbb{P})$, with the covariance operator $Q \in L(U)$. If $\operatorname{Tr} Q < +\infty$, then W is a genuine Wiener process, whereas if $Q = I$, W is a cylindrical process and in this case it has continuous paths in another Hilbert space U_1 larger than U, see Chapter 4. We assume that there exists a complete orthonormal system $\{e_k\}$ in U, a bounded sequence $\{\lambda_k\}$ of nonnegative real numbers such that

$$Qe_k = \lambda_k e_k, \quad k \in \mathbb{N},$$

and a sequence $\{\beta_k\}$ of real independent Brownian motions such that

$$\langle W(t), u \rangle = \sum_{k=1}^{\infty} \sqrt{\lambda_k}\, \langle u, e_k \rangle \beta_k(t), \quad u \in U,\ t \geq 0.$$

We will consider the following linear affine equation

$$\begin{cases} dX(t) = (AX(t) + f(t))dt + BdW(t) \\ X(0) = \xi \end{cases} \tag{5.1}$$

where $A\colon D(A) \subset H \to H$ and $B\colon U \to H$ are linear operators and f is an H-valued stochastic process. We will assume that the deterministic Cauchy problem

$$\begin{cases} u'(t) = Au(t), \\ u(0) = x \in H, \end{cases}$$

is uniformly well posed (see Definition A.1) and that B is bounded, that is, we have the following.

Hypothesis 5.1 *A generates a C_0-semigroup $S(\cdot)$ in H and $B \in L(U; H)$.*

It is also natural to require the following.

Hypothesis 5.2

(i) f is a predictable process with Bochner integrable trajectories on an arbitrary finite interval $[0, T]$.
(ii) ξ is $\mathscr{F}_0$-measurable.

Remark 5.1 If W is a Q-Wiener process in U, then $W_1 = BW$ is a BQB^*-Wiener process in H. So we could assume, without any restriction, that $U = H$. However, in some applications, for example wave or delay equations, it is convenient to have B different from the identity. □

An H-valued predictable process $X(t)$, $t \in [0, T]$, is said to be a *strong* solution to (5.1) if $X(t)$ takes values in $D(A)$, a.e.,

$$\int_0^T |AX(s)|ds < +\infty, \quad \mathbb{P}\text{-a.s.}$$

and for $t \in [0, T]$

$$X(t) = \xi + \int_0^t [AX(s) + f(s)]ds + BW(t), \quad \mathbb{P}\text{-a.s.}$$

This definition is meaningful only if W is a U-valued process and therefore requires that $\mathrm{Tr}\,[Q] < +\infty$. Note that a strong solution should necessarily have a continuous modification. An H-valued predictable process $X(t)$, $t \in [0, T]$, is said to be a *weak solution* of (5.1) if the trajectories of $X(\cdot)$ are $\mathbb{P}$-a.s. Bochner integrable and if for all $z \in D(A^*)$ and all $t \in [0, T]$ we have

$$\langle X(t), z\rangle = \langle \xi, z\rangle + \int_0^t [\langle X(s), A^*z\rangle + \langle f(s), z\rangle]ds + \langle BW(t), z\rangle, \quad \mathbb{P}\text{-a.s.} \quad (5.2)$$

This definition is meaningful for a cylindrical Wiener process because the scalar processes $\langle BW(t), z\rangle$, $t \in [0, T]$, are well defined random processes, (see Section 4.2.1). It is clear that a strong solution is also a weak one.

5.1.2 Stochastic convolution

It is of great importance in our study of linear and nonlinear equations to establish first the basic properties of the process

$$W_A(t) = \int_0^t S(t-s)B\,dW(s), \quad t \geq 0,$$

which is called *stochastic convolution*. The properties are collected in the following theorem. Note, however, that assertions (iii) below will not be used in this and in the next chapter.

Theorem 5.2 *Assume that Hypothesis 5.1 holds and*

$$\int_0^T \|S(r)B\|^2_{L_2^0}\, dr = \int_0^T \mathrm{Tr}\,[S(r)BQB^*S^*(r)]dr < +\infty. \tag{5.3}$$

Then

(i) the process $W_A(\cdot)$ is Gaussian, continuous in mean square and has a predictable version,
(ii) we have

$$\mathrm{Cov}(W_A(t)) = \int_0^t S(r)BQB^*S^*(r)dr, \quad t \in [0,T], \tag{5.4}$$

(iii) the trajectories of $W_A(\cdot)$ are $\mathbb{P}$-a.s. square integrable and the law $\mathscr{L}(W_A(\cdot))$ is a symmetric Gaussian measure on $\mathscr{H} = L^2(0,T;H)$ with the covariance operators

$$\mathscr{Q}\varphi(t) = \int_0^T G(t,s)\varphi(s)ds, \quad t \in [0,T], \tag{5.5}$$

where

$$G(t,s) = \int_0^{t\wedge s} S(t-r)BQB^*S^*(s-r)dr, \quad t,s \in [0,T] \tag{5.6}$$

and $t \wedge s = \min\{t,s\}$.

Proof To prove (i) fix $0 \leq s \leq t \leq T$, then

$$W_A(t) - W_A(s) = \int_s^t S(t-r)BdW(r) + \int_0^s [S(t-r) - S(s-r)]BdW(r).$$

Since the integrals are independent it follows that

$$\begin{aligned}\mathbb{E}|W_A(t) - W_A(s)|^2 &= \sum_{k=1}^\infty \lambda_k \int_0^{t-s} |S(r)Be_k|^2 dr \\ &\quad + \sum_{k=1}^\infty \lambda_k \int_0^s |(S(t-s+r) - S(r))Be_k|^2 dr.\end{aligned}$$

By (5.3) and the Lebesgue dominated convergence theorem, the mean square continuity follows. That the process is Gaussian follows easily from the definition of stochastic integral. Existence of a predictable version is a consequence of Proposition 3.6, and (5.4) follows from Proposition 4.28. It remains to prove (iii). For a measurable version of $W_A(\cdot)$ we have, by the Fubini theorem

$$\mathbb{E}\int_0^T |W_A(s)|^2 ds = \int_0^T \mathbb{E}|W_A(s)|^2 ds = \int_0^T \int_0^s \|S(r)B\|^2_{L_2^0}\, dr ds < +\infty,$$

so the initial part of (iii) is true and moreover $W_A(\cdot)$ can be regarded as an $\mathscr{H}$-valued random variable (see Proposition 3.18.) To show that $\mathscr{L}(W_A(\cdot))$ is symmetric and Gaussian on $\mathscr{H} = L^2(0, T; H)$ we apply Proposition 2.11 and consider the following family M of all functionals $(h \otimes a) \in \mathscr{H}^* = \mathscr{H},\ a \in H$

$$(h \otimes a)(\varphi) = \int_0^T h(t)\langle a, \varphi(t)\rangle dt, \quad \varphi \in \mathscr{H}.$$

We will need the following elementary lemma whose proof is left to the reader.

Lemma 5.3 *If $\xi(\cdot)$ is a real valued Gaussian process, mean square continuous, then for arbitrary $h \in L^2(0, T; \mathbb{R}^1)$, $\int_0^T h(s)\xi(s)ds$ is a Gaussian random variable $\mathscr{N}_{0,|h|^2}$.*

Since

$$(h \otimes a)(W_A) = \int_0^T h(t)\langle a, W_A(t)\rangle dt, \quad \varphi \in \mathscr{H}$$

and $\langle a, W_A(\cdot)\rangle$ is a real Gaussian process, mean square continuous, with mean 0, by Proposition 2.11 $\mathscr{L}(W_A(\cdot))$ is a symmetric Gaussian distribution on $\mathscr{H}$. If $\varphi, \psi \in \mathscr{H}$ then

$$\begin{aligned}\langle \mathscr{Q}\varphi, \psi\rangle_{\mathscr{H}} &= \mathbb{E}\left(\int_0^T \langle\varphi(s), W_A(s)\rangle ds \int_0^T \langle\psi(r), W_A(r)\rangle dr\right)\\ &= \int_0^T\int_0^T \mathbb{E}(\langle\varphi(s), W_A(s)\rangle\langle\psi(r), W_A(r)\rangle) ds\, dr.\end{aligned}$$

Since for $s > r$

$$\begin{aligned}&\mathbb{E}\left(\langle\varphi(s), W_A(s)\rangle\langle\psi(r), W_A(r)\rangle\right)\\ &= \mathbb{E}\int_0^T\int_0^T\int_0^s \langle\varphi(s), S(s-\sigma)BdW(\sigma)\rangle \int_0^r \langle\psi(r), S(r-\rho)BdW(\rho)\rangle ds\, dr\\ &= \int_0^T\int_0^T\int_0^{s\wedge r} \langle QB^*S^*(s-\sigma)\varphi(s), B^*S^*(r-\sigma)\psi(r)\rangle d\sigma\, ds\, dr\\ &= \int_0^T\int_0^T \langle G(s,r)\varphi(s), \psi(r)\rangle\, ds\, dr,\end{aligned}$$

where $G(s, r)$ is given by (5.6). Then (5.5) holds true. □

5.2 Existence and uniqueness of weak solutions

The main result of this section is the following.

Theorem 5.4 *Assume Hypotheses* 5.1, 5.2 *and* (5.3). *Then equation* (5.1) *has exactly one weak solution which is given by the formula*

$$X(t) = S(t)\xi + \int_0^t S(t-s)f(s)ds + \int_0^t S(t-s)BdW(s), \quad t \in [0,T]. \tag{5.7}$$

Formula (5.7) is a stochastic generalization of the classical variation of constants formula, see formula (A.17) in Appendix A.

Proof It easily follows from Proposition A.6 that a process X is a weak solution to (5.1) if and only if the process $\widetilde{X}$ given by the formula

$$\widetilde{X}(t) = X(t) - \left(S(t)\xi + \int_0^t S(t-s)f(s)ds \right), \quad t \in [0,T],$$

is a weak solution to

$$d\widetilde{X} = A\widetilde{X}dt + BdW, \quad \widetilde{X}(0) = 0.$$

So, we can assume, without any loss of generality, that $\xi = 0$ and $f \equiv 0$. To prove existence we show that equation (5.1) with $\xi = 0$ and $f \equiv 0$ is satisfied by the process $W_A(\cdot)$. We fix $t \in [0,T]$ and let $\zeta \in D(A^*)$. Note that

$$\int_0^t \langle A^*\zeta, W_A(s)\rangle ds = \int_0^t \left\langle A^*\zeta, \int_0^t 1\!\!1_{[0,s]}(r)S(s-r)BdW(r) \right\rangle ds$$

and consequently,

$$\begin{aligned}
\int_0^t \langle A^*\zeta, W_A(s)\rangle ds &= \int_0^t \left\langle \int_0^t 1\!\!1_{[0,s]}(r)B^*S^*(s-r)A^*\zeta ds, dW(r) \right\rangle \\
&= \int_0^t \left\langle \int_r^t B^*S^*(s-r)A^*\zeta ds, dW(r) \right\rangle \\
&= \int_0^t \left\langle \int_r^t \left(\frac{d}{ds} B^*S^*(s-r)\zeta \right) ds, dW(r) \right\rangle \\
&= \int_0^t \langle B^*S^*(t-r)\zeta), dW(r)\rangle - \int_0^t \langle B^*\zeta, dW(r)\rangle \\
&= \langle \zeta, W_A(t)\rangle - \langle \zeta, BW(t)\rangle.
\end{aligned}$$

Therefore $W_A(\cdot)$ is a weak solution.

To prove uniqueness we need the following lemma.

Lemma 5.5 *Let X be a weak solution of problem* (5.1) *with* $\xi = 0$, $f \equiv 0$. *Then, for arbitrary function* $\zeta(\cdot) \in C^1([0,T]; D(A^*))$ *and* $t \in [0,T]$, *we have*

$$\langle X(t), \zeta(t)\rangle = \int_0^t [\langle X(s), \zeta'(s) + A^*\zeta(s)\rangle]ds + \int_0^t \langle \zeta(s), BdW(s)\rangle.$$

Proof Consider first functions of the form $\zeta = \zeta_0\varphi(s)$, $s \in [0,T]$, where $\varphi \in C^1([0,T])$ and $\zeta_0 \in D(A^*)$. Let

$$F_{\zeta_0}(t) = \int_0^t \langle X(s), A^*\zeta_0\rangle ds + \langle BW(t), \zeta_0\rangle.$$

Applying Itô's formula to the process $F_{\zeta_0}(s)\varphi(s)$ we get

$$d[F_{\zeta_0}(s)\varphi(s)] = \varphi(s)dF_{\zeta_0}(s) + \varphi'(s)F_{\zeta_0}(s)ds.$$

In particular

$$F_{\zeta_0}(t)\varphi(t) = \int_0^t \langle \zeta(s), BdW(s)\rangle + \int_0^t [\varphi(s)(\langle X(s), A^*\zeta_0\rangle + \varphi'(s)\langle X(s), \zeta_0\rangle]ds.$$

Since $F_{\zeta_0}(\cdot) = \langle X(\cdot), \zeta_0\rangle$, $\mathbb{P}$-a.s. the lemma is proved for the special function $\zeta(t) = \zeta_0\varphi(t)$. Since these functions are linearly dense in $C^1([0,T]; D(A^*))$ the lemma is true in general. □

Let X be a weak solution and let $\zeta_0 \in D(A^*)$. Applying Lemma 5.5 to the function $\zeta(s) = S^*(t-s)\zeta_0$, $s \in [0,t]$, we have

$$\langle X(t), \zeta_0\rangle = \int_0^t \langle S(t-s)BdW(s), \zeta_0\rangle$$

and, since $D(A^*)$ is dense in H we find that $X = W_A$. The proof is complete. □

Example 5.6 *Delay equations.* In this example we use the notations of Section A.5.1. We are concerned with the problem

$$\begin{cases} dz(t) = \displaystyle\int_{-r}^0 a(d\theta)z(t+\theta)dt + f(t)dt + dW(t), \quad t \ge 0, \\ z(0) = h_0, \\ z(\theta) = h_1(\theta), \quad \theta \in [-r, 0], \ \mathbb{P}\text{-a.s.}, \end{cases} \tag{5.8}$$

where $a(\cdot)$ is an $N \times N$ matrix valued finite measure on $[-r, 0]$, $f\colon [0,+\infty) \to \mathbb{R}^N$ is a locally integrable function, $h_0 \in \mathbb{R}^N$, $h_1 \in L^2(-r, 0; \mathbb{R}^N)$ and r is a positive number representing the *delay*. In a similar way as in the deterministic case (see Section A.5.1), we can associate with the equation (5.8) a stochastic linear equation:

$$\begin{cases} dX = AXdt + Bf(t)dt + BdW(t), \\ X(0) = \begin{pmatrix} h_0 \\ h_1 \end{pmatrix} \end{cases} \tag{5.9}$$

on the space $H = \mathbb{R}^N \oplus L^2(-r, 0; \mathbb{R}^N)$, where the generator A is given by (A.40) and $B = \binom{I}{0}$. In the present situation U is equal to $\mathbb{R}^N$. Obviously (5.3) is fulfilled in this case and therefore the equation (5.9) as a unique weak solution. It has been shown by several authors [164, 256, 698] under different sets of assumptions that $X(t) = (z(t), z_t),\ t > 0$, where z and z_t are the solutions of the equation (5.8) and its segment respectively. This fact is important in the application of delay systems to control problems and stability.

Let us consider the special case when

$$a(\cdot) = a_0\delta_0(\cdot) + a_1\delta_{-r}(\cdot)$$

where a_0, a_1 are $N \times N$ matrices. Equation (5.9) with $f = 0$ can now be solved by successive steps. In particular, for $t \in [0, r]$

$$z(t) = e^{ta_0}h_0 + \int_0^t e^{(t-s)a_1}h_1(s-r)ds + \int_0^t e^{(t-s)a_0}dW(s). \tag{5.10}$$

Taking into account that trajectories of the stochastic convolution part in (5.10) are never absolutely continuous, we can see that in this case weak solution to (5.10) are never strong. □

Example 5.7 *Heat equation.* Let $U = H = L^2(\mathscr{O})$, where $\mathscr{O}$ is a bounded open set in $\mathbb{R}^N$ with a regular boundary $\partial\mathscr{O}$. Consider the problem

$$\begin{cases} d_tX(t,\xi) = \Delta_\xi X(t,\xi)dt + dW(t,\xi), & t \ge 0,\ \xi \in \mathscr{O}, \\ X(t,\xi) = 0, \quad t \ge 0,\ \xi \in \partial\mathscr{O}, \\ X(0,\xi) = 0, \quad \xi \in \mathscr{O}, \end{cases}$$

and let A be the realization of the Laplace operator in $L^2(\mathscr{O})$ with Dirichlet boundary conditions, defined in Section A.5.2. If $\mathrm{Tr}\ Q < +\infty,\ t \ge 0$, we have $\mathrm{Tr}\ Q_t < +\infty$ and (5.3) is fulfilled.

Assume now that $Q = I$ and that

$$Ae_k = -\mu_k e_k,$$

where $\mu_k > 0, k \in \mathbb{N}$. In this case (5.3) is fulfilled if and only if

$$\sum_{k=1}^{\infty} \frac{1}{\mu_k} < +\infty.$$

As easily seen, this conditions holds only for $N = 1$. □

Example 5.8 *Wave equation.* Let Λ be a strictly positive self-adjoint operator in U. Consider the problem

$$\begin{cases} dy_t = -\Lambda y(t)dt + dW(t), \\ y(0) = y \in U, \quad y_t(0) = z \in D(\Lambda^{-1/2}). \end{cases} \tag{5.11}$$

By $D(\Lambda^{-1/2})$ we mean the completion of U with respect to the norm

$$|x|_{D(\Lambda^{-1/2})} = |\Lambda^{-1/2}x|, \quad x \in U.$$

By proceeding as in Section A.5.4, we introduce the Hilbert space $H = U \oplus D(\Lambda^{-1/2})$ endowed with the inner product

$$\left\langle \begin{pmatrix} y \\ z \end{pmatrix}, \begin{pmatrix} y_1 \\ z_1 \end{pmatrix} \right\rangle = \langle y, y_1 \rangle + \langle \Lambda^{-1/2}z, \Lambda^{-1/2}z_1 \rangle.$$

Setting $X(t) = \begin{pmatrix} y(t) \\ y_t(t) \end{pmatrix}$ and $X_0 = \begin{pmatrix} y \\ z \end{pmatrix}$, we write problem (5.11) as

$$dX(t) = AX(t)dt + BdW(t), \quad X(0) = X_0,$$

where the linear operator A is defined by

$$\begin{cases} D(A) = D(\Lambda^{1/2}) \oplus U \\ A\begin{pmatrix} y \\ z \end{pmatrix} = \begin{pmatrix} 0 & 1 \\ -\Lambda & 0 \end{pmatrix} \begin{pmatrix} y \\ z \end{pmatrix}, \quad \forall \begin{pmatrix} y \\ z \end{pmatrix} \in D(A) \end{cases}$$

and $Bu = \begin{pmatrix} 0 \\ u \end{pmatrix}$. Then the stochastic convolution is given by

$$W_A(t) = \begin{pmatrix} \dfrac{1}{\sqrt{\Lambda}} \displaystyle\int_0^t \sin(\Lambda(t-s))dW(s) \\ \displaystyle\int_0^t \cos(\sqrt{\Lambda}(t-s))dW(s) \end{pmatrix}.$$

Thus, condition (5.3) holds, provided

$$\begin{aligned} &\text{(i)} \quad \int_0^T \mathrm{Tr}\left[\frac{\sin^2(\sqrt{\Lambda}s)}{\Lambda} Q\right] ds < +\infty \\ &\text{(ii)} \quad \int_0^T \mathrm{Tr}\left[\frac{\cos^2(\sqrt{\Lambda}s)}{\Lambda} Q\right] ds < +\infty. \end{aligned} \tag{5.12}$$

In particular (i) and (ii) hold if $\mathrm{Tr}\, Q < +\infty$. Let us assume now that $W(\cdot)$ is a cylindrical Wiener process with $Q = I$. Then the conditions (i) and (ii) are satisfied if and only if $\mathrm{Tr}\, \Lambda^{-1} < +\infty$.

In particular this holds if $U = L^2(0, \pi)$ and Λ is the realization of the second derivative with Dirichlet boundary conditions,

$$\Lambda x = -\frac{\partial^2 x}{\partial \xi^2}, \quad D(\Lambda) = H^2(0, \pi) \cap H_0^1(0, \pi).$$

Further regularity properties and their connection with the results in [702] are discussed in Section 5.4.

Note that if we take as basic state space $H = D(\Lambda^{1/2}) \oplus U$, with the energy norm

$$\left\| \begin{pmatrix} y \\ z \end{pmatrix} \right\|_H^2 = \|\Lambda^{1/2} y\|^2 + \|z\|^2, \quad \begin{pmatrix} y \\ z \end{pmatrix} \in H,$$

then condition (5.3) becomes

$$\text{(i)} \quad \int_0^T \text{Tr}[\sin^2(\sqrt{\Lambda} s) Q] ds < +\infty$$

$$\text{(ii)} \quad \int_0^T \text{Tr}[\cos^2(\sqrt{\Lambda} s) Q] ds < +\infty.$$

If $Q = I$ and Λ is an unbounded operator then (i) and (ii) never hold. The space $D(\Lambda^{1/2}) \oplus U$ is too small for the process $W_A(\cdot)$ to live in it. □

5.3 Continuity of weak solutions

In the previous section it was shown that, under natural assumptions, the weak solution exists, is unique and it is given by formula (5.7). It follows from the definition that a weak solution is a predictable process mean square continuous but, as one can expect, much more can be said about its properties.

We are concerned with the existence of a continuous version of the weak solution. We shall use the factorization method which we describe now.

5.3.1 Factorization formula

Proposition 5.9 *Assume that $p > 1$, $r \geq 0$, $\alpha > \frac{1}{p} + r$ and that E_1, E_2 are Banach spaces such that*

$$|S(t)x|_{E_1} \leq M t^{-r} |x|_{E_2}, \quad t \in [0, T], \ x \in E_2,$$

then G_α,

$$G_\alpha f(t) = \int_0^t (t-s)^{\alpha-1} S(t-s) f(s) ds, \quad t \in [0, T],$$

is a bounded linear operator from $L^p(0, T; E_2) =: L^p$ into $C([0, T]; E_1)$.

Proof We have

$$\left| \int_0^t (t-s)^{\alpha-1} S(t-s) f(s) ds \right|_{E_1} \leq M \int_0^t (t-s)^{\alpha-1-r} |f(s)| ds$$

$$\leq M \left(\int_0^t (t-s)^{(\alpha-1-r)q} ds \right)^{1/q} |f|_{L^p} \leq M \frac{t^{(\alpha-1-r)q+1}}{(\alpha-1-r)q+1} |f|_{L^p}.$$

Consequently, for a constant $C > 0$, independent of f,

$$\sup_{0\le t\le T} |G_\alpha f(t)|_{E_1} \le C|f|_{L^p}. \tag{5.13}$$

If $0 \le s \le t \le T$, then

$$\begin{aligned}
&G_\alpha f(t) - G_\alpha f(s)\\
&= \int_0^t \sigma^{\alpha-1} S(\sigma) f(t-\sigma)d\sigma - \int_0^s \sigma^{\alpha-1} S(\sigma) f(s-\sigma)d\sigma\\
&= \int_0^s \sigma^{\alpha-1} S(\sigma)(f(t-\sigma) - f(s-\sigma))d\sigma + \int_s^t \sigma^{\alpha-1} S(\sigma) f(t-\sigma)d\sigma.
\end{aligned}$$

Hence, there is a constant C_1 such that

$$\begin{aligned}
&|G_\alpha f(t) - G_\alpha f(s)|_{E_1}\\
&\le M\int_0^s \sigma^{\alpha-1-r}|f(t-\sigma)-f(s-\sigma)|_{E_2}d\sigma + M\int_s^t \sigma^{\alpha-1-r}|f(t-\sigma)|_{E_2}d\sigma\\
&\le M\left(\int_0^s \sigma^{(\alpha-1-r)q}d\sigma\right)^{1/q}\left(\int_0^s |f(t-\sigma)-f(s-\sigma)|^p_{E_2}d\sigma\right)^{1/p}\\
&\quad + M\left(\int_0^t \sigma^{(\alpha-1-r)q}d\sigma\right)^{1/q} |f|_{L^p}.
\end{aligned}$$

It is therefore clear that if $f \in C([0,T];E_2)$, then $G_\alpha f \in C([0,T];E_1)$. However, the space $C([0,T];E_2)$ is dense in L^p. So the result follows from (5.13). □

Assume now that U and H are Hilbert spaces and that W is a U-valued Wiener process. Denote

$$W_A(t) = \int_0^t S(t-s)\Phi(s)dW(s), \quad t \ge 0,$$

$$Y_\alpha(t) = \int_0^t (t-s)^{-\alpha} S(t-s)\Phi(s)dW(s), \quad t \ge 0.$$

The following result is a corollary of the stochastic Fubini theorem, Theorem 4.33.

Theorem 5.10 *Assume that for some $\alpha \in (0,1)$ and all $t \in [0,T]$,*

$$\int_0^t (t-s)^{\alpha-1}\left[\int_0^s (s-\sigma)^{-2\alpha}\mathbb{E}\left(\|S(t-\sigma)\Phi(\sigma)\|^2_{L_2^0}\right)d\sigma\right]^{1/2} ds < +\infty. \tag{5.14}$$

Then

$$W_A(t) = \frac{\sin\alpha\pi}{\pi}\int_0^t (t-s)^{\alpha-1} S(t-s) Y_\alpha(s)ds, \quad t \in [0,T]. \tag{5.15}$$

Condition (5.14) is precisely the condition (4.39) of Theorem 4.33 to exchange the deterministic integral of the right hand side of (5.15) with the stochastic integral Y_α.

Thus

$$\begin{aligned}
&\frac{\sin \alpha\pi}{\pi} \int_0^t (t-s)^{\alpha-1} S(t-s) Y_\alpha(s) ds \\
&= \frac{\sin \alpha\pi}{\pi} \int_0^t (t-s)^{\alpha-1} S(t-s) \left[\int_0^s (s-\sigma)^{-\alpha} S(s-\sigma) \Phi(\sigma) dW(\sigma) \right] ds \\
&= \frac{\sin \alpha\pi}{\pi} \int_0^t \left[\int_\sigma^t (t-s)^{\alpha-1} (s-\sigma)^{-\alpha} ds \right] S(t-\sigma) \Phi(\sigma) dW(\sigma).
\end{aligned}$$

Since

$$\int_\sigma^t (t-s)^{\alpha-1} (s-\sigma)^{-\alpha} ds = \frac{\pi}{\sin \alpha\pi}, \quad 0 \le \sigma \le t,\ \alpha \in (0,1),$$

the result follows. □

The main result of this section is the following, see [205].

Theorem 5.11 *If for some* $0 < \alpha < 1/2$,

$$\int_0^T t^{-2\alpha} \|S(t)B\|^2_{L_2^0}\, dt < +\infty, \tag{5.16}$$

then the weak solution of (5.1) has a continuous version.

Proof Condition (5.14) is satisfied by (5.16). To apply Proposition 5.9 (with $E_1 = E_2 = H$ and $r = 0$) we have to show that there exists $p > \frac{1}{\alpha}$ such that trajectories of Y_α are p-summable on $[0, T]$. By Theorem 4.36 for stochastic integrals, for $p > 0$,

$$\begin{aligned}
\mathbb{E}|Y_\alpha(t)|^p &= \mathbb{E}\left| \int_0^t (t-s)^{-\alpha} S(t-s) B dW(s) \right|^p \\
&\le c_p \left(\int_0^t \sigma^{-2\alpha} \|S(\sigma)B\|^2_{L_2^0}\, d\sigma \right)^{p/2}.
\end{aligned}$$

Consequently

$$\mathbb{E} \int_0^T |Y_\alpha(t)|^p dt \le T c_p \left(\int_0^T \sigma^{-2\alpha} \|S(\sigma)B\|^2_{L_2^0}\, d\sigma \right)^{p/2} < +\infty$$

and the result follows. □

The next theorem is a refinement of the previous one in the case when Tr $Q < +\infty$. In this case we can consider the continuous process

$$W_{A_n}(t) = \int_0^t e^{(t-s)A_n} dW(s), \quad t \in [0, T],$$

where A_n are the Yosida approximations of A. We have the following

Theorem 5.12 *If Tr* $Q < +\infty$ *then for arbitrary* $p > 2$,

$$\lim_{n \to \infty} \mathbb{E} \left(\sup_{t \in [0,T]} |W_A(t) - W_{A_n}(t)|^p \right) = 0. \tag{5.17}$$

Proof We choose $\alpha \in (0, 1/2)$ and $p > 1/\alpha$. We have

$$W_{A_n}(t) = \frac{\sin \pi\alpha}{\pi} \int_0^t e^{(t-s)A_n}(t-s)^{\alpha-1} Y_n(s)ds$$

where

$$Y_n(s) = \int_0^s e^{(s-\sigma)A_n}(s-\sigma)^{-\alpha} dW(\sigma).$$

Set

$$Y(s) = \int_0^s e^{(s-\sigma)A}(s-\sigma)^{-\alpha} dW(\sigma).$$

Thus, we can write

$$\begin{aligned} W_A(t) - W_{A_n}(t) = {} & \frac{\sin \pi\alpha}{\pi} \int_0^t [S(t-s) - e^{(t-s)A_n}](t-s)^{\alpha-1} Y(s)ds \\ & + \frac{\sin \pi\alpha}{\pi} \int_0^t e^{(t-s)A_n}(t-s)^{\alpha-1}[Y_n(s) - Y(s)]ds \\ =: {} & \frac{\sin \pi\alpha}{\pi} (\mathscr{K}_n Y(t) + \mathscr{J}_n(t)). \end{aligned}$$

Step 1 We have

$$\lim_{n\to\infty} \mathbb{E} \sup_{t\in[0,T]} |\mathscr{K}_n Y(t)|^p = 0, \tag{5.18}$$

where

$$\mathscr{K}_n Y(t) = \int_0^t [S(t-s) - e^{(t-s)A_n}](t-s)^{\alpha-1} Y(s)ds.$$

We first prove that the operators $\mathscr{K}_n$ are uniformly bounded in n as operators from the spaces of adapted processes in $L^p(\Omega \times [0, T]; H)$ into the spaces of adapted processes in $L^p(\Omega; L^\infty(0, T; H))$. In fact, by the Hölder inequality we have

$$|\mathscr{K}_n Y(t)| \le \left(\int_0^t \|S(t-s) - e^{(t-s)A_n}\|(t-s)^{\alpha-1} ds\right)^{1/q} \left(\int_0^t |Y(s)|^p\, ds\right)^{1/p},$$

from which

$$\sup_{t\in[0,T]} |\mathscr{K}_n Y(t)|^p \le \sup_{t\in[0,T]} (\|S(t)\| + \|e^{tA_n}\|)\left(\int_0^t s^{q(\alpha-1)}\, ds\right)^{1/q} \left(\int_0^t |Y(s)|^p\, ds\right)^{1/p}.$$

Since, by assumption $1/p < \alpha < 1$, the uniform boundedness follows.

We now prove Step 1 under the additional assumption that

$$\mathbb{E}\int_0^T |A^2 Y(s)|^p\, ds < +\infty.$$

In fact, recalling that by the Hille–Yosida theorem (Theorem A.3) the following estimate holds [(1)]

$$|S(t)x - e^{tA_n}x| \le M\frac{|A^2 x|}{n-\omega}, \quad n > \omega,\ t \in [0,T],\ x \in D(A^2),$$

we have

$$\begin{aligned}|\mathscr{K}_n Y(t)| &\le \frac{M}{n-\omega}\int_0^t (t-s)^{\alpha-1}|A^2 Y(s)|ds\\ &\le \frac{M}{n-\omega}\left(\int_0^t (t-s)^{(\alpha-1)q}\,ds\right)^{1/q}\left(\int_0^t |A^2 Y(s)|^p\,ds\right)^{1/p} \to 0,\end{aligned}$$

as $n \to \infty$. To conclude the proof of Step 1 we need a lemma.

Lemma 5.13 *Assume that* $\mathbb{E}\int_0^T |Y(s)|^p ds < \infty$ *and set*

$$\widetilde{Y}_m(s) = m^2(mI - A)^{-2}Y(s), \quad s \in [0,T],\ m > \omega.$$

Then

$$\mathbb{E}\int_0^T |A^2\widetilde{Y}_m(s)|^p ds < \infty, \quad m > \omega, \tag{5.19}$$

and

$$\lim_{m\to\infty}\mathbb{E}\int_0^T |Y(s) - \widetilde{Y}_m(s)|^p ds = 0. \tag{5.20}$$

Proof Since $mA(mI-A)^{-1} = m(I - m(mI-A)^{-1})$ therefore $A^2m^2(mI-A)^{-2}$ is a bounded operator and (5.19) follows. Moreover, since

$$\|m(mI-A)^{-1}\| \le \frac{M}{m-\omega}, \quad m > \omega$$

and $\lim_{m\to\infty} m(mI-A)^{-1}x = x$ for all $x \in H$ we have

$$\begin{aligned}\int_0^T |Y(s) - \widetilde{Y}_m(s)|^p ds &\le \int_0^T |(m^2(mI-A)^{-2} - I)Y(s)|^p ds\\ &\le \left(1 + \frac{M^2}{(m-\omega)^2}\right)^p \int_0^T |Y(s)|^p ds.\end{aligned}$$

Moreover,

$$\lim_{m\to\infty}|Y(s) - \widetilde{Y}_m(s)| = 0, \quad s \in [0,T]$$

[(1)] We assume that $\|S(t)\| \le Me^{\omega t},\ t \ge 0$.

and therefore, by the dominated convergence theorem,

$$\lim_{m\to\infty} \mathbb{E} \int_0^T |Y(s) - \widetilde{Y}_m(s)|^p ds = 0,$$

as required. □

To conclude the proof of Step 1 we notice that the operators $\mathscr{K}_n$ have uniformly bounded norms and $\mathscr{K}_n Y \to 0$ on a dense set of Y so the convergence $\mathscr{K}_n Y \to 0$ holds for all Y such that $\mathbb{E} \int_0^T |Y(s)|^p ds < \infty$.

Step 2 We have

$$\lim_{n\to\infty} \mathbb{E} \left(\sup_{t\in[0,T]} |\mathscr{I}_n(t)|^p \right) = 0. \tag{5.21}$$

The following estimate is proved as (5.13)

$$\sup_{t\in[0,T]} |\mathscr{I}_n(t)|^p \le C_1 \int_0^T |Y(s) - Y_n(s)|^p ds. \tag{5.22}$$

Let $N \ge 2$ with $N \in \mathbb{N}$ be such that $p \le N$. Then we have by the Hölder inequality

$$\mathbb{E}|Y(s) - Y_n(s)|^p \le \left(\mathbb{E}|Y(s) - Y_n(s)|^N\right)^{p/N}.$$

Now, we have

$$\mathbb{E}|Y(s) - Y_n(s)|^2 = \sum_{i=1}^{\infty} \lambda_i \int_0^T \sigma^{-2\alpha} |(S(\sigma) - e^{\sigma A_n})e_i|^2 d\sigma,$$

and, since $Y - Y_n$ is Gaussian, there exists $C_2 > 0$ (see Proposition 2.19), such that

$$\mathbb{E}|Y(s) - Y_n(s)|^N \le C_2 \left[\sum_{i=1}^{\infty} \lambda_i \int_0^T \sigma^{-2\alpha} |(S(\sigma) - e^{\sigma A_n})e_i|^2 d\sigma \right]^{N/2}.$$

Thus, by (5.22) and the dominated convergence theorem we obtain (5.21) and the theorem is proved. □

5.4 Regularity of weak solutions in the analytic case

5.4.1 Basic regularity theorems

In this section we shall assume that A is the infinitesimal generator of an analytic semigroup of negative type (see Section A.4.1 and (A.14)) and moreover that condition (5.16) holds. In this case we can prove additional regularity properties for the stochastic convolution $W_A(\cdot)$ and therefore for the weak solutions as well.

For any $\gamma \in (0,1)$ we shall denote by $(-A)^\gamma$ the fractional power of A. From (A.30) it follows that there exist constants M_k, $M_{k,\gamma}$, $k \in \mathbb{N}$, such that

$$\|A^k S(t)\| \le M_k t^{-k}, \quad t > 0,\ k = \{0\} \cup \mathbb{N}, \tag{5.23}$$

$$\|(-A)^\gamma A^k S(t)\| \le M_{k\gamma} t^{-k-\gamma}, \quad t > 0,\ k = \{0\} \cup \mathbb{N},\ \gamma \in (0,1). \tag{5.24}$$

We shall use the following result from [220, Proposition A.1.1].

Proposition 5.14 *Assume that $\alpha, \gamma \in (0, 1/2)$, $p > 1$ and let G_α be as in Proposition 5.9.*

(i) If $\alpha > \gamma + \frac{1}{p}$ then G_α is a bounded linear operator from $L^p(0,T;H)$ into $C^{\alpha-\gamma-\frac{1}{p}}([0,T]; D((-A)^\gamma))$.
(ii) If $\alpha > \frac{1}{p}$ then for arbitrary $\delta \in (0, \alpha - \frac{1}{p})$, G_α is a bounded linear operator from $L^p(0,T;H)$ into $C^\delta([0,T];H)$.

We will prove two kinds of regularity properties:

(i) *temporal regularity*, that is the trajectories of $W_A(\cdot)$ are Hölder continuous,
(ii) *spatial regularity*, that is the trajectories of $W_A(\cdot)$ are continuous as functions with values in the domain of a fractional power of A (see Section A.4.3).

Theorem 5.15 *Assume that $S(t)$, $t \ge 0$, is an analytic semigroup, $\alpha \in (0, 1/2)$ and that condition* (5.16) *holds.*

(i) For arbitrary $\delta \in (0,\alpha)$ trajectories of W_A are in $C^\delta([0,T];H)$.
(ii) Assume that $\beta, \gamma > 0$, and $\beta + \gamma < \alpha$, then trajectories of W_A are in $C^\beta([0,T]; D((-A)^\gamma))$.

Proof (i) We know from the proof of Theorem 5.11 that if $1/2 > \alpha > 1/p$ then trajectories of Y_α are p-integrable and the factorization formula is applicable. By Proposition 5.14(ii) applied for sufficiently large $p > 2$ we get point (i). We can always find $p > 2$ and

$$\beta' := \alpha - \gamma - \frac{1}{p} > \beta.$$

By Proposition 5.14(i) and the factorization formula we find that

$$W_A \in C^{\beta'}([0,T]; D((-A)^\gamma)).$$

Since $\beta' > \beta$ the result follows. □

If $B = I$ and W is an H-valued Wiener process then Theorem 5.15 holds true for $\alpha = 1/2$. In this case one can give more direct proofs.

Theorem 5.16 *Assume that $Tr\ Q < +\infty$ and $B = I$. Then Theorem 5.15 is true with $\alpha = 1/2$.*

Proof We give two proofs of independent interest. One is analytic and uses reduction to the deterministic case and the second one is based on the Kolmogorov regularity theorem. We start from the analytic one. We need the following lemma.

Lemma 5.17 *Assume* Tr $Q < +\infty$ *and set*

$$Y(t) = \int_0^t S(t-s)W(s)ds, \quad t \geq 0.$$

Then $Y(\cdot)$ *belongs to* $C^1([0,+\infty); H) \cap C([0,+\infty); D(A))$ $\mathbb{P}$*-a.s., and*

$$W_A(t) = W(t) + A\int_0^t S(t-s)W(s)ds. \tag{5.25}$$

Moreover

$$W_A(t) = \frac{d}{dt}\int_0^t S(t-s)W(s)ds = \frac{d}{dt}Y(t), \quad t \geq 0. \tag{5.26}$$

Proof Taking into account Theorem 5.12 we can assume (passing if necessary to a subsequence) that $\mathbb{P}$-a.s. $W_{A_n}(t) \to W_A(t)$ as $n \to \infty$, uniformly on bounded intervals. Moreover $W_{A_n}(\cdot)$ is the strong solution to the stochastic differential equation

$$\begin{cases} dW_{A_n}(t) = A_n W_{A_n}(t)dt + dW(t), \\ W_{A_n}(0) = 0, \end{cases}$$

so that

$$W_{A_n}(t) = \int_0^t A_n W_{A_n}(s)ds + W(t), \quad t \geq 0. \tag{5.27}$$

Setting $Z_n(t) = \int_0^t W_{A_n}(s)ds$ we have that $Z_n(\cdot)$ is the solution to the initial value problem

$$\frac{d}{dt}Z_n(t) = A_n Z_n(t) + W(t), \quad Z_n(0) = 0,$$

thus, we have

$$Z_n(t) = \int_0^t e^{(t-s)A_n} W(s)ds, \quad t \geq 0, \tag{5.28}$$

and so $\mathbb{P}$-a.s.

$$W_{A_n}(t) = A_n Z_n(t) + W(t), \quad t \geq 0. \tag{5.29}$$

Clearly

$$\lim_{n\to\infty} Z_n(t) = Z(t) = \int_0^t S(t-s)W(s)ds, \quad t \geq 0. \tag{5.30}$$

Now, recalling Theorem 5.12, (5.29) and Proposition A.4, we have

$$\lim_{n\to\infty} A_n Z_n(t) = \lim_{n\to\infty} AJ_n Z_n(t) = W_A(t) - W(t), \quad t \geq 0. \tag{5.31}$$

Since the operator A is closed, we conclude that $\mathbb{P}$-a.s $Z(t) \in D(A)$ and

$$AZ(t) = W_A(t) - W(t)$$

for all $t \geq 0$. Thus (5.25) is proved. The identity (5.26) follows from (5.25) and the definition of infinitesimal generator. □

First proof of Theorem 5.16 To show part (i) we use (5.25). Since by Theorem 3.3, $W(\cdot) \in C^\delta([0, T]; H)$, for arbitrary $\delta \in (0, 1/2)$ and $W(0) = 0$, the conclusion follows from Proposition A.24.

Let us prove (ii). From (5.25) we have $W_A = dY/dt$, $\mathbb{P}$-a.s., where Y is the solution to the initial value problem

$$\begin{cases} Y'(t) = AY(t) + W(t), & t \geq 0 \\ Y(0) = 0. \end{cases} \tag{5.32}$$

Fix $T > 0$. Since $\mathbb{P}$-a.s. $W(\cdot) \in C^\alpha([0, T]; H)$ for any $\alpha \in (0, 1/2)$, by Proposition A.25 and Remark A.16 we have

$$Y(\cdot) \in C^{1,\alpha}([0, T]; H) \cap C^\alpha([0, T]; D(A)) \cap C^{1,\alpha-\beta}([0, T]; D_A(\beta, \infty)),$$

$\mathbb{P}$-a.s. for all $\alpha \in (0, 1/2)$, $\beta \in (0, \alpha)$. Thus $W_A(\cdot) \in C^{\alpha-\beta}([0, T]; D_A(\beta, \infty))$. Since $D_A(\beta, \infty)$ is included, by Proposition A.15, in $D((-A)^{\beta-\varepsilon})$, the conclusion follows.

Second proof of Theorem 5.16 Fix $T > 0$, and let $T \geq t > s > 0$, then we have

$$\begin{aligned} \mathbb{E}|W_A(t) - W_A(s)|^2 &= \sum_{k=1}^\infty \lambda_k \int_s^t |S(t-\sigma)e_k|^2 d\sigma \\ &\quad + \sum_{k=1}^\infty \lambda_k \int_0^s |[S(t-\sigma) - S(s-\sigma)]\, e_k|^2\, d\sigma \\ &= I_1 + I_2. \end{aligned}$$

Next, we have $I_1 \leq M^2 \mathrm{Tr}\,(Q)(t-s)$ and

$$I_2 = \sum_{k=1}^\infty \lambda_k \int_0^s \left| \int_{s-\sigma}^{t-\sigma} AS(\rho)e_k d\rho \right|^2 d\sigma \leq M_1^2\ \mathrm{Tr}\,(Q) \int_0^s \left| \int_{s-\sigma}^{t-\sigma} \frac{d\rho}{\rho} \right|^2 d\sigma.$$

Let $\gamma \in (0, 1/2)$, then

$$\begin{aligned} I_2 &\leq M_1^2\ \mathrm{Tr}(Q) \int_0^s (s-\sigma)^{-2\gamma} \left| \int_{s-\sigma}^{t-\sigma} \rho^{\gamma-1} d\rho \right|^2 d\sigma \\ &\leq \frac{M_1^2 T^{1-2\gamma}}{\gamma^2(1-2\gamma)}\ \mathrm{Tr}\ Q(t-s)^{2\gamma}. \end{aligned}$$

Thus for any $\gamma \in (0, 1/2)$ we find

$$\mathbb{E}|W_A(t) - W_A(s)|^2 \leq \mathrm{Tr}(Q)\left[M^2(t-s) + \frac{M_1^2 T^{1-2\gamma}}{\gamma^2(1-2\gamma)}(t-s)^{2\gamma}\right], \quad (5.33)$$

and part (i) follows, since $W_A(t) - W_A(s)$ is Gaussian, compare Theorem 3.3.

Proof of part (ii) By (5.24) it follows that the stochastic integral

$$\int_0^t (-A)^\gamma S(t-s)dW(s),$$

is well defined. Since $(-A)^\gamma$ is closed, it follows that $W_A(t) \in D((-A)^\gamma)$ and

$$\int_0^t (-A)^\gamma S(t-s)dW(s) = (-A)^\gamma W_A(t).$$

Thus the process $(-A)^\gamma W_A(\cdot)$ is well defined.

We will now use the Kolmogorov test. Fix $T > 0$ and let $t \in [0, T]$, then

$$\begin{aligned}\mathbb{E}|(-A)^\gamma W_A(t) - (-A)^\gamma W_A(s)|^2 &= \sum_{k=1}^\infty \lambda_k \int_s^t |(-A)^\gamma S(t-\sigma)e_k|^2 d\sigma \\ &\quad + \sum_{k=1}^\infty \lambda_k \int_0^s |(-A)^\gamma [S(t-\sigma) - S(s-\sigma)]e_k|^2 d\sigma \\ &= J_1 + J_2.\end{aligned}$$

Now

$$|J_1| \leq \mathrm{Tr}(Q)\,\frac{M_\gamma^2}{1-2\gamma}\,(t-s)^{2\gamma}, \quad (5.34)$$

and

$$\begin{aligned}J_2 &= \sum_{k=1}^\infty \lambda_k \int_0^s \left|\int_{s-\sigma}^{t-\sigma} (-A)^\gamma A S(\rho) e_k d\rho\right|^2 d\sigma \\ &\leq M_\gamma^2 \; \mathrm{Tr}(Q) \int_0^s \left|\int_{s-\sigma}^{t-\sigma} \rho^{-1-\gamma} d\rho\right|^2 d\sigma.\end{aligned} \quad (5.35)$$

Choosing $\xi \in (\gamma, 1/2)$, we have

$$\begin{aligned}J_2 &\leq M_\gamma^2 \,\mathrm{Tr}(Q) \int_0^s (s-\sigma)^{-2\xi} \left|\int_{s-\sigma}^{t-\sigma} \rho^{\gamma-\xi-1} d\rho\right|^2 d\sigma \\ &\leq \mathrm{Tr}(Q)\,\frac{M_\gamma^2 T^{1-2\xi}}{(\xi-\gamma)^2(1-2\xi)}\,(t-s)^{2\xi-2\gamma}.\end{aligned} \quad (5.36)$$

Combining (5.34) and (5.36), we see that there exists $C_t > 0$ such that

$$\mathbb{E}|(-A)^\gamma W_A(t) - (-A)^\gamma W_A(s)|^2 \leq C_t(t-s)^{2\xi-2\gamma}.$$

The conclusion follows from the Kolmogorov test. □

5.4.2 Regularity in the border case

We consider now the border case. An example from [205] shows that, if the generator A is negative self-adjoint, then the process $(-A)^{1/2}W_A(t)$ may not have a pathwise continuous version. So the previous proposition cannot be extended to $\gamma = 1/2$. But the mean square continuity may take place. We are now going to investigate this question in some detail.

We consider first regularity of the stochastic convolution

$$W_A(t) = \int_0^t S(t-s)dW(s), \quad t \geq 0,$$

in the case when W is a Wiener process on H with nuclear covariance operator Q and A is a self-adjoint strictly negative operator. Thus $B := -A$ is a positive definite operator with spectrum located in $[\sigma_0, +\infty)$, $\sigma_0 > 0$. Then the domain $D((-A)^\alpha)$ of $(-A)^\alpha$, $\alpha > 0$, is given by the formula

$$D((-A)^\alpha) = \left\{ x \in H : \int_0^{+\infty} \lambda^\alpha \langle dE_\lambda x, x\rangle < +\infty \right\},$$

where E_λ is the spectral measure corresponding to B. We set

$$|x|_\alpha^2 = \int_0^{+\infty} \lambda^\alpha \langle dE_\lambda x, x\rangle, \quad x \in D((-A)^\alpha).$$

If $\{\lambda_j\}$ and $\{e_j\}$ are the eigenvalues and eigenvectors corresponding to Q, then the following measure μ_Q concentrated on $[\sigma_0, +\infty)$ is positive and finite

$$\mu_Q(d\lambda) = \sum_{j=1}^{\infty} \lambda_j \langle dE_\lambda e_j, e_j\rangle.$$

Note that

$$\mu_Q([\sigma_0, +\infty)) = \sum_{j=1}^{\infty} \lambda_j = \mathrm{Tr}Q.$$

We have the following explicit formula

Lemma 5.18 *For arbitrary $t, h, \alpha > 0$*

$$\mathbb{E}|W_A(t+h) - W_A(t)|_\alpha^2$$
$$= \frac{1}{2}\int_0^{+\infty} \lambda^{2\alpha-1}\left[(1-e^{-\lambda h})^2(1-e^{-2\lambda t}) + 1 - e^{-2\lambda h}\right]\mu_Q(d\lambda). \quad (5.37)$$

Proof Note that

$$\begin{aligned}\mathbb{E}|W_A(t+h) - W_A(t)|_\alpha^2 &= \mathbb{E}\left|\int_0^t (e^{-(t+h-\sigma)B} - e^{-(t-\sigma)B})dW(\sigma)\right|_\alpha^2 \\ &\quad + \mathbb{E}\left|\int_t^{t+h} e^{-(t+h-\sigma)B}dW(\sigma)\right|_\alpha^2 \\ &= \mathbb{E}\left|\int_0^t B^\alpha(e^{-(t+h-\sigma)B} - e^{-(t-\sigma)B})dW(\sigma)\right|^2 \\ &\quad + \mathbb{E}\left|B^\alpha\int_t^{t+h} e^{-(t+h-\sigma)B}dW(\sigma)\right|^2.\end{aligned}$$

Using now expansion of the Brownian motion in terms of eigenvectors $\{e_j\}$, from the definition of the measure μ_Q and the formula

$$|\psi(B)x|^2 = \int_0^{+\infty} \psi(\lambda)\langle dE_\lambda x, x\rangle,$$

after straightforward calculations we arrive at (5.37). □

Theorem 5.19 *For an arbitrary, strictly negative, self-adjoint generator A, the stochastic convolution W_A is mean square continuous in the domain $D((-A)^{-\alpha})$ of fractional powers of $-A$ for all $\alpha \in [0, 1/2]$. If $\alpha > 1/2$ one can construct an operator A and a one dimensional Wiener process such that $\mathbb{E}|W_A(t)|_\alpha^2 = +\infty$ for all $t > 0$.*

Proof If $\alpha \in [0, 1/2]$ then the function $\psi(\lambda) := \lambda^{2\alpha-1}$ is bounded on $[\sigma_0, +\infty)$ and then the result follows from (5.37) and the Lebesgue dominated convergence theorem. If $\alpha > 1/2$ the function $\psi(\lambda)$ is not bounded on $[\sigma_0, +\infty)$. Moreover, for arbitrary finite measure ν on $(0, +\infty)$ we can find H, A and an operator $Q \geq 0$ of rank one such that $\nu = \mu_Q$. In fact one can take $H = L^2(0, +\infty)$,

$$\mathbb{E}_\Gamma x(\xi) = 1\!\!1_\Gamma(\xi)x(\xi), \quad \xi \in (0, +\infty), \ \Gamma \in \mathscr{B}((0, +\infty))$$

and

$$Qx(\xi) = 1\!\!1_{[0,+\infty)}\langle 1\!\!1_{[0,+\infty)}, x\rangle, \quad x \in H.$$

This finishes the proof of the theorem. □

We now prove that the first part of Theorem 5.19 generalizes to a large class of stochastic convolution W_A, where A generates an analytic semigroup of negative type satisfying the following condition.

Hypothesis 5.3 *There exists $\varepsilon_0 \in (1/2, 1)$ such that $D_A(\varepsilon, 2)$ is isomorphic to $D((-A)^\varepsilon)$, for all $\varepsilon \in (0, \varepsilon_0]$.*

For a discussion of this hypothesis see Remark A.20. We only recall here that Hypothesis 5.3 is satisfied if $S(\cdot)$ is a contraction semigroup and so, in particular, if A is a variational or self-adjoint operator.

Theorem 5.20 *Under Hypothesis* 5.3 *the process* W_A *is mean-square continuous in* $D((-A)^{1/2})$.

Proof Assume that

$$W(t) = \sum_{k=1}^{\infty} \sqrt{\lambda_k}\beta_k(t)h_k, \quad t \geq 0,$$

where $\mathrm{Tr}\, Q = \sum_{k=1}^{\infty} \lambda_k < +\infty$, β_k are independent, standard Wiener processes and $\{h_k\}$ is an orthonormal basis in H. Denoting by $|\cdot|_\varepsilon$ the norm in $D_A(\varepsilon, 2)$ we have by Hypothesis 5.3 for a fixed $\varepsilon \in (0, \varepsilon_0 - 1/2)$,

$$\begin{aligned}
\mathbb{E}|W_A(t)|_{1/2}^2 &\leq c_1 \mathbb{E}|(-A)^{1/2} W_A(t)|_{1/2}^2 \\
&\leq c_1 \mathbb{E}|(-A)^{1/2+\varepsilon}(-A)^{-\varepsilon} W_A(t)|^2 \\
&\leq c_1 \mathbb{E}|(-A)^{1/2+\varepsilon} \widetilde{W}_A(t)|^2,
\end{aligned}$$

where

$$\widetilde{W}(t) = \sum_{k=1}^{\infty} \sqrt{\lambda_k}\beta_k(t)(-A)^{-\varepsilon}h_k, \quad t \geq 0.$$

Consequently

$$\mathbb{E}|W_A(t)|_{1/2}^2 \leq c_2 \mathbb{E}|\widetilde{W}_A(t)|_{1/2+\varepsilon}^2.$$

By the definition of $|\cdot|_\gamma$,

$$\begin{aligned}
\mathbb{E}|\widetilde{W}_A(t)|_{1/2+\varepsilon}^2 &= \mathbb{E} \int_0^{+\infty} \xi^{1-2(1/2+\varepsilon)} |AS(\xi)\widetilde{W}_A(t)|^2 d\xi \\
&= \sum_{k=1}^{\infty} \lambda_k \int_0^{+\infty} \xi^{-2\varepsilon} d\xi \int_0^t |AS(s+\xi)(-A)^{-\varepsilon}h_k|^2 ds \\
&\leq \sum_{k=1}^{\infty} \lambda_k \int_0^{+\infty} \xi^{-2\varepsilon} d\xi \int_\xi^{+\infty} |AS(\sigma)(-A)^{-\varepsilon}h_k|^2 d\sigma \\
&= \sum_{k=1}^{\infty} \frac{\lambda_k}{1-2\varepsilon} \int_0^{+\infty} \sigma^{1-2\varepsilon} d|AS(\sigma)(-A)^{-\varepsilon}h_k|^2 d\sigma \\
&\leq \frac{c_3}{1-2\varepsilon} \sum_{k=1}^{\infty} \lambda_k < +\infty.
\end{aligned}$$

Thus $W_A(t) \in D((-A)^{1/2})$. In a similar way one can prove continuity. If $t > s$ write

$$W_A(t) - W_A(s) = \int_0^s (S(t-\sigma) - S(s-\sigma))dW(\sigma) + \int_s^t S(t-\sigma)dW(\sigma).$$

Consequently

$$\begin{aligned}\mathbb{E}|W_A(t) - W_A(s)|_{1/2}^2 &= \mathbb{E}\left|\int_0^s (S(t-\sigma) - S(s-\sigma))dW(\sigma)\right|_{1/2}^2 \\ &\quad + \mathbb{E}\left|\int_s^t S(t-\sigma)dW(\sigma)\right|_{1/2}^2 := I_1 + I_2.\end{aligned}$$

Then

$$\begin{aligned}
I_1 &\le c_1\mathbb{E}\left|\int_0^s (-A)^{1/2+\varepsilon}(S(t-\sigma) - S(s-\sigma))d\widetilde{W}(\sigma)\right|^2 \\
&\le c_2\mathbb{E}\left|\int_0^s (S(t-\sigma) - S(s-\sigma))d\widetilde{W}(\sigma)\right|_{1/2+\varepsilon}^2 \\
&\le c_2\int_0^{+\infty} \xi^{-2\varepsilon}\mathbb{E}\left|AS(\xi)\int_0^s (S(t-\sigma) - S(s-\sigma))d\widetilde{W}(\sigma)\right|^2 d\xi \\
&= c_2\sum_{k=1}^{\infty}\lambda_k\int_0^{+\infty}\xi^{-2\varepsilon}d\xi\int_0^s |A(S(t-s+\xi+\sigma) - S(\xi+\sigma))(-A)^{-\varepsilon}h_k|^2 d\sigma \\
&= c_2\sum_{k=1}^{\infty}\lambda_k\int_0^{+\infty}\xi^{-2\varepsilon}d\xi\int_\xi^{s+\xi} |A(S(t-s+\sigma) - S(\sigma))(-A)^{-\varepsilon}h_k|^2 d\sigma \\
&\le c_2\sum_{k=1}^{\infty}\lambda_k\int_0^{+\infty}\xi^{-2\varepsilon}d\xi\int_\xi^{+\infty} |AS(\sigma)(S(t-s) - I)(-A)^{-\varepsilon}h_k|^2 d\sigma \\
&\le \frac{c_2}{1-2\varepsilon}\sum_{k=1}^{\infty}\lambda_k\int_0^{+\infty}\sigma^{1-2\varepsilon}|AS(\sigma)(S(t-s) - I)(-A)^{-\varepsilon}h_k|^2 d\sigma \\
&= \frac{c_2}{1-2\varepsilon}\sum_{k=1}^{\infty}\lambda_k|(S(t-s) - I)(-A)^{-\varepsilon}h_k|_\varepsilon^2 d\sigma \\
&\le c_3\sum_{k=1}^{\infty}\lambda_k|(S(t-s) - I)h_k|^2.
\end{aligned}$$

By the Lebesgue dominated convergence theorem it follows that $I_1 \to 0$ as $t - s \to 0$. Similarly $I_2 \to 0$ as $t - s \to 0$. □

5.5 Regularity of weak solutions in the space of continuous functions

In some applications we need continuity of the stochastic convolution $W_A,\ t \geq 0$, not in a Hilbert space H but in a smaller Banach space E. We restrict ourselves to a specific situation when $H = L^2(\mathscr{O})$ and $E = C(\overline{\mathscr{O}})$ where $\mathscr{O}$ is an open bounded subset of $\mathbb{R}^N$.

We are given a Q-Wiener process $W(\cdot)$ in H such that the following holds.

$$Qe_k = \lambda_k e_k, \quad k \in \mathbb{N},\ \lambda_k > 0, \tag{5.38}$$

where $\{e_k\}$ is a complete orthonormal basis in H. We do not assume that $\mathrm{Tr}\ Q < \infty$, and we set

$$W(t) = \sum_{k=1}^{\infty} \sqrt{\lambda_k}\ e_k \beta_k(t), \quad t \geq 0$$

where $\{\beta_k\}$ are independent Wiener processes.

We are also given a linear operator A in H such that the following holds.

Hypothesis 5.4 *A generates a strongly continuous semigroup $S(\cdot)$ in H such that $S(t)E \subset E$ for all $t \geq 0$ and $\{e_k\} \subset D(A)$.*

We set

$$W_A(t)(\xi) = W_A(t, \xi), \quad t \geq 0\ \xi \in \mathscr{O}.$$

In the next subsections we study regularity of $W_A(\cdot, \cdot)$ in both variables.

5.5.1 The case when A is self-adjoint

We are given a self-adjoint negative operator A in $L^2(\mathscr{O})$ such that

$$Ae_k = -\alpha_k e_k, \quad k \in \mathbb{N}, \tag{5.39}$$

where the sequence of positive numbers $\{\alpha_k\}$ satisfies

$$\alpha_k \geq \omega > 0, \qquad \sum_{k=1}^{\infty} \frac{\lambda_k}{\alpha_k} < +\infty. \tag{5.40}$$

Concerning the functions $\{e_k\}$ we shall assume that

$$\{e_k\} \subset C(\overline{\mathscr{O}}), \ |e_k(\xi)| \leq C, \ |\nabla e_k(\xi)| \leq C\alpha_k^{1/2}, \quad \forall\, \xi \in \mathscr{O},\ k \in \mathbb{N}, \tag{5.41}$$

for some positive constant C.

In order to find sufficient conditions for the stochastic convolution $W_A(t)$ to have an E-continuous version, we need a lemma.

Lemma 5.21 *Assume that (5.39)–(5.41) hold and that for some $\gamma \in (0,1)$*

$$\sum_{k=1}^{\infty} \frac{\lambda_k}{\alpha_k^{1-\gamma}} < +\infty. \tag{5.42}$$

Then there exists a constant $C_1 > 0$ such that

$$\mathbb{E}\left(|W_A(t,\xi) - W_A(t,\eta)|^2\right) \leq C_1 |\xi - \eta|^{2\gamma} \tag{5.43}$$

$$\mathbb{E}\left(|W_A(t,\xi) - W_A(s,\xi)|^2\right) \leq C_1 |t - s|^{\gamma} \tag{5.44}$$

for all $t, s \geq 0$ and $\xi, \eta \in \overline{\mathscr{O}}$.

Proof We have the following representation for $W_A(\cdot,\cdot)$

$$W_A(t,\xi) = \sum_{k=1}^{\infty} \sqrt{\lambda_k} e_k(\xi) \int_0^t e^{-\alpha_k(t-s)} d\beta_k(s), \quad t \geq 0, \tag{5.45}$$

with the series (5.45) converging in $L^2(\Omega, \mathscr{F}, \mathbb{P})$ for all $t \geq 0$ and $\xi \in \overline{\mathscr{O}}$. This is because we have, for arbitrary $n, p \in \mathbb{N}$

$$\mathbb{E}\left|\sum_{k=n+1}^{n+p} \sqrt{\lambda_k} e_k(\xi) \int_0^t e^{-\alpha_k(t-s)} d\beta_k(s)\right|^2$$
$$= \sum_{k=n+1}^{n+p} \frac{(1 - e^{-2\alpha_k t})\lambda_k}{2\alpha_k} |e_k(\xi)|^2 \leq \frac{C^2}{2} \sum_{k=n+1}^{n+p} \frac{\lambda_k}{\alpha_k}.$$

We note also that by (5.41) we have

$$|e_k(\xi) - e_k(\eta)| \leq C \alpha_k^{1/2} |\xi - \eta|, \quad k \in \mathbb{N},$$

and so

$$|e_k(\xi) - e_k(\eta)| \leq C 2^{1-\gamma} \alpha_k^{\gamma/2} |\xi - \eta|^{\gamma}, \quad k \in \mathbb{N}, \tag{5.46}$$

for all $\gamma \in [0,1]$ and $\xi, \eta \in \mathscr{O}$. To see (5.46) we note that simultaniously

$$\frac{1}{2C} |e_k(\xi) - e_k(\eta)| \leq 1, \quad \frac{1}{2C} |e_k(\xi) - e_k(\eta)| \leq \frac{1}{2}\, \alpha_k^{1/2} |\xi - \eta|.$$

Thus if $\frac{1}{2}\alpha_k^{1/2}|\xi - \eta| \leq 1$ we have $\frac{1}{2}\alpha_k^{1/2}|\xi - \eta| \leq (\frac{1}{2}\alpha_k^{1/2}|\xi - \eta|)^{\gamma}$ and if $\frac{1}{2}\alpha_k^{1/2}|\xi - \eta| > 1$ we have $(\frac{1}{2}\alpha_k^{1/2}|\xi - \eta|)^{\gamma} > 1$ so, (5.46) holds in both cases.

The estimate (5.43) follows now from the identity

$$\mathbb{E}|W_A(t,\xi) - W_A(t,\eta)|^2 = \sum_{k=1}^{\infty} \lambda_k \int_0^t e^{-2\alpha_k(t-s)} |e_k(\xi) - e_k(\eta)|^2 ds.$$

To show (5.44) fix $T \geq t > s > 0$. Then we have

$$\mathbb{E}\left(|W_A(t,\xi) - W_A(s,\xi)|^2\right) = \sum_{k=1}^{\infty} \lambda_k \int_s^t e^{-2(t-\sigma)\alpha_k} |e_k(\xi)|^2 d\sigma$$

$$+ \sum_{k=1}^{\infty} \lambda_k \int_0^s \left| \left[e^{-(t-\sigma)\alpha_k} - e^{-(s-\sigma)\alpha_k} \right] e_k(\xi) \right|^2 d\sigma$$

$$=: I_1(t,s,\xi) + I_2(t,s,\xi).$$

Next [(2)]

$$I_1(t,s,\xi) \leq C^2 \sum_{k=1}^{\infty} \lambda_k \int_s^t e^{-2(t-\sigma)\alpha_k} d\sigma$$

$$= C^2 \sum_{k=1}^{\infty} \frac{\lambda_k}{\alpha_k} \left(1 - e^{-2(t-s)\alpha_k}\right) \leq 2^\gamma C^2 \sum_{k=1}^{\infty} \frac{\lambda_k}{\alpha_k^{1-\gamma}} |t-s|^\gamma. \quad (5.47)$$

Moreover,

$$I_2(t,s,\xi) \leq C^2 \sum_{k=1}^{\infty} \lambda_k \int_0^s \left| e^{-(t-\sigma)\alpha_k} - e^{-(s-\sigma)\alpha_k} \right|^2 d\sigma$$

$$= C^2 \sum_{k=1}^{\infty} \frac{\lambda_k}{\alpha_k} [2(1 - e^{-(t-s)\alpha_k}) - (1 - e^{-2(t-s)\alpha_k}) - (e^{-t\alpha_k} - e^{-s\alpha_k})^2]$$

$$\leq 2C^2 \sum_{k=1}^{\infty} \frac{\lambda_k}{\alpha_k} (1 - e^{-(t-s)\alpha_k}) \leq 2C^2 \sum_{k=1}^{\infty} \frac{\lambda_k}{\alpha_k^{1-\gamma}} |t-s|^\gamma. \quad (5.48)$$

Collecting (5.47) and (5.48), the conclusion follows. □

We can prove now the following result.

Theorem 5.22 *Assume* (5.39)–(5.41), *and* (5.42) *for some* $\gamma \in (0,1)$. *Then the process* $W_A(t)$ *has a version* $W_A(t,\xi)$, $t \geq 0$, $\xi \in \overline{\mathscr{O}}$, *α-Hölder continuous with respect to* $t \geq 0$, $\xi \in \overline{\mathscr{O}}$ *and with any* $\alpha \in (0, \frac{\gamma}{2})$. *In particular the process* $W_A(t)$ *has an* E*-valued version with α-Hölder continuous paths.*

Proof It follows from Lemma 5.21 that there exists $C_2 > 0$ such that

$$\mathbb{E}|W_A(t,\xi) - W_A(s,\eta)|^2 \leq C_2 \left(|t-s| + |\xi - \eta|^2\right)^\gamma.$$

Since $W_A(t,\xi) - W_A(s,\eta)$ is a Gaussian random variable, therefore for constants C_m^1

$$\mathbb{E}|W_A(t,\xi) - W_A(s,\eta)|^{2m} \leq C_m^1 \left(|t-s| + |\xi - \eta|\right)^{m\gamma},$$

for $m \in \mathbb{N}$. Kolmogorov's test for random fields (see Theorem 3.5) now implies the result. □

(2) Note that, for arbitrary $\gamma \in [0,1]$ and all $x \geq 0$, $y \geq 0$, $|e^{-x} - e^{-y}| \leq |x-y|^\gamma$. Proof: if $0 \leq y \leq x$ then $|e^{-x} - e^{-y}| \leq (x-y)$. If in addition $x - y \leq 1$ then $x - y \leq (x-y)^\gamma$, so $|e^{-x} - e^{-y}| \leq |x - y|^\gamma$ and if $x - y > 1$ then also $|x-y|^\gamma \geq 1 \geq |e^{-x} - e^{-y}|$.

Remark 5.23 Condition (5.40) is necessary and sufficient for the process W_A to take values in $L^2(\mathscr{O})$. The stronger condition (5.42) with $\gamma = 2\alpha$ implies by Theorem 5.11 continuity of W_A as an $L^2(\mathscr{O})$-valued process.

For better regularity of W_A, additional conditions on eigenfunctions $\{e_k\}$ had to be imposed. This is natural if one takes into account that for a one dimensional Wiener process $W(t, \xi) = \beta(t)e(\xi),\ t \geq 0, \xi \in \mathscr{O}$, the spatial regularity of W_A is precisely that of the function $e(\xi)$. □

Example 5.24 Assume that $\mathscr{O}$ is the cube $[0, \pi]^N$ in $\mathbb{R}^N$ with the boundary $\partial\mathscr{O}$. If A is the linear operator

$$\begin{cases} D(A) = H^2(\mathscr{O}) \cap H_0^1(\mathscr{O}) \\ Au = \Delta u, \quad \forall\, u \in D(A) \end{cases}$$

where Δ represents the Laplace operator, then

$$\begin{aligned} g_{n_1,\dots,n_N}(\xi) &= (2/\pi)^{\frac{N}{2}} \sin(n_1\xi_1) \cdots \sin(n_N\xi_N). \\ |\nabla g_{n_1,\dots,n_N}(\xi)| &\leq (2/\pi)^{\frac{N}{2}} (|n_1| + \cdots + |n_N|) \\ &\leq (2/\pi)^{\frac{N}{2}} N^{1/2} \sqrt{n_1^2 + \ldots + n_N^2} \\ \alpha_{n_1,\dots,n_N} &= n_1^2 + \cdots + n_N^2 \end{aligned}$$

and the conditions (5.41) hold. Moreover if $Q = I$, then (5.42) holds if and only if $\gamma < 1 - N/2$. If $N = 1$ one has $\gamma < 1/2$. For $N \geq 2$ one can apply the results for appropriate powers of A. □

Elliptic operators of order $r \geq 2$ and with suitable boundary conditions, often generate analytic semigroups of negative type in $E = L^p(\mathscr{O})$ ($\mathscr{O} \subset \mathbb{R}^d$) for any $p \geq 1$. For such semigroups $S(t),\ t \geq 0$, and arbitrary $\gamma > 0$ and $T > 0$, there exist constants c_γ such that

$$|(-A)^\gamma S(t)x|_E \leq c_\gamma t^{-\gamma} |x|_E, \quad t \in [0, T].$$

The domain $D((-A)^\gamma)$ can often be identified with the Sobolev space $W^{r\gamma,p}(\mathscr{O})$ or some of its subspace. In the next theorem we will therefore require that

$$|S(t)x|_{W^{r\gamma,p}(\mathscr{O})} \leq c_\gamma t^{-\gamma} |x|_{L^p(\mathscr{O})}, \quad t \in [0, T]. \tag{5.49}$$

Theorem 5.25 *Assume the following.*

(i) For some $\alpha \in (0, 1/2)$

$$\sum_{k=1}^{\infty} \frac{\lambda_k}{\alpha_k^{1-2\alpha}} < +\infty.$$

(ii) For some constant $c > 0$

$$|e_k(\xi)| \le c, \quad k \in \mathbb{N}, \ \xi \in \overline{\mathscr{O}}.$$

(iii) Condition (5.49) *holds with some* $r \ge 1, \gamma > 0, \ p > 1$ *such that* $\alpha > 1/p + \gamma$.

Then the stochastic convolution W_A *is a continuous process with values in* $W^{r\gamma,p}(\mathscr{O})$*. If in addition,*

(iv) $p > d/r\gamma$*, then* W_A *is continuous with values in* $C(\overline{\mathscr{O}})$.

Proof By condition (i) the factorization formula is applicable. We will apply Proposition 5.9 with $E_1 = L^p(\mathscr{O})$, $E_2 = W^{r\gamma,p}(\mathscr{O})$. It is enough to check that the process

$$Y_\alpha(t) = \int_0^t (t-s)^{-\alpha} S(t-s) dW(s), \quad t \in [0,T],$$

with

$$W(t,\xi) = \sum_{k=1}^{\infty} \sqrt{\lambda_k} \beta_k(t) e_k(\xi), \quad t \in [0,T], \ \xi \in \mathscr{O},$$

has trajectories in $L^p(0,T; L^p(\mathscr{O}))$. Note that for $t \in [0,T], \ \xi \in \mathscr{O}$

$$Y_\alpha(t,\xi) = \sum_{k=1}^{\infty} \sqrt{\lambda_k} \left(\int_0^t (t-s)^{-\alpha} e^{-\alpha_k(t-s)} d\beta_k(s) \right) e_k(\xi),$$

is a Gaussian random variable with the second moment

$$\begin{aligned}
\mathbb{E}[Y_\alpha^2(t,\xi)] &= \sum_{k=1}^{\infty} \lambda_k \left(\int_0^t s^{-2\alpha} e^{-2\alpha_k s} ds \right) e_k^2(\xi) \\
&\le c^2 \sum_{k=1}^{\infty} \lambda_k \int_0^t s^{-2\alpha} e^{-2\alpha_k s} ds \\
&\le c_1 \sum_{k=1}^{\infty} \frac{\lambda_k}{\alpha_k^{1-2\alpha}}.
\end{aligned}$$

Since for each p and a constant c_p,

$$\mathbb{E}|Y_\alpha(t,\xi)|^p = c_p \left(\mathbb{E}|Y_\alpha(t,\xi)|^2 \right)^{p/2},$$

we have that

$$\mathbb{E} \int_0^T |Y_\alpha(t)|^p_{L^p(\mathscr{O})} dt \le T|\mathscr{O}| \left(\sum_{k=1}^{\infty} \frac{\lambda_k}{\alpha_k^{1-2\alpha}} \right)^{p/2} < +\infty.$$

This proves the first part of the theorem. If $p > d/(r\gamma)$ then $W^{r\gamma,p}(\mathscr{O}) \subset C(\overline{\mathscr{O}})$ continuously and the second part follows as well. □

Remark 5.26 For an elliptic generator A, condition (5.49) holds for arbitrary $\gamma > 0$ and $p > 1$; therefore one can choose p and γ such that both (iii) and (iv) hold.

Consequently, in such a situation the stochastic convolution W_A is continuous if only (i) and (ii) are satisfied. □

Remark 5.27 The uniform boundedness condition (ii) of Theorem 5.25 is not satisfied in general. One can show that if the boundary of $\mathscr{O} \subset \mathbb{R}^n$ is regular, there is a constant $C(\mathscr{O})$ such that

$$\|e_k\|_\infty \leq C(\mathscr{O})\alpha_k^{(n-1)/2}, \quad k \in \mathbb{N}^n.$$

Moreover this result is optimal, see Grieser [361]. □

Even if condition (ii) is not satisfied, in particular if $\mathscr{O}$ is a ball in $\mathbb{R}^d$, $d > 1$, and A is the Laplace operator with Dirichlet boundary conditions, the condition (5.39) implies that $e_k \in D(A)$ and

$$|Ae_k|_{L^2(\mathscr{O})} := \alpha_k < +\infty, \quad k \in \mathbb{N}.$$

Thus if $D(A)$ is included in the Sobolev space $W^{r,2}(\mathscr{O})$, then for some constant $c_1 > 0$

$$|e_k|_{W^{r,2}(\mathscr{O})} \leq c_1\alpha_k, \quad k \in \mathbb{N}$$

and if $2r > d$ then by Sobolev's embedding theorem, for some $c > 0$

$$|e_k(\xi)| \leq c\alpha_k, \quad k \in \mathbb{N}, \ \xi \in \mathscr{O}.$$

Considering higher powers $A^\gamma, \gamma \in \mathbb{N}$, of A for which $D(A^\gamma)$ is included in $W^{r\gamma,2}(\mathscr{O})$ one obtains, in the same way, that if $2r\gamma > d$ then for a constant $c > 0$,

$$|e_k(\xi)| \leq c\alpha_k^\gamma, \quad k \in \mathbb{N}, \ \xi \in \mathscr{O}. \tag{5.50}$$

With the same proof of as of Theorem 5.25 but replacing inequality (ii) by (5.50) we arrive at the following statement.

Theorem 5.28 *Assume that for some $\gamma \in \mathbb{N}$ such that $2r\gamma > d$ the space $D(A^\gamma)$ is included in $W^{r\gamma,2}(\mathscr{O})$ and, in addition, that for some $\alpha \in (0, 1/2)$*

$$\sum_{k=1}^{\infty} \frac{\lambda_k}{\alpha_k^{1-2(\alpha+\gamma)}} < +\infty.$$

Then the stochastic convolution W_A is a continuous process with values in $C(\overline{\mathscr{O}})$.

Estimate (5.50) is usually not optimal. The constant γ can be lowered, as in the case when $\mathscr{O}$ is a cube, as then, by Example 5.24, $\gamma = 0$ is equal to 0 and not to 1. Thus, even in the case of the Laplace operator A, the optimal γ depends on the geometry of the domain $\mathscr{O}$. In the case when $\mathscr{O} = \{x \in \mathbb{R}^2 : |x| < 1\}$ and A is the Laplace operator with Dirichlet boundary conditions, eigenvalues form a double sequence $\{\alpha_{k,l}\}$ determined by equations

$$J_k(\sqrt{\alpha_{k,l}}) = 0, \quad h \in \{0\} \cup \mathbb{N}, \ l \in \mathbb{N},$$

where J_k are the Bessel functions of order k. It follows from the theory of special functions see [636, Appendix 2, §2] that the optimal γ is the minimal positive integer for which there exists $c > 0$ such that

$$\alpha_{k,l}^{\gamma} |J_{k+1}(\sqrt{\alpha_{k,l}})| \ge c, \quad h \in \{0\} \cup \mathbb{N}, \ l \in \mathbb{N}.$$

The asymptotic behavior of J_{k+1} suggests that γ might be $1/4$.

5.5.2 The case of a skew-symmetric generator

Let $\mathscr{H} = L^2(\mathscr{O}) \oplus D(\Lambda^{-1/2})$ where $\Lambda = -A$ and A satisfies (5.39)–(5.41). Consider the skew-symmetric operator $\mathscr{A}$ in $\mathscr{H}$

$$\begin{cases} D(\mathscr{A}) = D(\Lambda^{1/2}) \oplus H \\ \mathscr{A}\begin{pmatrix} y \\ z \end{pmatrix} = \begin{pmatrix} 0 & 1 \\ -\Lambda & 0 \end{pmatrix}\begin{pmatrix} y \\ z \end{pmatrix}, \quad \forall \begin{pmatrix} y \\ z \end{pmatrix} \in D(\mathscr{A}). \end{cases}$$

If $B = \begin{pmatrix} 0 \\ I \end{pmatrix}$ and W is the same as in Section 5.5.1, then the stochastic convolution is of the form $W_A(\cdot) = \begin{pmatrix} W_A^1(\cdot) \\ W_A^2(\cdot) \end{pmatrix}$ where

$$W_A^1(t) = \frac{1}{\sqrt{\Lambda}} \int_0^t \sin(\sqrt{\Lambda}(t-s)) dW(s)$$

$$W_A^2(t) = \int_0^t \cos(\sqrt{\Lambda}(t-s)) dW(s).$$

The condition (5.40) implies (5.8) so that the process $W_A(\cdot)$ takes values in $\mathscr{H}$. However, the process $W_A^2(\cdot)$ does not take values in $L^2(\mathscr{O})$ but only in $D(\Lambda^{-1/2})$, so we cannot construct a version of the process $W_A(\cdot, \cdot)$ continuous in both variables. However, for the process $W_A^1(\cdot)$ we have the following result similar to Theorem 5.22.

Theorem 5.29 *Assume* (5.39)–(5.41)*, and* (5.42) *for some* $\gamma \in (0, 1)$. *Then the process* $W_A^1(t)$ *has a version* $W_A^1(t, \xi)$, $t \ge 0$, $\xi \in \overline{\mathscr{O}}$, *α-Hölder continuous with respect to* $t \ge 0$, $\xi \in \overline{\mathscr{O}}$ *with any* $\alpha \in (0, \frac{\gamma}{2})$. *In particular the process* $W_A^1(t)$ *has an* E*-valued version with α-Hölder continuous paths.*

Proof The proof is completely analogous to that of Theorem 5.22 and is based on the following lemma.

Lemma 5.30 *Assume that* (5.39)–(5.41) *and* (5.42) *hold for some* $\gamma \in (0, 1)$. *Then there exists a constant* $C_1 > 0$ *such that*

$$\mathbb{E}\left(|W_A^1(t, \xi) - W_A^1(t, \eta)|^2\right) \le C_1 |\xi - \eta|^{2\gamma} \tag{5.51}$$

$$\mathbb{E}\left(|W_A^1(t, \xi) - W_A^1(s, \xi)|^2\right) \le C_1 |t - s|^{\gamma} \tag{5.52}$$

for all $t, s \ge 0$ *and* $\xi, \eta \in \overline{\mathscr{O}}$.

Proof We have the following representation for $W_A^1(\cdot,\cdot)$

$$W_A^1(t,\xi) = \sum_{k=1}^{\infty} \sqrt{\frac{\lambda_k}{\alpha_k}} e_k(\xi) \int_0^t \sin(\sqrt{\alpha_k}(t-s))d\beta_k(s),\ t \geq 0. \tag{5.53}$$

Therefore

$$\mathbb{E}|W_A^1(t,\xi) - W_A^1(t,\eta)|^2 = \sum_{k=1}^{\infty} \frac{\lambda_k}{\alpha_k} \int_0^t \sin(\sqrt{\alpha_k}(t-s))]^2 |e_k(\xi) - e_k(\eta)|^2 ds$$

and [(3)]

$$\begin{aligned}
&\mathbb{E}\left(\left|W_A^1(t,\xi) - W_A^1(s,\xi)\right|^2\right) \\
&= \textstyle\sum_{k=1}^{\infty} \frac{\lambda_k}{\alpha_k} \int_s^t \sin(\sqrt{\alpha_k}(t-\sigma))]^2 |e_k(\xi)|^2 d\sigma \\
&+ \sum_{k=1}^{\infty} \frac{\lambda_k}{\alpha_k} \int_0^s \left|\left[\sin(\sqrt{\alpha_k}(t-\sigma)) - \sin(\sqrt{\alpha_k}(s-\sigma))\right] e_k(\xi)\right|^2 d\sigma \\
&=: I_1(t,s,\xi) + I_2(t,s,\xi).
\end{aligned}$$

Therefore the lemma follows easily in the same way as Lemma 5.21. □

Example 5.31 Take $H = L^2(0,\pi)$, $Q = I$ and let Λ be given by

$$\begin{cases} \Lambda x = -\dfrac{\partial^2 x}{\partial \xi^2} \\ D(\Lambda) = H^2(0,\pi) \cap H_0^1(0,\pi) \end{cases}$$

then $D(\Lambda^{-1/2})$ is isomorphic to $H^{-1}(0,\pi)$. Then $\lambda_k = 1$ and $\alpha_k = k^2$, so hypotheses (5.39)–(5.41) are fulfilled and therefore the process $W_A^1(\cdot,\cdot)$ has a continuous modification. This result was obtained earlier in [702]. □

5.5.3 Equations with spatially homogeneous noise

We are concerned with the stochastic heat equation,

$$\begin{cases} dX(t,\xi) = \Delta X(t,\xi)dt + dW_\Gamma(t,\xi), \\ X(0,\xi) = 0, \quad \xi \in \mathbb{R}^d, \end{cases} \tag{5.54}$$

where $W_\Gamma(t)$, $t \geq 0$, is a Wiener process with covariance kernel Γ introduced in Section 4.1.4. In the presentation we follow [441].

Although W_Γ is an $\mathscr{S}'(\mathbb{R}^d)$-valued Wiener process, the stochastic integral

$$\int_0^t \Psi(s)dW_\Gamma(s), \quad t \geq 0, \tag{5.55}$$

(3) Note that, for arbitrary $\gamma \in [0,1]$, there exists $c_\gamma > 0$ such that $|\sin x - \sin y| \leq c_\gamma |x-y|^\gamma$ for all $x \geq 0, y \geq 0$.

can be defined in the classical Hilbertian way by replacing the space U_0 by U_Γ, see the comments at the end of Section 4.2.1.

To study (5.54) consider first the heat equation

$$\begin{cases} \dfrac{\partial Z(t,\xi)}{\partial t} = \Delta Z(t,\xi), \quad t \ge 0, \\ Z(0,\xi) = z(\xi) \end{cases} \tag{5.56}$$

where $z \in \mathscr{S}'(\mathbb{R}^d)$.

Setting $\widehat{Z}(t) = \mathscr{F} Z(t),\ t \ge 0$, we have

$$\begin{cases} \dfrac{\partial \widehat{Z}(t,\lambda)}{\partial t} = -(2\pi)^2 \widehat{Z}(t,\lambda), \quad t \ge 0, \\ \widehat{Z}(0,\lambda) = \mathscr{F} z = \widehat{z} \end{cases}$$

and therefore

$$\widehat{Z}(t,\lambda) = e^{-(2\pi)^2 t |\lambda|^2} \widehat{z}.$$

Consequently, for arbitrary $z \in \mathscr{S}'_c(\mathbb{R}^d)$, equation (5.56) has a unique solution in $\mathscr{S}'_c(\mathbb{R}^d)$ and the solution is given by the formula

$$Z(t,\xi) = p(t,\xi) * z(\xi) = (\xi, p(t,\xi - \cdot)),$$

where

$$p(t,\xi) = (4\pi t)^{-d/2} e^{-|\xi|^2/4t}, \quad t \ge 0,\ x \in \mathbb{R}^d. \tag{5.57}$$

The family

$$S(t)z = p(t,\cdot) * z, \quad z \in \mathscr{S}'_c(\mathbb{R}^d),$$

forms a strongly continuous semigroup of operators in $\mathscr{S}'_c(\mathbb{R}^d)$.

As a solution of (5.54) we take

$$X(t) = \int_0^t S(t-s) dW_\Gamma(ds), \quad t \ge 0. \tag{5.58}$$

Our main concern here is to find conditions on Γ under which the process (5.58) is function valued. We say that the solution X is function valued if the integral (5.58) exists in $H = L^2_\rho(\mathbb{R}^d)$ for some positive, continuous, integrable weight ρ. We shall see that our characterization is independent of ρ.

Theorem 5.32 *Let Γ be a positive definite, tempered distribution on $\mathbb{R}^d$ with spectral measure μ. Then equation* (5.54) *has a function valued solution if and only if*

$$\int_{\mathbb{R}^d} \frac{1}{\lambda^2} \mu(d\lambda) < +\infty. \tag{5.59}$$

Proof It follows from the definition of stochastic integral that $X(t)$, given by (5.58), takes values in $L^2_\rho(\mathbb{R}^d)$ if and only if

$$\int_0^t \|S(r)\|^2_{L_2(U_\Gamma, L^2_\rho(\mathbb{R}^d))} dr < +\infty, \quad t \geq 0.$$

Let $\{f_k\}$ be an orthonormal basis in $L^2_{(s)}(\mathbb{R}^d, \mu)$. It follows from the proof of Theorem 4.14 that elements

$$g_k := \widehat{f_k \mu}, \quad k \in \mathbb{N},$$

form an orthonormal and complete basis on the reproducing kernel U_γ. Thus we have

$$\|S(r)\|^2_{L_2(U_\Gamma, L^2_\rho(\mathbb{R}^d))} = \sum_{k=1}^{\infty} |S(r)\widehat{f_k \mu}|^2_{L^2_\rho(\mathbb{R}^d)} = \sum_{k=1}^{\infty} \int_{\mathbb{R}^d} |p(r, \cdot) * \widehat{f_k \mu}(\xi)|^2 \rho(\xi) d\xi.$$

However, $p(r, \cdot) \in \mathscr{S}(\mathbb{R}^d)$ and therefore

$$p(r, \cdot) * f_k \mu(\xi) = (p(r, \xi - \cdot), \widehat{f_k \mu}) = (f_k \mu, \hat{p}(r, \xi - \cdot)).$$

Since

$$\hat{p}(r, \xi - \cdot)) = e^{i\langle \xi, \lambda \rangle} e^{-r|\lambda|^2},$$

therefore

$$(f_k \mu, \hat{p}(r, \xi - \cdot)) = (f_k \mu, e^{i\langle \xi, \cdot \rangle} e^{-r|\cdot|^2}).$$

Consequently

$$\begin{aligned}
\|S(r)\|^2_{L_2(U_\Gamma, L^2_\rho(\mathbb{R}^d))} &= \sum_{k=1}^{\infty} \int_{\mathbb{R}^d} |(f_k \mu, e^{i\langle \xi, \cdot \rangle} e^{-r|\cdot|^2})|^2 \rho(\xi) d\xi \\
&= \int_{\mathbb{R}^d} \left[\sum_{k=1}^{\infty} \left| \langle f_k, e^{-i\langle \xi, \cdot \rangle} e^{-r|\cdot|^2} \rangle_{L^2_{(s)}(\mathbb{R}^d, \mu)} \right|^2 \right] \rho(\xi) d\xi.
\end{aligned}$$

By the Parseval identity

$$\begin{aligned}
\sum_{k=1}^{\infty} \left| \langle f_k, e^{-i\langle \xi, \cdot \rangle} e^{-r|\cdot|^2} \rangle_{L^2_{(s)}} \right|^2 &= \int_{\mathbb{R}^d} |e^{-i\langle \xi, \lambda \rangle} e^{-r|\lambda|^2}|^2 \mu(d\lambda) \\
&= \int_{\mathbb{R}^d} e^{-2r|\lambda|^2} \mu(d\lambda).
\end{aligned}$$

Finally,

$$\begin{aligned}
\int_0^t \|S(r)\|^2_{L_2(U_\Gamma, L^2_\rho(\mathbb{R}^d))} dr &= \int_{\mathbb{R}^d} \rho(\xi) d\xi \int_{\mathbb{R}^d} e^{-2r|\lambda|^2} \mu(d\lambda) \\
&= \frac{1}{2} \int_{\mathbb{R}^d} \rho(\xi) d\xi \int_{\mathbb{R}^d} \frac{1 - e^{-2t|\lambda|^2}}{|\lambda|^2} \mu(d\lambda).
\end{aligned}$$

□

Before proving a second theorem we gather some properties of the resolvent kernel G_d of the d-dimensional Wiener process

$$G_d(\xi) = \int_0^\infty e^{-t} p(t,\xi)dt, \quad \xi \in \mathbb{R}^d,$$

where p is given by (5.57). It is easy to see that

$$G_d(\xi) = (2\pi)^{-d} \int_{\mathbb{R}^d} e^{-i\langle \xi,\lambda\rangle} \frac{1}{1+|\lambda|^2}\, d\lambda, \quad \xi \in \mathbb{R}^d.$$

The following formulae in the propositions below are well known, see for example [344, 351, 484].

Proposition 5.33 *The resolvent kernel G_d is given by,*

$$G_1(\xi) = \frac{1}{2}\, e^{-|\xi|}, \quad \xi \in \mathbb{R}^1,$$

$$G_d(\xi) = (2\pi)^{-d/2} |\xi|^{-\frac{d-2}{2}} K_{\frac{d-2}{2}}(|\xi|), \quad d \geq 2,\ \xi \in \mathbb{R}^d,$$

where $K_\gamma,\ \gamma > 0$ denotes the modified Bessel function of the third order.

Note that we have in particular

$$G_3(\xi) = \frac{1}{4\pi|\xi|}\, e^{-|\xi|}, \quad \xi \in \mathbb{R}^3.$$

Proposition 5.34 *The function G_d enjoys the following properties.*

(i) For $d \geq 1$, for $|\xi|$ bounded away from a neighborhood of 0 and for a constant $c > 0$ we have

$$G_d(\xi) \leq c|\xi|^{-\frac{d-2}{2}} e^{-|\xi|}, \quad \xi \in \mathbb{R}^d.$$

(ii) For $d \geq 3$ and for a constant $c > 0$, in a neighborhood of 0 we have

$$G_d(\xi) \sim c|\xi|^{2-d}, \quad \xi \in \mathbb{R}^d.$$

(iii) For $d = 2$ and for a constant $c > 0$, in a neighborhood of 0 we have

$$G_d(\xi) \sim -c \log |\xi|, \quad \xi \in \mathbb{R}^2.$$

We will also need the following lemma.

Lemma 5.35 *Assume that the positive definite distribution Γ is a measure. Then*

$$(\Gamma, G_d) = \int_{\mathbb{R}^d} G_d(\xi)\Gamma(d\xi) = (2\pi)^d \int_{\mathbb{R}^d} \frac{1}{1+|\lambda|^2} \mu(d\lambda).$$

Proof Since $\mu = \mathscr{F}^{-1}(\Gamma)$ and the function

$$\lambda \to \frac{e^{-t|\lambda|^2}}{1+|\lambda|^2|}$$

belongs to $\mathscr{S}(\mathbb{R}^d)$, we have

$$\int_{\mathbb{R}^d} \frac{e^{-t|\lambda|^2}}{1+|\lambda|^2|}\mu(d|\lambda) = \left(\mathscr{F}^{-1}(\Gamma), \frac{e^{-t|\cdot|^2}}{1+|\cdot|^2|}\right)$$

$$= \left(\Gamma, \mathscr{F}\left(\frac{e^{-t|\cdot|^2}}{1+|\cdot|^2|}\right)\right) = (2\pi)^{-d}(\Gamma, p(t,\cdot) * G_d.$$

Therefore

$$\lim_{t\to 0}(2\pi)^{-d}(\Gamma, p(t,\cdot) * G_d) = \int_{\mathbb{R}^d} \frac{1}{1+|\lambda|^2}\mu(d\lambda).$$

Moreover

$$p(s,\cdot) * G_d = \int_0^\infty e^{-t} p(t,\cdot) * p(s,\cdot) ds$$

$$= e^s \int_0^\infty e^{-(t+s)} p(t+s,\cdot) ds = e^s \int_s^\infty e^{-r} p(r,\cdot) dr.$$

So

$$e^{-s} p(s,\cdot) * G_d = \int_s^\infty e^{-r} p(r,\cdot) dr$$

and therefore

$$\lim_{s\to 0} e^{-s} p(s,\cdot) * G_d = G_d.$$

Hence, since Γ is a measure on $\mathbb{R}^d$, we have

$$\int_{\mathbb{R}^d} \frac{1}{1+|\lambda|^2}\mu(d\lambda) = \lim_{t\to 0} e^{-t}(\Gamma, p(t,\cdot) * G_d) = (2\pi)^{-d}(\Gamma, G_d).$$

□

Now we are ready to prove the following theorem.

Theorem 5.36 *Assume that the positive definite distribution Γ is a measure. Then the equation* (5.54) *has function valued solutions*

(i) for all Γ if $d = 1$,
(ii) for exactly those Γ for which $-\int_{\{|\lambda|\le 1\}} \log|\lambda|\Gamma(d\lambda) < +\infty$ if $d = 2$,
(iii) for exactly those Γ for which $\int_{\{|\lambda|\le 1\}} |\lambda|^{d-2}\Gamma(d\lambda) < +\infty$ if $d \ge 3$.

Proof It is well known that a measure Γ belongs to $\mathscr{S}'(\mathbb{R}^d)$ if and only if for some $r > 0$

$$\int_{\mathbb{R}^d} \frac{1}{|\xi|^r}\Gamma(d\xi) < +\infty. \tag{5.60}$$

Also for arbitrary $d \in \mathbb{N}$

$$\int_{\mathbb{R}^d} G_d(\xi)\Gamma(d\xi) = \int_{\{|\xi|\le 1\}} G_d(\xi)\Gamma(d\xi) + \int_{\{|\xi|>1\}} G_d(\xi)\Gamma(d\xi).$$

Then by Proposition 5.34(i)

$$\int_{\{|\xi|>1\}} G_d(\xi)\Gamma(d\xi) \le c \int_{\{|\xi|>1\}} e^{-|\xi|}\Gamma(d\xi),$$

and from (5.60)

$$\int_{\{|\xi|>1\}} G_d(\xi)\Gamma(d\xi) < +\infty,$$

and the theorem is true for $d = 1$.

If $d = 2$ then

$$\int_{\mathbb{R}^2} G_2(\xi)\Gamma(d\xi) < +\infty \Leftrightarrow \int_{\{|\xi|>1\}} e^{-|\xi|}\Gamma(d\xi) < +\infty.$$

But $G_2(\xi) \sim c \log\left(\frac{1}{|\xi|}\right)$ in a neighborhood of 0, so

$$\lim_{\xi\to 0} \frac{G_2(\xi)}{\log\left(\frac{1}{|\xi|}\right)} = 1.$$

Therefore for some $c_1, c_2 > 0$,

$$c_1 \log\left(\frac{1}{|\xi|}\right) \le G_2(\xi) \le c \log\left(\frac{1}{|\xi|}\right), \quad |\xi| \le 1.$$

Consequently

$$\int_{\mathbb{R}^2} G_2(\xi)\Gamma(d\xi) < +\infty \Leftrightarrow \int_{\{|\xi|>1\}} \log\left(\frac{1}{|\xi|}\right)\Gamma(d\xi) < +\infty.$$

If $d \ge 3$, in the same way,

$$\int_{\mathbb{R}^d} G_d(\xi)\Gamma(d\xi) < +\infty \Leftrightarrow \int_{\{|\xi|>1\}} |\xi|^{2-d}\Gamma(d\xi) < +\infty.$$

This completes the proof. □

One can prove a similar result for the stochastic wave equation

$$dX_t = \Delta X(t)dt + dW_\Gamma(t), \quad X(0) = X_t(0) = 0. \tag{5.61}$$

For the proof of the following result we refer to [441]. This paper also contains comments on the literature concerning problem (5.61).

Theorem 5.37 *Problem* (5.61) *has a function valued solution if and only if condition* (5.60) *holds.*

Thus equations (5.54) and (5.61) have a function valued solution under the same conditions.

5.6 Existence of strong solutions

We are concerned now with the existence of strong solutions to (5.1). Since $BW(t)$, $t \geq 0$, should be an H-valued process we will assume that $B = I$ and that W is an H-valued Wiener process with covariance operator Q with Tr $Q < +\infty$. Thus we are concerned with the equation

$$X(t) = x + \int_0^t (AX(s) + f(s))ds + W(t), \quad t \in [0, T]. \tag{5.62}$$

As strong solutions are also weak solutions we know that they should be of the form

$$X(t) = S(t)x + \int_0^t S(t-s)f(s)ds + \int_0^t S(t-s)dW(s), \quad t \in [0, T]. \tag{5.63}$$

Theorem 5.38 *Assume that*

(i) $Q^{1/2}(H) \subset D(A)$ *and* $AQ^{1/2}$ *is a Hilbert–Schmidt operator,*
(ii) $x \in D(A)$ *and* $f \in C^1([0, T]; H) \cap C([0, T]; D(A))$.

Then problem (5.1) has a strong solution.

Proof The result is true if $f = 0$, $W = 0$, by Proposition A.2. Assume now that $x = 0$, $W = 0$. For every $t \in [0, T]$ we have

$$\int_0^t |AS(t-\sigma)f(\sigma)|d\sigma \leq \int_0^t \|S(t-\sigma)\| \, |Af(\sigma)|d\sigma < +\infty,$$

and therefore, by Proposition A.7, $X(t) \in \int_0^t S(t-\sigma)f(\sigma)d\sigma \in D(A)$ and

$$AX(t) = \int_0^t AS(t-\sigma)f(\sigma)d\sigma, \quad t \in [0, T].$$

Moreover

$$\begin{aligned}
\int_0^t AX(s)ds &= \int_0^t \left(\int_0^s AS(s-\sigma)f(\sigma)d\sigma \right) ds \\
&= \int_0^t \left(\int_0^{t-\sigma} \frac{d}{ds} S(s)f(\sigma)ds \right) d\sigma \\
&= \int_0^t S(t-\sigma)f(\sigma)d\sigma - \int_0^t f(\sigma)d\sigma \\
&= X(t) - \int_0^t f(\sigma)d\sigma, \quad t \in [0, T].
\end{aligned}$$

Assume finally that $x = 0$, $f = 0$. Note that

$$\int_0^t \|AS(s)Q^{1/2}\|_{L_2}^2 ds = \int_0^t \|S(s)AQ^{1/2}\|_{L_2}^2 ds$$
$$\leq \|AQ^{1/2}\|^2 \int_0^t \|S(s)\|_{L_2}^2 ds < +\infty.$$

Therefore, by Proposition 4.30

$$W_A(t) = \int_0^t S(t-\sigma)dW(\sigma) \in D(A), \quad \mathbb{P}\text{-a.s.}$$

and for $t \in [0, T]$, $\mathbb{P}$-a.s.

$$AW_A(t) = \int_0^t AS(t-\sigma)dW(\sigma).$$

Since $W_A(\cdot)$ is a weak solution of (5.1), for $t \in [0, T]$ and $\zeta \in D(A^*)$, $\mathbb{P}$-a.s.

$$\langle W_A(t), \zeta\rangle = \int_0^t \langle W_A(s), A^*\zeta\rangle ds + \langle W(t), \zeta\rangle$$
$$= \int_0^t \langle AW_A(s), \zeta\rangle ds + \langle W(t), \zeta\rangle$$
$$= \langle A\int_0^t W_A(s)ds, \zeta\rangle + \langle W(t), \zeta\rangle.$$

Consequently

$$W_A(t) = \int_0^t AW_A(s)ds + W(t), \quad \mathbb{P}\text{-a.s.}$$

This finishes the proof. □

Example 5.39 Assume that W is a finite dimensional Wiener process, say

$$W(t) = \sum_{k=1}^m a_k\beta_k(t), \quad t \geq 0, \tag{5.64}$$

where $\beta_1, \dots, \beta_m$ are independent Wiener processes and vectors $a_1, \dots, a_m \in H$ are linearly independent. Then Theorem 5.38 says that if $a_1, \dots, a_m \in D(A)$ then W_A is a strong solution of (5.63) with $x = 0$, $f = 0$. Under additional conditions on the operator A the element $a_1, \dots, a_m$ can be less regular. This is the subject of the next theorem. □

Theorem 5.40 *Assume that A generates an analytic semigroup of negative type. Let in addition*

(i) for some $\beta \in (1/2, 1)$, $Q^{1/2}(H) \subset D((-A)^\beta)$ and $(-A)^\beta Q^{1/2}$ is a Hilbert–Schmidt operator,
(ii) $x \in D(A)$ and for some $\alpha \in (0, 1)$, $f \in C^\alpha([0, T]; H) \cup C([0, T]; D_A(\alpha, \infty))$.

Then problem (5.1) has a strong solution.

Proof By Proposition A.24, we have that

$$\widehat{X}(\cdot) = S(\cdot)\xi + \int_0^{\cdot} S(t-s)f(s)ds \in C^1([0,T];H) \cap C([0,T];D(A))$$

and moreover

$$\frac{d}{dt}\widehat{X}(t) = A\widehat{X}(t) + f(t), \quad t \in [0,T], \quad \widehat{X}(0) = x,$$

so, without any loss of generality, one can assume that $f = 0$ and $x = 0$.

We will only establish that AW_A has a continuous version. Define

$$W^\beta(t) = \sum_{j=1}^{\infty} \lambda_j(-A)^\beta e_j \beta_j(t), \quad t \geq 0. \tag{5.65}$$

By condition (i) formula (5.65) defines an H-valued Wiener process with trace class covariance. It is a version of the process $(-A)^\beta W$. Taking into account that $1 - \beta < 1/2$ we get by Theorem 5.15(ii) that

$$(-A)^{1-\beta} \int_0^t S(t-s)dW^\beta(s), \quad t \geq 0,$$

has a continuous version. The required continuity of AW_A now follows from the identity

$$-AW_A(t) = (-A)^{1-\beta} \int_0^t S(t-s)dW^\beta(s), \quad t \geq 0,$$

which can be justified by a finite dimensional approximation (see (5.64)). □

Example 5.41 If W is given by (5.64) and A generates an analytic semigroup of negative type, then a weak solution is the strong one provided that $a_1, \dots, a_m \in D((-A)^\beta)$ for some $\beta \in (1/2, 1)$. This is of course a weaker requirement than $a_1, \dots, a_m \in D(A)$. □

6
Linear equations with multiplicative noise

We start from definitions of strong, weak and mild solutions and establish their basic properties. Then we investigate regularity properties of stochastic convolutions with general random integrands, first for contraction semigroups, then for analytic ones. Using these results we prove the existence and uniqueness of mild solutions in several situations. At the end we derive the existence of strong solutions by reducing the stochastic evolution equation to a differential equation with stochastic coefficients.

6.1 Strong, weak and mild solutions

As in previous chapters, a probability space $(\Omega, \mathscr{F}, \mathbb{P})$ and a normal filtration $\mathscr{F}_t$, $t \geq 0$, are fixed. Let H and U be Hilbert spaces and Q a self-adjoint bounded nonnegative operator on U. Let $W(t)$, $t \geq 0$, be a Q-Wiener process on $U_1 \supset U$ and $U_0 = Q^{1/2}U$, see Chapter 4.

We recall that $\mathcal{N}_W^2(0, T; L_2^0)$, more simply $\mathcal{N}_W^2(0, T)$ or $\mathcal{N}_W^2$, will denote the set of all L_2^0-predictable processes Φ such that $|||\Phi|||_T < +\infty$ and $L_2^0 = L_2(U_0, H)$ is the space of all Hilbert–Schmidt operators from U_0 into H.

In this chapter we shall consider the following stochastic equation

$$\begin{cases} dX(t) = (AX(t) + f(t))dt + B(X(t))dW(t), \\ X(0) = \xi, \end{cases} \tag{6.1}$$

on a time interval $[0, T]$, where $A : D(A) \subset H \to H$ is the infinitesimal generator of a strongly continuous semigroup $S(\cdot)$, ξ is an $\mathscr{F}_0$-measurable H-valued random variable, f is a predictable process with local integrable trajectories and $B : D(B) \subset H \to L_2^0 = L_2(U_0; H)$ is a linear operator.

Let $\{g_j\}$ be a complete orthonormal system in U_0. Since for arbitrary $x \in D(B)$, $B(x)$ is a Hilbert–Schmidt operator from U_0 into H, therefore

$$\sum_{j=1}^{\infty} |B(x)g_j|^2 < +\infty, \quad \forall\, x \in D(B).$$

The operators

$$B_j x = B(x)g_j, \quad x \in D(B), \quad j \in \mathbb{N},$$

are linear and

$$B(x)u = \sum_{j=1}^{\infty} B_j x \langle u, g_j \rangle_{U_0}, \quad x \in D(B), \ u \in U_0. \tag{6.2}$$

Consequently if

$$W(t) = \sum_{j=1}^{\infty} \beta_j(t) g_j$$

then equation (6.1) can be written equivalently as

$$\begin{cases} dX(t) = (AX(t) + f(t))dt + \sum_{j=1}^{\infty} B_j X(t) d\beta_j(t), \\ X(0) = \xi. \end{cases} \tag{6.3}$$

Typical examples of operators B are the following.

Example 6.1 Fix $U = \mathbb{R}^m = U_0 = U_1$ and let $B_1, \ldots, B_m$ be linear operators on H with domains $D(B_1), \ldots, D(B_m)$ respectively. Define

$$\begin{cases} D(B) = \bigcap_{j=1}^{m} D(B_j), \\ B(x)u = \sum_{j=1}^{m} u_j B_j x, \quad (u_1, \ldots, u_m) \in \mathbb{R}^m, \ \forall\, x \in D(B). \end{cases} \tag{6.4}$$

Then B has the required form. □

Similarly as for additive noise, we define a *strong solution* of problem (6.1) as an H-valued predictable process $X(t)$, $t \in [0, T]$, which takes values in $D(A) \cap D(B)$, $\mathbb{P}$-a.s. such that

$$\begin{cases} \mathbb{P}\left(\int_0^T [|X(s)| + |AX(s)|]\, ds < +\infty \right) = 1, \\ \mathbb{P}\left(\int_0^T \|B(X(s))\|_{L_2^0}^2\, ds < +\infty \right) = 1, \end{cases}$$

and, for arbitrary $t \in [0, T]$ and $\mathbb{P}$-a.s.

$$X(t) = \xi + \int_0^t (AX(s) + f(s))ds + \int_0^t B(X(s))dW(s). \tag{6.5}$$

An H-valued predictable process $X(t)$, $t \in [0, T]$, is said to be a *weak solution* to (6.1) if X takes values in $D(B)$, $\mathbb{P}_T$-a.s.,

$$\mathbb{P}\left(\int_0^T |X(s)|ds < +\infty\right) = 1 \tag{6.6}$$

$$\mathbb{P}\left(\int_0^T \|B(X(s))\|_{L_2^0}^2 ds < +\infty\right) = 1 \tag{6.7}$$

and for arbitrary $t \in [0, T]$ and $\zeta \in D(A^*)$,

$$\langle X(t), \zeta\rangle = \langle \xi, \zeta\rangle + \int_0^t (\langle X(s), A^*\zeta\rangle + \langle f(s), \zeta\rangle)ds$$
$$+ \int_0^t \langle \zeta, B(X(s))dW(s)\rangle, \quad \mathbb{P}\text{-a.s.}$$

We also need the concept of the so called *mild solution* of (6.1). An H-valued predictable process $X(t)$, $t \in [0, T]$, is said to be a *mild solution* to (6.1) if X takes values in $D(B)$, $\mathbb{P}$-a.s., (6.6) and (6.7) hold and for arbitrary $t \in [0, T]$,

$$X(t) = S(t)\xi + \int_0^t S(t-s)f(s)ds + \int_0^t S(t-s)B(X(s)dW(s).$$

It is clear that a strong solution is also a weak solution. We will show now that a weak solution is also a mild solution. To do so we need some preliminary results on the *stochastic convolution* defined as,

$$W_A^\Phi(t) = \int_0^t S(t-s)\Phi(s)dW(s), \quad t \in [0, T], \ \Phi \in \mathcal{N}_W^2. \tag{6.8}$$

Proposition 6.2 *Assume that $A : D(A) \subset H \to H$ is the infinitesimal generator of a C_0 semigroup $S(\cdot)$ in H. Then for arbitrary $\Phi \in \mathcal{N}_W^2$ the process $W_A^\Phi(\cdot)$ has a predictable version.*

Proof By Proposition 4.31, for arbitrary $a > 0, b > 0$ and $t \in [0, T]$

$$\mathbb{P}(|W_A^\Phi(t)| > a) \le \frac{b}{a^2} + \mathbb{P}\left(\int_0^t \|S(t-s)\Phi(s)\|_{L_2^0}^2 ds > b\right).$$

If M is a constant such that $\|S(t)\| \le M$ for $t \in [0, T]$ then

$$\mathbb{P}(|W_A^\Phi(t)| > a) \le \frac{b}{a^2} + \mathbb{P}\left(\int_0^t \|\Phi(s)\|_{L_2^0}^2 ds > \frac{b}{M^2}\right). \tag{6.9}$$

Now, if Φ is an elementary process, then $W_A^\Phi(\cdot)$ has a predictable version by an easy inductive application of Theorem 5.2(i). If $\Phi \in \mathcal{N}_W^2$ is general then there exists a sequence $\{\Phi_n\}$ of elementary processes such that for arbitrary $c > 0$,

$$\mathbb{P}\left(\int_0^T \|\Phi(s) - \Phi_n(s)\|_{L_2^0}^2 ds > c\right) \to 0,$$

as $n \to \infty$. Inequality (6.9) implies that there exists a subsequence $\{\Phi_{n_k}\}$ of $\{\Phi_n\}$ such that, again for arbitrary $c > 0$,

$$\sup_{t\in[0,T]} \mathbb{P}\left(\|W_A^{\Phi}(t) - W_A^{\Phi_{n_k}}(t)\| > c\right) \to 0$$

as $k \to \infty$.

Now repeating the argument from the proof of Proposition 3.7(ii) one gets the required result. □

Proposition 6.3 *Assume that $A : D(A) \subset H \to H$ is the infinitesimal generator of a C_0-semigroup $S(\cdot)$ in H, that $\Phi \in \mathcal{N}_W^2$ and X is an H-valued predictable process with integrable trajectories. If for arbitrary $t \in [0, T]$ and $\zeta \in D(A^*)$,*

$$\langle X(t), \zeta\rangle = \int_0^t \langle X(s), A^*\zeta\rangle + \int_0^t \langle \zeta, \Phi(s)dW(s)\rangle, \quad \mathbb{P}\text{-a.s.,} \tag{6.10}$$

then we have $X(\cdot) = W_A^{\Phi}(\cdot)$.

Proof The proof is similar to the uniqueness part of Theorem 5.4. We show first that if (6.10) holds then for arbitrary $\zeta(\cdot) \in C^1([0, T]; D(A^*))$ and $t \in [0, T]$,

$$\langle X(t), \zeta(t)\rangle = \int_0^t \langle \zeta(s), \Phi(s)dW(s)\rangle + \int_0^t \langle X(s), \zeta'(s) + A^*\zeta(s)\rangle ds, \quad \mathbb{P}\text{-a.s.}$$

Applying this equality to the function $\zeta(s) = S^*(t - s)\zeta,\ s \in [0, t]$, and using the fact that $\Phi \in \mathcal{N}_W^2$, and the $D(A^*)$ is dense in H, we get (6.8). □

Conversely we have the following result.

Proposition 6.4 *Assume that $A : D(A) \subset H \to H$ is the infinitesimal generator of a C_0 semigroup $S(\cdot)$ in H and that $\Phi \in \mathcal{N}_W^2$.*

(i) The stochastic convolution W_A^{Φ} satisfies equation (6.10).
(ii) If in addition $\Phi(\cdot,\cdot)(U_0) \subset D(A)$, $\mathbb{P}_T$-a.s. and $A\Phi \in \mathcal{N}_W^2$, then we have:

$$W_A^{\Phi}(t) = \int_0^t AW_A^{\Phi}(s)ds + \int_0^t \Phi(s)dW(s). \tag{6.11}$$

In particular (6.11) *holds if A is bounded.*

Proof To prove (i), let $\zeta \in D(A^*)$ and note that $W_A^{\Phi}(\cdot)$ has integrable trajectories and

$$\int_0^t \langle W_A^{\Phi}(s), A^*\zeta\rangle ds = \int_0^t ds \int_0^s \langle S(s-\sigma)\Phi(\sigma)dW(\sigma), A^*\zeta\rangle. \tag{6.12}$$

It is straightforward to see that the stochastic Fubini theorem (Theorem 4.33) is applicable to the right hand side of (6.12) and yields,

$$\int_0^t ds \int_0^s \langle S(s-\sigma)\Phi(\sigma)dW(\sigma), A^*\zeta\rangle = \left\langle \int_0^t (S(t-\sigma) - I)\Phi(\sigma)dW(\sigma), \zeta\right\rangle$$

because

$$A\int_{\sigma}^{t} S(s-\sigma)ds = S(t-\sigma) - I.$$

We have

$$\int_0^t \langle W_A^{\Phi}(s), A^*\zeta\rangle ds = \int_0^t \langle S(t-s)\Phi(s)dW(s), \zeta\rangle - \int_0^t \langle \Phi(s)dW(s), \zeta\rangle$$
$$= \langle W_A^{\Phi}(t), \zeta\rangle - \int_0^t \langle \zeta, \Phi(s)dW(s)\rangle,$$

and (6.10) is proved. In order to prove (ii) note that, if A is a bounded operator, then identity (6.11) follows immediately from (6.10). Consequently if $A_n = nAR(n, A) = AJ_n$ are the Yosida approximations [(1)] of A and

$$W_{A,n}^{\Phi}(t) = \int_0^t e^{(t-s)A_n}\Phi(s)dW(s), \qquad n \in \mathbb{N},\ t \geq 0,$$

then

$$W_{A,n}^{\Phi}(t) = \int_0^t A_n W_{A,n}^{\Phi}(s)ds + \int_0^t \Phi(s)dW(s)), \qquad n \in \mathbb{N},\ t \geq 0.$$

Moreover, by Proposition 4.30

$$A_n W_{A,n}^{\Phi}(t) = J_n \int_0^t e^{(t-s)A_n} A\Phi(s)dW(s), \qquad n \in \mathbb{N},\ t \geq 0.$$

Now, by a direct computation, and by the Lebesgue dominated convergence theorem we have for any $T > 0$,

$$\lim_{n\to\infty} \sup_{t\in[0,T]} \mathbb{E}\left|W_{A,n}^{\Phi}(t) - W_A^{\Phi}(t)\right|^2 = 0$$
$$\lim_{n\to\infty} \sup_{t\in[0,T]} \mathbb{E}\left|A_n W_{A,n}^{\Phi}(t) - AW_A^{\Phi}(t)\right|^2 = 0.$$

This easily implies the result. □

Taking into account Propositions 6.2, 6.3 and 6.4 one gets the following result.

Theorem 6.5 *Assume that $A : D(A) \subset H \to H$ is the infinitesimal generator of a C_0 semigroup $S(\cdot)$ in H and that $\Phi \in \mathcal{N}_W^2$. Then a strong solution is a weak solution and a weak solution is always a mild solution of problem* (6.1). *Conversely if X is a mild solution of* (6.1) *and*

$$\mathbb{E}\int_0^T \|B(X(s))\|_{L_2^0}^2 ds < +\infty,$$

then X is also a weak solution of (6.1).

[(1)] See Appendix A.2 for basic proprties of A_n and J_n.

Remark 6.6 By the Hille–Yosida theorem (see Theorem A.2), there exist ω, $M \in \mathbb{R}^1$ such that:

$$\|S(t)\| \leq Me^{\omega t}, \quad \forall\, t \geq 0.$$

For $\lambda \in \mathbb{R}^1$ define $\widetilde{S}(t) = e^{-\lambda t}S(t)$, $\widetilde{f}(t) = e^{-\lambda t}f(t)$, $t \geq 0$, and assume that X is a solution of (6.1) and set $\widetilde{X}(t) = e^{-\lambda t}X(t)$. Then $\widetilde{X}$ is a solution to

$$\widetilde{X}(t) = \widetilde{S}(t)x + \int_0^t \widetilde{S}(t-s)f(s)ds \int_0^t \widetilde{S}(t-s)B(\widetilde{X}(s))ds.$$

Since $\widetilde{S}(\cdot)$ is the semigroup generated by $A - \lambda I$, for existence and regularity purposes, we can always assume that $\omega \leq 0$. □

6.1.1 The case when B is bounded

Although the case of the operator B bounded will be studied in a more generalí setting in Chapter 7, we discuss it here briefly, for illustration purposes. We first consider mild solutions.

Theorem 6.7 *Assume that A is the infinitesimal generator of a C_0-semigroup $S(\cdot)$ in H, $\mathbb{E}|\xi|^2 < +\infty$ and $B \in L(H; L_2^0)$. Then equation (6.1), has a unique mild solution $X \in \mathcal{N}_W^2(0,T;H)$, identical with a weak solution.*

Proof Denote by $\mathcal{H}$ the space of all H-valued predictable processes Y such that $|Y|_{\mathcal{H}} = \sup_{t\in[0,T]} \mathbb{E}|Y(t)|^2 < +\infty$ and for any Y define

$$\mathcal{K}(Y)(t) = S(t)\xi + \int_0^t S(t-s)f(s)ds + \int_0^t S(t-s)B(Y(s))dW(s), \quad t \in [0,T],$$

and

$$\mathcal{K}_1(Y)(t) = \int_0^t S(t-s)B(Y(s))dW(s), \quad t \in [0,T].$$

We can assume, see Remark 6.6, that $\|S(t)\| \leq M$, $t \geq 0$, and we have

$$\begin{aligned} |\mathcal{K}_1(Y)|_{\mathcal{H}} &\leq \sup_{t\in[0,T]} \mathbb{E}\left(\int_0^t \|S(t-s)B(Y(s)))\|^2_{L_2^0} ds\right)^{1/2} \\ &\leq M\|B\|_{L(H;L_2^0)}\sqrt{T}|Y|_{\mathcal{H}}, \quad t \in [0,T]. \end{aligned}$$

So, if T is sufficiently small, $\mathcal{K}$ is a contraction and it is easy to see that its unique fixed point can be identified as the solution to (6.1). The case of general T can be handled in a standard way. □

We now give an existence result on strong solutions.

Proposition 6.8 *Assume that the hypotheses of Theorem* 6.7 *hold,* $\xi = x \in D(A)$ *and* $f \equiv 0$*. Let moreover* $0 \in \rho(A)$ *and assume that* B_A*, given by*

$$B_A(x)u = AB(A^{-1}x)u, \quad x \in H,\ u \in U,$$

belongs to $L(H; L_2^0)$*. Then the equation* (6.1) *has a unique strong solution.*

Proof Let $x \in D(A)$, and let X and Y be the mild solutions of (6.1) and

$$\begin{cases} dY(t) = AY(t)dt + AB(A^{-1}Y(t))dW(t), \\ Y(0) = Ax, \end{cases} \tag{6.13}$$

which exists by Theorem 6.7. Consider the approximating problems

$$\begin{cases} dX_n(t) = A_nX_n(t)dt + B(X_n(t))dW(t), \\ X_n(0) = x, \end{cases}$$

and

$$\begin{cases} dY_n(t) = A_nY_n(t)dt + A_nB(A_n^{-1}Y_n(t))dW(t), \\ Y_n(0) = Ax, \end{cases}$$

where A_n are the Yosida approximations of A. We have clearly

$$Y_n(t) = A_nX_n(t), \quad t \geq 0,\ n \in \mathbb{N}$$

and so

$$X_n(t) = x + \int_0^t Y_n(s)ds + \int_0^t B(X_n(s))dW(s), \quad t \geq 0,\ n \in \mathbb{N}. \tag{6.14}$$

Moreover, it is easy to check that $X_n \to X$ and $Y_n \to Y$, as $n \to \infty$, in $\mathcal{N}_W^2(0, T; H)$. But the operator A is closed and this implies $X(t) \in D(A)$, a.s. and $Y(t) = AX(t)$. Now, letting n tend to infinity in (6.14), the conclusion follows. □

Example 6.9 (Stochastic wave equation) Let $\mathcal{O}$ be a bounded set of $\mathbb{R}^N$ with regular boundary $\partial\mathcal{O}$. Consider the wave equation with Dirichlet boundary conditions

$$\begin{cases} dy_t(t,\xi) = \Delta_\xi y(t,\xi)dt + a(\xi)\cdot\nabla_\xi y(t,\xi)d\beta(t), & t \in \mathbb{R}^1,\ \xi \in \mathcal{O}, \\ y(t,\xi) = 0, \quad t > 0,\ \xi \in \partial\mathcal{O}, \\ y(0,\xi) = x_0(\xi), \quad y_t(0,\xi) = x_1(\xi),\ \xi \in \mathcal{O}, \end{cases} \tag{6.15}$$

where $a \in C^1(\overline{\mathcal{O}})$ and $\beta(\cdot)$ is a one dimensional standard Wiener process.

According to notations introduced in Section A.5.4, let Λ be the positive self-adjoint operator defined by (A.63), let $H = D(\Lambda^{1/2}) \oplus L^2(\overline{\mathcal{O}})$ and let A be defined by (A.64). Moreover set $U = \mathbb{R}^1$ and define the operator $B \in L(H)$ by

$$B\begin{pmatrix} y \\ z \end{pmatrix} = \begin{pmatrix} 0 \\ a \cdot \nabla_\xi y \end{pmatrix} = \begin{pmatrix} 0 \\ B_1 y \end{pmatrix}.$$

Now, setting $X(t) = \begin{pmatrix} y(t) \\ y_t(t) \end{pmatrix}$, problem (6.15) reduces to problem (6.1), and a mild solution is a solution of the following integral equation

$$\begin{aligned} y(t,\cdot) &= \cos(\sqrt{\Lambda}t)x_0 + \frac{1}{\sqrt{\Lambda}}\sin(\sqrt{\Lambda}t)x_1 \\ &+ \int_0^t \frac{1}{\sqrt{\Lambda}}\sin(\sqrt{\Lambda}(t-s))B_1 y(s,\cdot)d\beta(s). \end{aligned} \tag{6.16}$$

By Theorem 6.7 it follows that, given $x_0 \in H_0^1(\mathscr{O})$ and $x_1 \in L^2(\mathscr{O})$, there exists a unique predictable solution y of (6.16) such that

(i) $y \in \mathcal{N}_W^2(0, T; H_0^1(\mathscr{O}))$,
(ii) $y_t \in \mathcal{N}_W^2(0, T; L^2(\mathscr{O}))$.

It will follow from the next section that the process Y is continuous as a $H^1(\mathscr{O})$-valued process, see Proposition 6.11 below. □

6.2 Stochastic convolution for contraction semigroups

In many circumstances weak and mild solutions coincide, see Theorem 6.7. Therefore for existence results it is of vital importance to study the properties of the mapping $\Phi \to W_A^\Phi$. With strong regularity properties of the mapping at hand, several existence and uniqueness results can be obtained by fixed point argument.

A generalization of the maximal inequality of martingales to stochastic convolution, when A generates a contraction semigroup, was obtained in [453]. From this result the continuity of a suitable modification follows. The next result was obtained independently in [679] and [454]. A simple proof of a maximal inequality in L^p spaces was obtained in [397]. Finally, another way to prove maximal inequalities is by using the *Factorization* method, see Section 5.3.1.

Theorem 6.10 *Assume that A generates a contraction semigroup and let $\Phi \in \mathcal{N}_W^2(0, T; L_2^0)$. Then the process $W_A^\Phi(\cdot)$ has a continuous modification and there exists a constant K such that*

$$\mathbb{E} \sup_{s\in[0,t]} |W_A^\Phi(s)|^2 \le K\mathbb{E}\int_0^t \|\Phi(s)\|_{L_2^0}^2 ds, \quad t \in [0, T]. \tag{6.17}$$

Proof To simplify the notation denote W_A^Φ by Z. Assume first that Φ is an elementary process such that $A\Phi \in \mathcal{N}_W^2(0, T; L_2^0)$. By Proposition 6.4 $Z(\cdot)$ has continuous trajectories, moreover the process $AZ(\cdot)$ has square integrable trajectories and

$$dZ(t) = AZ(t)dt + \Phi(t)dW(t), \quad t \in [0, T].$$

By Itô's formula

$$d|Z(t)|^2 = 2\langle Z(t), AZ(t)\rangle dt + 2\langle Z(t), \Phi(t)dW(t) + \|\Phi(s)\|^2_{L_2^0} ds. \quad (6.18)$$

Since A generates a contraction semigroup, therefore $\langle Ax, x\rangle \le 0$ for all $x \in D(A)$ (see Example D.8) and from (6.18) we deduce

$$|Z(t)|^2 \le 2\int_0^t \langle Z(s), \Phi(s)dW(s)\rangle + \int_0^t \|\Phi(s)\|^2_{L_2^0} ds.$$

In particular

$$\sup_{s\in[0,t]} |Z(s)|^2 \le 2 \sup_{s\in[0,t]} \left|\int_0^s \langle Z(\sigma), \Phi(\sigma)dW(\sigma)\rangle\right| + \int_0^t \|\Phi(s)\|^2_{L_2^0} ds,$$

and so

$$\mathbb{E} \sup_{s\in[0,t]} |Z(s)|^2 \le 2\mathbb{E} \sup_{s\in[0,t]} \left|\int_0^s \langle Z(r), \Phi(r)dW(r)\rangle\right| + \mathbb{E}\int_0^t \|\Phi(s)\|^2_{L_2^0} ds. \quad (6.19)$$

However, from Theorem 4.27 the process

$$M(t) = \int_0^t \langle Z(s), \Phi(s)dW(s)\rangle, \quad t \in [0, T],$$

is a square integrable martingale with quadratic variation

$$\langle\langle M(t)\rangle\rangle = \int_0^t |(\Phi(s)Q^{1/2})^* Z(s)|^2 ds, \quad t \in [0, T].$$

By Theorem 3.15

$$\mathbb{E} \sup_{s\in[0,t]} |M(s)| \le 3\mathbb{E}\left(\int_0^t |(\Phi(s)Q^{1/2})^* Z(s)|^2 ds\right)^{1/2}. \quad (6.20)$$

Taking into account (6.19) and (6.20) we have

$$\begin{aligned}
\mathbb{E} \sup_{s\in[0,t]} |Z(s)| &\le 6\mathbb{E}\left(\int_0^t |(\Phi(s)Q^{1/2})^* Z(s)|^2 ds\right)^{1/2} + \mathbb{E}\int_0^t \|\Phi(s)\|^2_{L_2^0} ds \\
&\le 6\mathbb{E}\left(\int_0^t \|\Phi(s)\|^2_{L_2^0}|Z(s)|^2 ds\right)^{1/2} + \mathbb{E}\int_0^t \|\Phi(s)\|^2_{L_2^0} ds \\
&\le 6\left(\mathbb{E} \sup_{s\in[0,t]} |Z(s)|^2\right)^{1/2}\left(\mathbb{E}\int_0^t \|\Phi(s)\|^2_{L_2^0} ds\right)^{1/2} + \mathbb{E}\int_0^t \|\Phi(s)\|^2_{L_2^0} ds.
\end{aligned}$$

For arbitrary $\varepsilon > 0$ it follows that

$$\mathbb{E} \sup_{s\in[0,t]} |Z(s)|^2 \le 12\left(\varepsilon\mathbb{E} \sup_{s\in[0,t]} |Z(s)|^2 + \frac{1}{\varepsilon}\mathbb{E}\int_0^t \|\Phi(s)\|^2_{L_2^0} ds\right) + \mathbb{E}\int_0^t \|\Phi(s)\|^2_{L_2^0} ds.$$

In particular, taking $\varepsilon = 1/24$ one obtains (6.17) for the considered class of elementary processes and the theorem is true in this special case. One can now use (6.17) to prove the theorem in the general case by a standard approximation argument. □

Here is a direct application of Theorem 6.10 to which we referred in Example 6.9.

Proposition 6.11 *If the conditions of Theorem* 6.10 *are fulfilled and* $S(\cdot)$ *is a contraction semigroup, then the solution* X *of* (6.1) *has a continuous modification.*

6.3 Stochastic convolution for analytic semigroups

We can obtain better regularity of the stochastic convolution $W_A^\Phi(\cdot)$ if we consider analytic semigroups. A first result, giving only continuity of $W_A^\Phi(\cdot)$, but with $A = A(t),\ t \in [0, T]$, was proved in [203].

We study here the maximal regularity result of $W_A^\Phi(\cdot)$ when $S(\cdot)$ is a general analytic semigroup by showing that, in a sense, the regularity one can obtain is the half of the corresponding regularity for the deterministic case, see Proposition A.26. This result was obtained in [186], see also [187], and it is maximal, see [186, Example 3.8]. When A is variational the quoted result is contained in [575]. Finally, a particular case, with A self-adjoint was studied by semigroups methods in [412].

We remark that equations of the form (6.1) with an unbounded operator B arise also in the abstract semigroup formulation of a general class of stochastic delay equations with delays in the coefficients of the noise. The techniques developed in the previous sections do not apply here (for instance, the semigroup $S(\cdot)$ associated to delay systems in the Hilbert space $\mathbb{R}^d \times L^2(-r, 0; \mathbb{R}^d)$ is not analytic or contraction). However, it is possible to give a very general regularity result for the stochastic convolution, which allows us to solve, in a generalized sense equation (6.1), see [295].

6.3.1 General results

Here we assume the following.

Hypothesis 6.1 *A generates an analytic semigroup of negative type.*

We denote by $D_A(\theta, 2)$ the real interpolation space, between $D(A)$ and X, see (A.33) and (A.34). Let $\{e_k\}$ be an orthonormal basis in U_0 and set

$$\varphi_k(s) = \Phi(s)e_k, \quad s \in [0, T],\ k \in \mathbb{N}.$$

For any Hilbert space K let $\mathcal{N}_W^2(0, T; K)$ denote the space of all K-valued predictable processes X such that

$$\|X\|^2_{\mathcal{N}_W^2(0,T;K)} = \int_0^T \mathbb{E}\|X(s)\|_K^2 ds < +\infty.$$

If $K = D_A(\theta, 2)$ and $T = +\infty$ then we write shortly $\|\cdot\|_\theta$ instead of $\|\cdot\|_{\mathcal{N}_W^2(0,+\infty;D_A(\theta,2))}$. In a similar way $\|\Phi\|_\theta$ stands for $\|\Phi\|_{\mathcal{N}_W^2(0,+\infty;L_2^{0,\theta})}$, where $L_2^{0,\theta} = L_2(U_0; D_A(\theta, 2))$. We notice that $\|\Phi\|_\theta$ is given by the formula

$$\|\Phi\|_\theta = \sum_{k=1}^{\infty} \int_0^{\infty} \mathbb{E}|\varphi_k(s)|_\theta^2 \, ds. \tag{6.21}$$

Important information is provided by the following theorem.

Theorem 6.12 *Assume Hypothesis* 6.1 *and that* $\Phi \in \mathcal{N}_W^2(0, +\infty; L_2^{0,\theta})$.

(i) If $\theta \in (0, \frac{1}{2})$ *then* $W_A^\Phi \in \mathcal{N}_W^2(0, +\infty; D_A(\theta + \frac{1}{2}, 2))$ *and*

$$\|W_A^\Phi\|^2_{\theta+1/2} = \frac{1}{1-2\theta} \|\Phi\|_\theta^2.$$

(ii) If $\theta = \frac{1}{2}$ *then* $W_A^\Phi \in \mathcal{N}_W^2(0, +\infty; D(A))$, *and*

$$\|AW_A^\Phi\|^2 = \|\Phi\|^2_{1/2}.$$

(iii) If $\theta \in (\frac{1}{2}, 1)$ *then* $AW_A^\Phi \in \mathcal{N}_W^2(0, +\infty; D_A(\theta - \frac{1}{2}, 2))$ *and*

$$\|AW_A^\Phi\|^2_{\theta-1/2} = \frac{1}{3-2\theta} \|\Phi\|_\theta^2.$$

In particular, for $t \geq 0$, $W_A^\Phi \in D(A)$, $\mathbb{P}$*-a.s.*

Proof (i) We have

$$\begin{aligned}\|W_A^\Phi\|^2_{\theta+1/2} &= \sum_{k=1}^{\infty} \mathbb{E} \int_0^{\infty} dt \int_0^{\infty} d\xi \xi^{-2\theta} \int_0^t |AS(t-s+\xi)\varphi_k(s)|^2 ds \\ &= \sum_{k=1}^{\infty} \mathbb{E} \int_0^{\infty} ds \int_0^{\infty} d\xi \xi^{-2\theta} \int_s^{\infty} |AS(t-s+\xi)\varphi_k(s)|^2 dt.\end{aligned}$$

By setting $t - s + \xi = \tau$, we obtain

$$\begin{aligned}\|W_A^\Phi\|^2_{\theta+1/2} &= \sum_{k=1}^{\infty} \mathbb{E} \int_0^{\infty} ds \int_0^{\infty} d\xi \xi^{-2\theta} \int_\xi^{\infty} |AS(\tau)\varphi_k(s)|^2 d\tau \\ &= \sum_{k=1}^{\infty} \mathbb{E} \int_0^{\infty} ds \int_0^{\infty} d\tau |AS(\tau)\varphi_k(s)|^2 \int_0^{\tau} d\tau \xi^{-2\theta} d\xi \\ &= \frac{1}{1-2\theta} \sum_{k=1}^{\infty} \mathbb{E} \int_0^{\infty} |\varphi_k(s)|_\theta^2 = \frac{1}{1-2\theta} \sum_{k=1}^{\infty} \mathbb{E}|\varphi_k(s)|_\theta^2.\end{aligned}$$

(ii) We have

$$
\begin{aligned}
\|AW_A^\Phi\|^2 &= \sum_{k=1}^\infty \mathbb{E}\int_0^\infty ds \int_s^\infty |AS(t-s)\varphi_k(s)|^2 dt \\
&= \sum_{k=1}^\infty \mathbb{E}\int_0^\infty ds \int_0^\infty |AS(\tau)\varphi_k(s)|^2 d\tau \\
&= \sum_{k=1}^\infty \mathbb{E}\int_0^\infty ds |\varphi_k(s)|_{1/2}^2.
\end{aligned}
$$

(iii) The proof is completely analogous to the proof of (i). □

As a direct corollary of Theorem 6.10 and the fact that the restriction of $S(\cdot)$ to $D_A(\theta, 2)$ is a contraction semigroup (see comments after (A.34)) we obtain the following.

Proposition 6.13 *If $\Phi \in \mathcal{N}_W^2(0, +\infty; L_2^{0,\theta})$ for some $\theta \in (0, 1)$, then W_A^Φ has a continuous version in $D_A(\theta, 2)$.*

In order to study the limit case $\theta = 0$, we need an additional assumption.

Hypothesis 6.2 *$D((-A)^\varepsilon)$ is isomorphic to $D_A(\varepsilon, \infty)$ for all $\varepsilon \in (0, 1)$.*

Hypothesis 6.2 is fulfilled in several situations, for example when $S(\cdot)$ is a contraction semigroup and, in particular, when A is a variational or a self-adjoint operator (see Remark A.20). We now prove the following theorem.

Theorem 6.14 *Assume Hypotheses 6.1 and 6.2 and let $\Phi \in \mathcal{N}_W^2(0, \infty; L_0^2)$. Then $W_A^\Phi \in \mathcal{N}_W^2(0, \infty; D_A(1/2, 2))$.*

Proof We first remark that, by the hypotheses, there exists a constant $C_\theta > 1, \theta \in (0, 1)$ such that

$$
\frac{1}{C_\theta}|(-A)^\theta x| \le |x|_\theta \le C_\theta |(-A)^\theta x|.
$$

Now fix $\varepsilon \in (0, 1/2)$. If $\Phi \in \mathcal{N}_W^2(0, \infty; H)$ we have:

$$
\begin{aligned}
\|W_A^\Phi\|_{1/2} = \|(-A)^\varepsilon W_A^{(-A)^{-\varepsilon}\Phi}\|_{1/2} &\le C_{1/2}\|(-A)^{\varepsilon+1/2} W_A^{(-A)^{-\varepsilon}\Phi}\| \\
&\le C_{1/2}C_{1/2+\varepsilon}\|W_A^{(-A)^{-\varepsilon}\Phi}\|_{1/2+\varepsilon}.
\end{aligned}
$$

By Theorem 6.12 it follows that

$$
\|W_A^\Phi\|_{1/2} \le \frac{C_{1/2}C_{1/2+\varepsilon}}{1-2\varepsilon}\|(-A)^\varepsilon\Phi\|_\varepsilon \le \frac{C_{1/2}C_{1/2+\varepsilon}C_\varepsilon}{1-2\varepsilon}\|\Phi\|,
$$

and the theorem is proved. □

Remark 6.15 Assume Hypothesis 6.2. Theorem 6.14 does not give any precise information about the constant:

$$\lambda_0 = \sup_{\Phi \in \mathcal{N}_W^2(0,\infty;H)} \frac{\| W_A^\Phi \|_{1/2}}{\|\Phi\|}. \tag{6.22}$$

As we will see in Section 6.5, knowledge of λ_0 is very important in order to solve stochastic equations. Thus it is convenient to study separately the cases when A is a variational operator or self-adjoint, because in these cases a characterization of λ_0 is possible. This will be done in the next subsections. □

6.3.2 Variational case

We consider here the framework of Section A.4.2.

Proposition 6.16 *Assume that A is a variational operator and let $\Phi \in \mathcal{N}_W^2(0, T; L_2^0)$. Then $W_A^\Phi \in \mathcal{N}_W^2(0, T; V)$ and*

$$\| W_A^\Phi \|^2_{\mathcal{N}_W^2(0,T;V)} \le \frac{1}{2} \| \Phi \|^2_{\mathcal{N}_W^2(0,T;L_2^0)}. \tag{6.23}$$

Proof Set

$$W_{A,n}^\Phi(t) = nR(n, A) W_A^\Phi(t), \quad \varphi_{k,n}(t) = nR(n, A)\varphi_k(t),$$

where $\varphi_k = \Phi e_k$. Since $\varphi_{k,n}(t) \in D(A)$ we have

$$\begin{aligned}\mathbb{E}\langle A W_{A,n}^\Phi(t), W_{A,n}^\Phi(t)\rangle &= \sum_{k=1}^\infty \int_0^t \langle AS(t-s)\varphi_{k,n}(s), S(t-s)\varphi_{k,n}(s)\rangle ds \\ &= \frac{1}{2} \sum_{k=1}^\infty \int_0^t \left[\frac{d}{dt} \left|S(t-s)\varphi_{k,n}(s)\right|^2\right] ds.\end{aligned}$$

By integrating in $[0, T]$ and by exchanging integrals, it follows that

$$\begin{aligned}\mathbb{E}\langle A W_{A,n}^\Phi(t), W_{A,n}^\Phi(t)\rangle &= \frac{1}{2} \sum_{k=1}^\infty \int_0^T ds \int_s^t \left[\frac{d}{dt} \left|S(t-s)\varphi_{k,n}(s)\right|^2\right] ds \\ &= \frac{1}{2} \sum_{k=1}^\infty \int_0^T \left[\left|S(t-s)\varphi_{k,n}(s)\right|^2 - \left|\varphi_{k,n}(s)\right|^2\right] ds.\end{aligned}$$

Thus

$$\mathbb{E} \int_0^T \left|W_{A,n}^\Phi(s)\right|_V^2 ds \le \frac{1}{2} \sum_{k=1}^\infty \int_0^T |\varphi_{k,n}(s)|^2 ds,$$

and the conclusion follows for $n \to \infty$. □

6.3.3 Self-adjoint case

We assume here the following hypothesis.

Hypothesis 6.3 $A : D(A) \subset H \to H$ *is selfadjoint negative and denote by* $\{E_\lambda\}_{\lambda \in R^1}$ *the spectral family associated to* A *(see for instance [721]).*

Lemma 6.17 *For all* $\theta \in (0, 1)$ *we have*

$$|x|_\theta^2 = 2^{2\theta-2}\Gamma(2-2\theta)|(-A)^\theta x|^2, \quad \forall\, x \in D_A(\theta, 2). \tag{6.24}$$

Proof For all $\xi > 0$ and $x \in H$, we have

$$|AS(\xi)x|^2 = \int_{-\infty}^0 \lambda^2 e^{2\lambda\xi} d|E_\lambda x|^2, \tag{6.25}$$

it follows that

$$\begin{aligned}
|x|_\theta^2 &= \int_{-\infty}^0 d|E_\lambda x|^2 \int_0^\infty \xi^{1-2\theta}\lambda^2 e^{2\lambda\xi} d\xi \\
&= 2^{2\theta-2}\Gamma(2-2\theta)\int_{-\infty}^0 \lambda^{2\theta} d|E_\lambda x|^2 \\
&= 2^{2\theta-2}\Gamma(2-2\theta)|(-A)^\theta x|^2.
\end{aligned} \tag{6.26}$$

□

Proposition 6.18 *Assume Hypothesis* 6.3 *and that, for some* $\theta \geq 0$, $(-A)^\theta \Phi \in \mathcal{N}_W^2(0, \infty; L_2^0)$. *Then* $W_A^\Phi \in \mathcal{N}_W^2(0, \infty; D_A(\theta + \frac{1}{2}; 2)$ *and*

$$\|(-A)^{\theta+1/2} W_A^\Phi\|^2 = \frac{1}{2}\sum_{k=1}^\infty \mathbb{E}|(-A)^\theta \varphi_k(s)|^2. \tag{6.27}$$

Proof Write

$$\begin{aligned}
\|(-A)^{\theta+\frac{1}{2}} W_A^\Phi\|^2 &= \sum_{k=1}^\infty \mathbb{E}\int_0^\infty dt \int_0^t ds \int_{-\infty}^0 (-\lambda)^{2\theta+1} e^{2\lambda(t-s)} dE_\lambda|\varphi_k(s)|^2 \\
&= \sum_{k=1}^\infty \mathbb{E}\int_0^\infty ds \int_{-\infty}^0 dE_\lambda|\varphi_k(s)|^2 \int_s^\infty (-\lambda)^{2\theta+1} e^{2\lambda(t-s)} dt \\
&= \sum_{k=1}^\infty \mathbb{E}\int_0^\infty ds \int_{-\infty}^0 (-\lambda)^{2\theta} dE_\lambda|\varphi_k(s)|^2 \\
&= \frac{1}{2}\sum_{k=1}^\infty \mathbb{E}|(-\lambda)^\theta \varphi_k(s)|^2.
\end{aligned}$$

□

6.4 Maximal regularity for stochastic convolutions in L^p spaces

We fix $U = \mathbb{R}^1 = U_0 = U_1$ for simplicity and consider a one dimensional Brownian motion in $(\Omega, \mathscr{F}, \mathbb{P})$ adapted to the filtration $(\mathscr{F}_t)_{t\geq 0}$. We denote by $\mathscr{N}_W^p(0, T; H)$, $p \geq 2$, the space of all H-valued predictable processes X such that

$$|X|^p_{\mathscr{N}_W^p(0,T;H)} = \int_0^T \mathbb{E}|X(s)|^p ds < +\infty.$$

We are here concerned with maximal regularity of the stochastic convolution

$$W_A(t) = \int_0^t e^{(t-s)A}\varphi(s)dW(s), \quad t \geq 0,$$

where $A : D(A) \subset X \mapsto X$ is the generator of an analytic semigroup e^{tA} in a Banach space X and $\varphi \in \mathscr{N}_W^p(0, T; H)$.

In this section we shall generalize, following [206], some of the previous regularity results to a wide class of Banach spaces, $p \geq 1$, and $0 \leq \theta < 1$. The class of Banach spaces that we consider are those spaces where the Burkholder inequality (see next section) holds. Such an inequality is known to be true in the 2-uniformly smooth Banach spaces, and so, in particular, in the Lebesgue spaces $L^q(\mathbb{R}^d)$ and in the Sobolev spaces $W^{k,q}(\mathbb{R}^d)$, $q \geq 2$. This follows from [601, Proposition 2.4] and [490, Lemma 1.1]).

In the case where A is the realization of a second order elliptic operator with regular coefficients in $L^p(\mathbb{R}^d)$ with $p \geq 2$, Krylov [466] proved that if $\varphi \in \mathscr{N}_W^p(0, T; W^{1,p}(\mathbb{R}^d))$ then $W_A \in \mathscr{N}_W^p(0, T; W^{2,p}(, R^d)) = \mathscr{N}_W^p(0, T; D(A))$. The method presented here does not allow us to prove such a result, since for $p \neq 2$, $W^{1,p}(\mathbb{R}^d)$ is not a real interpolation space between $L^p(\mathbb{R}^d)$ and $D(A) = W^{2,p}(\mathbb{R}^d)$. We have in fact $D_A(1/2, p) = B^1_{p,p}(\mathbb{R}^d)$, and $D_A(1, p) = B^2_{p,p}(\mathbb{R}^d)$ ($B^1_{p,p}(\mathbb{R}^d)$ and $B^2_{p,p}(\mathbb{R}^d)$ are Besov spaces, see [677]), so that we get an optimal regularity result in Besov spaces rather than in Sobolev spaces.

6.4.1 Maximal regularity

We recall the definition and some properties of the interpolation spaces wich will be used in the following, referring to [677] for an extensive treatment of interpolation theory.

Let X be a Banach space with norm $\|\cdot\|$ and let $A : D(A) \subset X \mapsto X$ generate an analytic semigroup e^{tA} in X. For any $x \in X$, $\theta \in (0, 1)$, and $p \geq 1$ we set

$$|x|^p_{D_A(\theta,p)} = \int_0^1 \|\xi^{1-\theta} A e^{\xi A} x\|^p \frac{d\xi}{\xi},$$

and

$$|x|^p_{D_{A^2}(\theta,p)} = \int_0^1 \|\xi^{2(1-\theta)} A^2 e^{\xi A} x\|^p \frac{d\xi}{\xi}.$$

The interpolation spaces $D_A(\theta, p)$, $D_A(\theta+1, p)$ and $D_{A^2}(\theta, p)$ are defined by

$$D_A(\theta, p) = \{x \in X : |x|_{D_A(\theta,p)} < +\infty\}, \quad \|x\|_{D_A(\theta,p)} = \|x\| + |x|_{D_A(\theta,p)},$$

$$D_A(\theta+1, p) = \{x \in D(A) : Ax \in D_A(\theta, p)\}, \quad \|x\|_{D_A(\theta+1,p)} = \|x\| + \|Ax\|_{D_A(\theta,p)},$$

$$D_{A^2}(\theta, p) = \{x \in X : |x|_{D_{A^2}(\theta,p)} < +\infty\}, \quad \|x\|_{D_{A^2}(\theta,p)} = \|x\| + |x|_{D_{A^2}(\theta,p)}.$$

There is a different notation for $\theta = 0$. Assume for simplicity that $0 \in \rho(A)$, so that the seminorm

$$x \mapsto |x|_{D_A(0,p)} = \left(\int_0^1 \|\xi A e^{\xi A} x\|^p \frac{d\xi}{\xi}\right)^{1/p}$$

is in fact a norm. The space

$$X_0 = \{x \in X : |x|_{D_A(\theta,p)} < +\infty\},$$

endowed with the norm $|\cdot|_{D_A(0,p)}$, is not complete in general. So, $D_A(0, p)$ is defined as the completion of X_0 in the norm $|\cdot|_{D_A(0,p)}$. If $0 \in \rho(A)$, for all $\omega \in \mathbb{R}$ such that $A - \omega I$ is of negative type, the spaces $D_{A-\omega I}(0, p)$ are equivalent. In this case we set $D_A(0, p) = D_{A-\omega I}(0, p)$. The semigroup e^{tA} has a natural extension to $D_A(0, p)$, which will be still denoted e^{tA}, and it turns out to be an analytic semigroup. For a detailed treatment of the spaces $D_A(0, p)$ see [260].

In the proof of our result the next proposition will play a key role, see [677].

Proposition 6.19 *For every $p \geq 1$, $\theta \in [0, 3/2)$,*

$$D_A\left(\theta + \frac{1}{2}, p\right) = D_{A^2}\left(\frac{\theta}{2} + \frac{1}{4}, p\right), \tag{6.28}$$

holds with equivalence of the respective norms.

We will also need the Burkhölder inequality. A class of Banach spaces in which it holds are the 2-uniformly convex spaces, see [601].

From now on we assume that X and $p \geq 1$ are such that the Burkhölder inequality holds, that is, we assume the following hypothesis.

Hypothesis 6.4 *There exists $C_p > 0$ such that for all $\varphi \in \mathcal{N}_W^p(0, T; X)$ we have*

$$\mathbb{E}\left(\left\|\int_0^T \varphi(s) dW(s)\right\|^p\right) \leq C_p \mathbb{E}\left[\left(\int_0^T \|\varphi(s)\|^2 ds\right)^{p/2}\right]. \tag{6.29}$$

We are now able to state our main result of this section

Theorem 6.20 *Let X be a Banach space, $p \geq 1$ be such that Hypothesis 6.4 holds. Let $A : D(A) \subset X \mapsto X$ generate an analytic semigroup e^{tA} in X. Then for every $\theta \in [0, 1)$ and $\varphi \in \mathcal{N}_W^p(0, T; D_A(\theta, p))$, $W_A(\varphi) \in \mathcal{N}_W^p(0, T; D_A(\theta + \frac{1}{2}, p))$, and there exists K independent of φ such that*

$$\|W_A(\varphi)\|_{\mathcal{N}_W^p(0,T;D_A(\theta+\frac{1}{2},p))} \leq K \|\varphi\|_{\mathcal{N}_W^p(0,T;D_A(\theta,p))}.$$

Proof Recalling (6.28), we have only to estimate

$$J := \mathbb{E}\int_0^T dt \int_0^1 \left\| \xi^{3/2-\theta} A^2 e^{\xi A} \int_0^t e^{(t-s)A}\varphi(s)d\beta(s)\right\|^p \frac{d\xi}{\xi}$$

$$= \mathbb{E}\int_0^T dt \int_0^1 \left\| \xi^{3/2-\theta} \int_0^t A^2 e^{(t-s+\xi)A}\varphi(s)d\beta(s)\right\|^p \frac{d\xi}{\xi}. \tag{6.30}$$

Using the Burkholder inequality (6.29) we get

$$J \le C_p\mathbb{E}\int_0^T dt \int_0^1 \xi^{(3/2-\theta)p} \left(\int_0^t \|A^2 e^{(t-s+\xi)A}\varphi(s)\|^2 ds\right)^{p/2} \frac{d\xi}{\xi}.$$

Splitting $A^2 e^{(t-s+\xi)A} = Ae^{\frac{(t-s+\xi)}{2}A}Ae^{\frac{(t-s+\xi)}{2}A}$ and using the estimate

$$\|\sigma Ae^{\sigma A}\|_{L(X)} \le M$$

for $0 < \sigma < (T+1)/2$ we get

$$J \le C_p(2M)^p\mathbb{E}\int_0^T dt \int_0^1 \xi^{(3/2-\theta)p}$$
$$\times \left(\int_0^t \frac{1}{(t-s+\xi)^2}\|Ae^{\frac{(t-s+\xi)}{2}A}\varphi(s)\|^2 ds\right)^{p/2} \frac{d\xi}{\xi},$$

so that by Hölder's inequality

$$J \le c\mathbb{E}\int_0^T dt \int_0^1 \xi^{(3/2-\theta)p}\frac{1}{\xi^{p/2+1}} \int_0^t \|Ae^{\frac{(t-s+\xi)}{2}A}\varphi(s)\|^p ds \frac{d\xi}{\xi}.$$

Now, setting $\tau = \xi + t - s$ and exchanging integrals, we find

$$J \le c\mathbb{E}\int_0^T ds \int_0^{T+1-s} \|Ae^{\tau A/2}\varphi(s)\|^p d\tau \int_s^{\tau+s} (\tau - t + s)^{(1-\theta)p-2}dt$$
$$= c'\mathbb{E}\int_0^T ds \int_0^{T+1-s} \tau^{(1-\theta)p-1}\|Ae^{\tau A/2}\varphi(s)\|^p d\tau$$
$$\le c'\mathbb{E}\int_0^T ds \int_0^{T+1} \tau^{(1-\theta)p-1}\|Ae^{\tau A/2}\varphi(s)\|^p d\tau$$
$$= c'\mathbb{E}\int_0^T ds \left(\int_0^{\min(2,T+1)} + \int_{\min(2,T+1)}^{T+1}\right)\tau^{(1-\theta)p-1}\|Ae^{\tau A/2}\varphi(s)\|^p d\tau$$
$$\le c''\|\varphi\|_{\mathcal{N}_W^p(0,T;D_A(\theta+\frac{1}{2},p))}.$$

□

6.5 Existence of mild solutions in the analytic case

6.5.1 Introduction

In this section we will give, by applying the regularity results of the previous sections, several existence and regularity results for a mild solution of the problem

$$\begin{cases} dX(t) = AX(t)dt + B(X(t))dW(t), \\ X(0) = x \in H, \end{cases} \tag{6.31}$$

that is for the integral equation

$$X(t) = S(t)x + \int_0^t S(t-s)B(X(s))dW(s). \tag{6.32}$$

Let us introduce some notation. We set $v(t) = S(t)x,\ t \geq 0$, and for any process y we denote by $\Gamma(y)$ the following process

$$\Gamma(y)(t) = \int_0^t S(t-s)B(y(s))dW(s), \quad t \in [0, T], \tag{6.33}$$

so that solving equation (6.31) is equivalent to finding a fixed point for the problem

$$X(t) = v(t) + \Gamma(X)(t), \quad t \in [0, T]. \tag{6.34}$$

6.5.2 Existence of solutions in the analytic case

We are here concerned with problem (6.31) under the hypothesis that A is the infinitesimal generator of an analytic semigroup $S(\cdot)$ in H. For $x \in D(B)$ and $\theta \in (0, 1)$ we denote by $\|B(x)\|_\theta$ the Hilbert–Schmidt norm of the operator $B(x)$ considered as operator from U_0 into $D_A(\theta, 2)$. If it does not act between those spaces we set $\|B(x)\|_\theta = +\infty$.

Theorem 6.21 *Assume that there exists $\theta \in (0, \frac{1}{2})$, $\eta \in (0, 1-2\theta)$ and $K > 0$ such that $B \in L(D_A(\theta + \frac{1}{2}, 2), L_2(U_0; D_A(\theta, 2)))$ and*

$$\|B(x)\|_\theta^2 \leq \eta |x|_{\theta+1/2}^2 + K|x|_\theta^2, \quad x \in D_A(\theta + 1/2, 2). \tag{6.35}$$

Then for any $x \in D_A(\theta, 2)$ equation (6.31) has a mild solution

$$X \in \mathcal{N}_W^2(0, T; D_A(\theta + 1/2, 2)),$$

identical with a weak solution. Moreover the solution has a continuous modification as a process with values in $D_A(\theta, 2)$.

Proof We introduce the space $Z_T = \mathcal{N}_W^2(0, T; D_A(\theta + 1/2, 2))$ endowed with the norm

$$\|y\|_Z^2 = \mathbb{E}\int_0^T |y(t)|_{\theta+1/2}^2\, dt + L\mathbb{E}\int_0^T |y(t)|_\theta^2\, dt,$$

where L is a positive number to be chosen later. The mapping Γ, defined by

$$\Gamma(Y)(t) = S(t)x + \int_0^t S(t-s)B(Y(s))dW(s)$$

is a well defined transformation from Z_T into Z_T and moreover

$$\begin{aligned}\mathbb{E}\int_0^T |\Gamma(y)(t)|^2_{\theta+1/2}\,dt &\le \frac{1}{1-2\theta}\,\mathbb{E}\int_0^T |B(y(t))|^2_\theta\,dt\\ &\le \frac{\eta}{1-2\theta}\,\mathbb{E}\int_0^T |y(t)|^2_{\theta+1/2}\,dt + K\frac{\eta}{1-2\theta}\mathbb{E}\int_0^T |y(t)|^2_\theta\,dt.\end{aligned} \tag{6.36}$$

On the other hand, we have

$$\mathbb{E}|\Gamma(y)(t)|^2_\theta \le \mathbb{E}\int_0^t |B(y(s))|^2_\theta\,ds \le \eta\mathbb{E}\int_0^t |y(s)|^2_{\theta+1/2}\,ds + K\mathbb{E}\int_0^t y(s)|^2_\theta\,ds. \tag{6.37}$$

Combining (6.36) and (6.37) we find

$$\begin{aligned}|\Gamma(y)|^2_Z \le& \left[\frac{\eta}{1-2\theta} + LT\eta\right]\mathbb{E}\int_0^T |y(s)|^2_{\theta+1/2}\,ds\\ &+K\left[\frac{1}{1-2\theta} + LT\right]\mathbb{E}\int_0^T |y(s)|^2_\theta\,ds.\end{aligned} \tag{6.38}$$

Now choose $L > \frac{K}{1-2\theta}$ and $T < \frac{1-2\theta-\eta}{L\eta(1-2\theta)}$ (this is possible because $\eta < 1-2\theta$). Then by (6.38) it follows that Γ is a contraction.

Moreover $v \in Z_T$ in virtue of Proposition A.23; therefore by the contraction principle, equation (6.31) has a unique solution X in Z_T. By standard arguments the restriction on T can be removed. Since the process $\Phi(t) = B(X(t))$, $t \in [0,T]$, belongs to $\mathcal{N}^2_W(0,T;H)$ therefore by Theorem 6.5 the process X is also a weak solution to (6.31). □

Corollary 6.22 *Assume that for $\theta \in (0,\frac{1}{2})$ and $x \in D_A(\theta,2)$*

$$\sup\left\{\|B(z)\|_\theta : |z|_{\theta+1/2} \le 1\right\} < +\infty. \tag{6.39}$$

Then, for sufficiently small $\varepsilon > 0$, the equation

$$\begin{cases} dX_\varepsilon(t) = AX_\varepsilon(t)dt + \varepsilon B(X_\varepsilon(t))dW(t),\\ X_\varepsilon(0) = x,\end{cases} \tag{6.40}$$

has a unique weak (and mild) solution $X_\varepsilon \in \mathcal{N}^2_W(0,\infty;D_A(\theta+1/2,2))$. Moreover one has

$$\lim_{\varepsilon\to 0} X_\varepsilon = S(\cdot)x \quad \text{in } \mathcal{N}^2_W(0,\infty;D_A(\theta+1/2,2)). \tag{6.41}$$

Example 6.23 Let $\mathscr{O}$ be a bounded open set of $\mathbb{R}^N$, with a regular boundary $\partial\mathscr{O}$. We set $H = L^2(\mathscr{O})$ and consider an elliptic operator

$$A_0 u = \sum_{i,j=1}^{N} a_{ij}(x) \frac{\partial^2 u}{\partial x_i \partial x_j},$$

and a differential operator of the first order

$$B_0 u - \sum_{i=1}^{N} b_i(x) \frac{\partial u}{\partial x_i}.$$

We assume that the coefficients a_{ij}, b_i are continuous in $\overline{\mathscr{O}}$, and that there exists $\nu > 0$ such that (ellipticity)

$$\sum_{i,j=1}^{N} a_{ij}(x)\xi_i\xi_j \geq \nu|\xi|^2, \quad \forall\, \xi \in \mathbb{R}^N. \tag{6.42}$$

We denote by A the realization of A_0 under the Dirichlet boundary conditions, defined by (A.44) (with $p = 2$), that is

$$\begin{cases} D(A) = H^2(\mathscr{O}) \cap H_0^1(\mathscr{O}), \\ Au = A_0 u, \quad \forall\, u \in D(A). \end{cases} \tag{6.43}$$

Then (compare Section A.5.2), A is the infinitesimal generator of an analytic semigroup in $L^2(\mathscr{O})$ and

$$D_A(\theta, 2) = \begin{cases} H^{2\theta}(\mathscr{O}) \ \text{ if } \theta \in (0, \frac{1}{4}), \\ H^{2\theta}(\mathscr{O}) \cap H_0^1(\mathscr{O}) \quad \text{if } \theta \in (\frac{1}{4}, 1). \end{cases} \tag{6.44}$$

Moreover we denote by B the operator B_0 with domain $H_0^1(\mathscr{O})$. By (6.44) it follows that

$$B \in L(D_A(\theta + 1/2; 2)), \quad \forall\, \theta \in (0, 1/4). \tag{6.45}$$

In order to apply Theorem 6.21 we have to compute the norm of B; this is not easy, in general. However, we can apply Corollary 6.22. □

We consider now the limit case $\theta = 0$. The proof of the following result is similar to that of Theorem 6.21, we have only to use Theorem 6.14 instead of Theorem 6.12.

Theorem 6.24 *Assume Hypothesis 6.1 and that there exists $\eta \in (0, 1/\lambda_0)$, where λ_0 is defined by* (6.22) *and $K > 0$ such that*

$$B \in L(D_A(1/2, 2), L_2(U_0; H))$$

and

$$\|B(z)\|^2_{L_2^0} \leq \eta |z|^2_{1/2} + K|z|^2, \quad z \in D_A(1/2, 2). \tag{6.46}$$

Then for any $x \in H$ equation (6.31) *has a mild solution*

$$X \in \mathcal{N}_W^2(0, T; D_A(1/2, 2)),$$

identical with a weak solution. Moreover the solution has a continuous modification as a process with values in H.

Example 6.25 Let $\mathscr{O}$ be a bounded open set of $\mathbb{R}^N$, with a regular boundary $\partial\mathscr{O}$. We consider the strongly damped wave equation (see Section A.5.5)

$$\begin{cases} d(y_t(t,\xi)) = (\Delta_{\xi} y(t,\xi) + \rho \Delta_\xi y_t(t,\xi))dt \\ \qquad\qquad + [a \cdot \nabla_\xi y(t,\xi) + b \cdot \nabla_\xi y_t(t,\xi)]d\beta(t), \quad t > 0, \ \xi \in \mathscr{O}, \\ y(t,\xi) = 0, \quad t > 0, \ \xi \in \partial\mathscr{O}, \\ y(0,\xi) = x_0(\xi), \quad y_t(0,\xi) = x_1(\xi), \quad \xi \in \mathscr{O}, \end{cases} \tag{6.47}$$

where a and b are fixed vectors in $\mathbb{R}^N$. Let us write problem (6.47) in abstract form. Let Λ be the positive self-adjoint operator defined by (A.63), and let C_a be the linear operator in $L^2(\mathscr{O})$,

$$\begin{cases} D(C_a) = H_0^1(\mathscr{O}), \\ C_a y = a \cdot \nabla y, \quad \forall y \in D(C_a), \end{cases}$$

and let C_b be defined similarly. We consider the Hilbert space $H = D(\Lambda^{1/2}) \oplus L^2(\mathscr{O})$ endowed with the inner product

$$\left\langle \begin{pmatrix} y \\ z \end{pmatrix}, \begin{pmatrix} y_1 \\ z_1 \end{pmatrix} \right\rangle = \langle \sqrt{\Lambda} y, \sqrt{\Lambda} y_1 \rangle + \langle z, z_1 \rangle.$$

Notice that $D(\sqrt{\Lambda}) = H^1(\mathscr{O})$. Let A be the linear operator in H defined by $(A.69)$ and let B be defined by

$$\begin{cases} D(B) = \left\{ \begin{pmatrix} y \\ z \end{pmatrix} : \ y \in D(\sqrt{\Lambda}), z \in L^2(\mathscr{O}) \right\} \\ B \begin{pmatrix} y \\ z \end{pmatrix} = \begin{pmatrix} 0 & 0 \\ C_a & C_b \end{pmatrix} \begin{pmatrix} y \\ z \end{pmatrix}, \quad \forall \begin{pmatrix} y \\ z \end{pmatrix} \in D(B). \end{cases}$$

Then A generates an analytic semigroup in H of contractions and $D_A(1/2, 2)$ is given by (A.70). It follows that $B \in L(D_A(1/2, 2); H)$ and moreover

$$\left\| B \begin{pmatrix} y \\ z \end{pmatrix} \right\|^2 = \|Cy\|^2 \le |a|^2 \| \sqrt{\Lambda} y\|^2.$$

Thus, if $|a| < 1$ problem (6.47) has a unique mild solution. □

We pass now to the variational situation.

Theorem 6.26 *We assume that A is a variational operator and that there exists $\eta \in (0, 1)$ and $K > 0$ such that*

$$\frac{1}{2}\|B(z)\|^2_{L_2^0} + \eta a(z, z) \leq K|z|^2, \quad \forall\, z \in V. \tag{6.48}$$

Then, for any $x \in H$ there exists a unique weak (and mild) solution $X \in \mathcal{N}^2_W(0, T; V)$ of (6.31). Moreover the solution has a continuous modification as a process with values in H.

Proof The proof is similar to the previous one but instead of Theorem 6.12 one has to use Proposition 6.16. Take as Z the space $\mathcal{N}^2_W(0, T; V)$ and note that

$$-\int_0^T a(y(s), y(s))ds = \eta \int_0^T \|y(s)\|^2_V ds.$$

□

Example 6.27 Let $\mathscr{O}$ be a bounded open set of $\mathbb{R}^d$, with a regular boundary $\partial\mathscr{O}$. We set $H = L^2(\mathscr{O})$ and consider an elliptic operator in variational form

$$A_0 u = \sum_{i,j=1}^{N} \frac{\partial}{\partial x_j} a_{ij}(x) \frac{\partial u}{\partial x_j},$$

and a differential operator of the first order

$$B_0 u = \sum_{i=1}^{N} b_i(x) \frac{\partial u}{\partial x_i}.$$

We assume that the coefficients a_{ij}, b_i are continuous in $\overline{\mathscr{O}}$, and that there exists $\nu > 0$ such that (superellipticity)

$$\sum_{i,j=1}^{N} \left(a_{ij}(x) - \frac{1}{2} b_i b_j \right) \xi_i \xi_j \geq \nu |\xi|^2, \quad \forall\, \xi \in \mathbb{R}^N. \tag{6.49}$$

Moreover we denote by A the realization of A_0 under the Dirichlet boundary conditions, and by B the operator B_0 with domain $H^1_0(\mathscr{O})$. Now one can easily realize that all hypotheses of Theorem 6.26 hold true. □

Remark 6.28 The assumption $\eta \in (0, 1)$ cannot be removed as the following example shows. Let $H = L^2(\mathbb{R}^1)$, $Au = u_{xx}$, $D(A) = H^2(\mathbb{R}^1)$, $Bu = bu_x$, $b \in \mathbb{R}^1$, $D(B) = H^1(\mathbb{R}^1)$, $U = \mathbb{R}^1$, $W(t) = \beta(t)$. In this case we can write the explicit solution of (6.31), namely

$$X(t, x) = e^{(1 - \frac{b^2}{2})tA} e^{B\beta(t)} x. \tag{6.50}$$

Thus, X is defined for all x in H and belongs to $\mathcal{N}^2_W(0, \infty; D_A(1/2, 2))$, provided $1 - \frac{b^2}{2} > 0$. □

6.6 Existence of strong solutions

Strong solutions exist very rarely. Here we study, following [202], a class of equations (6.31) for which this is the case. Their special feature is that they can be reduced to deterministic problems.

We are concerned with the problem

$$\begin{cases} dX = AXdt + \sum_{k=1}^{N} B_k X d\beta_k, \\ X(0) = x \in H, \end{cases} \tag{6.51}$$

where

$$A : D(A) \subset H \to H, \quad B_k : D(B_k) \subset H \to H, \ k = 1, 2, \dots, N$$

are generators of semigroups $S(t) = e^{tA}$ and $S_k(t) = e^{tB_k}$ respectively.

We concentrate on a finite number of independent, real Wiener processes $\beta_1(\cdot), \dots, \beta_N(\cdot)$ to simplify presentation, but generalizations to the case of infinitely many Wiener processes are possible. For the same reason we assume that $\xi = 0$ and $f \equiv 0$.

We will need the following conditions.

Hypothesis 6.5

(i) *Operators $B_1, \dots, B_N$ generate mutually commuting C_0-groups $e^{tB_1}, \dots, e^{tB_N}$ respectively.*

(ii) *For $k = 1, \dots, N$, $D(B_k^2) \supset D(A)$ and $\bigcap_{k=1}^{N} D((B_k^*)^2)$ is dense in H.*

(iii) *The operator*

$$C = A - \frac{1}{2} \sum_{k=1}^{N} B_k^2, \quad D(C) = D(A),$$

is the infinitesimal generator of a C_0-semigroup $S_0(t) = e^{tC}$, $t \geq 0$.

Remark 6.29 The commutativity Hypothesis 6.5(i) is very strong, and it is essential to use the method below. However, in some cases it can be removed, as shown in [680], by using *stochastic characteristics* introduced by Kunita, see [476]. □

Remark 6.30 When A is a second order elliptic operator in $\mathbb{R}^N$ and the B_i's are first or zero order differential operators, problem (6.51) arises in filtering theory, see introduction, and has been studied by PDE methods or control arguments by several authors. See for instance [36, 136, 137, 469, 565, 577, 578, 633]. □

In order to solve (6.51) define

$$U(t) = \prod_{k=1}^{N} S_k(\beta_k(t)), \quad v(t) = U^{-1}(t)X(t), \quad t \in [0, T],$$

and introduce the equation

$$\begin{cases} v'(t) = U^{-1}(t)CU(t)v(t), \\ v(0) = x, \end{cases} \tag{6.52}$$

which can be studied by analytical methods.

Proposition 6.31 *Assume Hypothesis* 6.5. *If X is a strong solution to* (6.51) *then the process v satisfies* (6.52). *Conversely if v is a predictable process whose trajectories are of class C^1 and satisfies* (6.52) *a.s., then the process $X(\cdot) = U(\cdot)v(\cdot)$ takes values in $D(C)$, $\mathbb{P}$-a.s. and it is a strong solution of* (6.51).

Proof For fixed $\zeta \in H$ define

$$z_\zeta(t) = \prod_{i=1}^{N} S_i^*(-\beta_i(t))\zeta = (U^{-1})^*(t)\zeta, \quad t \in [0, T].$$

We will show that if $\zeta \in \bigcap_{k=1}^N D((B_k^*)^2)$, then

$$dz_\zeta = \frac{1}{2}\sum_{i=1}^{N}(B_i^*)^2 z_\zeta dt - \sum_{i=1}^{N} B_i^* z_\zeta d\beta_i. \tag{6.53}$$

To do so let us fix $\eta \in H$ and apply Itô's formula to the process

$$\langle z_\zeta(t), \eta\rangle = \psi(\beta_1(t), \dots, \beta_N(t)), \quad t \in [0, T],$$

where

$$\psi(x_1, \dots, x_N) = \left\langle \zeta, \prod_{i=1}^{N} S_i(-x_i)\eta \right\rangle, \quad (x_1, \dots, x_N) \in \mathbb{R}^N.$$

Let $\zeta \in \bigcap_{i=1}^N D((B_i^*)^2)$. Since

$$\frac{\partial \psi}{\partial x_j} = -\left\langle B_j^* \prod_{i=1}^{N} S_i^*(-x_i)\zeta, \eta \right\rangle,$$

and

$$\frac{\partial^2 \psi}{\partial x_j^2} = \left\langle (B_j^*)^2 \prod_{i=1}^{N} S_i^*(-x_i)\zeta, \eta \right\rangle,$$

therefore

$$d\langle z_\zeta, \eta\rangle = -\sum_{i=1}^{N}\langle B_i^* z_\zeta, \eta\rangle d\beta_i + \frac{1}{2}\left\langle \sum_{i=1}^{N}(B_i^*)^2 z_\zeta, \eta \right\rangle dt,$$

and consequently (6.53) holds. Taking into account Hypothesis 6.5(i)(ii) and that

$$\langle v(t), \zeta\rangle = \langle X(t), z_\zeta(t)\rangle, \quad t \in [0, T],$$

one obtains (applying again Itô's formula)

$$\begin{aligned} d\langle v(t),\zeta\rangle &= \langle dX(t), z_\zeta(t)\rangle + \langle X(t), dz_\zeta(t)\rangle - \sum_{i=1}^N \langle B_i X, B_i^* z_\zeta(t)\rangle dt \\ &= \left\langle \left(A - \frac{1}{2}\sum_{i=1}^N B_i^2\right) X, z_\zeta \right\rangle dt \\ &= \langle U^{-1}(t) C U(t) v(t), \zeta\rangle dt, \end{aligned}$$

and (ii) holds. The proof of the converse assertion is analogous. □

We assume now that Hypothesis 6.5 holds and formulate some condition implying solvability of the equation (6.52). We set

$$\begin{cases} D(C(t)) = \{x \in H;\ U(t)x \in D(C)\}, \\ C(t)x = U(-t)CU(t)x \quad, \forall\, x \in D(C(t)). \end{cases} \tag{6.54}$$

It is easy to see that the resolvent set $\rho(C(t))$ of $C(t)$ coincides with the resolvent set of C and we have

$$R(\lambda, C(t)) = U^{-1}(t) R(\lambda, C) U(t), \quad \forall\, \lambda \in \rho(C). \tag{6.55}$$

Theorem 6.32 *Assume Hypothesis 6.5 and that there exists $\lambda_0 \in \mathbb{R}^1$ and bounded operators $K_1, \dots, K_N$ such that*

$$(\lambda_0 - C) B_k (\lambda_0 - C)^{-1} = B_k + K_k, \quad k = 1, 2, \dots, N. \tag{6.56}$$

If $x \in D(C)$ then problem (6.52) has a unique strict solution $v \in C^1([0,T];H) \cap C([0,T];D(C))$.

Proof We assume for simplicity $\lambda_0 = 0$ and remark that, by our hypotheses,

$$C e^{tB_k} C^{-1} = e^{t(B_k + K_k)}, \quad k = 1, \dots, N. \tag{6.57}$$

Moreover (by a straightforward computation), there is a constant $M > 0$, such that

$$\left\| e^{t(B_k+K_k)} - e^{tB_k} \right\| \le Mt, \quad t \ge 0,\ k = 1, \dots, N. \tag{6.58}$$

Now using (6.57)

$$\begin{aligned} C(t)C^{-1}(s) - I &= U^{-1}(t) C \textstyle\prod_{k=1}^N S_k(\beta_k(t) - \beta_k(s)) C^{-1}(s) \\ &\quad - U^{-1}(t) \prod_{k=1}^N S_k(\beta_k(t) - \beta_k(s)) U(s) \\ &= U^{-1}(t) \left(\prod_{k=1}^N e^{(B_k+K_k)(\beta_k(t)-\beta_k(s))} - \prod_{k=1}^N e^{B_k(\beta_k(t)-\beta_k(s))} \right) U(s). \end{aligned}$$

It follows from (6.58) by an easy induction argument that there exists a positive random variable η such that

$$\|C(t)C^{-1}(s) - I\| \le \eta \sum_{k=1}^{N} |\beta_k(t) - \beta_k(s)|.$$

Since the Brownian motion has α-Hölder continuous trajectories, for any $\alpha \in (0, \frac{1}{2})$ there exists a positive random variable η_α such that

$$\|C(t)C^{-1}(s) - I\| \le \eta_\alpha |t-s|^\alpha, \quad t, s \in [0, T].$$

Now the conclusion follows from [668, Theorem 5.2.1]. □

Example 6.33 (Zakaï equation) Let $H = L^2(\mathbb{R}^d)$, $d \in \mathbb{N}$, and

$$Ay = \sum_{i,j=1}^{d} a_{ij} \frac{\partial^2 y}{\partial \xi_i \partial \xi_j} + \sum_{i=1}^{d} q_i \frac{\partial y}{\partial \xi_i} + ry, \quad \forall\, y \in D(A) = H^2(\mathbb{R}^d),$$

$$By = \sum_{i=1}^{d} b_i \frac{\partial y}{\partial \xi_i} + cy, \quad \forall\, y \in D(B) = \{y \in L^2(\mathbb{R}^d) : \ By \in L^2(\mathbb{R}^d)\}.$$

We assume the following.

(i) The coefficients, a_{ij}, q_i, r, b_i, c are of class C^3 and bounded with derivatives.
(ii) There exists $\nu > 0$ such that

$$\sum_{i,j=1}^{d} \left(a_{ij} - \frac{1}{2} b_i b_j\right) \lambda_i \lambda_j \ge \nu \sum_{i=1}^{d} \lambda_i^2,$$

for all vectors $\lambda = (\lambda_1, \ldots, \lambda_d)$ of R^d.

Then all hypotheses of Theorem 6.32 are fulfilled and, by Proposition 6.31, the corresponding equation (6.31) has a strong continuous solution with values in $H^2(R^d)$. In fact in this case the operator C is of the form

$$Cy = \left(A - \tfrac{1}{2}B^2\right) y = \sum_{i,j=1}^{d} \left(a_{ij} - \tfrac{1}{2} b_i b_j\right) \frac{\partial^2 y}{\partial \xi_i \partial \xi_j}$$
$$+ \text{terms containing only } \frac{\partial y}{\partial \xi_i} \text{ and } y, \quad y \in D(C) = H^2(\mathbb{R}^d)$$

$$Cy = \left(A - \tfrac{1}{2}B^2\right) y = \sum_{i,j=1}^{d} \left(a_{ij} - \tfrac{1}{2} b_i b_j\right) \frac{\partial^2 y}{\partial \xi_i \partial \xi_j}$$
$$+ \text{terms containing only } \frac{\partial y}{\partial \xi_i} \text{ and } y, \ y \in D(C) = H^2(\mathbb{R}^d),$$

and therefore satisfies (6.54), see Section A.5.2. Moreover, assumption (6.53) is clearly fulfilled. To see that condition (6.52) holds as well, note that in the present situation $N = 1$ and the group $S_1(\cdot)$ can be easily constructed by a characteristic method.

In order to apply Theorem 6.32, it remains to check the validity of (6.54). To do this fix $\lambda_0 > 0$ in the resolvent set of the elliptic operator C and remark that the operator $L = (\lambda_0 - C)B - B(\lambda_0 - C)$ is a second order differential operator (since terms involving third order derivatives cancel). Thus we can write

$$(\lambda_0 - C)B - B(\lambda_0 - C) = L(\lambda_0 - C)^{-1}(\lambda_0 - C). \tag{6.59}$$

Now the linear operator $K = L(\lambda_0 - C)^{-1}$ is clearly bounded and, by multiplying in the right equality (6.59) by $(\lambda_0 - C)^{-1}$, (6.54) follows.

We notice that one can find more general results in [466]. □

7

Existence and uniqueness for nonlinear equations

We first extend to Hilbert spaces finite dimensional results on existence and uniqueness, under linear growth and Lipschitz conditions on the coefficients. We distinguish two cases according to which the Wiener process has nuclear or identity covariance. Then we relax the Lipschitz condition in two important situations: when the diffusion term is additive or multiplicative. In the former case the nonlinear drift term is defined on an embedded Banach subspace and it is either locally Lipschitz or dissipative. In the multiplicative case existence is obtained under a dissipativity assumption involving the drift and diffusion terms. Finally, existence of strong solutions is discussed.

7.1 Equations with Lipschitz nonlinearities

We proceed to study nonlinear equations

$$\begin{cases} dX = (AX + F(t, X))dt + B(t, X)dW(t), \\ X(0) = \xi, \end{cases} \tag{7.1}$$

starting from the case when $F(\cdot)$ and $B(\cdot)$ satisfy properly formulated Lipschitz and linear growth conditions.

We will assume that a probability space $(\Omega, \mathscr{F}, \mathbb{P})$ together with a normal filtration $\mathscr{F}_t,\ t \geq 0$, are given. As in the previous chapter, $\mathscr{P}$ and $\mathscr{P}_T$ will denote predictable σ-fields on $\Omega_\infty = [0, +\infty) \times \Omega$ and on $\Omega_T = [0, T] \times \Omega$ respectively. For any $T > 0$ we define $\mathbb{P}_T$ to be the product of the Lebesgue measure in $[0, T]$ and the measure $\mathbb{P}$.

We assume also that U and H are separable Hilbert spaces and that W is a Q-Wiener process on $U_1 \supset U$ and $U_0 = Q^{\frac{1}{2}}U$, see Chapter 4. Spaces $U,\ H$ and $L_2^0 = L_2(U_0, H)$ are equipped with Borel σ-fields $\mathscr{B}(U)$, $\mathscr{B}(H)$ and $\mathscr{B}(L_2^0)$. Morever ξ is an H- valued random variable $\mathscr{F}_0$- measurable.

We fix $T > 0$ and impose first the following conditions on coefficients A, F and B of the equation.

Hypothesis 7.1

(i) A is the generator of a C_0*-semigroup* $S(t) = e^{tA},\ t \geq 0$, *in* H.
(ii) The mapping

$$F : [0, T] \times \Omega \times H \to H,\ (t, \omega, x) \mapsto F(t, \omega, x)$$

is measurable from $(\Omega_T \times H, \mathscr{P}_T \times \mathscr{B}(H))$ *into* $(H, \mathscr{B}(H))$.
(iii) The mapping

$$B : [0, T] \times \Omega \times H \to L_2^0,\ (t, \omega, x) \mapsto B(t, \omega, x)$$

is measurable from $(\Omega_T \times H, \mathscr{P}_T \times \mathscr{B}(H)$ *into* $(L_2^0, \mathscr{B}(L_2^0))$.
(iv) There exists a constant $C > 0$ *such that for all* $x, y \in H,\ t \in [0, T],\ \omega \in \Omega$ *we have*

$$|F(t, \omega, x) - F(t, \omega, y)| + \|B(t, \omega, x) - B(t, \omega, y)\|_{L_2^0} \leq C|x - y|, \quad (7.2)$$

and

$$|F(t, \omega, x)|^2 + \|B(t, \omega, x)\|_{L_2^0}^2 \leq C^2(1 + |x|^2). \quad (7.3)$$

The following proposition, whose proof is left as an exercise, gives conditions implying the measurability assumption in Hypothesis 7.1(ii)(iii).

Proposition 7.1 *Assume Hypothesis* 7.1(iv) *and that for arbitrary* $x, h \in H, u \in U$ *the processes* $\langle F(\cdot, \cdot; x), h\rangle, \langle B(\cdot, \cdot, x)Q^{1/2}u, h\rangle$ *are predictable. Then Hypotheses* 7.1(ii)(iii) *are fulfilled.*

A predictable H-valued process $X(t)$, $t \in [0, T]$, is said to be a *mild solution* of (7.1) if

$$\mathbb{P}\left(\int_0^T |X(s)|^2 ds < +\infty\right) = 1, \quad (7.4)$$

and, for arbitrary $t \in [0, T]$, we have

$$X(t) = S(t)\xi + \int_0^t S(t - s)F(s, X(s))ds + \int_0^t S(t - s)B(s, X(s))dW(s),\ \mathbb{P}\text{-a.s.} \quad (7.5)$$

The condition (7.4) implies that the integrals in (7.5) are well defined. *Weak* and *strong* solutions of (7.1) are defined as in Chapters 5 and 6.

The main result of the present section is the following.

Theorem 7.2 *Assume that ξ is an $\mathscr{F}_0$- measurable H-valued random variable and Hypothesis 7.1 is satisfied.*

(i) There exists a mild solution X to (7.1) unique, up to equivalence, among the processes satisfying

$$\mathbb{P}\left(\int_0^T |X(s)|^2 ds < +\infty\right) = 1.$$

Moreover X possesses a continuous modification.

(ii) For any $p \geq 2$ there exists a constant $C_{p,T} > 0$ such that

$$\sup_{t\in[0,T]} \mathbb{E}|X(t)|^p \leq C_{p,T}(1 + \mathbb{E}|\xi|^p). \tag{7.6}$$

(iii) For any $p > 2$ there exists a constant $\widehat{C}_{p,T} > 0$ such that

$$\mathbb{E} \sup_{t\in[0,T]} |X(t)|^p \leq \widehat{C}_{p,T}(1 + \mathbb{E}|\xi|^p). \tag{7.7}$$

Proof We first prove uniqueness. We show that if $X_1(\cdot)$ and $X_2(\cdot)$ are two processes satisfying (7.4) and (7.5) then, for arbitrary $t \in [0, T]$, $\mathbb{P}(X_1(t) = X_2(t)) = 1$. For a fixed number $R > 0$ we define

$$\tau_i = \inf\left\{t \leq T : \int_0^t |F(s, X_i(s))|ds \geq R \quad \text{or} \quad \int_0^t \|B(s, X_i(s))\|^2_{L_2^0} ds \geq R\right\}, \quad i = 1, 2$$

and $\tau = \tau_1 \wedge \tau_2$. Let $\widehat{X}_i(t) = I_{[0,\tau]}(t)X_i(t),\ t \in [0, T],\ i = 1, 2$. Then for arbitrary $t \in [0, T]$, $\mathbb{P}$-a.s.

$$\begin{aligned}\widehat{X}_i(t) = I_{[0,\tau]}(t)S(t)\xi + I_{[0,\tau]}(t)\int_0^t I_{[0,\tau]}(s)S(t-s)F(s, \widehat{X}_i(s))ds\\ + I_{[0,\tau]}(t)\int_0^t I_{[0,\tau]}(s)S(t-s)B(s, \widehat{X}_i(s))dW(s).\end{aligned}$$

Consequently, for arbitrary $t \in [0, T]$, $\mathbb{P}$-a.s.

$$\begin{aligned}\mathbb{E}|\widehat{X}_1(t) - \widehat{X}_2(t)|^2 \leq 2\mathbb{E}\left\{\int_0^t |F(s, \widehat{X}_1(s)) - F(s, \widehat{X}_2(s))|ds\right\}^2\\ + 2\mathbb{E}\left\{\int_0^t \|B(s, \widehat{X}_1(s)) - B(s, \widehat{X}_2(s))\|^2_{L_2^0} ds\right\}.\end{aligned} \tag{7.8}$$

By the very definition of the stopping times τ, τ_1, τ_2 one finds that the right hand side and therefore also the left hand side of (7.8) is a bounded function on $t \in [0, T]$. Again by (7.2)–(7.3)

$$\mathbb{E}|\widehat{X}_1(t) - \widehat{X}_2(t)|^2 \leq 2C^2(T+1) \int_0^t \mathbb{E}|\widehat{X}_1(s) - \widehat{X}_2(s)|^2 ds.$$

The boundedness of $\mathbb{E}|\widehat{X}_1(t) - \widehat{X}_2(t)|^2$, $t \in [0, T]$, and the Gronwall lemma imply $\mathbb{E}|\widehat{X}_1(t) - \widehat{X}_2(t)|^2 = 0$. Therefore, for all $t \in [0, T]$, one has $\mathbb{P}(\widehat{X}_1(t) = \widehat{X}_2(t)) = 1$. So the predictable processes $\widehat{X}_1(\cdot)$, $\widehat{X}_2(\cdot)$ are $\mathbb{P}_T$-a.s. identical. Since this is true for arbitrary $R > 0$ therefore $X_1(\cdot)$ and $X_2(\cdot)$ are $\mathbb{P}_T$ -a.s. identical. Taking into account that X_1 and X_2 are solutions of the equation (7.5) one easily deduces that for arbitrary $t \in [0, T]$, $X_1(t) = X_2(t)$, $\mathbb{P}$-a.s.

The proof of existence is based on the classical fixed point theorem for contractions. Denote by $\mathscr{H}_p$, $p \geq 2$, the Banach space of all the H-valued predictable processes Y defined on the time interval $[0, T]$ such that

$$|||Y|||_p = \left(\sup_{t\in[0,T]} \mathbb{E}|Y(t)|^p \right)^{1/p} < +\infty.$$

If one identifies processes which are identical $\mathbb{P}_T$-a.s. then $\mathscr{H}_p$, with the norm $|||\cdot|||_p$, becomes a Banach space. Let $\mathscr{K}$ be the following transformation:

$$\begin{aligned}\mathscr{K}(Y)(t) &= S(t)\xi + \int_0^t S(t-s)F(s, Y(s))ds + \int_0^t S(t-s)B(s, Y(s))dW(s) \\ &= S(t)\xi + \mathscr{K}_1(Y)(t) + \mathscr{K}_2(Y)(t), \quad t \in [0, T],\ Y \in \mathscr{H}_p.\end{aligned}$$

We assume that $\mathbb{E}(|\xi|^p) < +\infty$ and show that $\mathscr{K}$ maps $\mathscr{H}_p$ into $\mathscr{H}_p$. As the composition of measurable mappings is measurable therefore, taking into account Hypothesis 7.1, one obtains that the transformations $\mathscr{K}_1$ and $\mathscr{K}_2$ are well defined. Moreover

$$\begin{aligned}|||\mathscr{K}_1(Y)|||_p^p &\leq M^p \mathbb{E} \left(\int_0^T |F(s, Y(s))| ds \right)^p \\ &\leq T^{p-1} M^p \mathbb{E} \int_0^T |F(s, Y(s))|^p ds \\ &\leq 2^{p/2-1} T^{p-1} M^p C^p \mathbb{E} \int_0^T (1 + |Y(s)|^p) ds \\ &\leq 2^{p/2-1} (TMC)^p (1 + |||Y|||^p),\end{aligned}$$

where $M = \sup_{t\in[0,T]} \|S(t)\|$. Consequently $\mathcal{K}_1$ maps $\mathcal{H}_p$ into $\mathcal{H}_p$. To show the same property for $\mathcal{K}_2$ we remark that, by Theorem 4.36 we find

$$\begin{aligned}
|||\mathcal{K}_2(Y)|||_p^p &\leq \sup_{t\in[0,T]} \mathbb{E}\left(\left|\int_0^t S(t-s)B(s,Y(s))dW(s)\right|^p\right)\\
&\leq M^p C_{p/2}\mathbb{E}\left(\int_0^T \|B(s,Y(s))\|_{L_2^0}^2\,ds\right)^{p/2}\\
&\leq M^p C_{p/2} C^p \mathbb{E}\left(\int_0^T (1+|Y(s)|^2)ds\right)^{p/2}\\
&\leq M^p C_{p/2} c T^{p/2-1}\mathbb{E}\int_0^T (1+|Y(s)|^2)^{p/2}ds\\
&\leq M^p C_{p/2} C T^{p/2-1} 2^{p/2-1}\mathbb{E}\int_0^T (1+|Y(s)|^p)ds\\
&\leq M^p C_{p/2} C(2T)^{p/2-1}(T+|||Y|||_p)^p.
\end{aligned}$$

Now let Y_1 and Y_2 be arbitrary processes from $\mathcal{H}_p$ then

$$\begin{aligned}
|||\mathcal{K}(Y_1)-\mathcal{K}(Y_2)|||_p &\leq |||\mathcal{K}_1(Y_1)-\mathcal{K}_1(Y_2)|||_p + |||\mathcal{K}_2(Y_1)-\mathcal{K}_2(Y_2)|||_p\\
&= I_1 + I_2,
\end{aligned}$$

and

$$\begin{aligned}
I_1^p &\leq \sup_{t\in[0,T]} \mathbb{E}\left\{\left|\int_0^t [S(t-s)(F(s,Y_1(s))-F(s,Y_2(s)))]ds\right|^p\right\}\\
&\leq M^p \sup_{t\in[0,T]} \mathbb{E}\left\{\int_0^t [|F(s,Y_1(s))-F(s,Y_2(s))|]ds\right\}^p\\
&\leq (MC)^p T^{p-1}\left\{\int_0^T \mathbb{E}|Y_1(s))-Y_2(s))|^p ds\right\}\\
&\leq (MC)^p T^p \sup_{t\in[0,T]} \mathbb{E}\left\{|Y_1(t)-Y_2(t)|^p\right\}\\
&\leq (MC)^p T^p |||Y_1-Y_2|||_p^p.
\end{aligned}$$

In a similar way, by Theorem 4.36 we have

$$\begin{aligned}
I_2^p &\leq C_{p/2}M^p\mathbb{E}\left\{\int_0^T \|B(s,Y_1(s))-B(s,Y_2(s))\|_{L_2^0}^2 ds\right\}^{p/2}\\
&\leq C_{p/2}(MC)^p T^{p/2-1}\mathbb{E}\left\{\|B(s,Y_1(s))-B(s,Y_2(s))\|_{L_2^0}^p\right\}ds\\
&\leq C_{p/2}(MC)^p T^{p/2}|||Y_1-Y_2|||_p^p.
\end{aligned}$$

Summing up the obtained estimates we have:

$$|||\mathcal{K}(Y_1)-\mathcal{K}(Y_2)|||_p \leq CM(T^p + C_{p/2}T^{p/2})^{1/p}|||Y_1-Y_2|||_p, \tag{7.9}$$

for all $Y_1, Y_2 \in \mathscr{K}$. Consequently if

$$MCT(1 + c_{p/2}T^{1/2})^{1/p} < 1, \tag{7.10}$$

then the transformation $\mathscr{K}$ has unique fixed point X in $\mathscr{H}_p$ which, as it is easy to see, is a solution of the equation (7.1). The extra condition (7.10) on T can be easily removed by considering the equation on intervals $[0, \widetilde{T}], [\widetilde{T}, 2\widetilde{T}], \dots$ with $\widetilde{T}$ satisfying (7.10). Thus we have proved assertion (ii) of the theorem since (7.6) follows easily by using Gronwall's lemma.

To construct a solution when $\mathbb{E}|\xi|^p = +\infty$, we show first that if ξ and η are two initial conditions satisfying $\mathbb{E}|\xi|^p < +\infty$, $\mathbb{E}|\eta|^p < +\infty$, and if $X, Y \in \mathscr{H}_p$ are the corresponding solutions of equation (7.1), then

$$I_\Gamma X(\cdot) = I_\Gamma Y(\cdot), \quad \mathbb{P}\text{-a.s.}, \tag{7.11}$$

where

$$\Gamma = \{\omega \in \Omega : \xi(\omega) = \eta(\omega)\}.$$

To see this define

$$X^0 = S(\cdot)\xi, \; X^{k+1} = \mathscr{K}(X^k), \quad t \in [0, T], \; k \in \mathbb{N}.$$

Thus for $t \in [0, T]$, $\mathbb{P}$-a.s.

$$X^{k+1}(t) = S(t)\xi + \int_0^t S(t-s)F(s, X^k(s))ds + \int_0^t S(t-s)B(s, X^k(s))dW(s).$$

Since I_Γ is an $\mathscr{F}_0$-measurable random variable, therefore $I_\Gamma B(\cdot, X^k(\cdot))$ is an L_0^2-predictable process and for $t \in [0, T]$,

$$\int_0^t S(t-s)I_\Gamma B(s, X^k(s))dW(s) = I_\Gamma \int_0^t S(t-s)B(s, X^k(s))dW(s).$$

Thus, for $t \in [0, T]$,

$$\begin{aligned} I_\Gamma X^{k+1}(t) = S(t)I_\Gamma\xi &+ \int_0^t S(t-s)I_\Gamma F(s, X^k(s))ds \\ &+ \int_0^t S(t-s)I_\Gamma B(s, I_\Gamma X^k(s))dW(s). \end{aligned} \tag{7.12}$$

If for a similarly defined sequence

$$Y^0(t) = S(t)\eta, \; Y^{k+1}(t) = \mathscr{K}(Y^k), \quad k \in \mathbb{N}, \; t \in [0, T],$$

and some k we have

$$I_\Gamma X^k(\cdot) = I_\Gamma Y^k(\cdot), \quad \mathbb{P}_T\text{-a.s.}$$

then also

$$I_\Gamma F(\cdot, X^k(\cdot)) = I_\Gamma F(\cdot, Y^k(\cdot)),\; I_\Gamma B(\cdot, X^k(\cdot)) = I_\Gamma B(\cdot, Y^k(\cdot)), \quad \mathbb{P}_T\text{-a.s.}$$

Consequently

$$I_\Gamma X^{k+1}(\cdot) = I_\Gamma Y^{k+1}(\cdot), \quad \mathbb{P}_T \text{ a.s.}$$

Since the processes X and Y are limits in the $|||\cdot|||_p$ norm of the sequences $\{X^k(\cdot)\}$ and $\{Y^k(\cdot)\}$ respectively, therefore (7.11) must be true. Moreover the process $I_\Gamma X(\cdot)$ satisfies the equation (7.1) with the initial condition $I_\Gamma \xi = I_\Gamma \eta$.

We now prove existence. Let us define, for $n \in \mathbb{N}$

$$\xi_n = \begin{cases} \xi & \text{if } |\xi| \le n, \\ 0 & \text{if } |\xi| > n, \end{cases}$$

and denote by $X_n(\cdot)$ the corresponding solution of (7.10). By the previous argument we have

$$X_n(t) = X_{n+1}(t) \;\text{ on }\; \{\omega \in \Omega : |\xi| \le n\}.$$

It is now easy to see that the process

$$X(t) = \lim_{n\to\infty} X_n(t),\; t \in [0, T],$$

is $\mathbb{P}$-a.s. well defined and satisfies the equation (7.1).

For proof of existence of continuous modification of the mild solution assume first that $\mathbb{E}|\xi|^{2r} < +\infty$ for some $r > 1$. From the first part of the theorem one knows that

$$\sup_{t\in[0,T]} \mathbb{E}\|X(t)\|^{2r} < +\infty. \tag{7.13}$$

Define

$$\Phi(t) = B(t, X(t)), \quad t \in [0, T],$$

and

$$I = \mathbb{E}\int_0^T \|\Phi(t)\|^{2r}_{L_2^0}\, dt = \mathbb{E}\int_0^T \|B(t, X(t))\|^{2r}_{L_2^0}\, dt.$$

By (7.2)–(7.3) we have

$$I \le C^{2r}\mathbb{E}\left(\int_0^T (1 + |X(t)|^2)^r\, dt\right) < +\infty.$$

Consequently Proposition 7.3 below implies that the process

$$\int_0^t S(t-s)B(s, X(s))dW(s), \quad t \in [0, T],$$

and therefore also $X(t)$, $t \in [0, T]$, has a continuous modification.

The case of initial conditions satisfying $\mathbb{E}|\xi|^{2r} = +\infty$ can be reduced to the case just considered by regarding initial conditions ξ_n

$$\xi_n = \begin{cases} \xi & \text{if } |\xi| \le n, \\ 0 & \text{if } |\xi| > n, \end{cases}$$

as in the proof of existence. Finally (7.7) follows again from Gronwall's lemma. The proof is complete. □

We consider now the approximating problem

$$\begin{cases} dX_n = (A_n X + F(t, X_n))dt + B(t, X_t)dW(t), \\ X_n(0) = \xi, \end{cases} \tag{7.14}$$

where A_n are the Yosida approximations of A, see Appendix A, Section A.2. Clearly problem (7.14) has a unique solution X_n for any random variable ξ, $\mathscr{F}_0$-measurable.
We will need the following result.

Proposition 7.3 *Let $p > 2$, $T > 0$ and let Φ be an L_2^0-valued predictable process such that $\mathbb{E}\left(\int_0^T \|\Phi(s)\|^p_{L_2^0} ds\right) < +\infty$. There exists a constant $C_T > 0$ such that*

$$\mathbb{E} \sup_{t\in[0,T]} \left| \int_0^t S(t-s)\Phi(s)dW(s) \right|^p \le C_T \mathbb{E}\left(\int_0^T \|\Phi(s)\|^p_{L_2^0} ds \right). \tag{7.15}$$

Moreover

$$\lim_{n\to\infty} \mathbb{E} \sup_{t\in[0,T]} |W_A^\Phi(t) - W_{A,n}(t)|^p = 0, \tag{7.16}$$

where $W_{A,n}^\Phi$ is defined as

$$W_{A,n}^\Phi(t) = \int_0^t e^{(t-s)A_n}\Phi(s)dW(s), \quad t \in [0, T],$$

and A_n are the Yosida approximations of A.
Finally $W_A^\Phi = \int_0^t S(t-s)\Phi(s)dW(s)$ has a continuous modification.

Proof We will use the factorization method, see the proof of Proposition 5.9. Let $\alpha \in (\frac{1}{p}, \frac{1}{2})$, the stochastic Fubini theorem implies that

$$W_A^\Phi(t) = \frac{\sin \pi\alpha}{\pi} \int_0^t (t-s)^{\alpha-1} S(t-s)Y(s)ds, \quad t \in [0, T],$$

where

$$Y(s) = \int_0^s (s-\sigma)^{-\alpha} S(s-\sigma)\Phi(\sigma)dW(\sigma), \quad s \in [0, T].$$

Since $\alpha > \frac{1}{p}$, applying Hölder's inequality one obtains that there exists a constant $C_{1,T} > 0$ such that

$$\sup_{t\in[0,T]} |W_A^\Phi(t)|^p \le C_{1,T} \int_0^T |Y(s)|^p ds. \tag{7.17}$$

Moreover, by Theorem 4.36, there exists a constant $C_{2,T} > 0$ such that

$$\mathbb{E}|Y(s)|^p \le C_{2,T}\mathbb{E}\left(\int_0^s (s-\sigma)^{-2\alpha}\|\Phi(\sigma)\|^2_{L_2^0}\, d\sigma\right)^{p/2}. \tag{7.18}$$

Now, using the Young inequality for convolutions [(1)] we obtain that

$$\begin{aligned}\int_0^T \mathbb{E}|Y(s)|^p ds &\le C_{2,T}\mathbb{E}\left(\int_0^T \sigma^{-2\alpha} d\sigma\right)^{p/2} \int_0^s \|\Phi(\sigma)\|^{2r}_{L_2^0} d\sigma\\ &\le C_{3,T}\mathbb{E}\left(\int_0^T \|\Phi(\sigma)\|^p_{L_2^0} d\sigma\right).\end{aligned}$$

This finishes the proof of (7.15) with $C = C_{1,T}C_{3,T}$.

We now prove (7.16). We have:

$$W_{A,n}^\Phi(t) = \frac{\sin\pi\alpha}{\pi}\int_0^t e^{(t-s)A_n}(t-s)^{\alpha-1}Y_n(s)ds,$$

where

$$Y_n(s) = \int_0^s e^{(s-\sigma)A_n}(s-\sigma)^{-\alpha}\Phi(\sigma)dW(\sigma).$$

Thus, we can write

$$\begin{aligned}W_A^\Phi(t) - W_{A,n}^\Phi(t) &= \frac{\sin\pi\alpha}{\pi}\int_0^t [S(t-s) - e^{(t-s)A_n}](t-s)^{\alpha-1}Y(s)ds\\ &= \frac{\sin\pi\alpha}{\pi}\int_0^t [S(t-s) - e^{(t-s)A_n}](t-s)^{\alpha-1}Y(s)ds\\ &\quad + \frac{\sin\pi\alpha}{\pi}\int_0^t e^{(t-s)A_n}(t-s)^{\alpha-1}[Y(s)-Y_n(s)]ds\\ &=: I_n(t) + J_n(t).\end{aligned}$$

We proceed now in two steps.

Step 1 Exactly as in Step 1 of the proof of Theorem 5.12 we show that

$$\lim_{n\to\infty}\mathbb{E}\sup_{t\in[0,T]}|I_n(t)|^p = 0. \tag{7.19}$$

Step 2 We have:

$$\lim_{n\to\infty}\mathbb{E}\sup_{t\in[0,T]}|J_n(t)|^p = 0. \tag{7.20}$$

(1) If $1+\frac{1}{r} = \frac{1}{p}+\frac{1}{q}$, $f\in L^p$, $g\in L^q$ then $f*g\in L^r$ and $|f*g|_{L^r} \le |f|_{L^p}\,|g|_{L^q}$.

The following estimate is proved as (7.17)

$$\sup_{t\in[0,T]} |J_n(t)|^p \le C_{2,T} \int_0^T |Y(s) - Y_n(s)|^p ds. \tag{7.21}$$

Our aim is to show that

$$\lim_{n\to\infty} \mathbb{E} \int_0^T |Y(s) - Y_n(s)|^p \, ds = 0.$$

We define the operators

$$\mathscr{K}_n\Phi(s) = \int_0^s (s-\sigma)^{-\alpha}(S(s-\sigma) - e^{A_n(s-\sigma)})\Phi(s) dW(s).$$

Thus $Y - Y_n = \mathscr{K}_n\Phi$. We will show that if $\mathbb{E}\int_0^T \|\Phi(s)\|^p_{L_2^0} ds < \infty$, then

$$\lim_{n\to\infty} \mathbb{E} \int_0^T |\mathscr{K}_n\Phi(s)|^p ds = 0. \tag{7.22}$$

It follows from considerations following (7.18) that the operators $\mathscr{K}_n$ have bounded norms as operators acting from $L^p(\Omega \times [0,T]; L_2^0)$ into $L^p(\Omega \times [0,T]; H)$. It is enough to prove (7.22) for a dense set of Φ such that

$$\mathbb{E} \int_0^T \|A^2\Phi(s)\|^p_{L_2^0} ds < \infty.$$

In fact the processes Y and Φ can by approximated as follows:

$$Y_m = (m(mI - A)^{-1})^2 Y, \quad \Phi_m = (m(mI - A)^{-1})^2\Phi.$$

By Theorem 4.36

$$\mathbb{E}|\mathscr{K}_n\Phi(s)|^p \le c_p \mathbb{E}\left[\int_0^s (s-\sigma)^{-2\alpha} \|(S(s-\sigma) - e^{A_n(s-\sigma)})\Phi(\sigma)\|^2_{L_2^0} d\sigma\right]^{p/2}.$$

However,

$$\|(S(s-\sigma) - e^{A_n(s-\sigma)})\Phi(\sigma)\|^2_{L_2^0} \le \left(\frac{M}{n-\omega}\right)^2 \|A\Phi(\sigma)\|^p_{L_2^0}$$

and therefore

$$\mathbb{E}|\mathscr{K}_n\Phi(s)|^p \le c_p \left(\frac{M}{n-\omega}\right)^p \mathbb{E}\left[\int_0^s (s-\sigma)^{-2\alpha} \|A\Phi(\sigma)\|^p_{L_2^0} d\sigma\right]^{p/2}.$$

By Young's inequality it follows that

$$\mathbb{E}\int_0^T |\mathscr{K}_n\Phi(s)|^p \, ds \le c_p \left(\frac{M}{n-\omega}\right)^p \left(\int_0^T \sigma^{-2\alpha}\right)^{2/p} \mathbb{E}\int_0^T \|A\Phi(\sigma)\|^p_{L_2^0} d\sigma,$$

and we get the required convergence.

Finally, the existence of a continuous modification of W_A^Φ now follows easily from (7.16). □

Proposition 7.4 *Under the hypotheses of Theorem 7.2, assume that $\xi \in L^p(\Omega, \mathscr{F}, \mathbb{P})$, with $p \geq 2$, and let X and X_n be the solutions of problems* (7.1) *and* (7.14) *respectively. Then we have*

$$\lim_{n\to\infty} \sup_{t\in[0,T]} \mathbb{E}(|X(t) - X_n(t)|^p) = 0. \tag{7.23}$$

Moreover, if $p > 2$

$$\lim_{n\to\infty} \mathbb{E} \sup_{t\in[0,T]} |X(t) - X_n(t)|^p = 0. \tag{7.24}$$

Proof The result follows from a straightforward application of the contraction principle depending on the parameter n, Theorem 4.36 and Proposition 7.3. □

7.1.1 The case of cylindrical Wiener processes

If $\mathrm{Tr}\, Q = +\infty$ then the identity mapping $B = I$ does not satisfy the condition (7.2) although the equation (7.1) might have a solution. To cover this important case we will introduce the following assumptions.

Hypothesis 7.2

(i) A is the infinitesimal generator of a strongly continuous semigroup $S(t),\ t \geq 0$, on H.

(ii) The mapping

$$F : [0, T] \times \Omega \times H \to H,\ (t, \omega, x) \mapsto F(t, \omega, x)$$

is measurable from $(\Omega_T \times H, \mathscr{P}_T \times \mathscr{B}(H))$ into $(H, \mathscr{B}(H))$.

Moreover there exists a constant $C > 0$ such that for all $x, y \in H,\ t \in [0, T],\ \omega \in \Omega$ we have

$$|F(t, \omega, x) - F(t, \omega, y)| \leq C|x - y|, \tag{7.25}$$

and

$$|F(t, \omega, x)| \leq C(1 + |x|). \tag{7.26}$$

(iii) B is a strongly continuous mapping from H into $L(U; H)$ [(2)] *such that for any $t > 0$ and $x \in H$, $S(t)B(x)$ belongs to $L_2^0 = L_2(U_0; H)$, and there exists a*

[(2)] This means that for any $u \in U$ the mapping $x \to B(x)u$ from H into H is continuous.

locally square integrable mapping

$$K : [0, +\infty) \to [0, +\infty), \quad t \mapsto K(t),$$

such that

$$\|S(t)B(x)\|_{L_2^0} \le K(t)(1 + |x|), \quad t > 0,\ x \in H, \tag{7.27}$$

and

$$\|S(t)B(x) - S(t)B(y)\|_{L_2^0} \le K(t)|x - y|, \quad t > 0,\ x, y \in H. \tag{7.28}$$

An $\mathscr{F}_t$-adapted process $X(t)$, $t \ge 0$, is said to be a *mild solution* of (7.1) if it satisfies the following integral equation,

$$\begin{aligned} X(t) = S(t)\xi &+ \int_0^t S(t-s)F(s, X(s))ds \\ &+ \int_0^t S(t-s)B(X(s))dW(s), \quad t \in [0, T]. \end{aligned} \tag{7.29}$$

We have the following result.

Theorem 7.5 *Assume Hypothesis* 7.2 *and let* $p \ge 2$. *Then for an arbitrary* $\mathscr{F}_0$-*measurable initial condition* ξ *such that* $\mathbb{E}|\xi|^p < +\infty$ *there exists a unique mild solution* X *of* (7.1) *in* $\mathscr{H}_p$ *and there exists a constant* C_T, *independent of* ξ, *such that*

$$\sup_{t \in [0,T]} \mathbb{E}|X(t)|^p \le C_T(1 + \mathbb{E}|\xi|^p). \tag{7.30}$$

Finally, if there exists $\alpha \in (0, 1/2)$ *such that*

$$\int_0^1 s^{-2\alpha} K^2(s)ds < +\infty, \tag{7.31}$$

where K *is the function from Hypothesis* 7.2(iii), *then the solution* $X(\cdot)$ *is continuous* $\mathbb{P}$-*a.s.*

Proof For arbitrary $\xi \in L^p(\Omega, H)$ and $X \in \mathscr{H}_p$ define a process $Y = \mathscr{K}(\xi, X)$ by the formula

$$\begin{aligned} Y(t) = S(t)\xi &+ \int_0^t S(t-s)F(s, X(s))ds \\ &+ \int_0^t S(t-s)B(X(s))dW(s), \quad t \in [0, T]. \end{aligned} \tag{7.32}$$

We will note first that by Theorem 4.36, $K(\xi, X) \in \mathcal{H}_p$ for arbitrary $X \in \mathcal{H}_p$. Moreover, setting $M_T = \sup_{t\in[0,T]} \|S(t)\|$, we have

$$\begin{aligned}\mathbb{E}|Y(t)|^p &\le 3^{p-1}\|S(t)\|^p\mathbb{E}|\xi|^p + 3^{p-1}\mathbb{E}\left[\left(\int_0^t |S(t-s)F(s, X(s))|ds\right)^p\right] \\ &\quad + 3^{p-1}\mathbb{E}\left[\left|\int_0^t S(t-s)B(X(s))dW(s)\right|^p\right] \\ &\le 3^{p-1}M_T^p\mathbb{E}|\xi|^p + 3^{p-1}T^{p-1}M_T^p\int_0^t \mathbb{E}|F(s, X(s))|^p ds \\ &\quad + c_p 3^{p-1}\left[\int_0^t \left(\mathbb{E}\|S(t-s)B(X(s))\|^p_{L_2^0}\right)^{2/p} ds\right]^{p/2}.\end{aligned}$$

In addition,

$$\int_0^t \mathbb{E}|F(s, X(s))|^p ds \le 2^{p-1}C^p \sup_{s\in[0,t]} (1 + \mathbb{E}|X(s)|^p)t,$$

and

$$\begin{aligned}&\left[\int_0^t \left(\mathbb{E}\|S(t-s)B(X(s))\|^p_{L_2^0}\right)^{2/p} ds\right]^{p/2} \\ &\quad\le 2^{p-1}\left(\int_0^t K^2(t-s)(1 + \mathbb{E}|X(s)|^p)^{2/p} ds\right)^{p/2} \\ &\quad\le 2^{p-1}\left(\int_0^t K^2(t-s)ds\right)^{p/2} \sup_{s\in[0,t]} (1 + \mathbb{E}|X(s)|^p).\end{aligned}$$

It is now clear that, for some constants c_1, c_2, c_3,

$$\sup_{t\in[0,T]} \mathbb{E}|Y(t)|^p \le c_1 + c_2\mathbb{E}|\xi|^p + c_3 \sup_{t\in[0,T]} \mathbb{E}|X(t)|^p. \tag{7.33}$$

Thus $Y \in \mathcal{H}_p$.

In exactly the same way, if $X_1, X_2 \in \mathcal{H}_p$ and $Y_1 = \mathcal{K}(\xi, X_1)$, $Y_2 = \mathcal{K}(\xi, X_2)$, then

$$\sup_{t\in[0,T]} \mathbb{E}|Y_1(t) - Y_2(t)|^p \le c_3 \sup_{t\in[0,T]} \mathbb{E}|X_1(t) - X_2(t)|^p.$$

It is easy to see that if T is small enough then $c_3 < 1$ and consequently, by the contraction principle, equation (7.1) has a unique solution in $\mathcal{H}_p$. The case of general $T > 0$ can be treated by considering the equation in intervals $[0, \widetilde{T}]$, $[\widetilde{T}, 2\widetilde{T}], \ldots$ with $\widetilde{T}$ such that $c_3(\widetilde{T}) < 1$. Moreover with such a $\widetilde{T}$ we get from (7.33) for the solution of (7.1) that

$$\sup_{t\in[0,T]} \mathbb{E}|X(t)|^p \le \frac{1}{1 - c_3(\widetilde{T})}[c_1 + c_2\mathbb{E}|\xi|^p]$$

which is inequality (7.30). The case of general $T > 0$ can be easily obtained by iteration as well.

Finally, the continuity of the solution can be proved by using factorization as before. □

Example 7.6 [596] Let us consider the nonlinear heat equation

$$\begin{cases} dX(t,\xi) = D_\xi^2 \, X(t,\xi)dt + b(X(t,\xi))dW(t,\xi), \\ X(t,0) = X(t,1) = 0, \quad t \geq 0, \\ X(t,\xi) = x(\xi), \quad \xi \in (0,1), \quad x \in H = L^2(0,1), \end{cases} \tag{7.34}$$

in which b is a real valued function and W is a cylindrical Wiener process on $U = H = L^2(0,1)$.

Let us write equation (7.34) in the abstract form:

$$\begin{cases} dX = AX(t) + B(X)dW(t), \\ X(0) = x \in H, \end{cases} \tag{7.35}$$

where A is the linear operator

$$Ax = D_\xi^2 x, \quad \forall\, x \in H^2(0,1) \cap H_0^1(0,1)$$

and B is given by

$$(B(x)u)(\xi) = b(x(\xi))u(\xi), \quad \xi \in (0,1), \; u \in H,$$

where b is a bounded and Lipschitz function. It is easy to see that B is strongly continuous.

It is well known that

$$Ae_k = -\pi^2 k^2 e_k, \quad k \in \mathbb{N},$$

where

$$e_k(\xi) = \sqrt{2/\pi}\; \sin(\pi k \xi), \quad \xi \in [0,1], \; k \in \mathbb{N}.$$

We take $W(t)$ of the form

$$\langle W(t), u\rangle = \sum_{k=1}^{\infty} \langle u, e_k\rangle \, \beta_k(t), \quad u \in H,$$

where (β_k) is a sequence of mutually independent real Brownian motions.

Let us compute $\|S(t)B(x)\|^2_{L_2^0}$. Write

$$\begin{aligned} \|S(t)B(x)\|^2_{L_2^0} &= \sum_{k=1}^{\infty} |S(t)(b(x)e_k)|^2_H = \sum_{h,k=1}^{\infty} e^{-2\pi^2 h^2 t} |\langle b(x)e_k, e_h\rangle|^2 \\ &= \sum_{h=1}^{\infty} e^{-2\pi^2 h^2 t} |b(x)e_h|^2_H \leq \frac{2}{\pi} \, \|b\|^2_\infty \sum_{h=1}^{\infty} e^{-2\pi^2 h^2 t}. \end{aligned}$$

Setting

$$K^2(t) = \sum_{h=1}^{\infty} e^{-2\pi^2 h^2 t},$$

it follows that assumption (7.27) is fulfilled. In a similar way one can check (7.28). So, Theorem 7.5 applies. □

7.2 Nonlinear equations on Banach spaces: additive noise

In many situations of interest, the nonlinear operators F and B are defined only on a subset of the Hilbert space H. A typical example is provided by the so called polynomial nonlinearities, see Example 7.8 (and continuation) below. One way of treating such cases is to consider equation (7.1) on a smaller state space E on which the nonlinear operators F and B are well defined and sufficiently regular, say locally Lipschitz continuous or simply continuous. This method requires that the initial condition takes values in the smaller space E. In Section 7.2.4 we show that in some important cases the concept of mild solution can be extended to all initial conditions in H.

In this section we restrict our attention to equations with additive noise. Thus we consider the problem

$$\begin{cases} dX = (AX + F(X))dt + dW(t), \\ X(0) = \xi \in E, \end{cases} \tag{7.36}$$

where E is a Banach space continuously, densely and as Borel subset, embedded in H, $U = H$ and A generates a C_0-semigroup in H. We denote by A_E the part of A in E. We will need the following assumptions.

Hypothesis 7.3 *either*

(i) A_E generates a C_0-semigroup $S(\cdot)$ on E or
(ii) A generates an analytic semigroup $S_E(\cdot)$ on E.

Moreover the stochastic convolution W_A has an E-continuous version.

We recall that if the semigroup S_E is analytic, then it is strongly continuous at 0 if and only if $D(A_E)$ is dense in E, see Theorem A.9(iii).

7.2.1 Locally Lipschitz nonlinearities

In this subsection we impose the following conditions on F.

Hypothesis 7.4

(i) $D(F) \supset E$, F maps E into E and the restriction F_E of F to E is locally Lipschitz continuous and bounded on bounded subsets of E.

(ii) There exists an increasing function $a : \mathbb{R}^1_+ \to \mathbb{R}^1_+$ *such that*

$$\langle A_{E,n}x + F(x+y), x^* \rangle \leq a(\|y\|)(1+\|x\|), \quad \forall\, x, y \in E,\ x^* \in \partial\|x\|,\ n \in \mathbb{N}.$$

$A_{E,n}$ denotes the Yosida approximations of A_E. Moreover $\langle \cdot, \cdot \rangle$ is the duality form on $E \times E^*$ [(3)] and $\partial\|x\|$ the subdifferential of the E-norm $\|\cdot\|$ at the point $x \in E$ (see for instance [29]).

The main result of this subsection is given by the following theorem from [213], which generalizes [288] and [521].

Theorem 7.7 *Assume that* A *generates a* C_0*-semigroup and that Hypotheses* 7.3, 7.4 *hold.*

(i) *If the condition* 7.3(i) *holds then equation* (7.36) *has a unique mild solution in* $C([0,+\infty); E)$.
(ii) *If the condition* 7.3(ii) *holds then equation* (7.36) *has a unique mild solution in* $C((0,+\infty); E) \cap L^\infty_{\text{loc}}(0,+\infty; E)$.

Before going to the proof we will discuss applicability of the imposed conditions in the case when

$$H = L^2(\mathscr{O}), \quad E = C(\overline{\mathscr{O}})$$

where $\mathscr{O}$ is a bounded open subset of $\mathbb{R}^N$ with regular boundary $\partial\mathscr{O}$.

Note that the continuity of the stochastic convolution in E was an object of Section 5.5. It is satisfied in several situations.

As far as Hypotheses 7.3 and 7.4 are concerned we consider the following specific example.

Example 7.8 Let A be defined as:

$$\begin{cases} Ay = \Delta y, & \forall\, y \in D(A),\\ D(A) = H^2(\mathscr{O}) \cap H^1_0(\mathscr{O}), \end{cases}$$

where Δ represents the Laplace operator. As we shall observe in Section A.5.2, A is the infinitesimal generator of an analytic semigroup $S(\cdot)$ of class C_0. As easily checked, the part A_E of A in E is given by

$$\begin{cases} D(A_E) = \{y \in C(\overline{\mathscr{O}}) : \ \Delta y \in C(\overline{\mathscr{O}}) \ \text{ and } \ y = 0 \ \text{ on } \ \partial\mathscr{O}\},\\ A_E y = \Delta y, \quad \forall\, y \in D(A_E), \end{cases}$$

where, in the definition of $D(A_E)$ the Laplacian is understood in the distributional sense.

As we observe in Section A.5.2 , A_E is the infinitesimal generator of an analytic semigroup $S_E(\cdot)$ on E. However, the closure $\overline{D(A_E)}$ of $D(A_E)$ consists of the set of

(3) E^* is the topological dual of E.

all continuous functions on $\overline{\mathcal{O}}$ vanishing on $\partial\mathcal{O}$ and is different from E. Therefore $S_E(\cdot)$ is not strongly continuous at 0 (see Theorem A.9(iii)). Thus Hypothesis 7.3(ii) is fulfilled but not 7.3(i).

Let us discuss Hypotheses 7.4. Note, see (A.53), that we have the estimate

$$\|S_A(t)\| \leq 1, \quad t \geq 0. \tag{7.37}$$

Therefore by the Hille–Yosida theorem (Theorem A.3) it follows that $\|e^{tA_{E,n}}\| \leq 1$ which implies $\langle A_{E,n}x, x^*\rangle \leq 0$ for all $x \in E$, $x^* \in \partial\|x\|$. Thus the inequality in 7.4(ii) reduces to

$$\langle F(x+y), x^*\rangle \leq a(\|y\|)(1+\|x\|), \quad \forall x, y \in E, \ x^* \in \partial\|x\|. \tag{7.38}$$

Let φ be a real continuous function of class C^1 such that there exists an increasing function $a : \mathbb{R}^1_+ \to \mathbb{R}^1_+$ such that

$$\varphi(\xi+\eta)\,\mathrm{sgn}(\xi) \leq a(|\eta|)(1+|\xi|), \quad \forall\, \xi, \eta \in \mathbb{R}^1, \tag{7.39}$$

where sgn $\alpha = 1$ if $\alpha \geq 0$, and sgn $\alpha = -1$ if $\alpha < 0$, and define F as the Nemytskii operator

$$F(x)(\xi) = \varphi(x(\xi)), \quad \forall\, x \in E, \ \xi \in \mathcal{O}.$$

Then, by using a known characterization of the subdifferential of the norm in E, see Examples D.3, D.7 and [647], it is not difficult to show that Hypothesis 7.4 holds true. □

Example 7.9 Similar considerations can be followed for the linear operator A corresponding to the Neumann problem:

$$\begin{cases} Ay = \Delta y, & \forall\, y \in D(A), \\ D(A) = \left\{ y \in H^2(\mathcal{O}) : \frac{\partial y}{\partial \nu} = 0 \right\}, \end{cases}$$

where ν is the outward normal to $\partial\mathcal{O}$. In this case the part A_E of A in E is given by

$$\begin{cases} A_E y = \Delta y, & \forall\, y \in D(A_E), \\ D(A_E) = \left\{ y \in C(\overline{\mathcal{O}}) : \ \Delta y \in C(\overline{\mathcal{O}}) \ \text{and} \ \frac{\partial y}{\partial \nu} = 0 \text{ on } \partial\mathcal{O} \right\}, \end{cases}$$

and $D(A_E)$ is dense in E. From Section A.5.2, A_E (respectively A) is the infinitesimal generator of an analytic semigroup of contractions on E (respectively H). So condition 7.4(i) is fulfilled in this case. □

Proof of Theorem 7.7 Define $v(t) = X(t) - z(t)$, $t \in [0, T)$, where $z(\cdot) = W_A(\cdot)$, and note that for E-valued processes X the mild version of (7.36) can be written as

$$v(t) = S_E(t)\xi + \int_0^t S_E(t-s)F_E(v(s)+z(s))ds, \quad t \in [0, T]. \tag{7.40}$$

We are going to solve (7.40) pathwise, assuming that $W_A(\cdot)$ is E-continuous.

We first prove the theorem when $D(A_E)$ is dense in E. In this case, for any $T > 0$ we set $Z_T = C([0, T]; E)$ and moreover

$$(\gamma(v))(t) = S_E(t)\xi + \int_0^t S_E(t-s)F_E(v(s) + z(s))ds. \tag{7.41}$$

Clearly γ maps Z_T into Z_T and, by using the local inversion theorem, it is easy to show that if T is small enough, then there exists a unique mild solution on $[0, T]$. Since F_E is bounded on bounded sets of E, to obtain global existence it is sufficient to deduce an a priori estimate for $\|v(\cdot)\|$. Let $v(\cdot)$ be a mild solution of (7.40) on a (stochastic) interval $[0, T_0]$ and let $\{v_n\}$ be the sequence in $C^1([0, T_0]; E) \cap C([0, T_0]; D(A_E))$ defined by

$$v_n(t) = nR(n, A_E)S_E(t)x + \int_0^t nR(n, A_E)F_E(v(s) + W_A(s))ds,$$

then it is easy to check that (use Proposition A.4),

$$v_n \to v, \qquad \frac{dv_n}{dt} - Av_n - F_E(v_n + z) = \delta_n \to 0,$$

uniformly on $[0, T_0]$ as $n \to \infty$. Now, for $t \geq 0$ and $x^*_{t,n} \in \partial\|v_n(t)\|$

$$\begin{aligned}\frac{d^-}{dt}\|v_n(t)\| &\leq \langle Av_n(t) + F_E(v_n(t) + z(t)), x^*_{t,n}\rangle + \langle \delta_n(t), x^*_{t,n}\rangle \\ &\leq a(\|z(t)\|)(1 + \|v_n(t)\|) + \|\delta_n(t)\|.\end{aligned}$$

Consequently

$$\|v_n(t)\| \leq e^{\int_0^{T_0} a(\|z(s)\|)ds}\|v_n(s)\| + \int_0^{T_0} e^{\int_s^t a(\|z(u)\|)du}[a(\|z(s)\|) + \|\delta_n(s)\|]ds.$$

Therefore, letting n tend to infinity, and taking into account continuous dependence of the solutions on n we get

$$\|v(t)\| \leq e^{\int_0^{T_0} a(\|z(s)\|)ds}\|v(0)\| + \int_0^{T_0} e^{\int_s^t a(\|z(u)\|)du}a(\|z(s)\|)ds.$$

This finishes the proof for case (i). The proof of case (ii) is similar, one has only to replace the Banach space Z_T with $\widetilde{Z}_T = C([0, T]; E) \cap L^\infty(0, T; E)$, endowed with the sup norm. □

Example 7.10 Set $N = 1$, $\mathscr{O} = (0, 1)$ and let A and A_E be defined as in Example 7.8. Consider the problem

$$\begin{cases} dy(t,\xi) = \left(D_\xi^2 y(t,\xi) + \varphi(y(t,\xi))\right)dt + dW(t), & t \geq 0,\ \xi \in [0,1], \\ y(t,0) = y(t,1) = 0, & t \geq 0,\ \xi \in [0,1], \\ y(0,\xi) = x(\xi), & \xi \in [0,1], \end{cases}$$

where φ is a polynomial of odd degree and negative leading coefficient. Also in this case all hypotheses of Theorem 7.7 are fulfilled. In fact 7.3(ii) was discussed in Example 7.8, and continuity of the stochastic convolution follows from Theorem 5.22. It remains to prove property (7.39), which will imply Hypothesis 7.4. We have in fact

$$\varphi(\xi+\eta) \le \sup_{\alpha\ge 0} \varphi(\alpha+\eta) < +\infty, \quad \forall\, \xi > 0,\ \eta \in \mathbb{R}^1,$$
$$-\varphi(\xi+\eta) \le \sup_{\alpha\le 0} \varphi(\alpha+\eta) < +\infty, \quad \forall\, \xi < 0,\ \eta \in \mathbb{R}^1.$$

Now it is easy to see that (7.39) is fulfilled. □

7.2.2 Dissipative nonlinearities

We first remark that the local Lipschitz condition imposed on the mapping F cannot be replaced by the continuity only. Godunov's theorem [353] says that on an arbitrary infinite dimensional Banach space E one can define a mapping F continuous and bounded such that the deterministic equation

$$X(t) = x + \int_0^t F(X(s))ds, \quad t \ge 0, \tag{7.42}$$

does not have a local solution. However, continuity and dissipativity of F do imply existence of a global solution of (7.42), see [528], and we show that this is true also in the stochastic case. We shall assume the following.

Hypothesis 7.5 *The mapping F_E is dissipative and uniformly continuous on bounded sets of E.* [(4)]

We now prove the main result of this subsection following [216]. Earlier results for the case of A equal to the second derivative, with Dirichlet boundary conditions in a bounded interval of $\mathbb{R}^1$, can be found in [288, 520, 521].

Theorem 7.11 *Assume that A generates a C_0-semigroup on H, that Hypotheses 7.3 and 7.5 hold and that $\|S_E(t)\| \le e^{\omega t}$, for some $\omega \in \mathbb{R}^1$ and all $t \ge 0$. Then*

(i) if the condition 7.3(i) holds then equation (7.36) has a unique mild solution in $C([0,+\infty); E)$,
(ii) if the condition 7.3(ii) holds then equation (7.36) has a unique mild solution in $C((0,+\infty); E) \cap L^\infty_{loc}(0,+\infty; E)$.

Proof We restrict our considerations to the case (i) and set $F = F_E$ for simplicity. As in the proof of Theorem 7.7, define

$$v(t) = X(t) - z(t),$$

(4) For the definition of dissipativity, see Appendix D.

where $z(\cdot) = W_A(\cdot)$, and consider equation (7.40). Let us introduce, for any $\alpha > 0$, the approximating equation

$$v^\alpha(t) = S_E(t)\xi + \int_0^t S_E(t-s)F_\alpha((v_\alpha(s) + z(s))ds, \tag{7.43}$$

where F_α are the Yosida approximations of F, see Section D.3. Let us check that assumption 7.4, with F replaced by F_α holds true. First of all 7.4(i) is fulfilled since F_α is Lipschitz continuous, by Proposition D.10. Moreover, for any $x, y \in E$ and some $x^* \in \partial\|x\|$, we have

$$\begin{aligned}\langle F_\alpha(x+y), x^*\rangle &= \langle F_\alpha(x+y) - F_\alpha(y), x^*\rangle + \langle F_\alpha(y), x^*\rangle \\ &\le \langle F_\alpha(y), x^*\rangle \le \|F(y)\|,\end{aligned}$$

and, recalling that $\|S_E(t)\| \le e^{\omega t}$, $t \ge 0$, we have $\langle A_{E,n}x, x^*\rangle \le \omega\|x\|$. Consequently

$$\langle A_{E,n}x + F_\alpha(x+y), x^*\rangle \le \omega\|x\| + \|F(y)\|,$$

and 7.4(ii) is also fulfilled.

Now by Theorem 7.7, equation (7.43) has a unique global solution v^α. Fix $T > 0$ and let $\{v_n^\alpha\} \subset C^1([0,T];E) \cap C([0,T];D(A_E))$ and $\{\delta_n^\alpha\} \subset C([0,T];E)$ be sequences such that, uniformly on $[0,T]$,

$$v_n^\alpha \to v_\alpha, \quad \frac{dv_n^\alpha}{dt} - Av_n^\alpha - F(v_n^\alpha + z) = \delta_n^\alpha \to 0,$$

see the proof of Theorem 7.7. Now, for some $x^*_{t,n,\alpha} \in \partial\|v_n^\alpha(t)\|$ we get the estimate

$$\begin{aligned}\frac{d^-}{dt}\|v_n^\alpha(t)\| &\le \langle Av_n^\alpha(t) + F_\alpha(v_n^\alpha(t) + z(t)), x^*_{t,n,\alpha}\rangle + \langle \delta_n^\alpha(t), x^*_{t,n,\alpha}\rangle \\ &\le \omega\|v_n^\alpha(t)\| + \|F_\alpha(z(t))\| + \|\delta_{n,\alpha}(t)\|.\end{aligned}$$

Since $\|F_\alpha(z)\| \le \|F(z)\|$ for all $z \in E$, therefore, letting $n \to \infty$, we have

$$\|v^\alpha(t)\| \le e^{\omega t}\|x\| + \int_0^t e^{\omega(t-s)}\|F(z(s))\|ds. \tag{7.44}$$

This shows that the sequence $\{v^\alpha(\cdot)\}$ is bounded uniformly on bounded sets. To show convergence of the sequence, we set for any $\alpha, \beta > 0$,

$$g^{\alpha,\beta} = v^\alpha - v^\beta, \quad u^\alpha = v^\alpha + z, \quad u^\beta = v^\beta + z.$$

Then $g^{\alpha,\beta}$ is a classical solution to the problem

$$\begin{cases}\dfrac{d}{dt}g^{\alpha,\beta}(t) = Ag^{\alpha,\beta}(t) + F_\alpha(u^\alpha(t)) - F_\beta(u^\beta(t)),\\ g^{\alpha,\beta}(0) = 0.\end{cases}$$

Let $y^*_{\alpha,\beta,t} \in \partial \|g^{\alpha,\beta}(t)\|$, then we have

$$\begin{aligned}\frac{d^-\|g^{\alpha,\beta}(t)\|}{dt} &\leq \omega\|g^{\alpha,\beta}(t)\| + \langle F_\alpha(u^\alpha(t)) - F_\beta(u^\beta(t)), y^*_{\alpha,\beta,t}\rangle \\ &\leq \omega\|g^{\alpha,\beta}(t)\| + \langle F(u^\alpha(t)) - F(u^\beta(t)), y^*_{\alpha,\beta,t}\rangle \\ &\quad + \langle F(J_\alpha(u^\alpha(t))) - F(u^\alpha(t)) - F(J_\beta(u^\beta(t))) + F(u^\beta(t)), y^*_{\alpha,\beta,t}\rangle \\ &\leq \omega\|g^{\alpha,\beta}(t)\| + \|F(J_\alpha(u^\alpha(t))) - F(u^\alpha(t))\| \\ &\quad + \|F(J_\beta(u^\beta(t))) - F(u^\beta(t))\|.\end{aligned}$$

Now by (7.44) and recalling that F is bounded on bounded subsets of E, for a fixed $T > 0$ there exists $R > 0$ such that,

$$\|u^\alpha(t)\| \leq R, \quad \|F(u^\alpha(t))\| \leq R, \quad \forall\, t \in [0, T],\ \forall\, \alpha \in (0, 1].$$

Moreover

$$\|u^\alpha(t) - J_\alpha(u^\alpha(t))\| \leq \alpha\|F(u^\alpha(t))\| \leq \alpha R,$$

and so

$$\|F(J_\alpha(u^\alpha(t))) - F(u^\alpha(t))\| + \|F(J_\beta(u^\beta(t))) - F(u^\beta(t))\| \leq \rho_F(\alpha R) + \rho_F(\beta R),$$

which implies

$$\|g_{\alpha,\beta}(t)\| \leq [\rho_F(\alpha R) + \rho_F(\beta R)] \int_0^t e^{\omega s}\, ds,$$

where ρ_F is the modulus of continuity [(5)] of F restricted to $B(0, R)$. This yields the convergence of the sequence $\{v^\alpha\}$ in $C([0, T]; E)$ to a function v. It is easily seen that v solves (7.41). This finishes the proof of existence of a solution to (7.40) on arbitrary time interval.

To show uniqueness, let $\widehat{X}$ be another solution to (7.36). Then we have

$$\frac{d^-\|X(t) - \widehat{X}(t)\|}{dt} \leq \langle A(X(t) - \widehat{X}(t)) + F(X(t)) - F(\widehat{X}(t)), \widehat{x}^*_t\rangle$$

for $t \in [0, T]$ and for an appropriate $\widehat{x}^*_t \in \partial\|X(t) - \widehat{X}(t)\|$. Consequently

$$\frac{d^-\|X(t) - \widehat{X}(t)\|}{dt} \leq \omega\|X(t) - \widehat{X}(t)\|, \quad t \in [0, T].$$

Since $X(0) = \widehat{X}(0)$ therefore $X(t) = \widehat{X}(t),\ t \in [0, T]$. The proof is complete. □

(5) Any function $\rho_F : [0, +\infty] \to [0, +\infty)$ such that (i) $\lim_{r\to 0} \rho_F(r) = 0$, (ii) $|F(x) - F(y)| \leq \rho_F(\|x - y\|)$ for all $x, y \in B(0, R)$, is called a continuity modulus of F restricted to $B(0, R)$.

7.2.3 Dissipative nonlinearities by Euler approximations

A crucial role in the previous section was played by Yosida approximations of the coefficients of the equation. Here we will present a different proof based on Euler approximations of the solutions. We start from a deterministic result.

Let us consider the problem

$$\begin{cases} y'(t) = Ay(t) + G(t, y(t)), & t \geq 0, \\ y(0) = x_0 \in H, \end{cases} \tag{7.45}$$

under the following hypothesis.

Hypothesis 7.6

(i) *$A : D(A) \subset E \to E$ generates a semigroup $S(t)$, $t \geq 0$, on E which is strongly continuous in $(0, +\infty)$.*
(ii) *There exists $\omega \in \mathbb{R}^1$ such that*

$$\|S(t)\| \leq e^{\omega t}, \quad t \geq 0.$$

(iii) *$G : [0, T] \times E \to E$ is continuous.*
(iv) *There exists $\eta \in \mathbb{R}^1$ such that $A + G(t, \cdot) - \eta$ is dissipative in E for any $t \in [0, T]$.*

We say that $u \in C([0, T]; E)$ is a *mild* solution of (7.45) if

$$u(t) = S(t)x_0 + \int_0^t S(t-s)G(s, u(s))ds.$$

To solve problem (7.45) we need a lemma about the construction of approximate solutions, which is a straightforward extension to the nonautonomous case of a result proved in [708], see also [185].

Lemma 7.12 *Assume that conditions* (i) *and* (ii) *of Hypothesis* 7.6 *hold and let $x_0 \in E$. Then there exists $T_0 \in (0, T]$ such that for arbitrary $\varepsilon > 0$ there exist $u_\varepsilon \in C([0, T_0]; E)$ and θ_ε piecewise continuous on $[0, T_0]$, such that*

$$u_\varepsilon(t) = S(t)x_0 + \int_0^t S(t-s)[G(s, u_\varepsilon(s)) + \theta_\varepsilon(s)]ds, \tag{7.46}$$

and

$$\|\theta_\varepsilon(t)\| \leq \varepsilon, \quad \forall\, t \in [0, T_0]. \tag{7.47}$$

Proof In the proof we set $\omega = \eta = 0$ for simplicity. Let us first define for arbitrary $\varepsilon > 0$, $t \in [0, T]$, $x \in E$

$$\rho_\varepsilon(t, x) = \sup\Big\{\delta > 0 : t_1, t_2 \in [0, T], |t_1 - t| < \delta, \ |t_2 - t| < \delta,$$
$$x_1, x_2 \in B(x, \delta) \Longrightarrow \|G(t_1, x_1) - G(t_2, x_2)\| < \varepsilon\Big\}.$$

As is easily checked we have,

$$|\rho_\varepsilon(t,x) - \rho_\varepsilon(s,y)| \le |t-s| + \|x-y\|, \quad \forall\, t,s \in [0,T], \forall\, x,y \in E.$$

So $\rho_\varepsilon : [0,T] \times E \to E$ is Lipschitz continuous. Moreover $\rho_\varepsilon(t,x) > 0$ for all $t \in [0,T]$ and $x \in E$, since G is continuous.

Again by the continuity of G there exist $r > 0$ and $M > 1$ such that

$$\|G(t,x)\| \le M, \quad \forall\, t \in [0,T], \ \forall\, x \in B(x_0,r). \tag{7.48}$$

We now define a function u_ε fulfilling (7.46) and (7.47). Set $t_0 = 0$ and let $\{t_k\}_{k\in\mathbb{N}}$ be positive numbers. Define by recurrence

$$u_\varepsilon(t) = S(t-t_{n-1})x_{n-1} + \int_{t_{n-1}}^t S(t-s)G(s,x_{n-1})ds, \quad t \in [t_{n-1},t_n], \tag{7.49}$$

where $x_{n-1} = u_\varepsilon(t_{n-1})$. Then setting

$$\theta_\varepsilon(t) = G(t,u_\varepsilon(t)) - G(t,x_{n-1}), \quad t \in [t_{n-1},t_n], \tag{7.50}$$

we have

$$u_\varepsilon(t) = S(t-t_{n-1})x_{n-1} + \int_{t_{n-1}}^t S(t-s)[G(s,u_\varepsilon(s)) + \theta_\varepsilon(s)]ds, \tag{7.51}$$

for $t \in [t_{n-1},t_n]$. We want now to show that it is possible to choose $\{t_k\}$ such that

$$\|\theta_\varepsilon(t)\| \le \varepsilon, \quad t \in [t_{n-1},t_n].$$

Suppose we have chosen t_{n-1}. To choose t_n note that by (7.49) we have

$$\|u_\varepsilon(t) - x_{n-1}\| \le \|S(t-t_{n-1})x_{n-1} - x_{n-1}\| + M(t-t_{n-1}), \quad t \in [t_{n-1},t_n].$$

Now let $t_n = s_n \wedge \lambda_n$ where

$$s_n - t_{n-1} = \frac{1}{2M}\rho_\varepsilon(t_{n-1},x_{n-1}) \tag{7.52}$$

and λ_n is such that

$$\sup_{t\in[t_{n-1},\lambda_n]} \|S(t-t_{n-1})x_{n-1} - x_{n-1}\| = \frac{1}{2}\rho_\varepsilon(t_{n-1},x_{n-1}). \tag{7.53}$$

Then if $\|u_\varepsilon(t) - x_0\| \le r$ on $[t_{n-1},t_n]$ we have

$$\|\theta_\varepsilon(t)\| = \|G(t,u_\varepsilon(t)) - G(t,x_{n-1})\| \le \varepsilon, \ t \in [t_{n-1},t_n].$$

It remains to show that the times $\{t_n\}$ can be chosen sufficiently large and such that $\|u_\varepsilon(t) - x_0\| \le r$ on $[0,t_n]$. Let us distinguish two cases.

First case There exists $\bar t > 0$ such that

$$\|u_\varepsilon(t) - x_0\| \le r, \quad \forall\, t \in [0,\bar t], \text{ and } \|u_\varepsilon(\bar t) - x_0\| = r.$$

By (7.49) we have

$$u_\varepsilon(t) = S(t)x_0 + \sum_{k=1}^{n-1} \int_{t_{k-1}}^{t_k} S(t-s)G(s, x_{k-1})ds + \int_{t_{n-1}}^{t} S(t-s)G(s, x_{n-1})ds. \tag{7.54}$$

It follows that

$$r = \|u_\varepsilon(\overline{t}) - x_0\| \le \|S(\overline{t})x_0 - x_0\| + M\overline{t}.$$

Thus there exists $T_0 > 0$, depending only on x_0, r, M, and not on ε such that $\overline{t} \ge T_0 > 0$.

Second case $t_n \uparrow t^*$. If $t^* = +\infty$ the conclusion is obvious. Let us assume that t^* is finite. Then by (7.54) we have

$$x_n - x_{n-1} = S(t_n)x_0 - S(t_{n-1})x_0 + \sum_{k=1}^{n-2} \int_{t_{k-1}}^{t_k} [S(t_n - s) - S(t_{n-1} - s)]G(s, x_{k-1})ds$$
$$+ \int_{t_{n-1}}^{t_n} S(t_n - s)G(s, x_{n-1})ds.$$

It follows that

$$\|x_n - x_{n-1}\| \le \|S(t_n)x_0 - S(t_{n-1})x_0\|$$
$$+ \sum_{k=1}^{n-2} \int_{t_{k-1}}^{t_k} \|[S(t_n - s) - S(t_{n-1} - s)]G(s, x_{k-1})\|ds + M(t_n - t_{n-1}).$$

Thus there exists $x^* \in E$ such that $x_n \to x^*$, and, recalling that ρ_ε is continuous, we have

$$\rho_\varepsilon(t_n, x_n) \to \rho_\varepsilon(t^*, x^*).$$

We notice finally that by (7.52)–(7.53) it follows that $\rho_\varepsilon(t^*, x^*) = 0$, a contradiction.

□

Proposition 7.13 *Assume that Hypothesis* 7.6 *holds. Then for any* $x_0 \in E$, *problem* (7.45) *has a unique mild solution* $y(\cdot, x)$ *in* $[0, +\infty)$. [6]

Proof For any $\varepsilon > 0$ let u_ε be the continuous function in $[0, T_0]$ defined in Lemma 7.12. Then u_ε is the mild solution to the problem

$$\begin{cases} u_\varepsilon'(t) = Au_\varepsilon(t) + F(t, u_\varepsilon(t)) + \theta_\varepsilon(t), & t \in [0, T_0] \\ u_\varepsilon(0) = x_0. \end{cases}$$

[6] In fact one can show that the solution is *strong*, that is for any $T > 0$ there exists a sequence $\{y_n\} \subset C^1([0, T]; E) \cap C([0, T]; D(A))$, such that $y_n \to y(\cdot, x)$, $\frac{d}{dt}y_n - Ay_n - F(y_n) \to 0$, in $C([0, T]; E)$.

For arbitrary $\varepsilon_1, \varepsilon_2 > 0$ we have, by Hypothesis 7.6,

$$\frac{d^+}{dt}\|u_{\varepsilon_1}(t) - u_{\varepsilon_2}(t)\| \leq \varepsilon_1 + \varepsilon_2.$$

Thus there exists $u \in C([0, T_0]; E)$ such that $u_\varepsilon \to u$ in $C([0, T_0]; E)$. Passing to the limit for $\varepsilon \to 0$ in (7.46) it follows that u is a solution to (7.45) in $[0, T_0]$. By a standard extension argument one can show that there is a solution in $[0, T]$. Finally, the uniqueness follows from the dissipativity assumptions. □

We pass now to the main result of the section and consider the problem

$$\begin{cases} dX = (AX + F(X))dt + dW(t), \\ X(0) = x \in E, \end{cases} \tag{7.55}$$

on a Banach space E (norm $\|\cdot\|$) $\subset H$. We assume the following.

Hypothesis 7.7

(i) *$A : D(A) \subset E \to E$ generates a semigroup $S(t)$, $t \geq 0$, on E which is strongly continuous in $(0, +\infty)$.*
(ii) *There exists $\omega \in \mathbb{R}^1$ such that*

$$\|S(t)\| \leq e^{\omega t}, \quad t \geq 0.$$

(iii) *$F : E \to E$ is continuous.*
(iv) *There exists $\eta \in \mathbb{R}^1$ such that $A + F - \eta$ is dissipative.*
(v) *$W(\cdot)$ is a cylindrical Wiener process on H such that the stochastic convolution $W_A(t)$, $t \geq 0$, belongs to $C([0, T]; E)$ for arbitrary $T > 0$.*

We say that $X \in C([0, T]; E)$ is a *mild* solution of (7.55) if

$$X(t) = S(t)x_0 + \int_0^t S(t-s)F(X(s))ds + W_A(t). \tag{7.56}$$

Theorem 7.14 *Assume that Hypothesis 7.7 holds. Then for any $x \in E$ problem (7.55) has a unique mild solution.*

Proof Setting as before $Y(t) = X(t) - W_A(t)$, equation (7.56) reduces to the problem

$$\begin{cases} Y'(t) = AY(t) + F(Y(t) + W_A(t)), \\ Y(0) = x \in E. \end{cases}$$

Now it suffices to set $G(t, z) = F(Y(t) + W_A(t))$ and to apply Proposition 7.13. □

7.2.4 Dissipative nonlinearities and general initial conditions

If $x \in H$ and the conditions of Theorem 7.11 are fulfilled, then we are not able to establish the existence of a mild solution of equation (7.36). However, under additional conditions, we can show, see Theorem 7.15, that there exists a generalized solution in the following sense. A similar result with $H = L^2(0, 1)$, was proved in [128] using a comparison method.

A process X is a *generalized solution* of (7.36) if for arbitrary sequence $\{x_n\} \subset E$ such that $\lim_{n\to\infty} |x - x_n|_H = 0$, the corresponding sequence of solutions $\{X_n\}$ converges to X in $C([0, T]; H)$, $\mathbb{P}$-a.s. Under additional hypotheses, we can identify the generalized solution as a solution of a related integral equation, see Theorem 7.17.

For the rest of Section 7.2 we will always assume, without further comments, that E is a Banach space continuously embedded into a Hilbert space H and $S(t)$, $t \geq 0$, is a C_0-semigroup in H, generated by an operator A. We assume that $S(t)$ acts on E, that the restriction $S_E(t)$ of $S(t)$ to E is either a C_0-semigroup or an analytic semigroup. We will require that for some $\omega \in \mathbb{R}^1$

$$\|S_E(t)\|_E \leq e^{\omega t}, \quad t \geq 0.$$

By $W_A(t)$, $t \geq 0$, we will denote the stochastic convolution

$$W_A(t) = \int_0^t S(t-s)dW(s), \quad t \geq 0.$$

Then we have the following theorem.

Theorem 7.15 *Assume that $F : E \to E$ is a dissipative mapping, uniformly continuous on bounded subsets of E. Let moreover*

$$\langle A_n(x-y) + F(x) - F(y), x - y\rangle_H \leq \eta |x-y|_H^2, \quad \forall\, x, y \in E \tag{7.57}$$

for n sufficiently large where $A_n = nA(n-A)^{-1}$ are the Yosida approximations of A. Assume moreover that W_A has an E-continuous version. Then, for arbitrary $x \in H$, there exists a generalized solution to equation (7.36).

Proof Let $\{x_k\}$ be a sequence in E that converges to x in H. By Theorem 7.11, for any positive integer k there exists a unique solution X_k to the equation

$$X_k(t) = S(t)x_k + \int_0^t S(t-s)F(X_k(s))ds + W_A(t), \quad t \geq 0.$$

Set $Z_{j,k} = X_j - X_k$, then $Z_{j,k}$ is the mild solution of the problem:

$$\begin{cases} \dfrac{d}{dt} Z_{j,k}(t) = AZ_{j,k}(t) + F(X_j(t)) - F(X_k(t)), \\ Z_{j,k}(0) = x_j - x_k. \end{cases} \tag{7.58}$$

Let $Z_{j,k,n} = X_{j,n} - X_{k,n}$, $n \in \mathbb{N}$, be the solution of (7.58) with A replaced by the Yosida approximation A_n. We have by (7.57)

$$\begin{aligned}\frac{1}{2}\frac{d}{dt}|Z_{j,k,n}(t)|^2 &= \langle A_n Z_{j,k,n}(t), Z_{j,k,n}(t)\rangle + \langle F(X_{j,n}(t)) - F(X_{k,n}(t)), Z_{j,k,n}(t)\rangle \\ &\le \eta |Z_{j,k,n}(t)|^2, \quad t \ge 0. \end{aligned} \tag{7.59}$$

Consequently

$$|Z_{j,k,n}(t)|^2 \le e^{\eta t}|x_j - x_k|^2, \quad t \ge 0.$$

The same inequality holds for $Z_{j,k}$. This way the existence of a generalized solution has been shown. Uniqueness follows from a standard argument based on dissipativity. □

Remark 7.16 If $A - \alpha I$ and $F - \beta I$ are dissipative in H for some constants α and β such that $\alpha + \beta < \eta$, then assumption (7.57) is fulfilled. □

Now we will identify the generalized solution. To avoid technicalities we shall assume, from now on, that

$$H = L^2(\mathscr{O}), \quad E = C(\overline{\mathscr{O}}),$$

where $\mathscr{O}$ is a bounded open set in $\mathbb{R}^N$. We will assume that there exists $q \in (1, 2]$ such that we have the following.

Hypothesis 7.8

(i) *$S(\cdot)$ has an extension to a C_0-semigroup $\widetilde{S}(\cdot)$ to $L^q(\mathscr{O})$.*
(ii) *F has a continuous extension $\widehat{F}$ acting from $L^p(\mathscr{O})$ into $L^q(\mathscr{O})$, where $\frac{1}{p} + \frac{1}{q} = 1$.*
(iii) *There exist $c > 0$ and $\beta \in \mathbb{R}$ such that*

$$-\langle F(x) - F(y), x - y\rangle \ge c|x - y|_p^p - \beta|x - y|^2, \quad \forall\, x, y \in E,$$

where $\langle \cdot, \cdot \rangle$ denotes the duality between $L^p(\mathscr{O})$ and $L^q(\mathscr{O})$ and $|\cdot|_p$ the norm on $L^p(\mathscr{O})$.

Theorem 7.17 *Assume that the hypotheses of Theorem 7.15 are satisfied with (7.57) replaced by*

$$\langle A_n x, x\rangle \le \eta |x|^2, \quad x \in E, \tag{7.60}$$

and that condition (7.8) holds. Then the generalized solution X of (7.36) is the unique predictable process on $[0, T]$ such that

$$X \in C([0, T]; L^2(\mathscr{O})) \cap L^p([0, T] \times \mathscr{O}), \quad \mathbb{P}\text{-a.s.}$$

and satisfies the integral equation

$$X(t) = S(t)x + \int_0^t \widetilde{S}(t-s)\widetilde{F}(X(s))ds + W_A(t), \quad t \in [0, T], \ \mathbb{P}\text{-a.s.} \tag{7.61}$$

Proof Proceeding as in the proof of Theorem 7.15 we arrive at the identity (7.59), and taking into account (7.8) we obtain

$$\begin{aligned} \frac{1}{2}\frac{d}{dt}|Z_{j,k}(t)|^2 + c|Z_{j,k}(t)|_p^p &\le \langle AZ_{j,k}(t), Z_{j,k}(t)\rangle + \beta|Z_{j,k}(t)|^2 \\ &\le (\eta+\beta)|Z_{j,k}(t)|^2, \end{aligned} \tag{7.62}$$

for $j, k \in \mathbb{N}$, $t \in [0, T]$. So $\{X_k\}$ is a Cauchy sequence not only in $C([0, T]; L^2(\mathscr{O}))$, as proved in the previous theorem, but also in $L^p([0, T] \times \mathscr{O})$, $\mathbb{P}$-a.s. The approximating processes X_k obviously satisfy the integral equation, with the initial condition x replaced by x_k, so we can pass in the equation to limit for $k \to \infty$, to obtain existence. The uniqueness follows again by a standard dissipativity argument. □

Example 7.8 (Continuation) We continue our discussion and examine applicability of condition 7.8 from Theorem 7.17. Condition 7.8(i) is always satisfied by results of Section A.4.2. Condition 7.8(ii) holds if φ is a polynomial of order p or smaller, by elementary calculations. Finally, condition 7.8(iii) holds if in addition φ is a polynomial of order p with negative leading coefficient. To see this note that the number $\widehat{\beta} = \sup_{\rho \in R^1} p'(\rho)$ is finite and it is enough to define:

$$c = \inf_{\sigma \neq \rho \in \mathbb{R}^1} \frac{|\varphi(\sigma) - \beta\sigma - \varphi(\rho) + \beta\rho|}{|\sigma-\rho|^p}.$$

The reader will check that $c > 0$. □

7.2.5 Dissipative nonlinearities and general noise

We will show that the method of the previous subsection can be used to treat more general equations

$$\begin{cases} dX = (AX + F(X))dt + \Phi dW(t), \\ X(0) = x \in H \end{cases} \tag{7.63}$$

where $\Phi \in \mathcal{N}_W^2(0, T; L_2^0)$. [7] The results are interesting in themselves and they will be used in the next subsection on nonlinear equations with multiplicative noise. Note that if the stochastic convolution W_A^Φ is E-continuous, then problem (7.63) can be treated in exactly the same way as problem (7.36), see Theorem 7.15. As in some

[7] See the definition of $\mathcal{N}_W^2(0, T; L_2^0)$ before Lemma 4.24.

applications this E-continuity property may not take place, we will consider the following approximating problems

$$\begin{cases} dX_n = (AX_n + F(X_n))dt + \Phi_n dW(t), \\ X_n(0) = x_n \in H, \end{cases} \tag{7.64}$$

where Φ_n and x_n are appropriate regularization of Φ and x.

Proposition 7.18 *Assume that the hypotheses of Theorem* 7.15 *are satisfied and that for some $k > 0$, $D(A^k)$ is continuously embedded in E. Define*

$$\Phi_n = J_n^{k+1}\Phi, x_n = J_n^{k+1}x,$$

where $J_n = nR(n, A)$, $n \in \mathbb{N}$. Then equation (7.64) *has a unique E-valued solution X_n, $n \in \mathbb{N}$, and there exists an H-valued, predictable mean square continuous process X, such that*

$$\lim_{n\to\infty} \sup_{t\in[0,T]} \mathbb{E}(|X(t) - X_n(t)|^2) = 0, \quad T > 0.$$

Proof First we show that the stochastic convolution $W_A^{\Phi_n}$ is E-continuous $n \in \mathbb{N}$. By Proposition 6.4(ii) the process $J_n W_A^{\Phi}$ is H-continuous. Moreover

$$D(A^k) = J_n^k(H), \quad W_A^{\Phi_n} = J_n^k(J_n W_A^{\Phi}).$$

By our assumptions on $D(A^k)$, $W_A^{\Phi_n}$ is E-continuous. Applying Itô's formula, [8] we have

$$\begin{aligned} \mathbb{E}|X_n(t) - X_m(t)|^2 &= |x_n - x_m|^2 + \mathbb{E}\int_0^t \|\Phi_n - \Phi_m\|^2_{L_2^0} ds \\ &+ 2\mathbb{E}\int_0^t \langle (A+F)X_n(s) - (A+F)X_m(s), X_n(s) - X_m(s)\rangle ds. \end{aligned} \tag{7.65}$$

We proceed now in exactly same way as in the proof of Theorem 7.15. □

To identify the limit X from the previous proposition as the solution of a suitable integral equation, we will assume

$$H = L^2(\mathscr{O}), \ E = C(\overline{\mathscr{O}}).$$

Theorem 7.19 *Assume that the hypotheses of Theorem* 7.17 *are satisfied and, that for some $k > 0$, $D(A^k)$ is continuously embedded in E. Then the process X given by Proposition* 7.18 *is the unique H-valued predictable, mean square continuous process on* $[0, T]$ *such that*

$$\mathbb{E}\int_0^T |X(t)|_p^p dt < +\infty$$

[8] Strictly speaking we cannot apply Itô's formula to the mild solution X_n, because A is unbounded. So we proceed as usual by replacing A with A_m, the Yosida approximations of A.

and satisfies the integral equation

$$X(t) = S(t)x + \int_0^t \widetilde{S}(t-s)\widetilde{F}(X(s))ds + W_A^{\Phi}(t), \quad t \in [0,T], \ \mathbb{P}\text{-a.s.} \quad (7.66)$$

Proof We will assume, to simplify the presentation, that $\beta = 0$ in Hypothesis 7.8(iii). By (7.65) and taking into account 7.8 and (7.60) we get first

$$\begin{aligned}\mathbb{E}|X_n(t) - X_m(t)|^2 &= |(x_n - x_m)x|^2 + \mathbb{E}\int_0^t \|(\Phi_n - \Phi_m)\|^2_{L_2^0} ds \\ &\quad + 2\mathbb{E}\int_0^t \langle (A+F)X_n(s) - (A+F)X_m(s), X_n(s) - X_m(s)\rangle ds,\end{aligned} \quad (7.67)$$

and then

$$\begin{aligned}&\mathbb{E}|X_n(t) - X_m(t)|^2 + c\int_0^t \mathbb{E}|X_n(s) - X_m(s)|_p^p ds \\ &\leq |x_n - x_m|^2 + \eta\int_0^t \mathbb{E}|X_n(s) - X_m(s)|^2 ds + \int_0^t \mathbb{E}|(\Phi_n - \Phi_m)|^2_{L_2^0} ds.\end{aligned}$$

By the Gronwall lemma it follows that

$$\mathbb{E}|X_n(t) - X_m(t)|^2 \leq |x_n - x_m|^2 e^{\eta t} + \int_0^t e^{\eta(t-s)}\mathbb{E}|(\Phi_n - \Phi_m)|^2_{L_2^0} ds =: \varepsilon_{n,M}(t),$$

and so

$$\begin{aligned}c\int_0^t \mathbb{E}|X_n(s) - X_m(s)|^p ds &\leq |x_n - x_m|^2 + \int_0^t \mathbb{E}|(\Phi_n - \Phi_m)|^2_{L_2^0} ds \\ &\quad + \eta\int_0^t \varepsilon_{n,m}(s)ds.\end{aligned}$$

Therefore the sequence $\{X_n\}$ converges in the required norms. Uniqueness follows once again by Itô's formula. □

7.3 Nonlinear equations on Banach spaces: multiplicative noise

The results of this section were essentially obtained in [575] by variational methods, see also [469]. We follow here the semigroup approach as in [298]. We set $H = L^2(\mathscr{O})$ and $E = C(\overline{\mathscr{O}})$, where $\mathscr{O}$ is an open bounded subset of $\mathbb{R}^N$.

Let us consider the following problem

$$\begin{cases} dX(t) = (AX(t) + F(X(t)))dt + B(X(t))dW(t), \\ X(0) = x \in H \end{cases} \quad (7.68)$$

where A and F are as in the previous section and B is a linear closed operator from its domain $D(B)$ ($D(A) \subset D(B) \subset H$) into L_2^0. In the formulation of the next theorem A_n are the Yosida approximations of A and $B_n(x) = B(J_n x)$, $x \in H$.

By $\Xi := \mathcal{N}_W^2(0, T; D(B))$ we denote the space of all predictable $D(B)$-valued processes $Y(t)$, $t \in [0, T]$ such that

$$\|Y\|_\Xi^2 = \mathbb{E} \int_0^T |Y(s)|_H^2 \, ds + \mathbb{E} \left| \int_0^T B(Y(s)) dW(s) \right|^2$$
$$= \|Y\|^2_{\mathcal{N}_W^2(0,T;D(B))} + \|B(Y)\|^2_{\mathcal{N}_W^2(0,T;L_2^0)}.$$

Theorem 7.20 *Assume that the hypotheses of Theorem 7.19 are satisfied and that there exists $\eta \in \mathbb{R}^1$ and $\delta > 1$ such that*

$$2\langle A_n(x-y), x-y\rangle + \delta \|B_n(x-y)\|^2_{L_2^0} \le \eta |x-y|^2, \qquad \forall\, x, y \in E,\ n \in \mathbb{N}. \tag{7.69}$$

Then there exists a unique predictable process X such that

(i) $X \in \mathcal{N}_W^2(0, T; D(B))$,
(ii) $X \in L^p([0, T] \times \Omega \times \mathcal{O})$,
(iii) *X fulfills the integral equation*

$$X(t) = S(t)x + \int_0^t \widetilde{S}(t-s)\widetilde{F}(X(s))ds + \int_0^t S(t-s)B(X(s))dW(s). \tag{7.70}$$

Proof Fix $T > 0$ and, for any $Y \in \mathcal{N}_W^2(0, T; D(B))$ denote by $Z = \Gamma(Y)$ the solution to

$$\begin{cases} dZ(t) = (AZ(t) + \widetilde{F}(Z(t)))dt + B(Y(t))dW(t), \\ Z(0) = x \in H \end{cases} \tag{7.71}$$

provided by Lemma 7.18. In order to prove the theorem we are going to show that Γ is a contraction in the space Ξ. Let $Y_i \in \mathcal{N}_W^2(0, T : D(B))$, $Z_i = \Gamma(Y_i)$ and $Z_{i,n}$ be solutions of (7.71) corresponding to $Y = Y_i$, $i = 1, 2$. We denote by $Z_{i,n,m}$ solutions to the following equation

$$\begin{cases} dZ_{i,n,m}(t) = [A_n Z_{i,n,m}(t) + \widetilde{F}(Z_{i,n,m}(t))]dt + J_n^m B_i(Y_i(t))dW(t), \\ Z_{i,n,m}(0) = x \in H. \end{cases} \tag{7.72}$$

By the hypotheses and Itô's formula, see previous footnote, we have:

$$\begin{aligned} &\mathbb{E}|Z_{1,n,m}(t) - Z_{2,n,m}(t)|^2 + \delta \mathbb{E} \int_0^t \|B(Z_{1,n,m}(s)) - B(Z_{2,n,m}(s))\|^2_{L_2^0} ds \\ &\quad + \mathbb{E} \int_0^t |Z_{1,n,m}(s) - Z_{2,n,m}(s))|_p^p ds \\ &\quad \le \mathbb{E} \int_0^t \|J_n^m B(Y_1(s)) - J_n^m B(Y_2(s))\|^2_{L_2^0} ds + \eta \mathbb{E} \int_0^t |Z_{1,m}(s) - Z_{2,m}(s))|^2 ds. \end{aligned} \tag{7.73}$$

In particular,

$$\mathbb{E}|Z_{1,n,m}(t) - Z_{2,n,m}(t)|^2 \leq \eta\mathbb{E}\int_0^t |Z_{1,n,m}(s) - Z_{2,n,m}(s)|^2 ds$$
$$+\mathbb{E}\int_0^t \|J_n^m B(Y_1(s)) - J_n^m B(Y_2(s))\|_{L_2^0}^2\, ds.$$

Using Gronwall's lemma and letting m and n tend to infinity, we find the following two estimates

$$\mathbb{E}|Z_1(t) - Z_2(t)|^2 \leq \int_0^t e^{(t-s)\eta}\mathbb{E}\|B(Y_1(s)) - B(Y_2(s))\|_{L_2^0}^2 ds,$$

and

$$\delta\mathbb{E}\int_0^t \|B(Z_1(s)) - B(Z_2(s))\|_{L_2^0}^2 ds \leq \mathbb{E}\int_0^t \|B(Y_1(s)) - B(Y_2(s))\|_{L_2^0}^2\, ds$$
$$+\eta T e^{\eta T}\mathbb{E}\int_0^T \|B(Y_1(s)) - B(Y_2(s))\|_{L_2^0}^2 ds.$$

It follows from the above inequalities that

$$\|Z_1 - Z_2\|^2_{\mathcal{N}_W^2(0,T;H)} \leq T e^{T\eta}\|B(Y_1) - B(Y_2)\|^2_{\mathcal{N}_W^2(0,T;L_2^0)},$$

$$\delta\|BZ_1 - BZ_2\|^2_{\mathcal{N}_W^2(0,T;H)} \leq (1 + \eta T e^{T\eta})\|B(Y_1) - B(Y_2)\|^2_{\mathcal{N}_W^2(0,T;L_2^0)}.$$

Therefore

$$\|\Gamma(Y_1) - \Gamma(Y_2)\|_{\Xi}^2 \leq \left[T e^{T\eta} + \frac{1 + \eta T e^{T\eta}}{\delta}\right]\|Y_1 - Y_2\|_{\Xi}^2.$$

Since $\delta > 1$, the transformation Γ is a contraction on Ξ, provided T is sufficiently small, and so it has a unique a fixed point Y in Ξ. Moreover, by using (7.73) it follows that Y is the unique solution with the required integrability properties. □

Remark 7.21 If the operator A is self-adjoint negative definite, then the condition (7.69) can be replaced by the following, simpler to handle, condition. There exists $\eta \in \mathbb{R}^1$ and $\delta > 1$ such that

$$2\langle Ax, x\rangle + \delta\|B(x)\|_{L_2^0}^2 \leq \eta|x|^2, \quad \forall\, x, y \in E. \tag{7.74}$$

To check that this is the case, it is enough to use the following consequence of spectral decomposition for self-adjoint operators:

$$\langle AJ_n x, J_n x\rangle \geq \langle AJ_n x, x\rangle, \quad \forall\, x \in D(A).$$

□

Example 7.22 Consider the stochastic equation:

$$\begin{cases} dy(t,\xi) = (\Delta y(t,\xi) + \varphi(y(t,\xi))dt + \langle v(\xi), \nabla y(t,\xi)\rangle dW(t), & t \geq 0,\ \xi \in \mathscr{O}, \\ y(t,\xi) = 0, \quad t \geq 0,\ \xi \in \partial\mathscr{O}, \\ y(0,\xi) = x(\xi), \quad \xi \in \mathscr{O},\ x \in L^2(\mathscr{O}) \end{cases} \tag{7.75}$$

with one dimensional Brownian motion $W(\cdot)$ and assume, similarly as in Example 7.8, that $\mathscr{O}$ is an open bounded set of $\mathbb{R}^N$ with a regular boundary $\partial\mathscr{O}$, and φ a polynomial of odd order p, $p \geq 2$, with negative leading coefficient. Moreover, let v be a regular vector field on $\mathscr{O}$ such that

$$\|v\| = \sup_{\xi\in\overline{\mathscr{O}}} |v(\xi)| < \sqrt{2}.$$

Defining

$$Ay = \Delta_\xi y + \beta y, \quad \forall\, y \in H^2(\mathscr{O}) \cap H_0^1(\mathscr{O}) = D(A),$$

$$By(\xi) = \langle v(\xi), \nabla y(\xi)\rangle_{\mathbb{R}^N}, \quad \forall\, y \in H_0^1(\mathscr{O}) = D(B),$$

$$F(y) = \varphi(y) - \beta y, \quad \forall\, y \in L^q(\mathscr{O}),\ 1/p + 1/q = 1,$$

and arguing as in Example 7.8, we see that Theorem 7.20 applies. Consequently equation (7.75) has a unique predictable solution y such that

$$\mathbb{E}\left(\int_0^T \int_{\mathscr{O}} (|\nabla_\xi y(t,\xi)|^2 + |y(t,\xi)|^p) dt\, d\xi\right) < +\infty.$$

□

7.4 Strong solutions

In this section we will choose as Hilbert space $H = L^2(\mathbb{R}^d)$, where d is a positive integer. We want to find strong solutions to the equation

$$\begin{cases} dX = (AX + F(X))dt + \sum_{k=1}^N B_k X d\beta_k, \\ X(0) = x \in H, \end{cases} \tag{7.76}$$

which in addition live in a Banach space of continuous functions. We shall use the framework of Section 6.6, in particular we shall assume that $A\colon D(A) \subset H \to H$, and $B_k : D(B_k) \subset H \to H$, $k \in \mathbb{N}$, are generators of semigroups $S(t) = e^{tA}$ and $S_k(t) = e^{tB_k}$ respectively, such that Hypothesis 6.5 holds. Moreover we shall introduce the Banach space E of all functions u on $\mathbb{R}^d$ which are uniformly continuous and bounded on bounded subsets and such that

$$\|u\| = \sup_{\xi\in\mathbb{R}^d} \frac{|f(\xi)|}{(1+|\xi|^2)^{d/2}} < +\infty.$$

Clearly $E \subset H = L^2(\mathbb{R}^d)$. We shall denote by $A_1, B_{E,1}, \ldots, B_{E,N}$ the parts of $A, B_1, \ldots, B_N$ in E and we shall assume the following.

Hypothesis 7.9

(i) The operators $B_{E,1}, \ldots, B_{E,N}$ generate mutually commuting C_0-groups $S_{E,k}(t)$, $t \in \mathbb{R}^1$, in E.

(ii) The operator

$$C_E = A_E - \frac{1}{2}\sum_{k=1}^{N} B_{E,k}^2, \quad D(C_E) = D(A_E)$$

is the infinitesimal generator of a C_0-semigroup $S_{E,0}(t) = e^{tC_E}$.

We set

$$U_E(t) = \prod_{k=1}^{N} S_{E,k}(\beta_k(t)).$$

Concerning the nonlinear mapping $F : D(F) \subset H \to H$ we shall assume the following.

Hypothesis 7.10 *$E \subset D(F)$ and the restriction F_E of F to E is uniformly continuous and bounded on bounded sets of E.*

By proceeding as in as in Section 6.6, see also [211], we set $v(t) = U_E^{-1}(t)X(t)$ and reduce problem (7.76) to the deterministic equation

$$\begin{cases} v'(t) = U_E^{-1}(t)C_E U_E(t)v(t) + U_E^{-1}(t)F_E(U_E(t)v(t)), \\ v(0) = x \in E, \end{cases} \tag{7.77}$$

which can be studied by analytic methods. By Proposition 6.31 we have the following result.

Theorem 7.23 *Assume Hypotheses* 7.9–7.10. *If X is a strong solution to* (7.76) *then the process v satisfies* (7.77). *Conversely, if v is a predictable process such that*

(i) trajectories of v are of class C^1 and satisfy (7.77),
(ii) the process $X(\cdot) = U_E(\cdot)v(\cdot)$ takes values in $D(C)$, $\mathbb{P}_T$-a.s.,

then the process X is a strong solution of (7.76).

8
Martingale solutions

This chapter is devoted to a proof of existence of solutions of equations with only continuous operators F and B and with the operator A generating a compact semigroup. Weak convergence of approximating solutions is proved by using the factorization procedure.

8.1 Introduction

Let us assume that A is the infinitesimal generator of a C_0-semigroup $S(\cdot)$ on a separable Hilbert space H, and F is a measurable transformation from H into H. Let Q be a symmetric positive definite operator on a Hilbert space U and B a measurable transformation from H into $L_2(U_0, H)$ where the space $U_0 = Q^{1/2}U$ is endowed with the inner product

$$\langle u, v\rangle_0 = \langle Q^{-1/2}u, Q^{-1/2}v\rangle, \quad u, v \in U_0.$$

If there exists a probability space $(\Omega, \mathscr{F}, \mathbb{P})$, with a filtration $\{\mathscr{F}_t\}$, a Q-Wiener process W and a mild solution X of the equation

$$\begin{cases} dX = (AX + F(X))dt + B(X)dW(t) \\ X(0) = x \end{cases} \tag{8.1}$$

then one says that the equation (8.1) has a *martingale solution*. Thus for given data

$$H, A, F, Q, U, B, x \tag{8.2}$$

one looks for probabilistic quantities

$$\Omega, \mathscr{F}, \mathbb{P}, \{\mathscr{F}_t\}, W, X, \tag{8.3}$$

such that X is a mild solution to (8.1). The sequence

$$(\Omega, \mathscr{F}, \mathbb{P}, \{\mathscr{F}_t\}, W, X)$$

is called a *martingale solution* to (8.1).

For stochastic equations in finite dimensional spaces, solutions of the introduced type are called *weak solutions*; very often they are also solutions to the so called *martingale problem*; see [665] and [428]. Since in PDEs and in the present book the word *weak* has a different meaning, we decided, to avoid ambiguities, to use the term *martingale solutions*.

In previous chapters we have formulated conditions under which for given deterministic and probabilistic data one can find a mild solution of (8.1). Now we proceed to the problem of the existence of a martingale solution to (8.1) in the situation where a mild solution on a fixed probability space does not exist in general.

There are basically two methods of constructing martingale solutions. The first one is the so called *compactness method* and we will treat it by the factorization method introduced earlier for regularity purposes. The second method is based on the *Girsanov theorem* and will be discussed in Section 10.3.

The first method, which will be used in this section, can be described as follows. We start with solutions X_n of equation (8.1) with regular coefficients F_n, B_n, $n \in \mathbb{N}$. These solutions can be constructed on a probability space $(\Omega, \mathscr{F}, \mathbb{P})$, with a filtration $\{\mathscr{F}_t\}$ and a Q-Wiener process W. Next we show, using the factorization procedure, that the sequence of laws $\{\mathscr{L}(X_n)\}$ is weakly convergent on $C([0, T]; H)$ to a measure μ. Then we construct a solution X of (8.1), with the law μ, on a new probability space $(\widetilde{\Omega}, \widetilde{\mathscr{F}_t}, \widetilde{\mathbb{P}})$, for a new filtration $\widetilde{\mathscr{F}_t}$, and with respect to a new Q-Wiener process $\widetilde{W}$. The main tools we use at this final step, are the Skorohod embedding theorem, see Section 2.1, and a representation theorem for martingales, see Section 8.2. This way we will arrive at following theorem.

Theorem 8.1 *Assume that the operators* $S(t)$, $t > 0$, *are compact. Let moreover* F *and* B *be continuous mappings from* H *into* H *and from* H *into* L_2^0 *respectively. If in addition* $\mathrm{Tr}Q < \infty$ *and* F *and* B *satisfy the linear growth condition*

$$|F(x)| + \|B(x)\|_{L_2^0} \leq C(1 + |x|), \quad x \in H, \tag{8.4}$$

then there exists a martingale solution to (8.1).

The idea to use the factorization procedure in the compactness part of the proof is due to [337]. It seems to simplify earlier proofs of [542, 695].

Continuity of F and B and the linear growth condition (8.4) are not enough for existence of martingale solutions. This is so even in the deterministic case. [(1)] Examples with F continuous and bounded $B = 0$, $A = 0$ can be found for instance in [261, page 287]. For deterministic results similar to Theorem 8.1 we refer to [587].

[(1)] Compare the beginning of Section 7.2.2.

8.2 Representation theorem

In the present section we show that a large class of continuous martingales $M \in \mathscr{M}_T^2(H)$ can be represented as stochastic integrals with respect to Wiener processes. This representation result will play an important role in the construction of martingale solutions.

Theorem 8.2 *Assume that $M \in \mathscr{M}_T^2(H)$ and*

$$\langle\langle M(t) \rangle\rangle = \int_0^t (\Phi(s)Q^{1/2})(\Phi(s)Q^{1/2})^* ds, \quad t \in [0, T],$$

where Φ is a predictable L_2^0-valued process [(2)] *and Q a given bounded, symmetric non-negative operator in U. Then there exists a probability space $(\widetilde{\Omega}, \widetilde{\mathscr{F}}, \widetilde{\mathbb{P}})$, a filtration $\{\widetilde{\mathscr{F}}_t\}$ and a Q-Wiener process W with values in U defined on $(\Omega \times \widetilde{\Omega}, \mathscr{F} \times \widetilde{\mathscr{F}}, \mathbb{P} \times \widetilde{\mathbb{P}})$ adapted to $\{\mathscr{F}_t \times \widetilde{\mathscr{F}}_t\}$,* [(3)] *such that*

$$M(t, \omega, \widetilde{\omega}) = \int_0^t \Phi(s, \omega, \widetilde{\omega}) dW(s, \omega, \widetilde{\omega}), \quad t \in [0, T], \ (\omega, \widetilde{\omega}) \in \Omega \times \widetilde{\Omega}, \quad (8.5)$$

where

$$M(t, \omega, \widetilde{\omega}) = M(t, \omega), \quad \Phi(t, \omega, \widetilde{\omega}) = \Phi(t, \omega), \quad t \in [0, T], \ (\omega, \widetilde{\omega}) \in \Omega \times \widetilde{\Omega}. \quad (8.6)$$

Proof It is enough to prove the theorem for $Q = I$. The general case can be obtained by finding first the representation

$$M(t) = \int_0^t \Phi(s)Q^{1/2} dW(s), \quad t \in [0, T],$$

with the cylindrical I-Wiener process and then defining the required Q-Wiener process as $Q^{1/2}W(\cdot)$.

The proof will be done into two steps. We will show first that M has a representation as a stochastic integral but different from the required one. The second step will consist in properly modifying the obtained representation.

Step 1 Applying Proposition 1.8 to the predictable process Q_Φ

$$Q_\Phi(t, \omega) = \Phi(t, \omega)\Phi^*(t, \omega), \quad (t, \omega) \in \Omega_T$$

treated as an $L_1(H)$-valued random variable defined on $(\Omega_T, \mathscr{P}_T)$, we get the following spectral decomposition

$$Q_\Phi(t, \omega) = \sum_{n=1}^{\infty} \lambda_n(t, \omega) g_n(t, \omega) \otimes g_n(t, \omega), \quad (t, \omega) \in \Omega_T. \quad (8.7)$$

(2) We recall that $L_2^0 = L_2(U_0, H)$.

(3) Without saying explicitly, we always assume that in addition increments $W(t) - W(s)$ are independent of $\mathscr{F}_s$, $t \geq s$.

Let us fix an orthonormal basis $\{f_n\}$ in U and define an operator valued process V by the formula

$$V(t,\omega) = \sum_{n=1}^{\infty} g_n(t,\omega) \otimes f_n, \quad (t,\omega) \in \Omega_T.$$

Then

$$V^*(t,\omega) = \sum_{n=1}^{\infty} f_n \otimes g_n(t,\omega), \quad (t,\omega) \in \Omega_T.$$

Note that the process $\Lambda(\cdot) = V^*(\cdot)Q_\Phi(\cdot)V(\cdot)$ has the diagonal decomposition

$$\Lambda(t,\omega) = \sum_{n=1}^{\infty} \lambda_n(t,\omega) f_n \otimes f_n, \quad (t,\omega) \in \Omega_T. \tag{8.8}$$

If we define

$$N(t) = \int_0^t V^*(s)dM(s), \quad t \in [0,T] \tag{8.9}$$

then, compare Section 4.7,

$$\langle\langle N(t) \rangle\rangle = \int_0^t V^*(s)Q_\Phi(s)V(s)ds = \int_0^t \Lambda(s)ds, \quad t \in [0,T].$$

Consequently the quadratic variation of the martingale N also has a diagonal decomposition and for the coordinate process $N_n(\cdot)$

$$\langle\langle N_n(t) \rangle\rangle = \int_0^t \lambda_n(s)ds, \quad t \in [0,T], \quad n \in \mathbb{N}. \tag{8.10}$$

Let $(\widehat{\Omega}, \widehat{\mathscr{F}}, \widehat{\mathbb{P}})$ be a new probability space with filtration $\{\widehat{\mathscr{F}_t}\}$ and let $\{\alpha_n\}$ be a sequence of independent normalized Wiener processes adapted to $\{\widehat{\mathscr{F}_t}\}$. Define the new probability space $(\Omega \times \widehat{\Omega}, \mathscr{F} \times \widehat{\mathscr{F}}, \mathbb{P} \times \widehat{\mathbb{P}})$, the filtration $\{\mathscr{F}_t \times \widehat{\mathscr{F}_t}\}$, and extend (trivially) the processes M, N and α_n to $\Omega \times \widehat{\Omega}$ by setting, for $t \in [0,T]$ and $(\omega, \widehat{\omega}) \in (\Omega \times \widehat{\Omega})$,

$$M(t,\omega,\widehat{\omega}) = M(t,\omega), \quad N(t,\omega,\widehat{\omega}) = N(t,\omega), \quad \alpha_n(t,\omega,\widehat{\omega}) = \alpha_n(t,\widehat{\omega}).$$

Obviously M and N are $\{\mathscr{F}_t \times \widehat{\mathscr{F}_t}\}$ martingales and α_n, $n \in \mathbb{N}$, are independent Wiener processes with respect to $\{\mathscr{F}_t \times \widehat{\mathscr{F}_t}\}$. Define in addition

$$\widehat{\beta}_n = \int_0^t \gamma_n(s)d\alpha_n(s) + \int_0^t \delta_n(s)dN_n(s), \quad t \in [0,T], \ n \in \mathbb{N}, \tag{8.11}$$

where

$$\begin{aligned} \gamma_n(t,\omega) &= \begin{cases} 0 & \text{if } \lambda_n(t,\omega) > 0 \\ 1 & \text{if } \lambda_n(t,\omega) = 0 \end{cases} \\ \delta_n(t,\omega) &= \begin{cases} (\lambda_n(t,\omega))^{-1/2} & \text{if } \lambda_n(t,\omega) > 0 \\ 1 & \text{if } \lambda_n(t,\omega) = 0. \end{cases} \end{aligned} \tag{8.12}$$

Since the processes α_n and N_m are mutually independent, we can check easily that, see Proposition 3.13,

$$\langle\langle \widehat{\beta}_n(t), \widehat{\beta}_m(t) \rangle\rangle = 0 \quad \text{if } n \neq m, \; \langle\langle \widehat{\beta}_n(t) \rangle\rangle = t.$$

So, by Theorem 4.6, the processes $\widehat{\beta}_n(\cdot)$, $n \in \mathbb{N}$, are independent real valued Wiener processes (with respect to $\{\mathscr{F}_t \times \widehat{\mathscr{F}_t}\}$), and therefore the process

$$\widehat{W}(t) = \sum_{n=1}^{\infty} \widehat{\beta}_n(t) f_n, \quad t \in [0, T]$$

is a cylindrical I-Wiener process. It follows from (8.12) that for $n \in \mathbb{N}$,

$$\int_0^t \lambda_n^{1/2}(s) d\widehat{\beta}_n(s) = \int_0^t \lambda_n^{1/2}(s) \delta_n(s) dN_n(s), \quad t \in [0, T].$$

Moreover from (8.10)–(8.12)

$$\left\langle\left\langle \int_0^t \lambda_n^{1/2}(s) \delta_n(s) dN_n(s) - N_n(t) \right\rangle\right\rangle$$
$$= \int_0^t [\lambda_n^{1/2}(s) \delta_n(s) - 1]^2 \lambda_n(s) ds = 0, \quad t \in [0, T].$$

Consequently, by Proposition 3.11,

$$N_n(t) = \int_0^t \lambda_n^{1/2}(s) d\widehat{\beta}_n(s), \quad t \in [0, T], \; n \in \mathbb{N}.$$

We have shown in this way that

$$N(t) = \int_0^t \Lambda^{1/2}(s) d\widehat{W}(s), \quad t \in [0, T].$$

We will show now that

$$M(t) = \int_0^t V(s) \Lambda^{1/2}(s) d\widehat{W}(s), \quad t \in [0, T]. \tag{8.13}$$

Note that

$$\int_0^t V(s) dN(s) = \int_0^t V(s) V^*(s) dM(s) = \int_0^t \Pi(s) dM(s), \quad t \in [0, T],$$

where $\Pi(s)$ is the projector [(4)] on the closure of the image of $Q_\Phi(s)$, $s \in [0, T]$. Let $\Pi^\perp(s) = I - \Pi_s$, $s \in [0, T]$, then

$$M(t) = \int_0^t \Pi^\perp(s) dM(s) + \int_0^t \Pi(s) dM(s) = M_0(t) + M_1(t), \quad t \in [0, T].$$

[(4)] By a projector we always mean an orthogonal projector.

Since

$$\langle\langle M_0(t)\rangle\rangle = \int_0^t \Pi^\perp(s)Q_\Phi(s)\Pi^\perp(s)ds = 0,$$

therefore $M_0 \equiv 0$ and consequently

$$M(t) = \int_0^t V(s)dN(s)$$

so (8.13) holds.

Step 2 Although (8.13) is a representation of M as a stochastic integral, the integrand $\widehat{\Phi}(s) = V(s)\Lambda^{1/2}(s)$, $s \in [0, T]$, is different from the required one. To construct the proper modification of the obtained representation let us remark that

$$\Phi(s)\Phi^*(s) = \widehat{\Phi}(s)\widehat{\Phi}^*(s), \quad s \in [0, T].$$

By Remark B.4

$$\widehat{\Phi}(s) = \Phi(s)J(s), \quad s \in [0, T],$$

where $J(s)J^*(s)$ is the projector onto $\mathrm{Ker}(\Phi(s))^\perp$. Moreover, from the explicit formula in Remark B.4, predictability of the operator valued processes $J(\cdot)$ and $J(\cdot)J^*(\cdot)$ follows.

Let us consider an additional probability space $(\widehat{\widehat{\Omega}}, \widehat{\widehat{\mathscr{F}}}, \widehat{\widehat{\mathbb{P}}})$ with filtration $\{\widehat{\widehat{\mathscr{F}_t}}\}$ and a cylindrical Wiener process $\widehat{\widehat{W}}$. Define

$$\widetilde{\Omega} = \widehat{\Omega} \times \widehat{\widehat{\Omega}}, \quad \widetilde{\mathscr{F}} = \widehat{\mathscr{F}} \times \widehat{\widehat{\mathscr{F}}}$$

$$\widetilde{\mathbb{P}} = \widehat{\mathbb{P}} \times \widehat{\widehat{\mathbb{P}}}, \quad \widetilde{\mathscr{F}_t} = \widehat{\mathscr{F}_t} \times \widehat{\widehat{\mathscr{F}}}_t$$

and extend (trivially) the processes $M(\cdot)$, $\Phi(\cdot)$, $\widehat{\Phi}(\cdot)$, $\widehat{W}(\cdot)$, $\widehat{\widehat{W}}(\cdot)$ onto $\Omega \times \widetilde{\Omega}$. Let

$$W(t) = \int_0^t J(s)d\widehat{W}(s) + \int_0^t K(s)d\widehat{\widehat{W}}(s),$$

where $K(s) = (J(s)J^*(s))^\perp$, $s \in [0, T]$. Since

$$\langle\langle W(t)\rangle\rangle = \int_0^t [J(s)J^*(s) + K(s)]ds = tI$$

therefore, by Theorem 4.6, $W(\cdot)$ is a cylindrical Wiener process on U. Moreover

$$\int_0^t \Phi(s)dW(s) = \int_0^t \Phi(s)J(s)d\widehat{W}(s) + \int_0^t \Phi(s)K(s)d\widehat{\widehat{W}}(s)$$

$$= \int_0^t \widehat{\Phi}(s)d\widehat{W}(s) = M(t), \ t \in [0, T].$$

This finishes the proof. □

8.3 Compactness results

Let us fix a complete orthonormal basis in H and let Π_n be the projectors on $H_n = \text{lin}\{e_1, \ldots, e_n\}$. It is clear that the mappings

$$\widehat{F_n}(x) = \Pi_n(F(\Pi_n x)), \quad \widehat{B_n}(x) = \Pi_n(B(\Pi_n x)), \quad x \in H,$$

satisfy the estimate (8.4). By a finite dimensional result, mappings $\widehat{F_n}(x)$, $\widehat{B_n}(x)$, $x \in H$, $n \in \mathbb{N}$, can be approximated uniformly by a Lipschitz continuous mapping satisfying (8.4) with possibly another constant c independent of n. Therefore, under the assumptions of the theorem, one can find sequences $\{F_n\}$ and $\{B_n\}$ such that

$$|F_n(x)| + \|B_n(x)\|_{L_2^0} \le C(1 + |x|), \quad x \in H, \ n \in \mathbb{N} \tag{8.14}$$

and for which the conditions of Theorem 7.2 are satisfied. Therefore on a given probability space $(\Omega, \mathscr{F}, \mathbb{P})$ with a filtration $\{\mathscr{F}_t\}$ and Q-Wiener process $W(\cdot)$ there exists a sequence of mild solutions $\{X_n\}$ of the problems

$$\begin{cases} dX_n = (AX_n + F_n(X_n))dt + B_n(X_n)dW(t), \\ X_n(0) = x \in H. \end{cases} \tag{8.15}$$

We proceed now to the main result of this section. Its proof follows [337].

Theorem 8.3 *Let $\{X_n\}$ be solutions of problem* (8.15) *with cofficients satisfying* (8.14). *Then the laws $\{\mathscr{L}(X_n)\}$ form a tight family of probability measures on $C([0, T]; H)$.*

We start from the proposition.

Proposition 8.4 *If $S(t)$, $t > 0$, are compact operators and $0 < \frac{1}{p} < \alpha \le 1$, then the operator G_α*

$$G_\alpha f(t) = \int_0^t (t - s)^{\alpha - 1} S(s) f(s) ds, \quad f \in L^p(0, T; H), \ t \in [0, T],$$

is compact from $L^p(0, T; H)$ into $C([0, T]; H)$.

Proof Denote by $|\cdot|_p$ the norm in $L^p(0, T; H)$. According to the infinite dimensional version of the Ascoli–Arzelá theorem one has to show the following.

(i) For arbitrary $t \in [0, T]$ the sets

$$\{G_\alpha f(t) : \ |f|_p \le 1\} \tag{8.16}$$

are relatively compact in H.

(ii) For arbitrary $\varepsilon > 0$ there exists $\delta > 0$ such that

$$|G_\alpha f(t) - G_\alpha f(s)| \le \varepsilon, \ \text{ if } |f|_p \le 1, \ |t - s| \le \delta, \quad t \in [0, T]. \tag{8.17}$$

Let us fix $t \in (0, T]$ and define for $\varepsilon \in (0, t)$ operators G_α^ε from $L^p(0, T; H)$ into H

$$G_\alpha^\varepsilon f = \int_0^{t-\varepsilon} (t-s)^{\alpha-1} S(t-s) f(s) ds, \quad f \in L^p(0, T; H).$$

Since

$$G_\alpha^\varepsilon f = S(\varepsilon) \int_0^{t-\varepsilon} (t-s)^{\alpha-1} S(t-\varepsilon-s) f(s) ds, \quad f \in L^p(0, T; H)$$

and $S(\varepsilon)$, $\varepsilon > 0$, is compact by assumption, therefore operators G_α^ε are compact. Moreover, setting $M = \sup_{t \in [0,T]} \|S(t)\|$, we have, using the Hölder inequality and setting $q = \frac{p}{p-1}$

$$\begin{aligned}
|G_\alpha f(t) - R_\alpha^\varepsilon f(t)| &= \left| \int_{t-\varepsilon}^t (t-s)^{\alpha-1} S(t-s) f(s) ds \right| \\
&\le \left(\int_{t-\varepsilon}^t (t-s)^{(\alpha-1)q} \|S(t-s)\|^q ds \right)^{1/q} \left(\int_{t-\varepsilon}^t |f(s)|^p ds \right)^{1/p} \\
&\le M \left(\frac{\varepsilon^{(\alpha-1)q+1}}{(\alpha-1)q+1} \right)^{1/q} |f|_p.
\end{aligned}$$

But $\frac{(\alpha-1)q+1}{q} = \alpha - \frac{1}{p} > 0$. Consequently $G_\alpha^\varepsilon \to G_\alpha$ in the operator norm so that G_α is compact and (i) follows immediately. Now, for $0 \le t \le t+u \le T$ and $|f|_p \le 1$, we have

$$\begin{aligned}
&|G_\alpha f(t+u) - G_\alpha f(t)| \\
&\quad \le \int_0^t |(t+u-s)^{\alpha-1} S(t+u-s) - (t-s)^{\alpha-1} S(t-s)||f(s)| ds \\
&\quad + \int_t^{t+u} |(t+u-s)^{\alpha-1} S(t+u-s) f(s)| ds.
\end{aligned}$$

So

$$\begin{aligned}
|G_\alpha f(t+u) - G_\alpha f(t)| &\le \left(\int_0^T |(u+s)^{\alpha-1} S(u+s) - s^{\alpha-1} S(s)|^q ds \right)^{1/q} |f|_p \\
&\quad + M \left(\int_0^T s^{(\alpha-1)q} ds \right) |f|_p \\
&\le \left(\int_0^T |(u+s)^{\alpha+1} S(u+s) - s^{\alpha-1} S(s)|^q ds \right)^{1/q} |f|_p \\
&\quad + M \frac{u^{\alpha-1/p}}{((\alpha-1)q+1)^{1/q}} |f|_p = I_1 + I_2.
\end{aligned}$$

It is clear that $I_2 \to 0$ as $u \to 0$. Since the semigroup $S(\cdot)$ is compact therefore $\|S(u+s) - S(s)\| \to 0$ as $u \to 0$ for arbitrary $s > 0$. Moreover

$$\|(u+s)^{\alpha+1} S(u+s) - s^{\alpha-1} S(s)\|^q \le 2M s^{(\alpha-1)q}, \quad s \in [0, T], u \ge 0.$$

Then by Lebesgue's dominated convergence theorem $I_2 \to 0$ as $u \to 0$. Thus the proof of (ii) and therefore the proof of the proposition is complete. □

Proof of Theorem 8.3 Let us fix numbers $p > 2$ and $\alpha \in (0, 1)$ such that $0 < \frac{1}{p} < \alpha < \frac{1}{2}$. It follows from the first part of the proof of Theorem 7.2 that there exists a constant $c_p > 0$ such that

$$\mathbb{E}|X_n(t)|^p \le c_p, \quad n \in \mathbb{N},\ t \in [0, T]. \tag{8.18}$$

By the stochastic Fubini theorem we have the following factorization formula

$$\int_0^t S(t-s)\Phi(s)dW(s) = \frac{\sin \pi\alpha}{\pi} G_\alpha Y(t), \quad t \in [0, T],$$

where

$$Y(s) = \int_0^s (s-r)^{-\alpha} S(s-r)\Phi(r)dW(r), \quad s \in [0, T].$$

Therefore

$$\begin{aligned} X_n(t) &= S(t)x + \int_0^t S(t-s)F_n(X_n(s))ds + \int_0^t S(t-s)B_n(X_n(s))dW(s) \\ &= S(t)x + G_0(F_n(X_n))(t) + G_\alpha(Y_n)(t) \end{aligned}$$

where

$$Y_n(t) = \int_0^t (t-s)^\alpha S(t-s)B_n(X_n(s))dW(s).$$

As in the proof of the basic inequality (4.48) one obtains that there exists a constant $\widehat{c}_p$ such that for all $n \in \mathbb{N}$

$$\mathbb{E}\int_0^T |Y_n(s)|^p ds \le \widehat{c}_p\, \mathbb{E}\left(\int_0^T \|B_n(X_n(s))\|^2_{L_2^0} ds\right)^{p/2}.$$

Taking into account (8.4) we see that for some constants a_p, b_p

$$\mathbb{E}\int_0^T |Y_n(s)|^p ds \le c^p \widehat{c}_p \mathbb{E}\int_0^T (1 + |X_n(s)|)^p ds \le a_p + b_p \mathbb{E}\int_0^T |X_n(s)|^p ds,$$

and by (8.18)

$$\mathbb{E}\int_0^T |Y_n(s)|^p ds \le a_p + T b_p c_p.$$

We are now ready to show tightness of $\mathscr{L}(X_n)$, $n \in \mathbb{N}$. It follows from (8.18) and from Chebishev's inequality that for $\varepsilon > 0$ one can find $r > 0$ such that for all $n \in \mathbb{N}$

$$\mathbb{P}\left(\left(\int_0^T |Y_n(s)|^p ds\right)^{1/p} \leq r \quad \text{and} \quad \left(\int_0^T |F_n(Y_n(s))|^p ds\right)^{1/p} \leq r\right) \leq 1 - \varepsilon. \tag{8.19}$$

By Proposition 8.4 the set

$$K = \{S(\cdot)x + G_\alpha f(\cdot) + G_0 g(\cdot) : \ \|f\|_p \leq r, \|g\|_p \leq r\}$$

is compact. It follows from (8.15) that

$$\mathscr{L}(X_n)(K) \geq 1 - \varepsilon, \quad n \in \mathbb{N}$$

and the tightness follows. □

8.4 Proof of the main theorem

We can assume that the sequence $\{\mathscr{L}(X_n)\}$ is weakly convergent to a measure μ on $C = C([0,T]; H)$. By the Skorohod theorem (Theorem 2.4) there exists a probability space $(\widetilde{\Omega}, \widetilde{\mathscr{F}}, \widetilde{\mathbb{P}})$ and C-valued random variables $\widetilde{X}, \widetilde{X}_n$, $n \in \mathbb{N}$, such that $\mathscr{L}(\widetilde{X}) = \mu$, $\mathscr{L}(\widetilde{X}_n) = \mathscr{L}(X_n)$, $n \in \mathbb{N}$, and $\widetilde{X}_n \to \widetilde{X}$ as $n \to \infty$, $\widetilde{\mathbb{P}}$-a.s. We will show now that for each $n \in \mathbb{N}$ the process

$$\widetilde{M}_n(t) = \widetilde{X}(t) - x - A\int_0^t \widetilde{X}_n(s)ds - \int_0^t F_n(\widetilde{X}_n(s))ds, \quad t \in [0,T], \tag{8.20}$$

is a square integrable martingale with respect to the filtration

$$\widetilde{\mathscr{F}}_n(t) = \sigma\{\widetilde{X}_n(s) : \ s \leq t\},$$

having the following quadratic variation

$$\langle\langle \widetilde{M}_n(t) \rangle\rangle = \int_0^t (B_n(\widetilde{X}_n(s))Q^{1/2})(B_n(\widetilde{X}_n(s))Q^{1/2})^* ds, \quad t \in [0,T]. \tag{8.21}$$

Both facts are true for the process

$$M_n(t) = X_n(t) - \left[x + A\int_0^t X_n(s)ds + \int_0^t F_n(X_n(s))ds\right]$$

because

$$M_n(t) = \int_0^t B_n(X_n(s))dW(s), \quad t \in [0,T].$$

Since

$$\mathscr{L}(\widetilde{M}_n) = \mathscr{L}(M_n), \tag{8.22}$$

and $\mathbb{E}|M_n(t)|^2 < +\infty$ also $\widetilde{\mathbb{E}}|\widetilde{M}_n(t)|^2 < +\infty$. Let moreover φ be a real valued bounded and continuous function on C and $0 \leq s \leq t \leq T$. Since $M_n(\cdot)$ is a $\sigma\{X_n(s) :$

$0 \le s \le t\}$ martingale therefore

$$\mathbb{E}\left([M_n(t) - M_n(s)]\varphi(X_n(\cdot))\right) = 0$$

and consequently

$$\mathbb{E}\left(\left[X_n(t) - X_n(s) - A\int_s^t X_n(u)du - \int_s^t F_n(X_n(u))du\right]\varphi(X_n(\cdot))\right) = 0.$$

So using (8.22)

$$\widetilde{\mathbb{E}}\left(\left[\widetilde{X}_n(t) - \widetilde{X}_n(s) - A\int_s^t \widetilde{X}_n(u)du - \int_s^t F_n(\widetilde{X}_n(u))du\right]\varphi(\widetilde{X}_n(\cdot))\right) = 0. \quad (8.23)$$

This implies that $\widetilde{M}_n(\cdot)$ is a martigale on $(\widetilde{\Omega}, \widetilde{\mathscr{F}}, \widetilde{\mathbb{P}})$. Similarly note that, for $a, b \in H$,

$$\begin{aligned}&\mathbb{E}[\langle M_n(t), a\rangle \,\langle M_n(t), b\rangle]\\&\quad - \mathbb{E}\left[\int_0^t \langle B_n(X_n(s))Q^{1/2}a, B_n(X_n(s))Q^{1/2}b\rangle ds\; \varphi(X_n(\cdot))\right] = 0.\end{aligned}$$

Therefore

$$\begin{aligned}&\widetilde{\mathbb{E}}\Bigg[\left\langle \widetilde{X}_n(t) - \left(x + A\int_0^t \widetilde{X}_n(s)ds + \int_0^t F_n(\widetilde{X}n(s))ds\right), a\right\rangle\\&\qquad \times \left\langle \widetilde{X}_n(t) - \left(x + A\int_0^t \widetilde{X}_n(s)ds + \int_0^t F_n(\widetilde{X}n(s))ds\right), b\right\rangle\Bigg]\\&\quad -\mathbb{E}\int_0^t \langle B_n(\widetilde{X}_n(s))Q^{1/2}a, B_n(\widetilde{X}_n(s))Q^{1/2}b\rangle ds\; \varphi(\widetilde{X}_n(\cdot)) = 0. \qquad (8.24)\end{aligned}$$

If we show that we can pass in (8.23) and (8.24) to the limit, as $n \to \infty$, then we will know that the process

$$\widetilde{M}(t) = \widetilde{X}(t) - \left(x + A\int_0^t \widetilde{X}(s)ds + \int_0^t F(\widetilde{X}(s))ds\right)$$

is an H-valued martingale with respect to the filtration $\widetilde{\mathscr{F}}_t = \sigma\{\widetilde{X}(s) : s \le t\}$, $t \in [0, T]$, having the quadratic variation

$$\langle\langle \widetilde{M}(t)\rangle\rangle = \int_0^t (B(\widetilde{X}(s))Q^{1/2})(B(\widetilde{X}(s))Q^{1/2})^* ds, \quad t \in [0, T].$$

By (2.22), $\widetilde{\mathbb{E}}|\widetilde{M}_n(t)|^p = \mathbb{E}|M_n(t)|^p,\ p > 2$, and by (8.18), (4.47),

$$\sup_{n\in\mathbb{N}} \widetilde{\mathbb{E}}|\widetilde{M}_n(t)|^p = \sup_{n\in\mathbb{N}} \mathbb{E}|M_n(t)|^p < +\infty.$$

Therefore the sequence $\{\widetilde{M}_n(t)\}$ is uniformly integrable and

$$\lim_{n\to\infty} \widetilde{E}|\widetilde{M}_n(t)|^2 = \widetilde{\mathbb{E}}|\widetilde{M}(t)|^2 < +\infty, \quad t \in [0, T].$$

So $\widetilde{M}(\cdot)$ is a square integrable process. To proceed further we can assume that the operator A^{-1} is everywhere defined and bounded (otherwise we replace A with $A - \lambda$ with λ sufficiently large) and define the martingales

$$\widetilde{N}_n(t) = A^{-1}\widetilde{X}_n(t) - A^{-1}x - \int_0^t \widetilde{X}_n(s)ds - \int_0^t A^{-1}F_n(\widetilde{X}_n(s))ds, \quad t \in [0, T].$$

By the same arguments as above, $\widetilde{N}_n(\cdot)$ is a square integrable continuous martingale on $(\widetilde{\Omega}, \widetilde{\mathscr{F}}, \widetilde{\mathbb{P}})$, with the quadratic variation

$$\langle\langle \widetilde{N}_n(t) \rangle\rangle = \int_0^t [A^{-1}B_n(\widetilde{X}_n(s))Q^{1/2}][A^{-1}B_n(\widetilde{X}_n(s))Q^{1/2}]^* ds, \quad t \in [0, T].$$

The limiting passages in equations (8.23) and (8.24) are straightforward if we replace martingales $\widetilde{M}_n(\cdot)$ with martingales $\widetilde{N}_n(\cdot)$. Consequently the process

$$\widetilde{N}(t) = A^{-1}\widetilde{X}(t) - A^{-1}x - \left(\int_0^t \widetilde{X}(s)ds + \int_0^t A^{-1}F_n(\widetilde{X}(s))ds\right), \quad t \in [0, T],$$

is a square integrable martingale with respect to $\{\widetilde{\mathscr{F}}_t\} - \sigma\{\widetilde{X}(s) : s \leq t\}$ for which

$$\langle\langle \widetilde{N}(t) \rangle\rangle = \int_0^t [A^{-1}B(\widetilde{X}(s))Q^{1/2}][A^{-1}\widetilde{X}(s))BQ^{1/2}]^* ds, \quad t \in [0, T].$$

By the representation Theorem 8.2 there exists a probability space $(\widetilde{\widetilde{\Omega}}, \widetilde{\widetilde{\mathscr{F}}}, \widetilde{\widetilde{P}})$, a filtration $\{\widetilde{\widetilde{\mathscr{F}}}_t\}$, a Q-Wiener process $\widetilde{\widetilde{W}}(\cdot)$ and a predictable continuous process $\widetilde{\widetilde{X}}(\cdot)$ such that

$$A^{-1}\widetilde{\widetilde{X}}(t) - \left[A^{-1}x + \int_0^t \widetilde{\widetilde{X}}(s)ds + \int_0^t A^{-1}F(\widetilde{\widetilde{X}}(s))ds\right]$$
$$= \int_0^t A^{-1}B(\widetilde{\widetilde{X}}(s))\, d\widetilde{\widetilde{W}}(s), \quad t \in [0, T]. \tag{8.25}$$

However,

$$\int_0^t A^{-1}B(\widetilde{\widetilde{X}}(s))\, d\widetilde{\widetilde{W}}(s) = A^{-1}\int_0^t B(\widetilde{\widetilde{X}}(s))\, d\widetilde{\widetilde{W}}(s),$$

and consequently, from (8.25), the process $A^{-1}\widetilde{\widetilde{X}}(\cdot)$ is continuous as a process with values in $D(A)$. One therefore has that

$$\widetilde{\widetilde{X}}(t) = x + A\int_0^t \widetilde{\widetilde{X}}(s)ds + \int_0^t F(\widetilde{\widetilde{X}}(s))ds + \int_0^t B(\widetilde{\widetilde{X}}(s))d\widetilde{\widetilde{W}}(s), \quad t \in [0, T].$$

This shows that $(\widetilde{\widetilde{\Omega}}, \widetilde{\widetilde{\mathscr{F}}}, \widetilde{\widetilde{P}}, \widetilde{\widetilde{\mathscr{F}}}_t, \widetilde{\widetilde{X}}(\cdot), \widetilde{\widetilde{W}}(\cdot))$ is a martingale solution of (8.25).

Example 8.5 Consider the equation

$$\begin{cases} dp(t,\xi) = \Delta p(t,\xi)dt + \sqrt{p_+(t,\xi)}dW_t, & \xi \in \mathscr{O}, ^{(5)} \\ p(t,\xi) = 0, \quad \xi \in \partial\mathscr{O}, \ t > 0 \\ p(0,\xi) = x(\xi), \quad x \in L^2(\mathscr{O}), \ \xi \in \mathscr{O}, \end{cases} \tag{8.26}$$

where $\mathscr{O}$ is a bounded open set in $\mathbb{R}^d$ with regular boundary $\partial\mathscr{O}$ and $W(\cdot)$ is a Q-Wiener process with values in $L^2(\mathscr{O})$, Tr $Q < +\infty$, compare example 0.9. If we set $H = U = L^2(\mathscr{O})$, $F \equiv 0$ and

$$\begin{cases} D(A) = W^{2,2}(\mathscr{O}) \cap W_0^{1,2}(\mathscr{O}) \\ Ax = \Delta x, \quad \forall\, x \in D(A), \end{cases}$$

$$(B(x)u)(\xi) = \sqrt{x_+(\xi)}u(\xi), \quad x,\ u \in H, \xi \in \mathscr{O},$$

then it is easily seen that all the assumptions of Theorem 8.1 are satisfied. □

(5) $p_+(\lambda) = \max\{0, p(\lambda)\}, \quad \lambda \in R^1.$

Part III

Properties of solutions

9
Markov property and Kolmogorov equation

We start from several results on regular dependence of solutions on initial data, needed in what follows. Then we establish Markov and strong Markov properties, first in the simplest and then in the most general formulation. We also study the existence and uniqueness of associated differential equations of parabolic type called *backward Kolmogorov equations*. We consider separately smooth and general initial functions.

9.1 Regular dependence of solutions on initial data

We are here concerned with equation (7.1) under Hypothesis 7.1 or Hypothesis 7.2. Let us remark that all the results on stochastic integration and stochastic equations obtained for the time interval $[0, T]$ can be generalized in a natural way to intervals $[s, T]$, $s \in [0, T]$, with the σ-field $\mathscr{F}_s$ playing the role of the σ-field $\mathscr{F}_0$, and $W(t) - W(s)$, $t \geq s$, the role of the Wiener process. In particular, for any $s \in [0, T]$ and for an arbitrary H-valued random variable ξ, $\mathscr{F}_s$-measurable, there exists exactly one solution $X(t)$, $t \in [s, T]$, of the equation

$$\begin{aligned} X(t) = S(t-s)\xi + \int_s^t S(t-r)F(r, X(r))dr \\ + \int_s^t S(t-r)B(r, X(r))dW(r), \quad t \in [s, T]. \end{aligned} \tag{9.1}$$

This solution will be denoted as $X(\cdot, s, \xi)$. If x is an element of H and $\xi = x$, $\mathbb{P}$-a.s. the solution of (9.1) is denoted as $X(t, s, x)$.

We can assume, by Theorem 7.2 or 7.5, that the process $X(\cdot, s, \xi)$ has continuous trajectories. It turns out that the family $X(\cdot, s, \xi)$ depends continuously on the initial data in the sense specified in the following theorem.

Theorem 9.1 *Under the assumptions of Theorem 7.2 or 7.5, there exists $C_T > 0$ such that, for arbitrary $\xi, \eta \in L^2(\Omega, \mathscr{F}_s, \mathbb{P})$ and $0 \leq s \leq s' \leq t \leq t' \leq T$, the following estimates hold.*

(i) $\mathbb{E}(|X(t,s,\xi)|^2) \le C_T(1+\mathbb{E}|\xi|^2)$.
(ii) $\mathbb{E}(|X(t,s,\xi) - X(t,s,\eta)|^2) \le C_T\mathbb{E}(|\xi-\eta|^2)$.
(iii) $\mathbb{E}(|X(t',s,\xi) - X(t,s,\xi)|^2) \le C_T(\mathbb{E}(|S(t'-t)X(t,s,\xi) - X(t,s,\xi)|^2) + |t'-t|)$.
(iv) $\mathbb{E}(|X(t,s',\xi) - X(t,s,\xi)|^2) \le C_T(\mathbb{E}(|S(s'-s)\xi - \xi|^2) + |s'-s|)$.

In addition, for arbitrary $x, y \in H,\ s \in [0,T]$,

$$\mathbb{E}\left(\sup_{r\in[s,T]} |X(r,s,x) - X(r,s,y)|^2\right) \le C_T\mathbb{E}(|x-y|^2). \tag{9.2}$$

Proof Assume, for instance, that the conditions of Theorem 7.5 hold and set $M_T = \sup_{t\in[0,T]} \|S(t)\|$. Then we have

$$\begin{aligned}
\mathbb{E}|X(t,s,\xi)|^2 &\le 3\|S(t)\|^2\mathbb{E}|\xi|^2 + 3\mathbb{E}\left[\left(\int_s^t |S(t-r)F(r,X(r,s,\xi))|dr\right)^2\right]\\
&\quad + 3\mathbb{E}\left[\left|\int_s^t S(t-r)B(X(r,s,\xi))dW(r)\right|^2\right]\\
&\le 3M_T^2\mathbb{E}|\xi|^2 + 3TM_T^2\int_s^t \mathbb{E}|F(r,X(r,s,\xi))|^2dr\\
&\quad + 3c_2\left[\int_s^t \left(\mathbb{E}\|S(t-r)B(X(r,s,\xi))\|^2_{L_2^0}\right)dr\right].
\end{aligned}$$

In addition,

$$\int_s^t \mathbb{E}|F(r,X(r,s,\xi))|^2dr \le 2C^2 \sup_{r\in[s,t]}(1+\mathbb{E}|X(r,s,\xi)|^2)t,$$

and

$$\begin{aligned}
&\left[\int_s^t \left(\mathbb{E}\|S(t-r)B(X(r,s,\xi))\|^2_{L_2^0}\right)dr\right]\\
&\quad\le 2\left(\int_s^t K^2(t-r)(1+\mathbb{E}|X(r,s,\xi)|^2)dr\right)\\
&\quad\le 2\left(\int_s^t K^2(t-r)dr\right)\sup_{r\in[s,t]}(1+\mathbb{E}|X(r,s,\xi)|^2).
\end{aligned}$$

Using Gronwall's lemma, formula (i) follows.

By proceeding as in the proof of formula (i) we see that

$$\mathbb{E}(|X(t,s,\xi) - X(t,s,\eta)|^2) \le C_T\mathbb{E}(|\xi-\eta|^2), \tag{9.3}$$

so formula (ii) follows as well. To prove (iii) we start from the identity

$$X(t', s, \xi) - X(t, s, \xi) = X(t', t, X(t, s, \xi)) - X(t, s, \xi)$$
$$= (S(t'-t) - I)X(t, s, \xi) + \int_t^{t'} S(t-r)F(r, X(r, s, \xi))dr$$
$$+ \int_t^{t'} S(t-r)B(r, X(r, s, \xi))dW(r).$$

It follows that

$$\mathbb{E}(|X(t', s, \xi) - X(t, s, \xi)|^2) \leq 3\mathbb{E}(|(S(t'-t) - 1)(X(t, s, \xi)|^2)$$
$$+ 3C^2 M_T^2 (T+1) \int_t^{t'} \mathbb{E}(|X(r, s, \xi)|^2)dr,$$

which implies (iii).

We now prove (iv). We have

$$X(t, s', \xi) - X(t, s, \xi) = X(t, s', \xi) - X(t, s', X(s', s, \xi)),$$

and so, by (9.3),

$$\mathbb{E}(|X(t, s', \xi) - X(t, s, \zeta)|^2) \leq C_T \mathbb{E}(|X(s', s, \xi) - \xi|^2). \tag{9.4}$$

But

$$X(s', s, \xi) - \xi = S(s'-s)\xi - \xi + \int_s^{s'} S(s'-r)F(r, X(r, s, \xi))dr$$
$$+ \int_s^{s'} S(s'-r)B(r, X(r, s, \xi))dW(r).$$

It follows that

$$\mathbb{E}(|X(s', s, \xi) - \xi|^2) \leq 3\mathbb{E}(|S(s'-s)\xi - \xi|^2)$$
$$+ 3C^2 M_T^2 (T+1) \int_s^{s'} \mathbb{E}(|X(r, s, \xi)|^2)dr,$$

which, along with (9.3) and (9.4), completes the proof of (iv)

Finally we prove (9.2). Fix $p > 2$. Using (7.7), we can show easily that there exists a constant $D_T > 0$ such that

$$\mathbb{E}(\sup_{r \in [s,T]} |X(r, s, x) - X(r, s, y)|^p) \leq D_T(|x-y|^p)$$
$$+ D_T \int_s^t \mathbb{E}(|X(r, s, X) - X(r, s, Y)|^p)dr,$$

and

$$\sup_{r \in [s,T]} \mathbb{E}(|X(r, s, x) - X(r, s, y)|^p) \leq D_T(|x-y|^p).$$

Consequently, by Hölder's inequality

$$\begin{aligned}
\mathbb{E} \sup_{r\in[s,T]} |X(r,s,x) - X(r,s,y)|^2 &\\
&\le [\mathbb{E} \sup_{r\in[s,T]} |X(r,s,x) - X(r,s,y)|^p]^{2/p} \\
&\le D_T(|x-y|^p)^{2/p} \le \widetilde{D}_T|x-y|^2,
\end{aligned}$$

for some constant $\widetilde{D}_T$, which, along with (9.3) implies (9.2). □

9.1.1 Differentiability with respect to the initial condition

We will now show that if the coefficients F and B are smooth enough then the solution X is also, in an appropriate sense, a differentiable function of the initial data.

As in Chapter 7, we denote by $\mathscr{H}_p$ the Banach space of all (equivalence classes of) H-valued predictable processes Y defined on the time interval $[0,T]$ with norm

$$|\!|\!| Y |\!|\!|_p = \left(\sup_{t\in[0,T]} \mathbb{E}|Y(t)|^p \right)^{1/p}.$$

It follows from Theorem 9.1(i) that the mapping

$$H \to \mathscr{H}_2, \quad x \mapsto X(\cdot, x)$$

is Lipschitz continuous. We assume Hypothesis 7.1. To simplify the notation, we take $s=0$ and we set $X(t,0,\xi) = X(t,\xi) = X(t)$. Let $\mathscr{K}$ be the mapping:

$$\mathscr{K} : H \times \mathscr{H}_p \to \mathscr{H}_p, \quad (x, X) \mapsto \mathscr{K}(x, X),$$

given by

$$\begin{aligned}
\mathscr{K}(x,X)(t) = S(t)x &+ \int_0^t S(t-s)F(s,X(s))ds \\
&+ \int_0^t S(t-s)B(s,X(s))dW(s).
\end{aligned}$$

We know, by the proof of Theorem 7.2, that if T is sufficiently small, then

$$|\!|\!| \mathscr{K}(x,X) - \mathscr{K}(x,Y) |\!|\!|_p \le \frac{1}{2} |\!|\!| X - Y |\!|\!|_p, \quad \forall\, x \in H,\ \forall\, X, Y \in \mathscr{H}_p. \tag{9.5}$$

Note that, for arbitrary $x \in H$ the process $X(\cdot, x)$ is the unique solution of the equation $X = \mathscr{K}(x, X)$ in $\mathscr{H}_p$. To establish the required regularity property, we need the following classical lemmas. The first one is called the *local inversion theorem*.

Lemma 9.2 *Let Λ be an open subset of a Banach space, E_0 a Banach space densely and continuously embedded into a Banach space E, and F a mapping: $F : \Lambda \times E_0 \to E_0$, $F : \Lambda \times E \to E$ such that, for an $\alpha \in [0,1)$*

$$\|F(\lambda,x) - F(\lambda,y)\|_{E_0} \le \alpha \|x-y\|_{E_0}, \quad \lambda \in \Lambda,\ x,y \in E_0, \tag{9.6}$$

$$\|F(\lambda,x) - F(\lambda,y)\|_{E} \le \alpha \|x-y\|_{E}, \quad \lambda \in \Lambda,\ x,y \in E. \tag{9.7}$$

(i) There exists a unique mapping

$$\varphi : \Lambda \to E_0, \quad \lambda \mapsto \varphi(\lambda)$$

such that

$$\varphi(\lambda) = F(\lambda, \varphi(\lambda)), \quad \lambda \in \Lambda. \tag{9.8}$$

(ii) If $F : \Lambda \times E_0 \to E_0$ is a continuous mapping with respect to the first variable, then φ is continuous as an E_0-valued mapping.

(iii) If there exists the directional derivative

$$D_\lambda F(\lambda, x; \mu), \quad \lambda \in \Lambda,\ x \in E_0,\ \mu \in \Lambda,$$

of F regarded as a mapping from $\Lambda \times E_0$ into E, and this is continuous in all variables, and there exists the directional derivative

$$D_x F(\lambda, x; y), \quad \lambda \in \Lambda,\ x, y \in E,$$

and this is continuous as a mapping from $\Lambda \times E \times E$ into E, then for arbitrary $\lambda, \mu \in \Lambda$ there exists the directional derivative $D_\lambda \varphi(\lambda; \mu)$ which satisfies the equation

$$D_\lambda \varphi(\lambda; \mu) = D_\lambda F(\lambda, \varphi(\lambda); \mu) + D_x F(\lambda, \varphi(\lambda); D_\lambda \varphi(\lambda; \mu)), \tag{9.9}$$

and is continuous with respect to all variables and is given by the formula

$$D_\lambda \varphi(\lambda; \mu) = [I - D_x F(\lambda, \varphi(\lambda))]^{-1} D_\lambda F(\lambda, \varphi(\lambda); \mu), \tag{9.10}$$

where $D_x F(\lambda, x)$, $\lambda \in \Lambda$, $x \in E$ is the Gateaux derivative of $F(\lambda, \cdot)$.

Proof Part (i) is a consequence of the contraction mapping theorem. To prove (ii) fix $\lambda_0 \in \Lambda$, then for sufficiently small $t > 0$

$$\begin{aligned}
&\varphi(\lambda_0 + t\mu_0) - \varphi(\lambda_0)\\
&\quad = F(\lambda_0 + t\mu_0, \varphi(\lambda_0 + t\mu_0)) - F(\lambda_0 + t\mu_0)\\
&\quad = [F(\lambda_0 + t\mu_0, \varphi(\lambda_0 + t\mu_0)) - F(\lambda_0, \varphi(\lambda_0 + t\mu_0))]\\
&\qquad + [F(\lambda_0, \varphi(\lambda_0 + t\mu_0)) - F(\lambda_0, \varphi(\lambda_0))]\\
&\quad = \int_0^1 D_\lambda F(\lambda_0 + \sigma t\mu_0, \varphi(\lambda_0 + t\mu_0); t\mu_0) d\sigma\\
&\qquad + \int_0^1 D_x F(\lambda_0, \varphi(\lambda_0) + \sigma(\varphi(\lambda_0 + t\mu_0) - \varphi(\lambda_0)); \varphi(\lambda_0 + t\mu_0) - \varphi(\lambda_0)) d\sigma.
\end{aligned}$$

Both integrals have a well defined meaning. Moreover the strong integral

$$\int_0^1 D_x F(\lambda_0, a + \sigma b) d\sigma$$

exists for arbitrary $a, b \in E$ and by (9.6) its operator norm is not greater than α. Consequently

$$\frac{1}{t}\left(\varphi(\lambda_0 + t\mu_0) - \varphi(\lambda_0)\right) = \Big[I - \int_0^1 D_x F(\lambda_0, \varphi(\lambda_0) + \sigma(\varphi(\lambda_0 + t\mu_0) - \varphi(\lambda_0)); \varphi(\lambda_0 + t\mu_0) - \varphi(\lambda_0)) d\sigma\Big]^{-1} \times \int_0^1 D_\lambda F(\lambda_0 + \sigma t\mu_0, \varphi(\lambda_0 + t\mu_0); t\mu_0) d\sigma. \quad (9.11)$$

To pass to the limit as $t \to 0$ in the identity above we need the following lemma.

Lemma 9.3 *Assume that a sequence of linear bounded operators (B_n) strongly converges to B. If $\|B_n\| \le \alpha$, $\|B\| \le \alpha$, for some $\alpha \in (0, 1)$ and all $n \in \mathbb{N}$ and $x_n \to x$ as $n \to \infty$, then*

$$\lim_{n\to\infty} (I - B_n)^{-1} x_n = (I - B_n)^{-1} x.$$

Proof Set $y_n = (I - B_n)^{-1} x_n$ and $y = (I - B_n)^{-1} x$. Then

$$(I - B_n)(y_n - y) = x_n - x + (B_n y - By),$$

from which

$$|y_n - y|_E \le \frac{1}{1-\alpha} (|x_n - x|_E + |B_n y - By|_E).$$

□

Continuation of the proof By Lemma 9.3 one can pass to the limit in (9.11) as $t \to 0$ to get existence of the directional derivative $D_\lambda \varphi(\lambda_0, \mu_0)$ and formula (9.10). Differentiating identity (9.8) in any direction $\mu \in \Lambda$ leads to (9.9). □

Remark 9.4 For calculating higher directional derivatives of φ it is important to notice that

$$\psi(\lambda) := D_\lambda \varphi(\lambda, \mu)$$

is a solution of a fixed point problem

$$\psi(\lambda) = G(\lambda, \psi(\lambda)), \quad (9.12)$$

where

$$G(\lambda, x) := D_\lambda F(\lambda, \varphi(\lambda); \mu) + D_x F(\lambda, \varphi(\lambda); x) \quad (9.13)$$

is a linear transformation in x. Moreover, from (9.6)

$$|G(\lambda, x) - G(\lambda, y)|_E \le \alpha |x - y|_E, \quad x, y \in E. \quad (9.14)$$

However, in the applications it will often be useful to treat (9.12) and (9.13) in some smaller space $E_0 \subset E$. The fact that G is linear in ψ will simplify calculations a lot.

For instance (9.12) implies that

$$\begin{aligned} D^2_{\lambda,\lambda}\varphi(\lambda;\mu,\eta) := D_\lambda\psi(\lambda;\eta) &= D_\lambda G(\lambda,\psi(\lambda);\eta) + G(\lambda, D_\lambda\psi(\lambda;\eta)) \\ &= D_\lambda G(\lambda,\psi(\lambda);\eta) + G(\lambda, D^2_\lambda\varphi(\lambda;\mu,\eta)). \end{aligned}$$

□

We now prove a lemma concerning the mapping $\mathscr{K}$.

Lemma 9.5 *Assume that Hypothesis 7.1 holds. Then if $F(t,\omega;\cdot)$, $B(t,\omega;\cdot)$ are continuously differentiable and there exists a constant $C_1 > 0$ such that for all $x, y \in H$, $t \in [0,T]$, $\omega \in \Omega$,*

$$|F_x(t,\omega;x)\cdot y| + \|B_x(t,\omega;x)\cdot y\|_{L_2^0} \le C_1|y|, \tag{9.15}$$

then

$$\begin{aligned} (\mathscr{K}_x(x,X)\cdot y)(t) &= S(t)y, \quad \forall\, y \in H,\ t \in [0,T], \\ (\mathscr{K}_X(x,X)\cdot Y)(t) &= \int_0^t S(t-s)F_x(s,X(s))\cdot Y(s)ds \\ &\quad + \int_0^t S(t-s)B_x(s,X(s))\cdot Y(s)dW(s), \quad \mathbb{P}\text{-a.s.} \end{aligned}$$

for all $X, Y \in \mathscr{H}_2$, $x \in H$, $t \in [0,T]$.

Proof (i) Note that the transformations $\mathscr{K}_x$ and $\mathscr{K}_X$ are well defined. We will check for instance that $\mathscr{K}_X(x,X)\cdot Y$ defines the directional derivative of $\mathscr{K}$ in the direction $Y \in \mathscr{H}_p$. Note also that

$$\begin{aligned} I_\sigma(t) &:= \frac{1}{\sigma}\,[\mathscr{K}(x, X+\sigma Y) - \mathscr{K}(x,Y) - \sigma\mathscr{K}_X(x,X)Y](t) \\ &= \int_0^t S(t-s)[\sigma^{-1}(F(s,X(s)+\sigma Y(s)) - F(s,X(s))) - F_x(s,X(s))Y(s)]ds \\ &\quad + \int_0^t S(t-s)[\sigma^{-1}(B(s,X(s)+\sigma Y(s)) - B(s,X(s))) \\ &\qquad\qquad -B_x(s,X(s))Y(s)]dW(s). \end{aligned}$$

Applying Theorem 4.36 we obtain for some constant $M > 0$ and all $t \in [0,T]$

$$\begin{aligned} &\mathbb{E}|I_\sigma(t)|^p \\ &\le M\int_0^T \mathbb{E}\left|\frac{1}{\sigma}\,(F(s,X(s)+\sigma Y(s)) - F(s,X(s))) - F_x(s,X(s))Y(s)\right|^p ds \\ &\quad + M\int_0^T \mathbb{E}\left\|\frac{1}{\sigma}\,(B(s,X(s)+\sigma Y(s)) - B(s,X(s))) - B_x(s,X(s))Y(s)\right\|^p_{L_2^0} ds. \end{aligned}$$

However,

$$\frac{1}{\sigma}\left(F(s, X(s)+\sigma Y(s))-F(s, X(s))=\int_0^1 F_x(s, X(s)+u\sigma Y(s))Y(s)du,\right.$$

$$\frac{1}{\sigma}\left(B(s, X(s)+\sigma Y(s))-B(s, X(s))=\int_0^1 B_x(s, X(s)+u\sigma Y(s))Y(s)du,\right.$$

and therefore, for a possibly different constant $M > 0$,

$$\begin{aligned}
&\sup_{0\le t\le T} \mathbb{E}|I_\sigma(t)|^p \\
&\quad\le M\mathbb{E}\int_0^T\int_0^1 |F_x(s, X(s)+u\sigma Y(s))Y(s)-F_x(s, X(s))Y(s)|^p du\, ds \\
&\quad\le M\mathbb{E}\int_0^T\int_0^1 \|B_x(s, X(s)+u\sigma Y(s))Y(s)-B_x(s, X(s))Y(s)\|^p_{L_2^0} du\, ds.
\end{aligned}$$

Taking into account that F_x and B_x are bounded and continuous and that

$$\sup_{0\le t\le T} \mathbb{E}|I_\sigma(t)|^p < +\infty,$$

we obtain by the dominate convergence theorem that

$$\lim_{\sigma\to 0}\sup_{0\le t\le T} \mathbb{E}|I_\sigma(t)|^p = 0.$$

Proposition 9.6 *Assume that F and B are differentiable with strongly continuous and bounded derivatives. Then for all $p \ge 2$ the derivative $\mathscr{K}_X(x, X)$ given by*

$$\begin{aligned}
(\mathscr{K}_X(x, X))Y(t) &= \int_0^t S(t-s)F_x(s, X(s))Y(s)ds \\
&\quad + \int_0^t S(t-s)B_x(s, X(s))Y(s)dW(s), \\
&\qquad t\in[0, T],\ x\in H,\ X, Y\in\mathscr{H}^p,
\end{aligned}$$

is continuous in x, X, Y.

We will need the following lemma.

Lemma 9.7 *If $X_n \to X$ in $\mathscr{H}^p$ then there exists a sequence $(n_k)\uparrow\infty$ such that*

$$\lim_{k\to\infty} X_{n_k}(s,\omega) = X(s,\omega), \quad \textit{for almost all } (s,\omega)\in[0, T]\times\Omega.$$

Proof It is clear that

$$\lim_{n\to\infty}\int_0^T \mathbb{E}|X_n(s)-X(s)|^p ds = 0.$$

Thus treating X_n as a measurable function on $([0, T]\times\Omega, dt\times d\mathbb{P})$ the result follows from a well known result about L^p spaces. □

Proof of Proposition 9.6 Assume that $X_n \to X$ in $\mathscr{H}^p$. Then by Theorem 4.36, there exists a constant $M > 0$ such that

$$\sup_{t\in[0,T]} \mathbb{E}|\mathscr{K}_X(x, X_n)Y(t) - \mathscr{K}_X(x, X)Y(t)|^p$$

$$\leq M\mathbb{E} \int_0^T |(F_x(s, X_n(s)) - F_x(s, X))Y(s)|^p ds$$

$$+ M\mathbb{E} \int_0^T \|(B_x(s, X_n(s)) - B_x(s, X))Y(s)\|^p_{L_2^0} ds$$

$$\leq M(I_n^1 + I_n^2).$$

Assume, by contradiction, that (I_n^1) is not converging to 0 and for some sequence m_k and $\epsilon > 0$, $I_{m_n} \geq \epsilon$. Let (n_k) be a sequence such that

$$\lim_{k\to\infty} X_{m_{n_k}}(s, \omega) = X(s, \omega).$$

Then, for almost all $(s, \omega) \in [0, T] \times \Omega$ the integrated functions in the definition of $I^1_{m_{n_k}}$ converge to zero. Boundedness of F_x and B_x and finiteness of the expression

$$\mathbb{E} \int_0^T |Y(s)|^p ds$$

allow us to apply the dominate convergence theorem to get a contradiction. The same arguments apply to the sequence (I_n^2). □

We can now prove the following result.

Theorem 9.8 *Assume that Hypothesis* 7.1 *holds and* $F(t, \omega, \cdot)$, $B(t, \omega, \cdot)$ *are Gateaux differentiable on* H *with the derivatives*

$$F_x(t, \omega, x) \cdot y, \quad B_x(t, \omega, x) \cdot y, \quad x, y \in H$$

continuous in x, y *and such that for a constant* $C > 0$

$$|F_x(t, \omega, x) \cdot y| + \|B_x(t, \omega, x) \cdot y\|_{L_2^0} \leq C|y|, \quad x, y \in H.$$

Then the solution $X(\cdot, x)$, $x \in H$, *to problem* (7.1) *is Gateaux differentiable as a mapping from* H *into* $\mathscr{H}_p$ *for every* $p \geq 2$. *Moreover, for any* $h \in H$ *the process*

$$\zeta^h(t) = X_x(t, x) \cdot h, \quad t \in [0, T],$$

is the mild solution of the following equation

$$\begin{cases} d\zeta^h(t) = (A\zeta^h(t) + F_x(t, X) \cdot \zeta^h dt + B_x(t, X) \cdot \zeta^h dW(t) \\ \zeta^h(0) \quad = h. \end{cases} \tag{9.16}$$

Proof The proof is a direct consequence of Lemma 9.2(ii) and Proposition 9.6. □

Similar results also hold true for higher derivatives. We present and prove the result for second derivatives but the method will work for higher derivatives as well, see Remark 9.4.

Theorem 9.9 *In addition to the assumptions of Theorem 9.8, assume that $F(t, \omega, \cdot)$, $B(t, \omega, \cdot)$ also have second order Gateaux derivatives*

$$F_{xx}(t, \omega, x)(y, z), \quad B_{xx}(t, \omega, x)(y, z) \quad x, y, z \in H$$

continuous in x, y, z and such that for a constant $C > 0$

$$|F_{xx}(t, \omega, x)(y, z)| + \|B_{xx}(t, \omega, x)(y, z)\|_{L_2^0} \leq C|y||z|,$$

for all $x, y, z \in H$, $t \in [0, T]$, $\omega \in \Omega$. Then the solution $X(\cdot, x)$, $x \in H$, to problem (7.1) also has second order Gateaux derivatives as a mapping from H into $\mathscr{H}_p$ for every $p \geq 2$. Moreover, for any $g, h \in H$ the process

$$\zeta^h(t) = X_x(t, x) \cdot h, \quad t \in [0, T],$$

is the mild solution of the following equation

$$\begin{cases} d\eta^{h,g}(t) = (A\eta^{h,g}(t) + F_x(t, X) \cdot \eta^{h,g})dt + B_x(t, X) \cdot \eta^{h,g} dW(t) \\ \qquad\qquad + F_{xx}(t, X) \cdot (\zeta^h(t), \zeta^g(t))dt + B_{xx}(t, X) \cdot (\zeta^h, \zeta^g)dW(t) \\ \eta^{h,g}(0) \; = 0. \end{cases} \tag{9.17}$$

Proof Denote by $L(x, Z)$, $x \in H$, $Z \in \mathscr{H}^{2p}$, $p \geq 2$, the operator acting from $H \times \mathscr{H}^{2p}$ into $\mathscr{H}^{2p}$ given by the formula

$$L(x, Z) = S(\cdot)h + \mathscr{K}_X(x, X(x, \cdot)) \cdot Z.$$

By making the time interval $[0, T]$ small enough and then repeating the considerations on intervals $[T, 2T], \ldots$, we can assume that the transformations $L(x, Z)$, $x \in H$, have norms smaller than $\alpha \in (0, 1)$. As in the proof of Proposition 9.6 we can show that the transformation $L(x, Z)$, $x \in H$, $Z \in \mathscr{H}^{2p}$ is Gateaux differentiable in x and

$$\begin{aligned}(D_x L(x, Z) \cdot g)(t) = &\int_0^t S(t-s)F_{xx}(s, X(s, x))(X_x(s, x)g, Z(s))ds \\ &+ \int_0^t S(t-s)B_{xx}(s, X(s, x))(X_x(s, x)g, Z(s))dW(s).\end{aligned}$$

Moreover the derivatives take values in $\mathscr{H}^p$ and are continuous in $x \in H$, $Z \in \mathscr{H}^{2p}$ and

$$\zeta^h(\cdot, x) = L(x, \zeta^h(\cdot, x)), \quad x, h \in H.$$

We can apply Lemma 9.2 with $E_0 = \mathscr{H}^{2p}$, $E = \mathscr{H}^p$ and $F = L$ to deduce that $\zeta^h(\cdot, x)$, $x \in H$ has directional derivative denoted by $\eta^{h,g}(\cdot, x)$ and

$$\eta^{h,g}(\cdot, x) = D_x(L(x, \zeta^h(\cdot, x))) \cdot g + D_Z(L(x, \zeta^h(\cdot, x))) \cdot \zeta^h(\cdot, x).$$

The proof is complete. □

Assume that the processes $X_n(t,x),\ t \in [0,T], x \in H,\ n \in \mathbb{N}$ satisfy the equation

$$\begin{cases} dX_n(t) = (A_n X_n(t) + F(t, X_n(t)))dt + B(t, X_n(t))dW(t), \\ X_n(0) = x, \end{cases} \tag{9.18}$$

where A_n are the Yosida approximations of A. By Proposition 7.4, Theorem 9.8 and Theorem 9.9 we arrive at the following approximation result.

Theorem 9.10 *Let $X_n(t,\cdot)$ be the solution to (9.18) and $\zeta_n^h(t,x) = X_{n,x}(t,x)\cdot h$ and $\eta_n^{h,g}(t,x) = X_{n,xx}(t,x)\cdot(h,g)$, $h, g \in H,\ t \in [0,T]$. Then we have*

$$\begin{aligned} &\lim_{n\to\infty} \sup_{t\in[0,T]} \mathbb{E}|\zeta^h(t,x) - \zeta_n^h(t,x)| = 0, \\ &\lim_{n\to\infty} \sup_{t\in[0,T]} \mathbb{E}|\eta^{h,g}(t,x) - \eta_n^{h,g}(t,x)| = 0. \end{aligned} \tag{9.19}$$

Proof Assume that the assumptions of Theorem 9.9 are satisfied, that $X_n,\ n \in \mathbb{N}$, are solutions to (9.18) and

$$\zeta_n^h(t) = X_{n,x}(t,x)\cdot h, \quad \eta_n^{h,g}(t) = X_{n,x}(t,x,x)\cdot h, \quad h, g \in H,\ t \in [0,T].$$

Then for arbitrary $p \geq 2$ we have

$$\begin{aligned} &\lim_{n\to\infty} \sup_{t\in[0,T]} \mathbb{E}|\zeta^h(t) - \zeta_n^h(t)|^p = 0, \\ &\lim_{n\to\infty} \sup_{t\in[0,T]} \mathbb{E}|\eta^{h,g}(t) - \eta_n^{h,g}(t)|^p = 0. \end{aligned} \tag{9.20}$$

As we have noticed before, the process $X(\cdot, x)$ is the unique solution of the equation

$$\mathscr{K}(x, X) = X,$$

in $\mathscr{H}_2$. Moreover, by Proposition 7.4 and Lemma 9.5, conclusions (i) and (ii) of Lemma 9.2 hold. Moreover, part (iii) of Lemma 9.2 implies validity of the final part of the theorem. □

Remark 9.11 Let us change the formulation of the conditions (9.15), by replacing the space L_2^0 and its norm $\|\cdot\|_2$ by the space L with the operator norm. Then the theorem remains true under the conditions of Theorem 7.5. □

9.1.2 Comments on stochastic flows

Let $X(t,s,\omega;x)$, $t \geq s \geq 0$, $\omega \in \Omega$, $x \in H$ be a mild solution of (7.1) with the initial condition $X(s,s;x) = x$ given at the time $s \geq 0$ and assume that all the conditions of Theorem 7.5 hold. It follows from the uniqueness that for arbitrary $t \geq r \geq s,\ x \in H$

$$X(t,r,\omega; X(r,s,\omega;x)) = X(t,s,\omega;x), \quad \mathbb{P}\text{-a.s.} \tag{9.21}$$

Consequently the formula

$$\Phi(t, s, \omega)(x) = X(t, s, \omega; x), \quad x \in H$$

defines mappings $\Phi(t, s, \omega)$ from H into H. If for all $0 \le s \le t \le T$ and $\mathbb{P}$-a.e. $\omega \in \Omega$, $\Phi(t, s, \omega)$ are continuous, then we say that the equation (7.1) defines the stochastic flow $\Phi(t, s)$. It follows immediately from Theorem 9.1 that transformations $\Phi(t, s)$ are continuous in mean square:

$$\lim_{x_n \to x} \mathbb{E}|\Phi(t, s)x_n - \Phi(t, s)x|^2 = 0. \tag{9.22}$$

Therefore the solution X defines an "almost" stochastic flow. In the finite dimensional case ($\dim H < \infty$), under the conditions of Theorem 7.5, one can show, see [146, 418, 476] that equation (7.1) defines a stochastic flow. Surprisingly enough a similar result is not true in the infinite dimensional case even for linear equations, as the following example due to [650] shows. For more examples and related discussions, we refer to [551].

Example 9.12 Let $H = \ell^2,\ U = \ell^2,\ A = 0,\ F = 0$ and

$$B(x)u = \{\xi_n \gamma_n\}, \quad x = \{\xi_n\}, u = \{\gamma_n\}.$$

Note that $\|B(x)\|_2 = |x|,\ x \in \ell^2$, and therefore all the conditions of Theorem 7.5 hold. In fact the solution $X(t) = \{x_n(t)\}$ of the corresponding equation

$$dX = B(X)dW, \quad X(0) = x = \{\xi_n\}, \tag{9.23}$$

has the following explicit form

$$X(t) = \left\{ e^{\beta_n(t) - \frac{1}{2}t} \xi_n,\ t \ge 0 \right\},$$

where $W(t) = \{\beta_n(t)\},\ t \ge 0$. But for $\mathbb{P}$-a.s. $\omega \in \Omega$ the sequence $\{\beta_n(t, \omega)\}$ is unbounded and therefore the matrix operator

$$\Phi(t, 0) = \begin{pmatrix} e^{\beta_1(t)-\frac{1}{2}t} & 0 & 0 & 0 & \dots \\ 0 & e^{\beta_2(t)-\frac{1}{2}t} & 0 & 0 & \dots \\ 0 & 0 & e^{\beta_n(t)-\frac{1}{2}t} & 0 & \dots \\ 0 & 0 & 0 & 0 & \dots \\ \dots & \dots & \dots & \dots & \end{pmatrix}$$

is an unbounded and therefore discontinuous mapping. So equation (9.23) does not define a stochastic flow. □

Remark 9.13 In a number of cases, however, one can show the existence of stochastic flows. In particular, they exist for stochastic equations with additive noise and regular coefficients, because they can be reduced to deterministic integral equations. For equations with nonregular coefficients, existence of stochastic flows on a smaller space follows from Theorem 7.7 and Theorem 7.11. In the case of equations with multiplicative noise, stochastic flows exist when one is able to solve the problem pathwise, see Theorems 6.32 and 7.23. A generalization of these results is given by the so called method of *stochastic characteristic*, see [680]. Another approach similar to the previous one is based on the representation of solutions by means of the formula of Feynman–Kac, see [309]. Finally, a stochastic flow exists for equations with multiplicative noise when it is possible to construct a regular stochastic evolution operator, see [297, 634]. □

9.2 Markov and strong Markov properties

9.2.1 Case of Lipschitz nonlinearities

We continue here the study of problem (7.1) but assuming that the coefficients F and B depend only on x, and not on ω, and that Hypothesis 7.1 or Hypothesis 7.2 is fulfilled. It is not difficult to see, taking into account the proof of existence of Theorem 7.2 or Theorem 7.5, that the process $X(\cdot, s, x)$ is measurable with respect to the σ-fields $\sigma\{W(u) - W(s) : u \geq s\}$ and therefore it is independent of $\mathscr{F}_s$. If $\varphi \in B_b(H)$, [(1)] then we define for $0 \leq s \leq t \leq T$ and $x \in H$

$$P_{s,t}\varphi(x) = \mathbb{E}[\varphi(X(t, s, x))]. \tag{9.24}$$

From Theorem 9.1 it follows that, for arbitrary $\varphi \in C_b(H)$, the function

$$[0, +\infty) \times [0, +\infty) \times H, \quad (t, s, x) \mapsto P_{s,t}\varphi(x)$$

is continuous. In particular $P_{s,t}\varphi(\cdot)$ is bounded and continuous for any $\varphi \in C_b(H)$; this property is called the *Feller property*. The function

$$P(s, x; t, \Gamma) = P_{s,t}\mathbb{1}_\Gamma(x), \quad 0 \leq s \leq T, \; x \in H, \; \Gamma \in \mathscr{B}(H),$$

is called the *transition function* corresponding to the solution of equation (9.1). We set

$$\mathscr{L}(X(t, s, x))(dy) = P(s, x; t, dy), \quad y \in H.$$

We leave it as an exercise to prove that the operators $P_{s,t}$, $0 \leq s \leq t \leq T$, do not depend on the probability space $(\Omega, \mathscr{F}, P)$ and on the specific Q-Wiener process taken for the definition of the stochastic integral but are functionals of the operator A, the coefficients F and B and the operator Q. We will show now, following [665], that the processes $X(t, u, \xi)$, $t \in [u, T]$, are Markov with transition operators $P_{s,t}$, $0 \leq s \leq t \leq T$. According to the general definition (see Section 2.2.5), it is enough to prove the following.

(1) We denote by $B_b(H)$ the Banach space of all real bounded Borel functions, endowed with the sup norm.

Theorem 9.14 *Assume Hypothesis* 7.1 *or Hypothesis* 7.2 *holds and let* $X(\cdot, s, \xi)$ *be the solution of* (9.1). *For arbitrary* $\varphi \in B_b(H)$ *and* $0 \le u \le s \le t \le T$, *we have*

$$\mathbb{E}[\varphi(X(t,u,\xi))|\mathscr{F}_s] = P_{s,t}(\varphi)(X(s,u,\xi)), \quad \mathbb{P}\text{-a.s.} \tag{9.25}$$

Proof We restrict the proof to the condition of the first Hypothesis 7.1, as the second one can be handled similarly. Note that both sides of (9.25) have a well defined meaning. The uniqueness part of Theorem 7.5 implies that

$$X(t,u,\xi) = X(t,s,X(s,u,\xi)), \quad \mathbb{P}\text{-a.s.}$$

Denote $X(s,u,\xi)$ by η. Then (9.25) can be written as

$$\mathbb{E}[\varphi(X(t,s,\eta))|\mathscr{F}_s] = P_{s,t}(\varphi)(\eta), \quad \mathbb{P}\text{-a.s.} \tag{9.26}$$

It is enough to show that (9.26) holds for an arbitrary $\mathscr{F}_s$-measurable random variable η. Note that if (9.26) holds for all $\varphi \in C_b(H)$, then it holds also for $\varphi = \mathbb{1}_\Gamma$ where Γ is an arbitrary closed set of H. By Proposition 1.2 it holds for $\varphi = \mathbb{1}_\Gamma$ where Γ is an arbitrary Borel set of H and thus for all $\varphi \in B_b(H)$. Therefore, without any loss of generality, we can assume that $\varphi \in C_b(H)$. We know that if $\eta = x$, $\mathbb{P}$-a.s., then the random variable $X(t,s,x)$ is independent of $\mathscr{F}_s$ and therefore

$$\mathbb{E}[\varphi(X(t,s,x))|\mathscr{F}_s] = \mathbb{E}[\varphi(X(t,s,x)] = P_{s,t}\varphi(x), \quad \mathbb{P}\text{-a.s.}$$

If η takes on only a finite number of values, say

$$\eta = \sum_{j=1}^{N} x_j \mathbb{1}_{\Gamma_j},$$

where $\Gamma_1, \Gamma_2, \ldots, \Gamma_N \subset \mathscr{F}_s$ is a partition of Ω and $x_1, x_2, \ldots, x_N$ are some elements in H, then

$$X(t,s,\eta) = \sum_{j=1}^{N} X(t,s,x_j)\mathbb{1}_{\Gamma_j}, \quad \mathbb{P}\text{-a.s.}$$

and consequently

$$\mathbb{E}[\varphi(X(t,s,\eta))|\mathscr{F}_s] = \sum_{j=1}^{N} \mathbb{E}[\varphi(X(t,s,x_j))\mathbb{1}_{\Gamma_j}|\mathscr{F}_s], \quad \mathbb{P}\text{-a.s.}$$

Taking into account that the random variables $X(t,s,x_j)$ are independent of $\mathscr{F}_s$ and I_{Γ_j} are $\mathscr{F}_s$-measurable, $j = 1, \ldots, N$, one deduces that

$$\mathbb{E}[\varphi(X(t,s,\eta))|\mathscr{F}_s] = \sum_{j=1}^{N} P_{s,t}(\varphi)(x_j)I_{\Gamma_j} = P_{s,t}(\varphi)(\eta), \quad \mathbb{P}\text{-a.s.}$$

More generally, if $\mathbb{E}|\eta|^2 < \infty$ then there exists a sequence of simple random variables $\{\eta_n\}$ for which (9.26) holds and $\mathbb{E}|\eta - \eta_n|^2 \to 0$. But then for an appropriate subsequence, $\eta_n \to \eta$, and $X(t,s,\eta_n) \to X(t,s,\eta)$, $\mathbb{P}$-a.s., as $n \to \infty$; so, by

Theorem 9.1, we can pass in the identity (9.26), with η replaced by η_n, to the limit and (9.26) holds if $\mathbb{E}|\eta|^2 < \infty$. □

Corollary 9.15 *For arbitrary $\varphi \in B_b(H)$, $\varphi : H \to \mathbb{R}^1$ and $0 \le u \le s \le t \le T$*

$$P_{u,s}(P_{s,t}\varphi)(x) = P_{u,t}\varphi(x), \quad x \in H. \tag{9.27}$$

In particular for each Borel set Γ of H

$$P(u, x, t, \Gamma) = \int_H P(u, x, s, dy)P(s, y, t, \Gamma) \quad x \in H,\ u \le t. \tag{9.28}$$

Equation (9.28) is called the *Chapman–Kolmogorov* equation.

Proof Note that, in virtue of (9.25)

$$\begin{aligned} P_{u,t}(\varphi)(x) &= \mathbb{E}[\varphi(X(t, u, x))] = \mathbb{E}\left[\mathbb{E}[\varphi(X(t, u, x))|\mathscr{F}_s]\right] \\ &= \mathbb{E}(P_{s,t}\varphi(X(s, u, x))) = P_{u,s}(P_{s,t}\varphi)(x). \end{aligned}$$

Now (9.28) follows by setting $\varphi = 1\!\!1_\Gamma$ in (9.27). □

Corollary 9.16 *If the coefficients F and B do not depend on time then*

$$P_{s,t} = P_{0,t-s}, \quad s \in [0, t]. \tag{9.29}$$

Proof The processes $X(h, 0, x)$ and $X(t + h, t, x)$, $h \ge 0$, have the same distribution. This is because

$$\begin{aligned} X(t + h, t, x) &= S(h)x + \int_t^{t+h} S(t + h - u)F(X(u, t, x))du \\ &\quad + \int_t^{t+h} S(t + h - u)B(X(u, t, x))dW(u) \\ &= S(h)x + \int_0^h S(h - u)F(X(t + u, t, x))du \\ &\quad + \int_0^h S(h - u)B(X(t + u, t, x))dW^t(u), \quad h \ge 0, \end{aligned}$$

where $W^t(u) = W(t + u) - W(t)$, $u \ge 0$. Since $W^t(u)$ is again a Q-Wiener process, therefore, the process $X(t + h, t, x)$, $h \ge 0$, satisfies the same equation as $X(h, 0, x)$ but with a different Q-Wiener process, so the identity of distributions follows. This, of course, implies (9.29). □

When F and B do not depend on time, we define

$$P_t = P_{0,t}, \quad P(t, x; \Gamma) = P(0, x; t; \Gamma).$$

In the applications we need generalizations of the identity (9.25). We start with the following one.

Proposition 9.17 *Assume Hypothesis* 7.1 *or Hypothesis* 7.2 *and let $X(t,s,\xi)$ be the solution of* (9.1). *For arbitrary $\varphi_1,\dots,\varphi_n\in B_b(H)$, $0\le u\le s$ and $0\le h_1\le h_2\le \dots\le h_n$ we have*

$$\begin{aligned}&\mathbb{E}[\varphi_1(X(s+h_1,u,\xi))\varphi_2(X(s+h_2,u,\xi))\cdots\varphi_n(X(s+h_n,u,\xi))|\mathscr{F}_s]\\&\quad=Q^{\varphi_1,\dots,\varphi_n}_{h_1,\dots,h_n}(s,X(s,u,\xi)),\quad \mathbb{P}\text{-a.s.}\end{aligned}\tag{9.30}$$

where $Q^{\varphi_1,\dots,\varphi_n}_{h_1,\dots,h_n}(\cdot,\cdot)$ is a real Borel function on $[0,T]\times H$ given by

$$\begin{aligned}Q^{\varphi_1,\dots,\varphi_n}_{h_1,\dots,h_n}(s,x)&=\int_E P(s,x;s+h_1,dy_1)\varphi_1(y_1)\int_E P(s+h_1,y_1;s+h_2,dy_2)\varphi_2(y_2)\\&\quad\times\cdots\times\int_E P(s+h_{n-1},y_{n-1};s+h_n,dy_n)\varphi_n(y_n),\\&\qquad s\in[0,T],\ x\in H.\end{aligned}\tag{9.31}$$

Proof We prove (9.30) by induction. By Theorem 9.14 the result is true for $n=1$. Assume that the identity (9.30) holds for n and let $h_{n+1}\ge h_n$. Since $\mathscr{F}_s\subset\mathscr{F}_{s+h_n}$, therefore, $\mathbb{P}$-a.s.,

$$\begin{aligned}&\mathbb{E}[\varphi_1(X(s+h_1,u,\xi))\cdots\varphi_n(X(s+h_n,u,\xi))\varphi_{n+1}(X(s+h_{n+1},u,\xi))|\mathscr{F}_s]\\&\quad=\mathbb{E}[\mathbb{E}(\varphi_1(X(s+h_1,u,\xi))\cdots\varphi_{n+1}(X(s+h_{n+1},u,\xi))|\mathscr{F}_{s+h_n})|\mathscr{F}_s]\\&\quad=\mathbb{E}[\mathbb{E}(\varphi_{n+1}(X(s+h_{n+1},u,\xi))|\mathscr{F}_{s+h_n})\varphi_1(X(s+h_1,u,\xi))\\&\qquad\cdots\varphi_n(X(s+h_n,u,\xi))|\mathscr{F}_s]\\&\quad=\mathbb{E}[\varphi_1(X(s+h_1,u,\xi))...\varphi_n(X(s+h_n,u,\xi))\Psi_n(X(s+h_n,u;\xi))|\mathscr{F}_s]\end{aligned}$$

where

$$\begin{aligned}\Psi_n(x)&=\int_H P(s+h_n,x,s+h_{n+1},s+h_{n+1},dy_{n+1})\varphi_{n+1}(y_{n+1})\\&=P_{s+h_n,s+h_{n+1}}(\varphi_{n+1})(x),\ x\in H.\end{aligned}$$

Applying (9.30) with φ_n replaced by $\varphi_n\Psi_n$, we see that formula (9.30) is true for $n+1$. □

We now give the announced generalization of (9.25). For any $x\in H$ and $s\ge 0$ we denote by $\mathbb{P}^{s,x}$ the law of the process $X(s+\cdot,s,x)$ in $(C,\mathscr{B}(C))$, where $C=C([0,+\infty);H)$. Thus $\mathbb{P}^{s,x}$ is defined as follows

$$\mathbb{P}^{s,x}(\mathscr{A})=\mathbb{P}(X(s+\cdot,s,x)\in\mathscr{A}),\quad \forall\,\mathscr{A}\in\mathscr{B}(C).$$

Recall that $\mathbb{P}^{s,x}$ is determined by its values on the cylindrical subsets

$$\mathscr{I}(h_1,\dots,h_n;\Gamma_1,\dots,\Gamma_n)=\{f\in C: f(h_j)\in\Gamma_j,\ j=1,\dots,n\},$$

where $0 \le h_1 \le ... \le h_n$ and $\Gamma_j \in \mathscr{B}(H)$, $j = 1, \dots, n$, by the formula

$$\mathbb{P}^{s,x}(\mathscr{I}) = \mathbb{P}(X(s+h_1, s, x) \in \Gamma_1, \dots, X(s+h_n, s, x) \in \Gamma_n).$$

Notice that, by the Chapman–Kolmogorov equation, it follows that

$$\mathbb{P}^{s,x}(\mathscr{I}(h_1, \dots, h_n; \Gamma_1, \dots, \Gamma_n)) = \Phi_{h_1,\dots,h_n}^{\chi_{\Gamma_1},\dots,\chi_{\Gamma_n}}(s, x),$$

see formula (9.31). Now (9.30) can be written as

$$\mathbb{P}(X(s+\cdot, u, \xi) \in \mathscr{I} | \mathscr{F}_s) = \mathbb{P}^{s, X(s,u,\xi)}(\mathscr{I}), \quad \mathbb{P}\text{-a.s.} \tag{9.32}$$

with $\mathscr{I} = \mathscr{I}(h_1, \dots, h_n; \Gamma_1, \dots, \Gamma_n)$. Let $\mathscr{K}$ be the family of all cylindrical subsets of C. Then $\mathscr{K}$ is a π-system. Moreover the family $\mathscr{G}$ of all Borel subsets $\mathscr{I}$ of C for which (9.32) holds satisfies all conditions of Proposition 1.2. Therefore (9.32) holds for all Borel subsets of C. More generally we arrive at the following theorem.

Theorem 9.18 *Under the hypothesis of Theorem* 9.14, *for an arbitrary nonnegative Borel function* Ψ *on* $C([0, +\infty); H)$ *we have*

$$\mathbb{E}([\Psi(X(s+\cdot, u, \xi)]|\mathscr{F}_s) = \mathbb{E}^{s, X(s,u,\xi)}(\Psi), \quad \mathbb{P}\text{-a.s.}, \ s \ge u \ge 0. \tag{9.33}$$

Proof If Ψ is the indicator function of a set $\mathscr{I} \in \mathscr{B}(C)$ then (9.33) is identical with (9.32). Therefore (9.33) holds for simple functions and, as a consequence, for all monotone limits of simple nonnegative functions and thus for all nonnegative Borel functions. □

Corollary 9.19 *Under the hypothesis of Theorem* 9.18, *for arbitrary* $A \in \mathscr{F}_s$, $\Gamma \in \mathscr{B}(C[0, +\infty); H))$ *and Borel function*

$$\Psi : (C[0, +\infty); H) \to [0, +\infty)$$

we have

$$\mathbb{P}(X(s+\cdot, u, \xi) \in \Gamma \cap A) = \mathbb{E}[\mathbb{P}^{s, X(s,u,\xi)}(\Gamma) \mathbb{1}_A], \tag{9.34}$$

$$\mathbb{E}[\Psi(X(s+\cdot, u, \xi)) \mathbb{1}_A)] = \mathbb{E}[\mathbb{E}^{s, X(s,u,\xi)}(\Psi) \mathbb{1}_A], \quad s \ge u \ge 0. \tag{9.35}$$

In the time homogeneous case (see Corollary 9.16), measures $\mathbb{P}^{s,x}$ do not depend on s and we denote them as $\mathbb{P}^x$, $x \in H$.

We finish our considerations by showing that the solutions of (9.1) are even *strong Markov* processes in the sense that for an arbitrary stopping time $\tau \ge u$

$$\mathbb{E}[\Psi(X(\tau+\cdot, u, \xi))|\mathscr{F}_\tau] = \mathbb{E}^{\tau, X(\tau,u,\xi)}(\Psi), \quad \mathbb{P}\text{-a.s. on } \tau < \infty. \tag{9.36}$$

Let us recall that $\mathscr{F}_\tau$ denotes the σ-field consisting of all events $A \in \mathscr{F}$ such that $\{\tau \le t\} \cap A \in \mathscr{F}_t$, $t \ge 0$. We can easily show that if $\tau \le \sigma$ are two stopping times then $\mathscr{F}_\tau \subset \mathscr{F}_\sigma$.

Theorem 9.20 *Under the same conditions as in Theorem* 9.18 *the solutions of equation* (9.1) *are strong Markov processes.*

Proof One has to show that for arbitrary $A \in \mathscr{F}_\tau$

$$\mathbb{E}\left[\Psi(X(\tau+\cdot,u,\xi))1\!\!1_{A\cap\{\tau<+\infty\}}\right] = \mathbb{E}\left[\mathbb{E}^{\tau,X(\tau,u,\xi)}(\Psi)1\!\!1_{A\cap\{\tau<+\infty\}}\right]. \quad (9.37)$$

By Corollary 9.19 the identity (9.37) holds for stopping time τ taking on only a finite number of values. For arbitrary stopping times τ define $\tau_n = 2^{-n}([2^n\tau]+1)$, $n \in \mathbb{N}$. Then $\tau_n \downarrow \tau$ as $n \uparrow \infty$. Since $\tau_n \geq \tau$ therefore $\mathscr{F}_{\tau_n} \supset \mathscr{F}_\tau$, and $A \in \mathscr{F}_{\tau_n}$, $n \in \mathbb{N}$. Consequently

$$\mathbb{E}\left[\Psi(X(\tau_n+\cdot,u,\xi))1\!\!1_{A\cap\{\tau<+\infty\}}\right] = \mathbb{E}\left[\mathbb{E}^{\tau_n,X(\tau_n,u,\xi)}(\Psi)1\!\!1_{A\cap\{\tau<+\infty\}}\right]. \quad (9.38)$$

We show that one can pass in (9.38) to the limit for arbitrary $\Psi : C \to \mathbb{R}^1$ of the form

$$\psi(f) = \varphi(f(h)), \quad f \in C,$$

where $\varphi \in C_b(H)$ and $h \in [s,T]$. In this special case (9.38) becomes

$$\mathbb{E}\left[\varphi(X(\tau_n+h,u,\xi))1\!\!1_{A\cap\{\tau<+\infty\}}\right] = \mathbb{E}\left[(P_{\tau_n,\tau_n+h})(X(\tau_n,u,\xi))1\!\!1_{A\cap\{\tau<+\infty\}}\right]. \quad (9.39)$$

Now by Theorem 9.1, for any $t > 0$ the function

$$[0,+\infty) \times H \to \mathbb{R}^1, \quad h \mapsto P_{t,t+h}\varphi(x),$$

is continuous. Taking into account that the process $X(\cdot,u,\xi)$ has continuous trajectories, one can pass in (9.39) to the limit as $n \to \infty$ to obtain

$$\mathbb{E}\left[\varphi(X(\tau+h,u,\xi))1\!\!1_{A\cap\{\tau<+\infty\}}\right] = \mathbb{E}\left[(P_{\tau,\tau+h})(X(\tau_n,u,\xi))1\!\!1_{A\cap\{\tau<+\infty\}}\right],$$

for arbitrary bounded continuous φ and therefore for arbitrary Borel nonnegative φ. The proof that (9.37) holds for general Ψ is now very similar to the proof of (9.33) and therefore is omitted. □

9.2.2 Markov property for equations in Banach spaces

Assume that E is a Banach space embedded as a Borel subset into H. If for arbitrary E-valued, $\mathscr{F}_s$-measurable initial conditions ξ, the solution $X(t,s,\xi)$ is E-continuous then it is also a strong Markov process on E with the transition operators $P_{s,t}$ restricted to E. More generally, the previous techniques used to prove the Markov property can be adapted to solutions on E of different types of equations studied in Section 7.2. Instead of giving detailed formulations and proofs we present here a direct derivation of the Markov property for equations

$$\begin{cases} dX = (AX + F(X))dt + dW, \\ X(0) = \xi, \end{cases} \quad (9.40)$$

studied in Section 7.2.1; in the following we use notation from this section.

Theorem 9.21 *Under the hypothesis of Theorem 7.7, equation* (9.40) *has a strong Markov solution.*

Proof For simplicity we set here $F_E = F$ and $S_E = S$. Let us fix $t \geq 0$. For any $f \in C([0, +\infty); E)$ denote by $u(\cdot, f)$ the solution to the integral equation

$$v(t) = \int_0^t S(t-s)F(v(s))ds + f(t), \quad t \geq 0. \tag{9.41}$$

Such a solution does exist under the hypothesis of Theorem 7.7. Since

$$v(t+h) = S(h)v(t) + \int_0^h (t-s)F(v(t+s))ds + f(t+h) - S(h)f(t), \tag{9.42}$$

we have

$$u(t+h, f) = u(h, S(\cdot)u(t, f) + f(t+\cdot) - S(\cdot)f(t)), \quad h \geq 0.$$

Moreover the solution of (9.40) is given by the formula

$$X(t;\xi) = u(t, S(\cdot)\xi + W_A(\cdot)).$$

Consequently

$$X(t+h,\xi) = X(h, S(\cdot)u(t,\xi) + \widehat{W}_A(\cdot)), \quad h \geq 0 \tag{9.43}$$

where

$$\widehat{W}_A(s) = W_A(t+s) - S(s)W_A(t) = \int_t^{t+h} S(t+h-s)dW(s), \quad s \geq 0.$$

Moreover, the random process $\widehat{W}_A(\cdot)$ is independent of $\mathscr{F}_t$ and its law is identical to the law of $W_A(\cdot)$. Also $X(t,\xi)$ is $\mathscr{F}_t$-measurable. Consequently, applying Proposition 1.12, with $E_1 = E$ and $E_2 = C([0, +\infty); E)$, we obtain by (9.43) that for any $\varphi \in B_b(E)$

$$\begin{aligned}\mathbb{E}[\varphi(X(t+h,\xi))|\mathscr{F}_t] &= \mathbb{E}[\varphi(u(h, S(\cdot)X(t,\xi)) + \widehat{W}_A(\cdot)))|\mathscr{F}_t] \\ &= P_h\varphi(X(t,\xi)), \quad \mathbb{P}\text{-a.e.}\end{aligned}$$

This way we have shown that

$$\mathbb{E}[\varphi(X(t+h,\xi))|\mathscr{F}_t] = P_h\varphi(X(t,\xi)), \quad \mathbb{P}\text{-a.e.}, \tag{9.44}$$

so the process $X(\cdot,\xi)$ is Markov. Using the same technique as in the previous subsection we can also generalize (9.44) to obtain (9.33). □

9.3 Kolmogorov's equation: smooth initial functions

This and the next section are devoted to an analytical characterization of the transition family $P_{s,t}$, $0 \leq s \leq t \leq T$, defined in Section 9.2. To simplify the presentation we restrict our considerations to the time homogeneous case, and assume that the coefficients F and B of the equation (7.1) depend only on $x \in H$. It turns out that $v(t,x) = P_t\varphi(x)$, $t \geq 0$, $x \in H$, is, under some conditions, a solution of the following parabolic type equation called *Kolmogorov's backward equation*:

$$\begin{cases} v_t(t,x) = \dfrac{1}{2}\,\mathrm{Tr}[v_{xx}(t,x)(B(x)Q^{1/2})(B(x)Q^{1/2})^*] \\ \qquad\qquad + \langle Ax + F(x), v_x(t,x)\rangle, \quad t > 0,\ x \in D(A), \\ v(0,x) = \varphi(x), \quad x \in H. \end{cases} \tag{9.45}$$

Throughout this section we shall assume that Hypothesis 7.1 holds, so that W is a Q-Wiener process and Q is of trace class. More general Kolmogorov equations can be found in the monograph [222].

Let us introduce some notation. By $C_b^n(H)$ (respectively $C_b^n(H,H)$), $n \in \mathbb{N}$, we denote the space of all functions $\varphi : H \to \mathbb{R}^1$ (respectively $\varphi : H \to H$) that are n-times continuously Fréchet differentiable with all derivatives up to order n bounded. If $n < \infty$ the space $C_b^n(H)$ (respectively $C_b^n(H,H)$) is endowed with its natural norm which will be denoted by $\|\cdot\|_n$. Moreover by $C_b^{k,n}([0,T] \times H)$ (respectively $C_b^{k,n}([0,T] \times H, H)$), $n,\ k \in \mathbb{N}$, we denote the space of all functions $\varphi : [0,T] \times H \to \mathbb{R}^1$ (respectively $\varphi : [0,T] \times H \to H$) that are k-times continuously Fréchet differentiable with respect to t and n times continuous Garteaux differentiable with respect to x and with all their partial derivatives continuous and bounded in $[0,T] \times H$.

9.3.1 Bounded generators

We assume here that A is bounded and look for *strict solutions* to the problem (9.45).

Definition 9.22 *We say that v is a* strict solution *to the problem* (9.45) *if*

(i) $v \in C_b^{1,2}([0,T] \times H)$,
(ii) equation (9.45) *is fulfilled for all $x \in H$, $t \geq 0$.*

Theorem 9.23 *We assume the following.*

(i) Hypothesis 7.1 *is satisfied with $A \in L(H)$ and F and B nonrandom and independent on t.*
(ii) F and B fulfill hypotheses of Theorems 9.8 *and* 9.9.

Let $\varphi \in C_b^2(H)$, then problem (9.45) *has a unique strict solution $v \in C_b^{1,2}([0,T] \times H)$. Moreover v is given by the formula*

$$v(t,x) = \mathbb{E}[\varphi(X(t,x))] = P_t\varphi(x), \quad t \geq 0,\ x \in H. \tag{9.46}$$

Proof Step 1 Existence. We will prove first that the function v given by (9.46) satisfies (9.45) for $t = 0$. Since A is bounded, the process $X(t) = X(t,x)$, $t \geq 0$, is a strong solution of the stochastic equation (7.1) and by Itô's formula

$$d\varphi(X(t,x)) = \langle \varphi_x(X(t,x)), dX(t,x)\rangle + \frac{1}{2}\,\mathrm{Tr}\,[\varphi_{xx}(X(t,x))(B(X(t,x))Q^{1/2})(B(X(t,x))Q^{1/2})^*]dt.$$

Consequently

$$\begin{aligned} v(t,x) &= \mathbb{E}[\varphi(X(t,x))] \\ &= \varphi(x) + \mathbb{E}\int_0^t \langle AX(s,x) + F(X(s,x)), \varphi_x(X(s,x))\rangle ds \\ &\quad + \frac{1}{2}\,\mathbb{E}\int_0^t \mathrm{Tr}\,[\varphi_{xx}(X(s,x))(B(X(s,x))Q^{1/2})(B(X(s,x))Q^{1/2})^*]ds. \end{aligned} \tag{9.47}$$

By Lebesgue's dominated convergence theorem

$$\begin{aligned} v_t^+(0,x) &= \lim_{t\downarrow 0} \frac{1}{t}\,(v(t,x) - \varphi(x)) \\ &= \frac{1}{2}\,\mathrm{Tr}\,[\varphi_{xx}(x)\cdot(B(x)Q^{1/2})(B(x)Q^{1/2})^*] + \langle Ax + F(x), \varphi_x(x)\rangle. \end{aligned} \tag{9.48}$$

Now we remark that from Theorem 9.9, $v(t,\cdot)$ is two times differentiable for any $t \geq 0$, and we have

$$v_{xx}(t,x)\cdot(h,g) = \mathbb{E}[\varphi_{xx}(X(t,x))\cdot(X_x(t,x)\cdot h, X_x(t,x)\cdot g)], \quad h,g \in H.$$

Fix $s > 0$. By Corollary 9.15 it follows that $v(t+s,x) = P_t(v(s,\cdot))$. By applying the previous argument with $v(s,\cdot)$ replacing φ, we obtain that

$$D_t^+ v(s,x) = \frac{1}{2}\,\mathrm{Tr}\,[v_{xx}(s,x)(B(x)Q^{1/2})(B(x)Q^{1/2})^*] + \langle Ax + F(x), v_x(s,x)\rangle.$$

Since the right hand side of the above equality is a continuous function in $[0,T]\times H$, it follows, from a well known result, that $v(\cdot,\cdot)$ is continuously differentiable in t and satisfies (9.45).

Step 2 Uniqueness.

Assume that $\widehat{v}(\cdot,\cdot) \in C_b^{1,2}([0,T]\times H)$ is a strict solution of (9.45). Let us fix $t \in [0,T]$, $n > \omega$ and apply Itô's formula to the process $\widehat{v}(t-s, X(s,x))$, $s \in [0,t]$.

Then

$$
\begin{aligned}
& d\widehat{v}(t-s, X(s,x)) \\
&\quad = -D_t\widehat{v}(t-s, X(s,x))ds + \langle \widehat{v}_x(t-s, X(s,x)), AX(s,x) + F(X(s,x))\rangle ds \\
&\qquad + \frac{1}{2}\,\mathrm{Tr}\,[\widehat{v}_{xx}(t-s, X(s,x))(B(X(s,x))Q^{1/2})(B(X(s,x))Q^{1/2})^*] \\
&\qquad + \langle \widehat{v}_x(t-s, X(s,x)), B(X(s,x))\rangle dW(s), \quad s \in [0,t].
\end{aligned}
$$

Therefore, integrating over the interval $[0, t]$, and taking expectation one obtains

$$
\widehat{v}(t,x) = \mathbb{E}[\varphi(X(t,x))]
$$

and the theorem is proved. □

9.3.2 Arbitrary generators

The Kolmogorov equation with unbounded operators A has been studied by several authors, in particular by Daleckij [184]. However, the case of arbitrary generator A covered by the theorem below seems to be new. We first give a definition of the solution of problem (9.45). We assume here that Q is of trace class.

Definition 9.24 *We say that v is a* strict solution *to the problem* (9.45) *if*

(i) $v \in C_b^{0,0}([0,T] \times H)$,
(ii) $v(t,\cdot) \in C_b^2(H)$ *for any* $t \geq 0$ *and* $v_x,\ v_{x,x}$ *are continuous on* $[0,T] \times H$,
(iii) equation (9.45) *is fulfilled for any* $x \in D(A),\ t \geq 0$.

Theorem 9.25 *We assume the following.*

(i) Hypothesis 7.1 *is satisfied with F and B nonrandom and independent on t.*
(ii) F and B fulfill hypotheses of Theorems 9.8 *and* 9.9.

Let $\varphi \in C_b^2(H)$, then problem (9.45) *has a unique strict solution and v is given by the formula* (9.46).

Proof **Step 1** Existence. Set $v(t,x) = \mathbb{E}[\varphi(X(t,x))]$; to see that v is a solution it suffices to check the first of equation (9.45) for any $x \in D(A)$. Let $v_n(t,x) = \mathbb{E}[\varphi(X_n(t,x))]$ where X_n is the solution to (7.14). By Theorem 9.23 we have

$$
\begin{aligned}
D_t v_n(t,x) = &\ \frac{1}{2}\,\mathrm{Tr}\,[v_{n,xx}(t,x)(B(x)Q^{1/2})(B(x)Q^{1/2})^*] \\
& + \langle A_n x + F(x), v_{n,x}(t,x)\rangle, \quad t \geq 0,\ x \in H.
\end{aligned}
$$

Let us fix $x, y, z \in H$, by Theorems 9.8, 9.9 it is not difficult to check that

$$\lim_{n\to\infty} \langle D_x v_n(t,x), y\rangle = \langle v_x(t,x), y\rangle$$

and

$$\lim_{n\to\infty} \langle D_x^2 v_n(t,x)y, z\rangle = \langle v_{xx}(t,x), y\rangle$$

uniformly in $t \in [0, T]$ and $\mathbb{P}$-a.s. Thus, if $x \in D(A)$, it follows that

$$\begin{aligned}\lim_{k\to\infty} D_t v_{n_k}(t,x) = \frac{1}{2}\, &\mathrm{Tr}\,[v_{xx}(t,x)(B(x)Q^{1/2})(B(x)Q^{1/2})^*]\\ &+ \langle Ax + F(x), v_x(t,x)\rangle, \quad t \geq 0,\ x \in D(A)\end{aligned}$$

uniformly in t. This implies that

$$\lim_{k\to\infty} D_t v_{n_k}(t,x) = D_t v(t,x)$$

uniformly in $t \in [0, T]$. This finishes the proof of Step 1.

Step 2 Uniqueness.

We assume that v is a strict solution of (9.45). We are going to show that

$$v(t,x) = \mathbb{E}[\varphi(X(t,x))], \quad \forall\, t \geq 0,\ x \in H, \tag{9.49}$$

where X is the mild solution of the stochastic equation (7.1).

The classical proof of uniqueness is based on the computation of Itô's differential with respect to s of $v(t-s, X(s,x))$. This is impossible in the present situation for two reasons: v is not of class $C^{1,2}$ and X is not a strong solution of equation (7.1). For this reason we proceed as in [729] by approximating v by the function

$$v_n(t,x) := v(t, J_n x), \quad J_n = (nI - A)^{-1}, \quad t \geq 0,\ x \in H,$$

and introducing an auxiliary equation

$$\begin{cases} dX_n = (AX_n + J_n F(X_n))dt + J_n B(X_n)dW(t),\\ X_n(0) = J_n x,\end{cases} \tag{9.50}$$

where again $J_n = n(n-A)^{-1}$ (Warning: X_n is not the solution to (7.14).) The advantage of introducing (9.50) is that it possesses a strong solution, because one can easily check that $Y_n = AX_n$ is the mild solution of

$$\begin{cases} dY_n = (AY_n + AJ_n F(X_n))dt + AJ_n B dW(t),\\ Y_n(0) = AJ_n x.\end{cases} \tag{9.51}$$

Moreover one can easily check, by a fixed point argument, that $X_n(t) \to X(t)$ as $n \to \infty$ uniformly in t on any interval $[0, T]$.

Now we can apply Itô's formula to $v_n(t - x, X_n(s))$ obtaining

$$
\begin{aligned}
&d_s v_n(t - s, X_n(s)) \\
&\quad = -D_t v_n(t - s, X_n(s))ds \\
&\qquad + \frac{1}{2}\operatorname{Tr}[D_x^2 v_n(t - s, X_n(s))(J_n B(X_n(s))Q^{1/2})(J_n B(X_n(s))Q^{1/2})^*]ds \\
&\qquad + \langle AX_n(s) + J_n F(X_n(s)), D_x v_n(t - s, X_n(s))\rangle ds \\
&\qquad + \langle J_n B(X_n(s))dW(s), D_x v_n(t - s, X_n(s))\rangle. \qquad (9.52)
\end{aligned}
$$

On the other hand,

$$
D_x v_n(t - s, X_n(s)) = J_n^* D_x v(t - s, J_n X_n(s))
$$

and

$$
D_x^2 v_n(t - s, X_n(s)) = J_n^* D_x v(t - s, J_n X_n(s))J_n.
$$

Substituting into (9.52) yields

$$
\begin{aligned}
&d_s v_n(t - s, X_n(s)) \\
&\quad = -D_t v(t - s, J_n X_n(s))ds \\
&\qquad + \frac{1}{2}\operatorname{Tr}[J_n^* D_x^2 v(t - s, J_n X_n(s))J_n(J_n B(X_n(s))Q^{1/2})(J_n B(X_n(s))Q^{1/2})^*]ds \\
&\qquad + \langle AJ_n X_n(s) + J_n^2 F(X_n(s)), D_x v(t - s, J_n X_n(s))\rangle ds \\
&\qquad + \langle J_n^2 B(X_n(s))dW(s), D_x v(t - s, J_n X_n(s))\rangle. \qquad (9.53)
\end{aligned}
$$

Now we substitute $D_t v(t - s, J_n X_n(s))$ from (9.45) and obtain

$$
\begin{aligned}
&d_s v_n(t - s, X_n(s)) \\
&\quad = -\frac{1}{2}\operatorname{Tr}[D_x^2 v(t - s, J_n X_n(s))(B(J_n X_n(s))Q^{1/2})(B(J_n X_n(s))Q^{1/2})^*]ds \\
&\qquad - \langle AJ_n X_n(s) + F(J_n X_n(s)), D_x v(t - s, J_n X_n(s))\rangle ds \\
&\qquad + \frac{1}{2}\operatorname{Tr}[J_n^* D_x^2 v(t-s, J_n X_n(s))J_n(J_n B(J_n X_n(s))Q^{1/2})(J_n B(J_n X_n(s))Q^{1/2})^*]ds \\
&\qquad + \langle AJ_n X_n(s) + J_n^2 F(X_n(s)), D_x v(t - s, J_n X_n(s))\rangle ds \\
&\qquad + \langle J_n^2 B(X_n(s))dW(t), D_x v(t - s, J_n X_n(s))\rangle. \qquad (9.54)
\end{aligned}
$$

We deduce using (9.54) that

$$
\begin{aligned}
&d_s v_n(t-s, X_n(s))\\
&\quad = \frac{1}{2}\,\mathrm{Tr}[J_n^* D_x^2 v(t-s, J_nX_n(s))J_n(J_nB(J_nX_n(s))Q^{1/2})(J_nB(J_nX_n(s))Q^{1/2})^*]ds\\
&\quad\quad - \frac{1}{2}\,\mathrm{Tr}[D_x^2 v(t-s, J_nX_n(s))(B(J_nX_n(s))Q^{1/2})(B(J_nX_n(s))Q^{1/2})^*]ds\\
&\quad\quad + \langle J_n^2 F(X_n(s)), D_x v(t-s, J_nX_n(s))\rangle ds\\
&\quad\quad - \langle F(J_nX_n(s)), D_x v(t-s, J_nX_n(s))\rangle ds\\
&\quad\quad + \langle J_n^2 B(X_n(s))dW(t), D_x v(t-s, J_nX_n(s))\rangle.
\end{aligned}
\tag{9.55}
$$

Taking expectation and integrating from 0 to t we get

$$
\mathbb{E}[\varphi(J_nX_n(t))] - v_n(t, J_nx) = o(n),
$$

where $\lim_{n\to\infty} o(n) = 0$ (notice that the unbounded operator A does not appear in $o(n)$). Now the conclusion follows letting $n \to \infty$. The proof is complete. □

9.4 Further regularity properties of the transition semigroup

The aim of this section is to show that the requirements on the initial function φ imposed in the previous section can sometimes be relaxed. In fact we prove that in several important cases the function $v(t,x) = \mathbb{E}[\varphi(X(t,x))]$, $t \geq 0$, $x \in H$, is smooth in x although φ may be only a Borel function. In the case of diagonal operators A and Q this was also noticed in [341].

We divide the material into two subsections devoted respectively to linear and nonlinear equations.

9.4.1 Linear case

We assume here $F = 0$ and we limit ourselves to equations with additive noise only

$$
\begin{cases}
dX = AXdt + dW(t),\\
X(0) = x,
\end{cases}
\tag{9.56}
$$

where W is a Q-Wiener process (Q can be the identity) and

$$
\mathrm{Tr}\ Q_t = \int_0^t \mathrm{Tr}[S(s)QS^*(s)]ds < +\infty.
\tag{9.57}
$$

As we know, equation (9.56) in mild form reads as follows

$$
X(t) = S(t)x + W_A(t), \quad t \geq 0,
\tag{9.58}
$$

where $W_A(t)$ is the stochastic convolution

$$W_A(t) = \int_0^t S(t-s)dW(s),$$

introduced before. In particular, under assumption (9.57) we have seen that the process $W_A(t), t \geq 0$, is well defined as an H-valued Gaussian process.

To formulate our first theorem we fix $t > 0$ and introduce the following assumption [(2)]

$$S(t)(H) \subset Q_t^{1/2}(H). \tag{9.59}$$

If (9.59) holds then the operator $\Gamma(t) = Q_t^{-1/2} S(t)$ is a well defined bounded operator in H.[(3)] To simplify the presentation we will assume that Ker $Q_t = \{0\}$. The condition (9.59) is equivalent to the fact that all transition probability $P(t, x, \cdot)$, $x \in H$, are mutually absolutely continuous. This follows directly from Theorem 2.23 as $P(t, x, \cdot)$ are Gaussian measures with mean vector $S(t)x$ and covariance operator Q_t, that is $P(t, x, \cdot) = \mathcal{N}(S(t)x, Q_t)$. The condition (9.59) also has a control theoretic interpretation, see [207, 727] and Appendix B. More precisely, condition (9.59) holds if and only if the deterministic system

$$\dot{y} = Ay + Q^{1/2}u, \quad y(0) = x \in H \tag{9.60}$$

is null controllable. We recall also that the minimal energy required by the system (9.60) to drive an element $x \in H$ to 0 in time t is given by $|\Gamma(t)x|^2$. Thus

$$|\Gamma(t)x|^2 = \inf\left\{\int_0^t |u(s)|^2 ds : y(t) = 0, \quad y \text{ satisfies } (9.60)\right\}. \tag{9.61}$$

Theorem 9.26 *Assume that the conditions* (9.57) *and* (9.59) *hold. Then for arbitrary $\varphi \in B_b(H)$ the function $v(t, \cdot) := P_t(\varphi)$ belongs to $C_b^\infty(H)$. Its nth directional derivatives at point x and at directions $h_1, \ldots, h_n$ are given by the formula*

$$D_x^n v(t, x)(h_1, \ldots, h_n) = \frac{\partial^n}{\partial \alpha_1 \cdots \partial \alpha_n} \Bigg[e^{-\frac{1}{2}\left|\Gamma(t)\left(\sum_{j=1}^n \alpha_j h_j\right)\right|^2} \times \int_H \varphi(S(t)x + z) e^{\sum_{j=1}^n \alpha_j \langle \Gamma(t)h_j, Q_t^{-1/2}z\rangle} \mathcal{N}(0, Q_t)(dz)\Bigg]_{\alpha_1 = \cdots = \alpha_n = 0}. \tag{9.62}$$

In particular

$$\langle D_x v(t, x), h_1\rangle = \int_H \langle \Gamma(t)h_1, Q_t^{-1/2}z\rangle \varphi(S(t)x + z) \mathcal{N}(0, Q_t)(dz) \tag{9.63}$$

(2) This assumption will also be introduced in Chapter 11.

(3) In the definition of $\Gamma(t)$, $Q_t^{-1/2}$ stands for the pseudo-inverse of $Q_t^{1/2}$.

and

$$
\begin{aligned}
&D_x^2 v(t,x)(h_1,h_2)\\
&\quad = -\langle \Gamma(t)h_1, \Gamma(t)h_2\rangle\\
&\qquad + \int_H \langle \Gamma(t)h_1, Q_t^{-1/2}z\rangle\langle \Gamma(t)h_2, Q_t^{-1/2}z\rangle \varphi(S(t)x+z)\mathcal{N}(0,Q_t)(dz). \qquad (9.64)
\end{aligned}
$$

Proof Define

$$\psi(\alpha_1,\dots,\alpha_n) = v(t, x+\alpha_1 x_1+\cdots+\alpha_n x_n).$$

Then

$$D_x^n v(t,x)(h_1,\dots,h_n) = \frac{\partial^n \psi}{\partial\alpha_1\cdots\partial\alpha_n}(0,\dots,0).$$

By the Cameron–Martin formula, Theorem 2.23 we have

$$
\begin{aligned}
&\psi(\alpha_1,\dots,\alpha_n)\\
&\quad = \int_H \varphi(z) e^{-\frac12 \left|\Gamma(t)\left(x+\sum_{j=1}^n \alpha_j h_j\right)\right|^2 + \left\langle \Gamma(t)\left(x+\sum_{j=1}^n \alpha_j h_j\right), Q_t^{-1/2}z\right\rangle} \mathcal{N}(0,Q_t)(dz)\\
&\quad = \int_H \varphi(z) e^{\sum_{j=1}^n \alpha_j \langle \Gamma(t)h_j, Q_t^{-1/2}(z-S(t)x)\rangle}\\
&\qquad \times e^{-\frac12 \left|\sum_{j=1}^n \alpha_j h_j\right|^2} e^{-\frac12 |\Gamma(t)|^2 + \langle \Gamma(t)x, Q_t^{-1/2}z\rangle} \mathcal{N}(0,Q_t)(dz).
\end{aligned}
$$

And again the Cameron–Martin formula yields

$$
\begin{aligned}
&\psi(\alpha_1,\dots,\alpha_n)\\
&\quad = \int_H \varphi(z) e^{\sum_{j=1}^n \alpha_j \langle \Gamma(t)h_j, Q_t^{-1/2}(z-S(t)x)\rangle} e^{-\frac12 \left|\sum_{j=1}^n \alpha_j h_j\right|^2} \mathcal{N}(0,Q_t)(dz).
\end{aligned}
$$

Changing the variables $z - S(t) = y$, we arrive at the formula

$$
\begin{aligned}
&\psi(\alpha_1,\dots,\alpha_n)\\
&\quad = e^{-\frac12 \left|\sum_{j=1}^n \alpha_j h_j\right|^2} \int_H \varphi(S(t)x+z) e^{\sum_{j=1}^n \alpha_j \langle \Gamma(t)h_j, Q_t^{-1/2}(z-S(t)x)\rangle} \mathcal{N}(0,Q_t)(dz).
\end{aligned}
$$

Note that the vector

$$(\langle \Gamma(t)h_1, Q_t^{-1/2}z\rangle, \dots, \langle \Gamma(t)h_n, Q_t^{-1/2}z\rangle)$$

is Gaussian on $(H, \mathscr{B}(H), \mathcal{N}(0,Q_t))$, therefore the differentiability of ψ an arbitrary number of times follows from the following lemma.

Lemma 9.27 *Assume that* $(\xi_1,\dots,\xi_n)$ *is a Gaussian vector and* ζ *a bounded Gaussian variable. Then*

$$h(\alpha_1,\dots,\alpha_n) = \mathbb{E}\left(\zeta e^{\sum_{j=1}^n \alpha_j \xi_j}\right), \quad (\alpha_1,\dots,\alpha_n) \in \mathbb{R}^n,$$

is a C^∞ function and for arbitrary indices $k_1, \ldots, k_n$,

$$\frac{\partial^{k_1+\cdots+k_n}}{\partial\alpha_1^{k_1}\cdots\partial\alpha_n^{k_n}}(\alpha_1,\ldots,\alpha_n) = \mathbb{E}\left(\zeta e^{\sum_{j=1}^n \alpha_j\xi_j}\prod_{j=1}^n \xi_j^{k_j}\right), \quad (\alpha_1,\ldots,\alpha_n)\in\mathbb{R}^n. \tag{9.65}$$

Proof Since $(\xi_1, \ldots, \xi_n)$ is Gaussian, the function on the right hand side of (9.65) is well defined. We show that we can go with the differentiation of the right hand side with respect to say α_1, under the expectation sign. Note that

$$\begin{aligned} I_\sigma &:= \frac{1}{\sigma}\,\mathbb{E}\left(\zeta\left(e^{(\alpha_1+\sigma)\xi_1} - e^{\alpha_1\xi_1}\right)e^{\alpha_2\xi_2+\cdots+\alpha_n\xi_n}\xi_1^{k_1}\cdots\xi_n^{k_n}\right)\\ &= \mathbb{E}\left(\zeta e^{\alpha_1(\sigma)\xi_1+\alpha_2\xi_2+\cdots+\alpha_n\xi_n}\xi_1^{k_1+1}\cdots\xi_n^{k_n}\right), \end{aligned}$$

where, for sufficiently small σ, $|\alpha_1(\sigma)| \le M$, we can assume that $|\alpha_k| \le M$, $k = 2, \ldots, n$ and $\lim_{\sigma\to 0}\alpha_1(\sigma) = \alpha$. The family of random variables under the expectation sign is uniformly integrable because for $\epsilon > 0$ and σ such that $|\alpha_1(\sigma)| \le M$,

$$\begin{aligned} &\mathbb{E}\left|\zeta e^{\alpha_1(\sigma)\xi_1+\alpha_2\xi_2+\cdots+\alpha_n\xi_n}\xi_1^{k_1+1}\cdots\xi_n^{k_n}\right|^{1+\epsilon}\\ &\le \left(\mathbb{E}\left|\zeta\xi_1^{k_1+1}\cdots\xi_n^{k_n}\right|^{2(1+\epsilon)}\right)^{1/2}\left(\mathbb{E}e^{2M\left(\sum_{k=1}^n|\xi_k|^2\right)}\right)^{1/2}, \end{aligned}$$

with the right hand side finite. □

We now derive formulae (9.63) and (9.64). Set

$$\psi_1(\alpha_1) := \int_H \varphi(S(t)x+z)e^{-\frac{1}{2}\alpha_1^2|\Gamma(t)h_1|^2+\alpha_1\left\langle\Gamma(t)h_1,Q_t^{-1/2}z\right\rangle}\mathscr{N}(0,Q_t)(dz).$$

Then

$$\frac{\partial\psi_1}{\partial\alpha_1}(0) = \int_H \varphi(S(t)x+z)\left\langle\Gamma(t)h_1, Q_t^{-1/2}z\right\rangle\mathscr{N}(0,Q_t)(dz),$$

as required.

Similarly if

$$\begin{aligned} \psi_2(\alpha_1,\alpha_2) := &\int_H \varphi(S(t)x+z)e^{-\frac{1}{2}\alpha_1^2|\Gamma(t)h_1|^2-\alpha_1\alpha_2\langle\Gamma(t)h_1,\Gamma(t)h_2\rangle-\frac{1}{2}\alpha_2^2|\Gamma(t)h_2|^2}\\ &\times e^{\alpha_1\left\langle\Gamma(t)h_1,Q_t^{-1/2}z\right\rangle+\alpha_2\left\langle\Gamma(t)h_2,Q_t^{-1/2}z\right\rangle}\mathscr{N}(0,Q_t)(dz), \end{aligned}$$

we have

$$\begin{aligned} \frac{\partial\psi_2}{\partial\alpha_1}(0,0) = &\int_H \varphi(S(t)x+z)e^{-\frac{1}{2}\alpha_2^2|\Gamma(t)h_2|^2+\alpha_2\left\langle\Gamma(t)h_2,Q_t^{-1/2}z\right\rangle}\\ &\times\left[\left\langle\Gamma(t)h_1,Q_t^{-1/2}z\right\rangle - \alpha_2\langle\Gamma(t)h_1,\Gamma(t)h_2\rangle\right]\mathscr{N}(0,Q_t)(dz) \end{aligned}$$

and

$$\frac{\partial \psi_2}{\partial \alpha_1}(0,0) = \int_H \varphi(S(t)x+z) e^{-\frac{1}{2}\alpha_2^2 |\Gamma(t)h_2|^2 + \alpha_2 \left\langle \Gamma(t)h_2, Q_t^{-1/2}z\right\rangle}$$
$$\times \left[\left\langle \Gamma(t)h_1, Q_t^{-1/2}z\right\rangle - \alpha_2 \langle \Gamma(t)h_1, \Gamma(t)h_2\rangle\right] \mathcal{N}(0, Q_t)(dz)$$

and

$$\frac{\partial^2 \psi_2}{\partial \alpha_1 \partial \alpha_2}(0,0)$$
$$= \int_H \varphi(S(t)x+z) \left\langle \Gamma(t)h_1, Q_t^{-1/2}z\right\rangle \left\langle \Gamma(t)h_2, Q_t^{-1/2}z\right\rangle \mathcal{N}(0, Q_t)(dz)$$
$$- \langle \Gamma(t)h_1, \Gamma(t)h_2\rangle \int_H \varphi(S(t)x+z) \mathcal{N}(0, Q_t)(dz),$$

as required.

We show finally that for arbitrary $\varphi \in B_b(H)$, $P_t\varphi$ is Fréchet differentiable and the derivative is Lipschitz continuous. We show first that for $t > 0$, $P_t\varphi$ is Lipschitz continuous. In fact from the mean value theorem we have that

$$P_t\varphi(x_2) - P_t\varphi(x_1) = \int_0^1 \langle D_x P_t\varphi(x_1 + \sigma(x_2 - x_1)), x_2 - x_1\rangle d\sigma$$
$$\times \int_0^1 \int_H \varphi(S(t)(x_1 + \sigma(x_2 - x_1) + z)$$
$$\times \left\langle \Gamma(t)(x_2 - x_1), Q_t^{-1/2}z\right\rangle \mathcal{N}(0, Q_t)(dz) d\sigma.$$

Consequently

$$|P_t\varphi(x_2) - P_t\varphi(x_1)| \le \|\varphi\|_\infty \left[\int_H \left|\left\langle \Gamma(t)(x_2 - x_1), Q_t^{-1/2}z\right\rangle\right|^2 \mathcal{N}(0, Q_t)(dz)\right]^{1/2}$$
$$\le \|\varphi\|_\infty \|\Gamma(t)\| \, |x_2 - x_1|$$

so, Lipschitzianity of $P_t\varphi$ follows. To show that $P_t\varphi$ is Fréchet differentiable for any $\varphi \in B_b(H)$ it is enough to show that this holds for φ Lipschitz. Indeed, if $\epsilon \in (0, t)$ then $P_t\varphi = P_{t-\epsilon} P_\epsilon \varphi$ and $P_\epsilon\varphi$ is Lipschitz by the above argument. We have

$$\langle D_x P_t\varphi(x_1) - D_x P_t\varphi(x_2), h\rangle$$
$$\times \int_H [\varphi(S(t)x_1 + z) - \varphi(S(t)x_2 + z)] \left\langle \Gamma(t)h, Q_t^{-1/2}z\right\rangle \mathcal{N}(0, Q_t)(dz) d\sigma.$$

Taking into account that φ is Lipschitz and proceeding as above we get that for a constant $C > 0$

$$|\langle D_x P_t\varphi(x_1) - D_x P_t\varphi(x_2), h\rangle| \le C|x_1 - x_2| \, |h|$$

and the result follows. We pass to the second derivative. Only the continuity in x of the bilinear form

$$W(x,h_1,h_2)=\int_H \varphi(S(t)x+z)\Big\langle \Gamma(t)h_1, Q_t^{-1/2}z\Big\rangle\Big\langle \Gamma(t)h_2, Q_t^{-1/2}z\Big\rangle \mathscr{N}(0,Q_t)(dz),$$

should be checked. Note that

$$\begin{aligned}
&|W(x,h_1,h_2)-W(x_1,h_1,h_2)|\\
&\quad\le \int_H |\varphi(S(t)x+z)-\varphi(S(t)x_1+z)|\\
&\qquad\times \Big|\Big\langle \Gamma(t)h_1, Q_t^{-1/2}z\Big\rangle\Big\langle \Gamma(t)h_2, Q_t^{-1/2}z\Big\rangle\Big| \mathscr{N}(0,Q_t)(dz).
\end{aligned}$$

We can again assume that φ is Lipschitz continuous. Then we have for a constant $C>0$

$$\begin{aligned}
&|W(x,h_1,h_2)-W(x_1,h_1,h_2)|\\
&\quad\le C|x-x_1|\left[\int_H \Big|\Big\langle \Gamma(t)h_1, Q_t^{-1/2}z\Big\rangle\Big|^2 \mathscr{N}(0,Q_t)(dz)\right]^{1/2}\\
&\qquad\times \left[\int_H \Big|\Big\langle \Gamma(t)h_2, Q_t^{-1/2}z\Big\rangle\Big|^2 \mathscr{N}(0,Q_t)(dz)\right]^{1/2}\\
&\quad\le C|x-x_1|\,|\Gamma(t)h_1|\,|\Gamma(t)h_2|.
\end{aligned}$$

Therefore, for a different constant $C>0$

$$\sup_{|h_1|\le 1,\ |h_2|\le 1} |W(x,h_1,h_2)-W(x_1,h_1,h_2)|\le C|x-x_1|$$

and Lipschitz continuity of the form follows. □

Theorem 9.28 *Assume that the conditions* (9.57) *and* (9.59) *are satisfied.*

(i) If $\Gamma(t)A$ has a bounded extension to the whole H denoted $\Gamma_1(t)$, then $A^*v_x\in C_b(H,H)$ and

$$\|A^*v_x(t)\|_0\le \|\Gamma_1(t)\|\,\|\varphi\|_0.$$

(ii) If $\Gamma(t)$ is a Hilbert–Schmidt operator then

$$\|\mathrm{Tr}\, v_{xx}(t)\|_0\le \|\Gamma(t)\|_2^2\|\varphi\|_0.$$

Proof (i) For $h\in D(A)$ we have

$$\begin{aligned}
v_x(t,x)(Ah)&=\int_H \langle \Gamma(t)Ah, Q_t^{-1/2}y\rangle\varphi(S(t)x+y)\mathscr{N}(0,Q_t)(dy)\\
&=\int_H \langle \Gamma_1(t)h, Q_t^{-1/2}y\rangle\varphi(S(t)x+y)\mathscr{N}(0,Q_t)(dy).
\end{aligned}$$

Since the operator $\Gamma_1(t)$ is bounded we have $v_x(t,x) \in D(A^*)$ and

$$\langle A^* v_x(t,x), h\rangle = \int_H \langle \Gamma_1(t)h, Q_t^{-1/2}y\rangle \varphi(S(t)x+y)\mathcal{N}(0,Q_t)(dy).$$

Changing the variable as in the proof of the previous theorem one gets the conclusion.

(ii) Note that for $h \in H$

$$v_{xx}(t,x)(h,h) = \int_H \langle \Gamma(t)h, Q_t^{-1/2}y\rangle^2 \varphi(S(t)x+y)\mathcal{N}(0,Q_t)(dy),$$

and therefore, for an orthonormal and complete basis $\{e_n\}$,

$$\begin{aligned}
|\mathrm{Tr}\, v_{xx}(t,x)| &= \left|\sum_{n=1}^{\infty} v_{xx}(t,x)(e_n,e_n)\right| \\
&= \left|\int_H \sum_{n=1}^{\infty} \langle \Gamma(t)e_n, Q_t^{-1/2}y\rangle^2 \varphi(S(t)x+y)\mathcal{N}(0,Q_t)(dy)\right| \\
&\le \|\varphi\|_0 \sum_{n=1}^{\infty} \int_H |\langle \Gamma(t)e_n, Q_t^{-1/2}y\rangle|^2 \mathcal{N}(0,Q_t)(dy) \\
&\le \|\varphi\|_0 \sum_{n=1}^{\infty} \int_H \|\Gamma(t)e_n\|^2 \\
&\le \|\varphi\|_0 \|\Gamma(t)\|_2^2.
\end{aligned}$$

The proof is complete. □

Remark 9.29 If the condition (9.59) does not hold then there exists a sequence of elements $x_n \to 0$ such that all the measures $P(t,x_n,\cdot)$ are singular with respect to $P(t,0,\cdot)$. Consequently, there exists a bounded function φ such that $P_t\varphi(x_n) = 0$ and $P_t\varphi(0) = 1$. So the condition (9.59) is also necessary for the smoothness property to hold. We remark that one can construct an example that shows that even for very regular data, A self-adjoint, Q a trace class operator and $\varphi \in C_b(H)$, the function $v(t,\cdot)$ may fail to be Fréchet differentiable, see [214]. □

Corollary 9.30 *Assume* (9.57) *and that*

$$S(t)(H) \subset Q^{1/2}(H) \text{ and } \int_0^t \|Q^{-1/2}S(s)x\|^2 ds < +\infty, \quad \forall\, x \in H. \tag{9.66}$$

Then the conclusions of Theorem 9.26 *are true and we have*

$$\|\Gamma(t)\| \le \frac{1}{t}\left(\int_0^t \|Q^{-1/2}S(s)\|^2 ds\right)^{1/2}. \tag{9.67}$$

Proof Consider the control system (9.60). If (9.66) holds then the control $u(s) = -\frac{1}{t}Q^{-1/2}S(s)x$, $s \in [0,t]$ transfers x to 0. So the control system is null controllable and (9.67) holds. □

Corollary 9.31 *Assume that $Q = I$ then* (9.66) *holds and for a constant $c > 0$*

$$\|\Gamma(t)\| \le \frac{c}{\sqrt{t}}. \tag{9.68}$$

Other concrete situations for which the conditions of Theorems 9.26 and 9.28 are satisfied will be discussed in Section 9.5 devoted to specific examples.

9.4.2 Nonlinear case

In this section we present an infinite dimensional generalization of an important formula discovered by Bismut, Li and Elworthy, see Bismut [74], Li [495] and Elworthy [281].

We are concerned with the problem

$$\begin{cases} dX(t) = (AX + F(X))dt + B(X)dW(t), \\ X(0) = x \in H, \end{cases} \tag{9.69}$$

assuming the following.

Hypothesis 9.1

(i) A is the infinitesimal generator of a strongly continuous semigroup $S(t),\ t \geq 0$, on H.

(ii) F is a Lipschitz continuous mapping from H into H.

(iii) B is a Lipschitz continuous mapping from H into $L(H)$. Moreover, for all $z \in H$, $B(z)$ is invertible and there exists $K > 0$ such that

$$\|B^{-1}(z)\| \leq K, \quad \forall\, z \in H.$$

(iv) We have $S(t) \in L_2(H)$ for all $t > 0$, and

$$\int_0^1 \|S(t)\|_{HS}^2\, dt < +\infty.$$

(v) W is a cylindrical Wiener process.

The following proof is basically from Peszat and Zabczyk [596].

Theorem 9.32 *Assume that Hypothesis 9.1 holds. Then for any $T > 0$ there exists a constant $C_T > 0$ such that for all $\psi \in B_b(H)$ and $t \in [0, T]$*

$$|P_t\psi(x) - P_t\psi(y)| \leq \frac{C_T}{\sqrt{t}}\, \|\psi\|_0\, |x - y|, \quad x, y \in H. \tag{9.70}$$

In particular $P_t,\ t \geq 0$, is a strong Feller semigroup for all $t > 0$. [4]

Proof The proof will consist of three steps. In the first step we prove (9.70) under the additional hypothesis that F, B and ψ are regular. In the second and third steps we show how to dispense with that assumption.

[4] For the definition of strong Feller semigroup see Section 11.2.2 below.

Step 1 Here we assume that $\psi \in C_b^2(H)$, F and B are Lipschitz and of class C^2 with bounded second derivatives.

We have divided this step into a sequence of lemmas.

Lemma 9.33 *Assume that $\psi \in C_b^2(H)$, F and B are Lipschitz and of class C^2 with bounded second derivatives. Then*

$$\psi(X(t,x)) = P_t\psi(x) + \int_0^t \langle D_x P_{t-s}\psi(X(s,x)), B(X(s,x))dW(s)\rangle, \quad \mathbb{P}\text{-a.s.} \tag{9.71}$$

Proof Let $\{e_n\}$ be a complete orthonormal system in H. For each n let $X_n(\cdot, x)$ be the solution of the problem

$$\begin{cases} dX_n(t) = (A_n X_n + F(X_n))dt + B(X_n)P_n dW(t), \\ X_n(0) = x, \end{cases}$$

where $A_n = nA(n - A)^{-1}$ is the Yosida approximation of A and P_n is the orthogonal projection of H onto the linear subspace of H spanned by $\{e_1, \ldots, e_n\}$. It follows from Theorem 9.23 that the function

$$v_n(t,x) = \mathbb{E}(\psi(X_n(t,x))), \quad (t,x) \in (0, +\infty) \times H,$$

is a strict solution to the Kolmogorov equation

$$\begin{cases} D_t v^n(t,x) = \frac{1}{2}\,\mathrm{Tr}\,[B^*(x)P_n B(x) v^n_{xx}(t,x)] + \langle A_n x + F(x), v^n_x(t,x)\rangle, \\ v^n(0,x) \quad = \psi(x), \quad x \in H, \quad t \geq 0. \end{cases}$$

Applying Itô's formula (Theorem 4.32) to the process $v^n(t-s, X_n(s,x))$, $s \in [0,t]$, we obtain that

$$\psi(X_n(t,x)) = v^n(t,x) + \int_0^t \langle v^n_x(t-s, X_n(s,x)), B(X_n(s,x))P_n dW(s)\rangle, \quad \mathbb{P}\text{-a.s.}$$

Letting $n \to +\infty$ gives the desired result. □

Now we extend the finite dimensional Bismut–Elworthy–Li formula, see Elworthy [281].

Lemma 9.34 *Assume that $\psi \in C_b^2(H)$, F and B are Lipschitz and of class C^2 with bounded second derivatives. Then the directional derivatives $\langle D_x P_t \psi(x), h\rangle$*

are given by

$$\langle D_x P_t \psi(x), h\rangle = \frac{1}{t}\mathbb{E}\left[\psi(X(t,x))\int_0^t \langle B^{-1}(X(s,x))X_x(s,x)h, dW(s)\rangle\right], \quad \mathbb{P}\text{-a.s.} \tag{9.72}$$

Proof Fix $h \in H$ and set $v(t,x) = P_t\psi(x),\ t \geq 0,\ x \in H$. Multiplying both sides of (9.71) by

$$\int_0^t \langle B^{-1}(X(s,x))X_x(s,x)h, dW(s)\rangle,$$

and taking the expectation we get

$$\begin{aligned}
&\mathbb{E}\left(\psi(X(t,x))\int_0^t \langle B^{-1}(X(s,x))X_x(s,x)h, dW(s)\rangle\right)\\
&= \mathbb{E}\int_0^t \langle B^*(X(s,x))v_x(t-s, X(s,x)), B^{-1}(X(s,x))X_x(s,x)h\rangle ds\\
&= \int_0^t \langle D_x\mathbb{E}(D_x P_{t-s}\psi(X(s,x))), h\rangle\, ds\\
&= \int_0^t \langle v_x(t-s, X(s,x)), h\rangle\, ds = \int_0^t \langle D_x P_t\psi(x), h\rangle ds\\
&= t\langle D_x P_t\psi(x), h\rangle,
\end{aligned}$$

which yields (9.72). □

Lemma 9.35 *Assume that $\psi \in C_b^2(H)$, F and B are Lipschitz and of class C^2 with bounded second derivatives. Then the estimate* (9.70) *holds true.*

Proof Fix $T > 0$, then by (9.72) we have

$$\begin{aligned}
|\langle D_x P_t\psi(x), h\rangle|^2 &\leq \frac{1}{t^2}\,\|\psi\|_0^2\,\mathbb{E}\left[\int_0^t |B^{-1}(X(s,x))X_x(s,x)h|^2 ds\right]\\
&\leq \frac{K^2}{t^2}\,\|\psi\|_0^2\,\mathbb{E}\int_0^t |X_x(s,x)h|^2 ds,
\end{aligned}$$

and the conclusion follows from Remark 9.11. □

Step 2 We first show that if (9.70) holds for $\psi \in C_b^2(H)$ then it holds for all $\psi \in B_b(H)$. This follows from the next lemma.

Lemma 9.36 *Let $P_t\varphi(x) = \mathbb{E}[\varphi(X(t,x))],\ t \geq 0,\ \varphi \in B_b(H)$ and let $c > 0$ and $t > 0$ be fixed. Then the following conditions are equivalent.*

(i) *For all $\varphi \in C_b^2(H)$ and for all $x, y \in H$ we have*

$$|P_t\varphi(x) - P_t\varphi(y)| \leq c\|\varphi\|_0|x - y|.$$

(ii) For all $\varphi \in B_b(H)$ and for all $x, y \in H$ we have

$$|P_t\varphi(x) - P_t\varphi(y)| \le c\|\varphi\|_0|x-y|.$$

(iii) For all $x, y \in H$ we have

$$\mathrm{Var}(P_t(x,\cdot) - P_t(y,\cdot)) \le c\|\varphi\|_0|x-y|.$$

Proof Let

$$\mathscr{K}_1 = \{\varphi \in C_b(H) : \ \|\varphi\|_0 \le 1\},$$

and

$$\mathscr{K}_2 = \{\varphi \in C_b^2(H) : \ \|\varphi\|_0 \le 1\}.$$

Since each bounded continuous function on H can be approximated pointwise by functions of $C_b^2(H)$ we have

$$\sup_{\varphi\in\mathscr{K}_1} |P_t\varphi(x) - P_t\varphi(y)| = \sup_{\varphi\in\mathscr{K}_2} |P_t\varphi(x) - P_t\varphi(y)|,$$

for all $x, y \in H$. As a simple consequence of the Hahn decomposition theorem, we have

$$\sup_{\varphi\in\mathscr{K}_1} |P_t\varphi(x) - P_t\varphi(y)| = \mathrm{Var}(P_t(x,\cdot) - P_t(y,\cdot)).$$

Therefore (i) implies (iii). However, if (iii) holds then for all $\varphi \in B_b(H)$

$$\begin{aligned} |P_t\varphi(x) - P_t\varphi(y)| &\le \left| \int_H \varphi(z)(P_t(x,dz) - P_t(y,dz)) \right| \\ &\le \|\varphi\|_0 \ \mathrm{Var}(P_t(x,\cdot) - P_t(y,\cdot)) \\ &\le \|\varphi\|_0|x-y|, \end{aligned}$$

and we see that (ii) is true. Obviously (ii) implies (i) and the proof of the lemma is complete. □

Step 3 Conclusion.

By Step 2 it follows that, to prove the theorem in the general case, we can assume that $\varphi \in C_b^2(H)$. We will now construct suitable approximations F_n of F and B_n of B. For this we choose a negative self-adjoint operator $\beta : D(\beta) \subset H \to H$ such that β^{-1} is of trace class and consider the linear equation

$$dX = \beta X dt + dW(t), \quad X(0) = x,$$

and the corresponding transition semigroup

$$R_t\psi(x) = \mathbb{E}[\psi(X(t,x)], \quad \psi \in B_b(H).$$

Then we set

$$\langle F_n(x), h\rangle = R_{1/n}\langle F(x), h\rangle, \quad n \in \mathbb{N},\ x, h \in H \tag{9.73}$$

and

$$\langle B_n(x)z, h\rangle = R_{1/n}\langle B(x)z, h\rangle, \quad n \in \mathbb{N},\ x, z, h \in H. \tag{9.74}$$

By Theorem 9.26, $R_t\psi$ is of C^∞ class. Moreover, it is easy to check the following.

(i) $F_n,\ B_n,\ n \in \mathbb{N}$, are twice Fréchet differentiable with bounded and continuous derivatives.
(ii) $F_n,\ B_n,\ n \in \mathbb{N}$, satisfy a Lipschitz condition with the same Lipschitz constant.
(iii) The operators B_n are invertible and

$$\|B_n^{-1}(x)\| \le 2K, \quad \forall\, x \in H.$$

(iv) $\lim_{n\to\infty} |F(z) - F_n(z)| = 0$ and $\lim_{n\to\infty} \|B(z) - B_n(z)\| = 0$.

We check only (iii). Let us first notice that by the very definition (9.74) we have

$$B_n(x) = \int_H B(e^{\beta/n}x + y)\mathcal{N}_{Q_n}(dy),$$

where $Q_n = -\frac{1}{2}\beta^{-1}(1 - e^{2\beta/n})$. Moreover

$$\sup_n \sup_{x\in H} \|B^{-1}(e^{\beta/n}x)\| \le \sup_{x\in H} \|B^{-1}(x)\| \le K.$$

Consequently, if L is a Lipschitz constant of B, we have

$$\|B(e^{\beta/n}x) - B_n(x)\| \le L \int_H |y|\mathcal{N}_{Q_n}(dy) \to 0$$

as $n \to \infty$, uniformly on $x \in H$. Now choose n_0 such that

$$\|B(e^{\beta/n}x) - B_n(x)\| \le \frac{1}{2K}, \quad \forall\, n > n_0$$

and recall that if $S, T \in L(H)$, $\|T^{-1}\| \le K$, $\|S - T\| \le \frac{1}{2K}$ we have $S^{-1} \in L(H)$, $S^{-1} = T^{-1}(I - (T - S)T^{-1})^{-1}$ and $\|S^{-1}\| \le 2K$. Setting $T = B(e^{\beta/n}x)$ and $S = B_n(x)$ we see that if $n > n_0$, $\|B_n(x)^{-1}\| \le 2K$ as claimed.

Now we can finish the proof of the theorem. Let $X_n(\cdot, x)$ be the solution of the equation

$$\begin{cases} dX_n(t) = (AX_n + F_n(X_n))dt + B_n(X_n)dW(t), \\ X_n(0) = x, \end{cases}$$

and let $P_{n,t},\ t \ge 0$, be the corresponding transition semigroup. Fix $t > 0$, $\psi \in C_b^2(H)$. Then Lemma 9.36 applied to $P_{n,t},\ t \ge 0$, shows that

$$|P_{n,t}\psi(x) - P_{n,t}\psi(y)| \le K\|\psi\|_0|x - y|, \quad x, y \in H.$$

Letting $n \to \infty$ we have

$$P_{n,t}\psi(x) = \mathbb{E}[\psi(X_n(t, x))] \to \mathbb{E}(\psi(X(t, x))].$$

The proof follows. □

9.5 Mild Kolmogorov equation

We are concerned here with the Kolmogorov equation

$$\begin{cases} v_t(t,x) = \dfrac{1}{2}\,\mathrm{Tr}\,[Qv_{xx}(t,x)] + \langle Ax + F(x), v_x(t,x)\rangle, \\ v(0,x) = \varphi(x), \quad t \ge 0,\ x \in D(A), \end{cases} \tag{9.75}$$

under the following assumptions.

Hypothesis 9.2

(i) $A : D(A) \subset H \to H$ *is the infinitesimal generator of a* C_0*-semigroup* $S(\cdot)$.

(ii) $\displaystyle\int_0^\infty \mathrm{Tr}\,[S(t)QS(t)^*]dt < \infty.$

(iii) $S(t)(H) \subset Q_t^{1/2}(H)$, *where* $Q_t x = \displaystyle\int_0^t S(s)QS(s)^* ds$ *and there exists* $C > 0$ *and* $\alpha \in [1/2, 1)$ *such that*

$$\|\Gamma(t)\| \le \frac{C}{t^\alpha}, \quad t \in [0,T], \tag{9.76}$$

where $\Gamma(t) = Q_t^{-1/2} S(t)$.

(iv) $F \in C_b(H,H)$.

Definition 9.37 *We say that* v *is a* mild solution *to* (9.75) *if*

(i) $v \in C((0,T]; C_b^1(H))$,

(ii) $\displaystyle\sup_{t\in[0,T]} t^\alpha \|v(t,\cdot)\|_1 < \infty$,

(iii) v *fulfills the integral equation*

$$v(t,x) = R_t\varphi(x) + \int_0^t R_{t-s}\left(\langle F(\cdot), v_x(s,\cdot)\rangle\right)(x)ds, \quad t \in [0,T], \tag{9.77}$$

where $R_t\varphi(x) = \mathbb{E}[\varphi(S(t)x + W_A(t))]$ *and* $W_A(t)$, $t \ge 0$, *is the stochastic convolution.*

We are going to show in Section 9.5.1 that equation (9.75) has a unique mild solution. Then in Section 9.5.2, under the additional assumption that $F \in C_b^1(H)$, we shall identify $v(t,\cdot)$ with the transition semigroup $P_t\varphi$ corresponding to the problem

$$\begin{cases} dX = (AX + F(X))dt + \sqrt{Q}\,dW(t), \\ X(0) = x, \end{cases} \tag{9.78}$$

W being a cylindrical Wiener process.

9.5.1 Solution of (9.75)

Let us define, for given $n \in \mathbb{N}$, $T > 0$ the Banach space,

$$\Lambda^n_{T,\alpha} = \left\{ v \in C([0,T]; C_b(H)) \cap C((0,T]; C^n_b(H)) : \sup_{t\in[0,T]} t^\alpha \|v(t,\cdot)\|_n < \infty \right\}$$

endowed with the norm

$$|\!|\!| v |\!|\!|_n = \sup_{t\in[0,T]} \|v(t,\cdot)\|_0 + t^\alpha \|v(t,\cdot)\|_n.$$

Theorem 9.38 *Assume that Hypothesis 9.2 is fulfilled. Then the following statements hold.*

(i) If $\varphi \in B_b(H)$ then equation (9.77) has a unique solution in $\Lambda^1_{T,\alpha}$.
(ii) If $n \ge 2$, $F \in C^{n-1}_b(H,H)$ and $\varphi \in B_b(H)$ then equation (9.77) has a unique solution in $\Lambda^n_{T,\alpha}$.

Proof (i) Let $u(t,x) = R_t\varphi(x)$, $t > 0$, $x \in H$. It follows from (9.76) that

$$\|u_x(t,\cdot)\|_0 \le \|\Gamma(t)\| \, \|\varphi\|_0 \le \frac{C}{t^\alpha} \|\varphi\|_0, \quad t \ge 0.$$

Since $\|u(t,\cdot)\|_0 \le \|\varphi\|_0$, therefore

$$|\!|\!| u |\!|\!|_1 \le (T^\alpha + 1)\|\varphi\|_0 < +\infty,$$

and consequently $u(\cdot) \in \Lambda^1_{T,\alpha}$. Define the transformation γ

$$(\gamma(v))(t,\cdot) = \int_0^t R_{t-s}\left(\langle F(\cdot), v_x(s,\cdot)\rangle\right)(x)ds, \quad t \in [0,T].$$

It is easy to check that γ maps $\Lambda^1_{T,\alpha}$ into itself. Moreover for each $v \in \Lambda^1_{T,\alpha}$ and $t \in (0,T]$ we have

$$\begin{aligned} t^\alpha \|\gamma(v)(t,\cdot)\|_1 &= t^\alpha \left(\|\gamma(v)(t,\cdot)\|_0 + \|\gamma(v)_x(t,\cdot)\|_0\right) \\ &\le t^\alpha \left(\|F\|_0 \int_0^t \|v_x(s,\cdot)\|_0 + c(t-s)^{-\alpha}\|v_x(s,\cdot)\|_0 \right) ds \\ &\le t^\alpha \left(\|F\|_0 \int_0^t s^{-\alpha}(1 + c(t-s)^{-\alpha}) ds \, |\!|\!| v |\!|\!|_1 \right). \end{aligned}$$

Therefore for a different constant $C_1 > 0$

$$|\!|\!| \gamma(v) |\!|\!|_1 \le C_1 T^{1-\alpha} |\!|\!| v |\!|\!|_1.$$

Consequently, if $T < C_1^{-\frac{1}{1-\alpha}}$, γ is a contraction on $\Lambda^1_{T,\alpha}$ and the problem (9.77) has a unique solution v. The case of arbitrary $T > 0$ can be treated by the usual methods of steps.

(ii) It follows from (9.76) that

$$\|D^n u(t,\cdot)\|_0 \le \|\Gamma(t)\| \, \|S(t)\|^{n-1} \|D^{n-1}\varphi\|_0, \quad t \ge 0. \tag{9.79}$$

Therefore $u(t,\cdot) = R_t\varphi,\ t \in (0,T]$, belongs to $\Lambda^n_{T,\alpha}$. If $v \in \Lambda^n_{T,\alpha}$ and $t \in (0,T]$ then, for a constant $C_{n,1}$,

$$
\begin{aligned}
t^\alpha \|\gamma(v)(t,\cdot)\|_n &= t^\alpha \sum_{j=0}^{n} \|D^j \gamma(v)(t,\cdot)\|_0 \\
&\leq t^\alpha \|D^n \gamma(v)(t,\cdot)\|_0 + t^\alpha C_{n,1} \|F\|_{n-1} \int_0^t \sum_{j=0}^{n} \|D^j \gamma(v)(s,\cdot)\|_0 ds.
\end{aligned}
\tag{9.80}
$$

Moreover, compare (9.79), there exists $C_{n,2} > 0$ such that

$$
\begin{aligned}
\|D^n\gamma(v)(t,\cdot)\|_0 &\leq \int_0^t \|D^n R_{t-s}(\langle F(\cdot), D^1 v(s,\cdot)\rangle)\|_0 ds \\
&\leq C_{n,2} \int_0^t (\|\Gamma(t-s)\|\ \|S(t-s)\|^{n-1}\ \|F_{n-1}\| \|v(s,\cdot)\|_n) ds.
\end{aligned}
\tag{9.81}
$$

Taking into account (9.5.1) and (9.81) one can see that for a constant $C_n > 0$

$$
|||\gamma(v)|||_n \leq C_n T^{1-\alpha} |||v|||_n.
$$

So, as in part (i), the problem (9.77) has a unique solution in $\Lambda^n_{T,\alpha}$. □

If $\varphi \in B_b(H)$, define

$$
(\widetilde{P}_t\varphi)(x) = v(t,x), \quad t \in [0,T], \tag{9.82}
$$

where v is the unique solution of (9.77). The uniqueness implies that the family $\{\widetilde{P}_t\}$ has the semigroup property.

Theorem 9.39 *Assume that Hypothesis* 9.2 *holds.*

(i) If $F \in C_b(H,H)$ and $\varphi \in B_b(H)$ then $\widetilde{P}_t\varphi \in C^1_b(H),\ t \in (0,T]$.
(ii) If, in addition, $F \in C^{n-1}_b(H,H),\ n \geq 1$, then $\widetilde{P}_t\varphi \in C^n_b(H),\ t \in (0,T]$.

Proof Part (i) follows from Theorem 9.28(i). To show (ii) it is enough to remark that for arbitrary $\varepsilon \in (0,t)$, $\widetilde{P}_\varepsilon\varphi \in C^1_b(H)$. Consequently, by Theorem 9.26(ii) we have

$$
\widetilde{P}_t\varphi = \widetilde{P}_{t-\varepsilon}(\widetilde{P}_\varepsilon\varphi) \in C^1_b(H).
$$

So the result follows by an obvious induction argument in n. □

Remark 9.40 The above theorems could be regarded as a nonlinear counterpart of Theorem 9.26. In a similar way, requiring in addition that

$$
\|\Gamma_1(t)\| \leq \frac{c}{t^\alpha}, \quad t \geq 0, \tag{9.83}
$$

or

$$
\|\Gamma_1(t)\|_2^2 \leq \frac{c}{t^\alpha}, \quad t \geq 0, \tag{9.84}
$$

one could prove a complete analog of Theorem 9.28. However, the conditions (9.76) and (9.83) as well as (9.84) are not satisfied in many applications, therefore such a result would be of only very limited value. A real analog of Theorem 9.28, for the nonlinear equation (9.75), requires different methods which have to be invented. □

9.5.2 Identification of $v(t,\cdot)$ with $P_t\varphi$

For the sake of simplicity we shall prove the identification for a particular case. It is not difficult to generalize this proof in other specific situations.

We shall assume the following.

Hypothesis 9.3

(i) $A: D(A) \subset H \to H$ *is symmetric negative and there exists an orthonormal basis* (e_k) *in* H *and a sequence of positive numbers* (α_k) *such that*

$$Ae_k = -\alpha_k e_k, \quad k \in \mathbb{N}.$$

(ii) There exists $r \in (0,1]$ *such that* $Q = (-A)^{-r}$.

(iii) $F \in C_b^1(H,H)$.

Under these assumptions one can easily check that Hypothesis 9.2 is fulfilled and moreover that

$$\Gamma(t)e_k = \frac{1}{\sqrt{2}}\,\alpha_k^{\frac{1+r}{2}}\,\frac{e^{-\alpha_k t}}{(1-e^{-2\alpha_k t})^{1/2}}, \quad k \in \mathbb{N} \tag{9.85}$$

(see also Section 9.6.) It follows that

$$\|\Gamma(t)\| = \sup_{k\in\mathbb{N}} \frac{1}{\sqrt{2}}\,\alpha_k^{\frac{1+r}{2}}\,\frac{e^{-\alpha_k t}}{(1-e^{-2\alpha_k t})^{1/2}} = c_r\, t^{-\frac{1+r}{2}}, \quad t>0, \tag{9.86}$$

where

$$c_r = \sup_{z>0} \frac{1}{\sqrt{2}}\, z^{\frac{1+r}{2}}\,\frac{e^{-z}}{(1-e^{-2z})^{1/2}}.$$

We denote by $X(\cdot,x)$ the unique mild solution of (9.78) and by P_t the corresponding transition semigroup

$$P_t\varphi(x) = \mathbb{E}[\varphi(X(t,x))], \quad t \ge 0,\ x\in H,\ \varphi \in B_b(H).$$

It is convenient to consider an approximation of equation (9.78):

$$dX_N(t) = (AP_NX_N(t) + P_NF(X_N(t))dt + P_N\sqrt{Q}\,dW(t), \quad X_N(0) = P_Nx, \tag{9.87}$$

where $N \in \mathbb{N}$ and P_N is the orthogonal projector on the linear span of $\{e_1,\ldots,e_N\}$. It is clear that equation (9.87) has a unique solution $X_N(\cdot,x)$.

Lemma 9.41 *Assume Hypothesis* 9.3 *and and let* $x \in H$. *Then*

$$\lim_{n\to\infty} X_N(t,x) = X(t,x) \quad \text{in } \mathcal{H}_2(0,T),\ T>0. \tag{9.88}$$

Moreover $X_N(\cdot, x)$ is differentiable in any direction $h \in H$ and

$$\lim_{n\to\infty} DX_N(t,x)\cdot h = DX(t,x)\cdot h \quad in\ \mathscr{H}_2(0,T),\ T>0. \tag{9.89}$$

Proof Set $S(t) = e^{tA}$ and $S_N(t) = e^{tAP_N}$. Write

$$\begin{aligned}
X(t) - X_N(t) = {} & S(t)x - S_N(t)P_N x \\
& + \int_0^t [S(t-s) - S_N(t-s)]F(X(s))ds \\
& + \int_0^t S_N(t-s)(F(X(s)) - F(X_N(s)))ds \\
& + \int_0^t (I - P_N)S_N(t-s)F(X_N(s))ds \\
& + \int_0^t [S(t-s)\sqrt{Q} - S_N(t-s)P_N\sqrt{Q}]dW(s).
\end{aligned}$$

Therefore

$$\begin{aligned}
|X(t) - X_N(t)| \le {} & |S_N(t)x - S_N(t)x| \\
& + \left|\int_0^t [S(t-s) - S_N(t-s)]F(X(s))ds\right| \\
& + L\int_0^t |F(X(s)) - F(X_N(s))|ds \\
& + \left|\int_0^t (I - P_N)S_N(t-s)F(X_N(s))ds\right| \\
& + \left|\int_0^t [S(t-s)\sqrt{Q} - S_N(t-s)P_N\sqrt{Q}]dW(s)\right|,
\end{aligned}$$

where L is the Lipschitz constant of F. Since

$$\mathbb{E}\left|\int_0^t [S(t-s)\sqrt{Q} - S_N(t-s)P_N\sqrt{Q}]dW(s)\right|^2 \le \int_0^t \mathrm{Tr}\,[(I - P_N)QS(2s)]ds$$

the conclusion of the first part of the theorem follows from the Gronwall lemma. The proof of the second part is straighforward, repeating arguments from the proof of Theorem 9.8. □

Next we consider an approximating Kolmogorov equation.

$$v_N(t,x) = R_t^N\varphi(x) + \int_0^t R_{t-s}^N\left(\langle F(\cdot), D_x v_N(s,\cdot)\rangle\right)(x)ds, \quad t\in[0,T], \tag{9.90}$$

where

$$R_t^N\varphi(x) = \mathbb{E}[\varphi(S_N(t)x + W_A^N(t))], \quad \varphi\in B_b(H),\ t\ge 0,\ x\in H$$

and

$$W_A^N(t) = \int_0^t S_N(t-s)P_N Q dW(s).$$

Lemma 9.42 *Assume Hypothesis* 9.3 *and let* $N \in \mathbb{N}$. *If* $\varphi \in B_b(H)$ *then equation* (9.90) *has a unique solution* $v_N(t,\cdot)$ *in* $\Lambda^1_{T,\alpha}$. *Moreover,*

$$\lim_{N\to\infty} v_N(t,x) = v(t,x), \quad t \in [0,T],\ x \in H. \tag{9.91}$$

Proof Existence and uniqueness follow from Theorem 9.38, noticing that, thanks to (9.86) we have

$$\|\Gamma_N(t)\| \le c_r t^{-\frac{1+r}{2}},$$

(with c_r independent of N) where

$$\Gamma_N(t) = (Q_t^N)^{-1/2} S_N(t), \quad t > 0,\ N \in \mathbb{N},$$

and

$$Q_t^N = \int_0^t S_N(s) P_N Q ds.$$

Let us prove (9.91). We know that

$$v(t,\cdot) = (\gamma(v))(t,\cdot), \quad t \in [0,T],$$

and

$$v_N(t,\cdot) = (\gamma_N(v_N))(t,\cdot), \quad t \in [0,T],$$

where

$$(\gamma(v))(t,\cdot) = \int_0^t R_{t-s}\left(\langle F(\cdot), v_x(s,\cdot)\rangle\right)(x) ds, \quad t \in [0,T],$$

and

$$(\gamma_N(v))(t,\cdot) = \int_0^t R^N_{t-s}\left(\langle F(\cdot), v_x(s,\cdot)\rangle\right)(x) ds, \quad t \in [0,T].$$

We may assume that γ_N and γ are Lipschitz in Λ^1_T with constant $\frac{1}{2}$. Since we do not have $\gamma_N \to \gamma$ in Λ^1_T we cannot conclude, using the classical contraction principle depending on a parameter that $v_N \to v$ in Λ^1_T. We shall prove the weaker result (9.91) arguing as in [135]. Set

$$v^0 = 0, \quad v_N^0 = 0,$$

and define

$$v^M = \gamma^M(v^0), \quad v_N^M = \gamma_N^M(v^0), \quad M \in \mathbb{N}.$$

By the classical contraction principle, we have

$$\lim_{M\to\infty} v^M = v, \quad \lim_{M\to\infty} v_N^M = v_N \quad \text{in } \Lambda_T^1,\ N \in \mathbb{N}.$$

Moreover

$$\|v - v^M\|_{\Lambda_T^1} \le \sum_{k=M}^{\infty} 2^{-k} \|\gamma(v^0)\|_{\Lambda_T^1}, \quad \|v_N - v_N^M\|_{\Lambda_T^1} \le \sum_{k=M}^{\infty} 2^{-k} \|\gamma_N(v^0)\|_{\Lambda_T^1}.$$

Now fix $x \in H$, then for all $t \in [0, T]$

$$|v(t,x) - v_N(t,x)| \le |v(t,x) - v^M(t,x)| + |v^M(t,x) - v_N^M(t,x)| + |v_N^M(t,x) - v_N(t,x)|. \tag{9.92}$$

Given $\varepsilon > 0$ there exists $M_\varepsilon \in \mathbb{N}$ such that

$$\sum_{k=M}^{\infty} 2^{-k} \left[\|\gamma(v^0)\|_{\Lambda_T^1} + \|\gamma_N(v^0)\|_{\Lambda_T^1}\right] \le \frac{\varepsilon}{2}, \tag{9.93}$$

for all $M \ge M_\varepsilon$ and all $N \in \mathbb{N}$. By (9.94) and (9.93) it follows that

$$|v(t,x) - v_N(t,x)| < \frac{\varepsilon}{2} + |v^{M_\varepsilon}(t,x) - v_N^{M_\varepsilon}(t,x)|, \quad \forall\, t \in [0, T].$$

The conclusion follows letting $\varepsilon \to 0$. □

We are now ready to prove the following.

Theorem 9.43 *Assume Hypothesis 9.3. Then for any $\varphi \in B_b(H)$ we have*

$$v(t,x) = \mathbb{E}[\varphi(X(t,x))], \quad t \ge 0,\ x \in H,$$

where v is the mild solution of (9.77) *and X is the mild solution of* (9.78).

Proof It is enough to show the conclusion when $\varphi \in C_b^2(H)$. By Lemma 9.42 we have

$$\lim_{N\to\infty} v_N(t,x) = v(t,x), \quad t \in [0,T],\ x \in H,$$

where v_N is the solution of the mild Kolmogorov equation (9.91). On the other hand, by Theorem 9.23 the Kolmogorov equation

$$\begin{cases} z_t^N(t,x) = \dfrac{1}{2}\,\mathrm{Tr}\,[P_N Q z_{xx}^N(t,x)] + \langle AP_N x + F(x), z_x^N(t,x)\rangle, & t > 0, \\ z^N(0,x) = \varphi(x), \quad x \in H, \end{cases} \tag{9.94}$$

has a unique strict solution given by

$$z^N(t,x) = \mathbb{E}[\varphi(X_N(t,x))].$$

But, by the variation of constants formula we have $z^N = v_N$. So, the conclusion follows from Lemma 9.42. □

9.6 Specific examples

Results obtained in the previous sections are applicable in the important case when the operator A is self-adjoint and commutes with Q. More precisely, we assume that there exists a complete orthonormal basis $\{e_n\}$ in H such that

$$Ae_n = -\alpha_n e_n, \quad Qe_n = \lambda_n e_n, \tag{9.95}$$

where $\{\alpha_n\}$ and $\{\lambda_n\}$ are sequences of positive numbers, $\alpha_n \to +\infty$ and $\{\lambda_n\}$ is bounded. By a direct calculation one gets the following proposition.

Proposition 9.44 *Assume* (9.95), *then the following hold.*

(i) Condition (9.57) *is equivalent to*

$$\sum_{n=1}^{\infty} \frac{\lambda_n}{\alpha_n} < \infty. \tag{9.96}$$

(ii) Condition (9.59) *is equivalent to the boundedness of the sequence*

$$\left\{\sqrt{\frac{\alpha_n}{\lambda_n}} e^{-\alpha_n t}\right\}.$$

(iii) There exists $c > 0$ *and* $\alpha \in [1/2, 1)$ *such that*

$$\|\Gamma(t)\| \le \frac{c}{t^\alpha}, \quad t \in (0, T] \tag{9.97}$$

if and only if there exists $c > 0$ *and* $\alpha \in [1/2, 1)$ *such that*

$$\lambda_n \alpha_n^{2\alpha-1} \ge c, \quad n \in \mathbb{N}. \tag{9.98}$$

Proof (i) is clear, let us show (ii). Since

$$Q_t e_n = \frac{\lambda_n}{2\alpha_n}(1 - e^{-2t\alpha_n})e_n, \quad n \in \mathbb{N},$$

we have

$$\Gamma(t)e_n = \sqrt{\frac{2\alpha_n}{\lambda_n}} e^{-t\alpha_n}(1 - e^{-2t\alpha_n})^{-1/2},$$

so that condition (9.59) is fulfilled, provided that for any $t > 0$,

$$k_t := \sup_n \sqrt{\frac{2\alpha_n}{\lambda_n}} e^{-t\alpha_n}(1 - e^{-2t\alpha_n})^{-1/2} < \infty.$$

Since $\alpha_n \to +\infty$ as $n \to \infty$, this happens if and only if $\left\{\frac{2\alpha_n}{\lambda_n} e^{-t\alpha_n}\right\}$ is bounded.

Finally, let us show (iii). The eigenvalues of the operator

$$\Gamma^2(t) = Q_t^{-1} S(2t), \quad t > 0,$$

are

$$\frac{2\alpha_n}{\lambda_n}(1 - e^{-2t\alpha_n})^{-1/2} e^{-2t\alpha_n}.$$

So, there exists $c > 0$ such that (9.97) holds if and only if there exists $c > 0$ such that

$$\sup_n \sup_{t\in(0,T]} \left\{ \frac{2t^{2\alpha}\alpha^n}{\lambda_n} \frac{1}{e^{2t\alpha_n - 1}} \right\} \leq c^2. \tag{9.99}$$

□

Lemma 9.45 *Let $\alpha > 0,\ \beta > 0$. Then the following hold.*

(i) $\sup_{t\in(0,T]} \dfrac{t^\beta}{e^{t\alpha} - 1} < +\infty$ *if and only if $\beta \geq 1$.*

(ii) $\sup_{t\in(0,T]} \dfrac{t}{e^{t\alpha} - 1} = \dfrac{1}{\alpha}$.

(iii) If $\beta > 1$, denote by z_β the unique positive solution of the equation $z = \beta(1 - e^{-z})$. Then the supremum of the function $t \to \frac{t^\beta}{e^{t\alpha}-1}$, $t > 0$, is attained at $t = \frac{z_\beta}{\alpha}$ and is equal to $(\frac{z_\beta}{\alpha})^\beta \frac{1}{e^{z_\beta}-1}$.

Proof We check only (iii). If

$$\psi(t) = \tfrac{t^\beta}{e^{t\alpha}-1},$$

then $\psi'(t) = 0$ if and only if $\beta(1 - e^{-\alpha t}) = t\alpha$. So (iii) follows. □

Proof of Proposition 9.44(ii) (Continuation) Let $2\alpha \geq 1$. Putting $\beta = 2\alpha = 1$ and $\alpha = 2\alpha_n$ we get that the supremum in (9.99) with respect to $t \in (0, T]$ is $2\alpha_n$. So, (9.99) becomes

$$\sup_n \frac{1}{\lambda_n} < +\infty,$$

as required. Take now $\beta = 2\alpha > 1$ and $\alpha = 2\alpha_n$. Then the supremum in (9.99) with respect to $t \in (0, T]$ is attained at β/α_n. Since $\alpha_n \to +\infty$ we can assume that $\beta/\alpha_n \leq T$. Then the supremum in (9.99) with respect to $t \in (0, T]$ is equal to

$$\frac{1}{(2\alpha_n)^{\beta-1}\lambda_n} \frac{(2\beta)^\beta}{e^{2\beta} - 1}.$$

So, (9.99) becomes

$$\sup_n \left[\frac{1}{(2\alpha_n)^{\beta-1}\lambda_n} \frac{(2\beta)^\beta}{e^{2\beta} - 1} \right] \leq c^2$$

and we get 9.44(ii). □

Remark 9.46 The condition (9.66) from Corollary 9.30 implying smoothness of the mild solution is equivalent, under (9.95), to the boundnedess of the sequence $\{\frac{1}{\lambda_n\alpha_n}\}$.

This is a stronger condition than boundnedess of $\{\frac{1}{\lambda_n}\alpha_n e^{-\alpha_n t}\}$. However, it is always satisfied in the case of cylindrical noise $Q = I$, see Corollary 9.31. □

Example 9.47 Assume $\alpha_n = n^\gamma,\ \lambda_n = n^{-\beta},\ n \in \mathbb{N},\ \gamma > 0,\ \beta \geq 0$. Then (9.96) holds if and only if $\gamma + \beta > 1$ and (9.76) holds for some $c > 0$ and $\alpha \in (0, 1)$ if and only if $\beta < \gamma$. □

Example 9.48 Assume that $H = L^2(0, \pi)$. Denote by $x = x(\xi),\ \xi \in [0, \pi]$ elements of H and define

$$Ax = \partial_\xi^2 x, \quad \forall\, x \in D(A) = H^2(0, \pi) \cap H_0^1(0, \pi). \tag{9.100}$$

Let $Q = I$, then we have $\gamma = 2$ and $\beta = 0$, so $\gamma + \beta > 1$ and therefore there exists a C^∞ mild solution of the equation

$$\begin{cases} u_t(t, x) = \dfrac{1}{2}\ \mathrm{Tr}\,[u_{xx}(t, x)] + \left\langle \partial_\xi^2 x, u_x(t, x)\right\rangle, \\ u(0, x) = \varphi(x), \quad t \geq 0,\ x \in H^2(0, \pi) \cap H_0^1(0, \pi). \end{cases} \tag{9.101}$$

Moreover (9.96) is satisfied and therefore if $f \in C_b^1(\mathbb{R}^1)$ then there exists a C^2 mild solution of

$$\begin{cases} v_t(t, x) = \dfrac{1}{2}\ \mathrm{Tr}\,[v_{xx}(t, x)] + \left\langle \partial_\xi^2 x + f(x), v_x(t, x)\right\rangle, \\ v(0, x) = \varphi(x), \quad t \geq 0,\ x \in H^2(0, \pi) \cap H_0^1(0, \pi). \end{cases} \tag{9.102}$$

□

Example 9.49 More generally, consider the case when

$$\alpha_{n_1,\ldots,n_N} = (n_1^2 + \cdots + n_N^2)^{\gamma/2}, \quad \lambda_{n_1,\ldots,n_N} = (n_1^2 + \cdots + n_N^2)^{-\beta/2}, \tag{9.103}$$

then (9.96) holds if and only if $\gamma + \beta > N$ and (9.76) holds for some $c > 0$ and $\alpha \in (0, 1)$ if and only if $\beta < \gamma$. □

Example 9.50 Let $\mathscr{Q}$ be the unitary cube in $\mathbb{R}^N$ and let $\partial\mathscr{Q}$ be its boundary. Set $H = L^2(\mathscr{Q})$ and denote by $x = x(\xi),\ \xi \in \mathscr{Q}$ elements of H. Choose a positive integer m and set

$$\begin{cases} D(A) = \{x \in H^{2m}(Q);\ x = \frac{\partial x}{\partial \nu} = \cdots = \frac{\partial^{m-1} x}{\partial^{m-1} \nu} = 0 \ \text{ on } \ \mathscr{Q}\}, \\ Au = \Delta_\xi u, \quad \forall\, u \in D(A), \end{cases}$$

where ν is the outward normal derivative to $\partial\mathscr{Q}$ and Δ_ξ is the Laplace operator. Let $Q = I$, then $\gamma = 2m, \beta = 0$ so $\gamma + \beta > N$ means $N < 2m$. If $N < 2m$ then there exists a C^∞ classical solution of the equation

$$\begin{cases} u_t(t, x) = \dfrac{1}{2}\ \mathrm{Tr}\,[u_{xx}(t, x)] + \langle \Delta_\xi x, u_x(t, x)\rangle, \\ u(0, x) = \varphi(x), \quad t > 0,\ x \in D(A). \end{cases}$$

Moreover if $f \in C_b^1(\mathbb{R}^1)$ then there exists a C^2 mild solution of

$$\begin{cases} v_t(t,x) = \dfrac{1}{2}\,\mathrm{Tr}\,[v_{xx}(t,x)] + \langle \Delta_\xi x + f(x), v_x(t,x)\rangle, \\ v(0,x) = \varphi(x), \quad t > 0,\ x \in D(A). \end{cases}$$

□

Remark 9.51 Examples involving nonself-adjoint operators A can be discussed by using perturbation results of Section 5.6.3 and Corollary 9.25. □

10

Absolute continuity and the Girsanov theorem

We start from necessary and sufficient conditions for absolute continuity of laws of solutions of two linear differential stochastic equations with additive noise. The main tool for our investigation is the Feldman–Hajek theorem. Then we prove an infinite dimensional version of Girsanov's theorem and derive from it sufficient conditions for absolute continuity of solutions of nonlinear equations. We also apply Girsanov's theorem to prove the existence of martingale solutions for a class of equations with measurable coefficients.

10.1 Absolute continuity for linear systems

Let X and $\widetilde{X}$ be mild solutions of two different equations:

$$dX = (AX + F(X))dt + B(X)dW(t), \quad X(0) = x \tag{10.1}$$

$$d\widetilde{X} = (\widetilde{A}\widetilde{X} + \widetilde{F}(\widetilde{X}))dt + \widetilde{B}(\widetilde{X})dW(t), \quad \widetilde{X}(0) = x \tag{10.2}$$

with the same initial condition $x \in H$, and assume that both equations satisfy the hypotheses of either Theorem 7.2 or Theorem 7.5. Then the solutions have continuous modifications and therefore define two probability laws on $C([0, T]; H)$. One of our prime objectives in this chapter is to characterize equations (10.1)–(10.2), which have absolutely continuous laws. If the laws are absolutely continuous then each property which holds almost surely for the process X must also hold for the process $\widetilde{X}$ and vice versa. We examine first the linear case and follow basically [733] and [734]; Theorem 10.11 is taken from [592].

Assume that X and $\widetilde{X}$ are weak solutions of respectively

$$dX = AXdt + BdW(t), \quad X(0) = x \tag{10.3}$$

$$d\widetilde{X} = \widetilde{A}\widetilde{X}dt + \widetilde{B}dW(t), \quad \widetilde{X}(0) = x \tag{10.4}$$

where W is a cylindrical Wiener process (on a Hilbert space U), with the incremental covariance I. We denote by $\{e_k\}$ a complete orthonormal system in U and by $\{\beta_k\}$ a sequence of independent real Wiener processes such that

$$\langle W(t), x\rangle = \sum_{k=1}^{\infty} \beta_k(t)\langle x, e_k\rangle, \quad x \in H.$$

As we know, weak and mild solutions $X, \widetilde{X}$ do exist in H if and only if

$$\int_0^t \|S(r)B\|^2_{L_2^0}\, dr < +\infty, \quad \int_0^t \|\widetilde{S}(r)\widetilde{B}\|^2_{L_2^0}\, dr < +\infty, \quad t \geq 0. \tag{10.5}$$

The laws of X and $\widetilde{X}$ are Gaussian measures $\mu, \widetilde{\mu}$ on the Hilbert space $\mathscr{H} = L^2(0, T; H)$, see Theorem 5.2, and their covariance operators $\mathscr{Q}$ and $\widetilde{\mathscr{Q}}$ are given by (5.5) and (5.6). General necessary and sufficient conditions for the absolute continuity of μ and $\widetilde{\mu}$ follow from the Feldman–Hajek theorem, Theorem 2.23.

Proposition 10.1 *Assume* (10.5)*, and let X and $\widetilde{X}$ be the solutions of* (10.4) *with $x = 0$ and μ and $\widetilde{\mu}$ the corresponding laws. Then μ and $\widetilde{\mu}$ are equivalent if and only if*

$$\mathscr{Q}^{1/2}(\mathscr{H}) = \widetilde{\mathscr{Q}}^{1/2}(\mathscr{H}) = \mathscr{H}_0 \tag{10.6}$$

and the operator

$$(\widetilde{\mathscr{Q}}^{-1/2}\mathscr{Q}^{1/2})(\widetilde{\mathscr{Q}}^{-1/2}\mathscr{Q}^{1/2})^* - \mathscr{I} \tag{10.7}$$

is Hilbert–Schmidt on $\overline{\mathscr{H}_0}$. The identity on $\mathscr{H}$ is $\mathscr{I}$.

Unfortunately only in very special cases are explicit expressions for the operators $\mathscr{Q}^{1/2}$ and $\widetilde{\mathscr{Q}}^{1/2}$ known. Our basic aim is to replace the conditions (10.6) and (10.7) by more explicit ones. For this purpose we introduce the associated control systems

$$y' = Ay + Bu, \quad y(0) = 0 \tag{10.8}$$

$$\widetilde{y}' = \widetilde{A}\widetilde{y} + \widetilde{B}u, \quad \widetilde{y}(0) = 0 \tag{10.9}$$

and denote by L and $\widetilde{L}$ the corresponding input-output transformations acting from $L^2(0, T; U) = \mathscr{U}$ into $L^2(0, T; H) = \mathscr{H}$.

They are given by the formulae (see Appendix B)

$$Lu(t) = \int_0^t S(t-s)Bu(s)ds, u \in \mathscr{U}, t \in [0, T], \tag{10.10}$$

$$\widetilde{L}u(t) = \int_0^t \widetilde{S}(t-s)\widetilde{B}u(s)ds, u \in \mathscr{U}, t \in [0, T]. \tag{10.11}$$

To avoid notational complications we assume also that

$$\text{Ker } B = \text{Ker } \widetilde{B} = \{0\}. \tag{10.12}$$

Theorem 10.2 *Assume that* (10.5) *and* (10.12) *hold. Then the measures* μ *and* $\widetilde{\mu}$ *are equivalent if and only if*

$$L(U) = \widetilde{L}(U) \tag{10.13}$$

and the operator

$$(\widetilde{L}^{-1}L)(\widetilde{L}^{-1}L)^* - \mathscr{I} \tag{10.14}$$

is Hilbert–Schmidt.

Remark 10.3 In (10.14), $\widetilde{L}^{-1}$ stands for the pseudo-inverse of $\widetilde{L}$ (see Appendix B.2), identical by (10.12) with the inverse. By the closed graph theorem, if (10.13) holds, the operator

$$\mathscr{M} = \widetilde{L}^{-1}L$$

is linear and bounded (see Appendix B). □

Proof of Theorem 10.2 We show that (10.11) and (10.12) are equivalent respectively to (10.6) and (10.7). To show the first equivalence it is enough to prove that

$$\mathscr{Q}^{1/2}(\mathscr{H}) = L(U), \quad \widetilde{\mathscr{Q}}^{1/2}(\mathscr{H}) = \widetilde{L}(U).$$

By direct computation

$$\mathscr{Q} = LL^*. \tag{10.15}$$

Therefore

$$|\mathscr{Q}^{1/2}\varphi|^2_{\mathscr{H}} = \langle \mathscr{Q}\varphi, \varphi\rangle_{\mathscr{H}} = \langle LL^*\varphi, \varphi\rangle_{\mathscr{H}} = |L^*\varphi|^2_{\mathscr{U}}.$$

Consequently by Proposition B.1 we have $\mathscr{Q}^{1/2}(\mathscr{H}) = L(U)$.

Also from the equalities $LL^* = \mathscr{Q}$, $\widetilde{L}\widetilde{L}^* = \widetilde{\mathscr{Q}}$ it follows that

$$|\mathscr{Q}^{-1/2}h| = |L^{-1}h|, \quad |\widetilde{\mathscr{Q}}^{-1/2}h| = |\widetilde{L}^{-1}h|, \quad h \in L(U).$$

So the transformations $P_1 = \mathscr{Q}^{-1/2}L$, and $P_2 = \widetilde{\mathscr{Q}}^{-1/2}\widetilde{L}$ are isometries of $\mathscr{U}$ onto $\mathscr{H}$, compare Proposition B.1(ii). Moreover

$$\widetilde{\mathscr{Q}}^{-1/2}\mathscr{Q}^{1/2} = P_2\widetilde{L}LP_1^{-1}$$

and

$$\begin{aligned}(\widetilde{\mathscr{Q}}^{-1/2}\mathscr{Q}^{1/2})(\widetilde{\mathscr{Q}}^{-1/2}\mathscr{Q}^{1/2})^* &= P_2\widetilde{L}^{-1}LP_1^{-1}(P_2\widetilde{L}^{-1}LP_1^{-1})^* \\ &= P_2\mathscr{M}P_1^{-1}(P_1^{-1})^*\mathscr{M}^*P_2^* = P_2\mathscr{M}\mathscr{M}^*P_2^{-1}\end{aligned}$$

since $P_1^{-1}(P_1^{-1})^* = I$. Thus we have proved that the operators

$$(\widetilde{\mathscr{Q}}^{-1/2}\mathscr{Q}^{1/2})(\widetilde{\mathscr{Q}}^{-1/2}\mathscr{Q}^{1/2})^* - \mathscr{I}, \quad \text{and } \mathscr{M}\mathscr{M}^* - \mathscr{I}$$

are isometrically equivalent. So one of them is Hilbert–Schmidt if and only if the other one is. In addition $|\mathscr{M}^* u| \neq 0$ for $u \neq 0$. The proof of the theorem is therefore complete. □

Remark 10.4 The condition (10.13) and the operator $\mathscr{M}$ have clear control theoretic interpretations (see Appendix B). The condition (10.13) holds if and only if the control systems (10.8) and (10.9) produce the same set of trajectories. Moreover $\mathscr{M} u = \widetilde{u}$ if and only if $Lu = \widetilde{L}\widetilde{u}$. □

We intend to give an even more explicit version of condition (10.14) by calculating $\mathscr{M}$. We start from an auxiliary result of independent interest.

Proposition 10.5

(i) If (10.13) *holds then*

$$B(U) = \widetilde{B}(U). \tag{10.16}$$

(ii) If the laws μ *and* $\widetilde{\mu}$ *are absolutely continuous and the set* $D((A^*)^2) \cap D((\widetilde{A}^*)^2)$ *is dense in* H *then*

$$BB^* = \widetilde{B}\widetilde{B}^*. \tag{10.17}$$

Remark 10.6 Condition (10.16) is equivalent to the existence of constants $k_1 > 0$, $k_2 > 0$ such that

$$k_1 \widetilde{B}\widetilde{B}^* \leq BB^* \leq k_2 \widetilde{B}\widetilde{B}^*.$$

So (10.17) is almost the same as (10.16). □

Proof of Proposition 10.5 (i) If (10.16) holds then for some $k > 0$ and all $\varphi \in \mathscr{H}$: $|L^*\varphi|^2 \leq k|\widetilde{L}^*\varphi|^2$. Equivalently

$$\int_0^T \left| \int_t^T B^* S^*(s-t)\varphi(s)ds \right|^2 dt \leq k \int_0^T \left| \int_t^T \widetilde{B}^* \widetilde{S}^*(s-t)\varphi(s)ds \right|^2 dt. \tag{10.18}$$

For arbitrary $\delta \in (0, T)$ and $a \in H$ define

$$\varphi_\delta(s) = \sqrt{3}\delta^{-3/2} I_{[0,\delta]}(s)a, \quad s \in [0, T].$$

Inserting φ_δ into (10.18) and taking the limit passing as $\delta \downarrow 0$ one obtains that

$$|B^* a|^2 \leq k|\widetilde{B}^* a|^2, \quad a \in H.$$

Therefore $B(U) \subset \widetilde{B}(U)$. In the same way one shows that $B(U) \supset \widetilde{B}(U)$.

(ii) Let $v \in D((A^*)^2)$. By Theorem 5.4 we have $\mathbb{P}$-a.s.

$$\langle X(t), v\rangle = \int_0^t \langle X(s), A^* v\rangle ds + \langle B^* v, W(t)\rangle, \quad \forall\, t \in [0, T]. \tag{10.19}$$

Since $v \in D((A^*)^2)$, it follows that the process $\langle X(\cdot), A^*v\rangle$ is continuous. By the law of the iterated logarithm

$$\mathbb{P}\left(\limsup_{t\downarrow 0}\left(2t\log\log\frac{1}{t}\right)^{-1/2}\langle W(t), B^*v\rangle = |B^*v|\right) = 1$$

and by (10.19)

$$\mathbb{P}\left(\limsup_{t\downarrow 0}\left(2t\log\log\frac{1}{t}\right)^{-1/2}\langle W(t), B^*v\rangle = |B^*v|\right) = 1.$$

Since the laws $\mathscr{L}(X(\cdot))$ and $\widetilde{\mathscr{L}}(X(\cdot))$ are equivalent, therefore

$$\mathbb{P}\left(\limsup_{t\downarrow 0}\left(2t\log\log\frac{1}{t}\right)^{-1/2}\langle W(t), B^*v\rangle = |B^*v|\right) = 1$$

for all $v \in D((A^*)^2)$. If in addition $v \in D((A^*)^2) \cap D((\widetilde{A}^*)^2)$,

$$\mathbb{P}\left(\limsup_{t\downarrow 0}\left(2t\log\log\frac{1}{t}\right)^{-1/2}\langle W(t), B^*v\rangle = |\widetilde{B}^*v|\right) = 1.$$

Consequently for v from a dense subset of H

$$\langle BB^*v, v\rangle = |B^*v|^2 = |\widetilde{B}^*v|^2 = \langle \widetilde{B}\widetilde{B}^*v, v\rangle.$$

So $BB^* = \widetilde{B}\widetilde{B}^*$. □

Conjecture If laws $\mathscr{L}(X(\cdot))$ and $\mathscr{L}(\widetilde{X}(\cdot))$ are equivalent then (10.17) holds. □

Theorem 10.7 *Assume that* (10.12), (10.13) *and* (10.16) *hold; then for all* $u \in W^{1,2}(0,T;U)$ *and almost all* $t \in [0,T]$

$$\mathscr{M}u(t) = \widetilde{B}^{-1}(A-\widetilde{A})\int_0^t S(t-s)Bu(s)ds + \widetilde{B}^{-1}Bu(t). \tag{10.20}$$

Proof Assume that $u \in W^{1,2}(0,T;U)$, then by Proposition A.7(i), $y \in C^1([0,T:H) \cap C([0,T];D(A))$. If (10.14) holds then there exists $\widetilde{u} \in \mathscr{U}$ such that $y = \widetilde{L}\widetilde{u}$. Consequently, for almost all $t \in [0,T]$

$$\widetilde{B}\widetilde{u}(t) = (A-\widetilde{A})y(t) + Bu(t)$$

and by (10.12) and (10.16)

$$\widetilde{u}(t) = \widetilde{B}(A-\widetilde{A})y(t) + \widetilde{B}^{-1}Bu(t),$$

the required result. □

10.1.1 The case $B = \widetilde{B} = I$

In this paragraph we assume (10.5). In our present situation the operator $\mathscr{M}$ is of the form:

$$\mathscr{M}u(t) = (A - \tilde{A}) \int_0^t S(t-s)u(s)ds + u(t) = \mathscr{K}u(t) + u(t), \quad t \in [0, T].$$

If in addition (10.13) holds then $\mathscr{K}$ has an extension from $W^{1,2}(0, T; H)$ in the whole $\mathscr{H}$, which we denote also by $\mathscr{K}$. As a corollary of previous results we have the following.

Theorem 10.8 *The laws $\mathscr{L}(X)$ and $\mathscr{L}(\tilde{X})$ are equivalent if and only if (10.13) holds and the operator $\mathscr{K} + \mathscr{K}^* + \mathscr{K}\mathscr{K}^*$ is Hilbert–Schmidt.*

Remark 10.9 Using elementary arguments one can prove that (10.13) implies that domains D(A) and $D(\tilde{A})$ are equal, more generally $L(U) \subset \tilde{L}(U)$ implies that $D(A) \subset D(\widetilde{A})$. Indeed, if $x \in D(A)$, $x \neq 0$, then the function $y(t) = tx$, $t \geq 0$, is an absolutely continuous solution of the equation (10.8) where $u(t) = x - tAx$, $t \geq 0$. Consequently $y(\cdot)$ is also a strong solution of the equation (10.9), for some $v \in \mathscr{H}$. In particular for almost all $t \geq 0$, $tx \in D(\tilde{A})$. □

Let us consider the following assumption.

Hypothesis 10.1

(i) $D(A) = D(\widetilde{A})$.

(ii) For all $r \in [0, T]$, the operator $(A - \tilde{A})S(r) : D(A) \to H$ has an extension to H and

$$\int_0^T \|(A - \widetilde{A})S(t)\|^2_{L_2^0} dt < \infty.$$

Lemma 10.10 *Hypothesis* 10.1 *holds if and only if* (10.13) *is fulfilled and the operator $\mathscr{K}$ is Hilbert–Schmidt.*

Proof Let $\{e_n\}$ be an orthonormal basis in H and $\{f_m\}$ be an orthonormal basis in $L_2(0, T; \mathbb{R}^1)$. Since $D(A) = D(\widetilde{A})$ we may assume that for each $n \in \mathbb{N}$, $e_n \in D(A) \cap D(\tilde{A})$. Moreover, we assume that for each $m \in \mathbb{N}$, f_m is smooth. Functions $\{f_m e_n\}$, $n, m \in \mathbb{N}$ form an orthonormal basis in $\mathscr{H}$ and

$$\begin{aligned}\mathscr{K} f_m e_n(t) &= (A - \widetilde{A}) \int_0^t S(t-s) f_m(s) e_n ds \\ &= \int_0^t (A - \widetilde{A}) S(t-s) f_m(s) e_n ds.\end{aligned}$$

Hence

$$\sum_{n,l=1}^{\infty}\sum_{m,d=1}^{\infty}\langle \mathscr{K} f_m e_n, f_d e_l\rangle_{\mathscr{H}}^2$$
$$= \sum_{n,l=1}^{\infty}\sum_{m,d=1}^{\infty}\left(\int_0^T\int_0^t \langle (A-\tilde{A})S(t-s)e_n, e_l\rangle_H f_m(s) f_d(t) ds\, dt\right)^2$$
$$= \int_0^T\int_0^t \|(A-\tilde{A})S(t-s)\|_{L_2^0}^2 ds\, dt$$
$$\le T\int_0^T \|(A-\widetilde{A})S(t-s)\|_{L_2^0}^2 dt < +\infty.$$

This means that $\mathscr{K}$ has a Hilbert–Schmidt extension if and only if (10.1) holds. Since $\widetilde{L}(\mathscr{K}+\mathscr{I}) = L$, we have $L(U)\subset \widetilde{L}(U)$. Since $D(A) = D(\widetilde{A})$, the inclusion $L(U)\subset \widetilde{L}(U)$ can be proved by replacing A with $\widetilde{A}$. □

In [592] it is shown that for a variety of classes of generators the operator $\mathscr{M}$ is Hilbert–Schmidt if and only if $\mathscr{K}$ is Hilbert–Schmidt. Hence, according to Theorem 10.8 and Lemma 10.10, the laws $\mathscr{L}(X)$ and $\mathscr{L}(\widetilde{X})$ are equivalent if and only if the condition 10.1 holds. For simplicity we restrict our considerations to the case when A and $\widetilde{A}$ are self-adjoint. The following theorem is a special case of a more general result; see [592].

Theorem 10.11 *Assume that* (10.5) *holds and A, $\widetilde{A}$ are self-adjoint. Then the laws $\mathscr{L}(X)$ and $\mathscr{L}(\widetilde{X})$ are equivalent if and only if Hypothesis* 10.1 *is fulfilled.*

Proof According to the previous lemma it suffices to prove only that if the operator $\mathscr{M}$ is Hilbert–Schmidt, then the operator $\mathscr{K}$ is Hilbert–Schmidt.

Note first that if the laws of the processes X and $\widetilde{X}$ corresponding to the generators A and $\widetilde{A}$ are equivalent then, for each $\alpha, \beta\in\mathbb{R}^1$, also the laws of the processes X_1 and $\widetilde{X}_1$ corresponding to the generators $A+\alpha I$ and $\widetilde{A}+\beta I$ are equivalent. Moreover, it is not difficult to prove that the condition in Hypothesis 10.1 still holds if we replace A by $A+\alpha I$ and $\widetilde{A}$ by $\widetilde{A}+\beta I$. Therefore, without any loss of generality, we can assume that A is invertible and $\|I-\widetilde{A}A^{-1}\| < 2$. In this situation

$$K := A-\widetilde{A} = (I-\widetilde{A}A^{-1})A := BA,$$

where $\|B\| < 2$. Since A is self-adjoint, we have

$$Z(s) := \int_0^s S(s-r)(KS(s-r))^* dr$$
$$= \int_0^s S(s-r)(AS(s-r))^* dr\, B^*$$
$$= \frac{1}{2}\,(S(2s)-I)B^*.$$

Therefore, taking if necessary T small enough,

$$\sup\{\|Z(s)\|: \ 0 < s \leq T\} = \frac{1}{2}\, \|B\| < 1. \tag{10.21}$$

As in the proof of Lemma 10.10, let $\{e_n\}$ and $\{f_m\}$ be orthonormal bases in H and $L^2(0, T; \mathbb{R}^1)$, such that $e_n \in D(A) = D(\widetilde{A})$ and f_m is smooth for each $n, m \in \mathbb{N}$. Obviously for all $n \in \mathbb{N}$

$$\int_0^T \left(|KS(t)e_n|_H^2 |(KS(t))^* e_n|_H^2\right) dt < \infty.$$

Let

$$q_{n,l}^1(t,s) = \mathbb{1}_{\{s<t\}} \langle KS(t-s)e_n, e_l\rangle_H$$

$$q_{n,l}^2(t,s) = \mathbb{1}_{\{s>t\}} \langle KS(s-t)^* e_n, e_l\rangle_H$$

$$q_{n,l}^3(t,s) = \int_0^{t\wedge s} \left\langle (KS(s-r))^* e_n, (KS(t-r))^* e_l\right\rangle_H dr.$$

After simple calculations we have

$$\begin{aligned}&\left\langle \mathscr{K} + \mathscr{K}^* + \mathscr{K}\mathscr{K}^* f_m e_n(t), f_d(t) e_l\right\rangle \\ &\quad = \int_0^T \left\{q_{n,l}^1(t,s) + q_{n,l}^2(t,s) + q_{n,l}^3(t,s)\right\} f_m(s) f_d(t) ds.\end{aligned}$$

Using similar arguments as in the proof of Lemma 10.10, we have

$$\begin{aligned}&\sum_{m,d=1}^{\infty} \left\langle \mathscr{K} + \mathscr{K}^* + \mathscr{K}\mathscr{K}^* f_m e_n(t), f_d(t) e_l\right\rangle_{\mathscr{H}_T}^2 \\ &\quad = \int_0^T \int_0^T [q_{n,l}^1(t,s) + q_{n,l}^2(t,s) + q_{n,l}^3(t,s)]^2 ds\, dt.\end{aligned}$$

Since $\mathscr{M} = \mathscr{K} + \mathscr{K}^* + \mathscr{K}\mathscr{K}^*$ is Hilbert–Schmidt, we have

$$\sum_{n,l=1}^{\infty} \int_0^T \int_0^t \left\{q_{n,l}^1(t,s) + q_{n,l}^2(t,s) + q_{n,l}^3(t,s)\right\}^2 ds\, dt \leq \|M\|_2^2 < \infty.$$

This implies that

$$\begin{aligned}&\sum_{n,l=1}^{\infty} \int_0^T \int_0^t \left\langle KS(t-s)\left(e_n + \int_0^s S(s-r)KS(s-r)e_n dr\right), e_l\right\rangle^2 ds\, dt \\ &\quad = \sum_{n,l=1}^{\infty} \int_0^T \int_0^t \langle KS(t-s)(e_n + Z(s)e_n), e_l\rangle^2 ds\, dt \\ &\quad = \int_0^T \int_0^t \|KS(t-s)(I + Z(s))\|_{L_2^0}^2 ds\, dt < +\infty. \end{aligned}\tag{10.22}$$

From (10.21) the operators $I + Z(s)$, $s \in [0, T]$, are invertible and

$$\sup\{(I + Z(s))^{-1} : 0 \le s \le T\} = c_1 < +\infty.$$

Therefore

$$\begin{aligned}
&\int_0^T \int_0^t \|KS(t-s)\|^2_{L_2^0} \, ds \, dt \\
&= c_1^3 \int_0^T \int_0^t \|KS(t-s)(I + Z(s))(I + Z(s))^{-1}\|^2_{L_2^0} \, ds \, dt \\
&\le \int_0^T \int_0^t \|KS(t-s)(I + Z(s))\|^2_{L_2^0} \, ds \, dt < +\infty
\end{aligned}$$

and finally

$$\int_0^T \|KS(t)\|^2_{L_2^0} \, dt < \infty.$$

The proof is complete. □

For a self-adjoint, negative definite generator A and an arbitrary $\widetilde{A}$, Hypothesis 10.1 can be formulated in a more convenient form. Note first that A satisfies (10.5) if and only if A^{-1} is nuclear. Let $\{-\lambda_k, e_k\}$ be the sequence of all eigenvalues and the corresponding, normalized eigenvectors of A. We can assume that $0 < \lambda_1 \le \lambda_2 \ldots$. The semigroup S has the following form

$$S(t)x = \sum_{k=1}^{\infty} \exp(-\lambda_k t)\langle x, e_k\rangle e_k.$$

Lemma 10.12 *The pair of generators A and $\widetilde{A}$ satisfies condition* 10.1 *if and only if $D(A) = D(\widetilde{A})$ and the operator $(A - \widetilde{A})(-A)^{-1/2}$ defined on $D(A)$ has a Hilbert–Schmidt extension to the whole space H.*

Proof Let $F := \widetilde{A} - A$. For $k \in \mathbb{N}$ and $0 \le t$ we have $FS(t)e_k = \exp(-\lambda_k t)Fe_k$. Hence

$$\begin{aligned}
\int_0^T \|FS(t)\|^2_{L_2^0} \, dt &= \sum_{k=1}^{\infty} \int_0^T \exp(-2\lambda_k t) dt |Fe_k|^2 \\
&= \frac{1}{2} \sum_{k=1}^{\infty} (1 - \exp(-2\lambda_k T)) \, \lambda_k^{-1} |Fe_k|^2 \\
&= \frac{1}{2} \sum_{k=1}^{\infty} (1 - \exp(-2\lambda_k T)) \, |F(-A)^{-1/2} e_k|^2 \\
&\le \frac{1}{2} \|F(-A)^{-1/2}\|^2_{L_2^0} \le (1 - \exp(-2\lambda_1))^{-1} \int_0^1 \|FS(t)\|^2_{L_2^0} dt.
\end{aligned}$$

This completes the proof. □

Remark 10.13 Assume that A and $\widetilde{A}$ are the realizations of uniformly elliptic operators of order $2m$ end $2\tilde{m}$ respectively on a bounded subset $\mathscr{O}$ of $\mathbb{R}^d$. In [461] it is proved that the condition 10.1 is fulfilled if and only if

(i) $m = \widetilde{m}$ and the boundary conditions in the definition of A and $\widetilde{A}$ are the same,
(ii) the order of the differential operator $A - \widetilde{A}$ is less than m. □

10.2 Girsanov's theorem and absolute continuity for nonlinear systems

As the Feldman–Hajek theorem was an appropriate tool to treat absolute continuity in the linear case, Girsanov's theorem allows us to obtain sufficient conditions for absolute continuity in the nonlinear case.

10.2.1 Girsanov's theorem

The following result is due to [53] and [460]. Here $U_0 = Q^{1/2}U$, $|\cdot|_0 = |\cdot|_{U_0}$.

Theorem 10.14 *Assume that $\psi(\cdot)$ is a U_0-valued $\mathscr{F}_t$-predictable process such that*

$$\mathbb{E}\left(e^{\int_0^T \langle \psi(s), dW(s)\rangle_0 - \frac{1}{2}\int_0^T |\psi(s)|_0^2 ds}\right) = 1. \tag{10.23}$$

Then the process

$$\widehat{W}(t) = W(t) - \int_0^t \psi(s)ds, \quad t \in [0, T], \tag{10.24}$$

is a Q-Wiener process with respect to $\{\mathscr{F}_t\}_{t\geq 0}$ on the probability space $(\Omega, \mathscr{F}, \widehat{\mathbb{P}})$ where

$$d\widehat{\mathbb{P}}(\omega) = e^{\int_0^T \langle \psi(s), dW(s)\rangle_0 - \frac{1}{2}\int_0^T |\psi(s)|_0^2 ds} d\mathbb{P}(\omega). \tag{10.25}$$

We need the following lemma.

Lemma 10.15 *Assume that $\psi(t)$, $t \in [0, T]$, is a U_0-valued, $\mathscr{F}_t$-predictable process such that*

$$\mathbb{P}\left(\int_0^T |\psi(s)|_0^2 ds < +\infty\right) = 1.$$

Then there exists a real valued Wiener process $\beta(t)$, $t \in [0, T]$, with respect to $\{\mathscr{F}_t\}_{t\geq 0}$, normalized and such that $\mathbb{P}$-a.s.

$$\int_0^t |\psi(s)|_0 d\beta(s) = \int_0^t \langle \psi(s), dW(s)\rangle_0, \quad t \in [0, T].$$

Proof Let us fix $a \in U_0$, $\|a\|_0 = 1$ and define a process $\widetilde{\psi}$

$$\widetilde{\psi}(s) = \begin{cases} \frac{\psi(s)}{|\psi(s)|_0} & \text{if } \psi(s) \neq 0, \\ a & \text{if } \psi(s) = 0. \end{cases}$$

The process $\widetilde{\psi}$ is also predictable and moreover $|\widetilde{\psi}(t)| = 1$ for $t \in [0, T]$. Consequently the process

$$\beta(t) = \int_0^t \langle \widetilde{\psi}(s), dW(s) \rangle_0, \quad t \in [0, T], \tag{10.26}$$

is a square integrable martingale with the quadratic variation process $\langle\langle \beta(t) \rangle\rangle_t = t, t \in [0, T]$ (see Section 3.4). By Lévy's theorem, see Proposition 3.11, $\beta(\cdot)$ is a normalized real valued Wiener process with respect to $\{\mathscr{F}_t\}$. It follows from (10.26) that

$$\begin{aligned} \int_0^t |\widetilde{\psi}|_0 d\beta(s) &= \int_0^t |\widetilde{\psi}|_0 \langle \widetilde{\psi}(s), dW(s) \rangle_0 \\ &= \int_0^t \langle \psi(s), dW(s) \rangle_0, \quad t \in [0, T] \end{aligned}$$

as required. □

Proof of Theorem 10.14 Assume first that there exists $K > 0$ such that $|\psi(t)| \leq K,\ t \geq 0$. Let $g(t),\ t \in [0, T]$, be a Borel measurable, bounded function with values in U_0. Then

$$\int_0^T \langle g(t), d\widehat{W}(t) \rangle_0 = \int_0^T \langle g(t), dW(t) \rangle_0 - \int_0^T \langle g(t), \psi(t) \rangle_0 dt.$$

We now show that

$$\widehat{\mathbb{E}} \left(e^{\int_0^T \langle g(t), d\widehat{W}(t) \rangle_0} \right) = e^{\frac{1}{2} \int_0^T |g(t)|_0^2 dt}. \tag{10.27}$$

By the very definition of the measure $\widehat{\mathbb{P}}$

$$\begin{aligned} \widehat{\mathbb{E}} \left(e^{\int_0^T \langle g(t), d\widehat{W}(t) \rangle_0} \right) &= \mathbb{E}(e^{\int_0^T \langle \psi(t), dW(t) \rangle_0 - \frac{1}{2} \int_0^T |\psi(t)|_0^2 dt + \int_0^T \langle g(t), dW(t) \rangle_0 - \int_0^T \langle g(t), \psi(t) \rangle_0 dt}) \\ &= e^{\frac{1}{2} \int_0^T |g(t)|_0^2 dt} \mathbb{E} \left(e^{\int_0^T \langle g(t) + \psi(t), dW(t) \rangle_0 - \frac{1}{2} \int_0^T |g(t) + \psi(t)|_0^2 dt} \right). \end{aligned}$$

The expectation in the last line is equal to 1. This follows from Lemma 10.15, and the fact that the process $\gamma(t) = |g(t) + \psi(t)|_0$ is bounded by a constant, and from the observation that for a bounded process γ and a real valued Wiener process β,

$$\mathbb{E} \left(e^{\int_0^T \gamma(t) d\beta - \frac{1}{2} \int_0^T \gamma^2(t) dt} \right) = 1.$$

It follows from (10.27) that for real numbers λ

$$\widehat{\mathbb{E}}\left(e^{\lambda\int_0^T\langle g(t),d\widehat{W}(t)\rangle_0}\right)=e^{\frac{\lambda^2}{2}\int_0^T|g(t)|_0^2dt}. \tag{10.28}$$

For an arbitrary complex number z define a function h

$$h(z)=\widehat{\mathbb{E}}\left(e^{z\int_0^T\langle g(t),d\widehat{W}(t)\rangle_0}\right),\quad z\in\mathbb{C}.$$

From the fact that h is finite for all real numbers z it easily follows that h is well defined for all complex numbers and is continuously differentiable with respect to the complex variable z. Therefore h is analytic on $\mathbb{C}$ and

$$h(z)=e^{\frac{z^2}{2}\int_0^T\|g(t)\|_0^2dt},\quad z\in\mathbb{C}.$$

In particular

$$\widehat{\mathbb{E}}\left(e^{i\lambda\int_0^T\langle g(t),d\widehat{W}(t)\rangle_0}\mathbb{1}_\Gamma\right)=e^{\frac{-\lambda^2}{2}\int_0^T|g(t)|_0^2dt}\,\widehat{\mathbb{P}}(\Gamma),\quad \Gamma\in\mathscr{F}_0,\ \lambda\in\mathbb{R}^1.$$

Therefore random variables $\int_0^T\langle g(t),d\widehat{W}(t)\rangle_0$ are Gaussian with covariances $\int_0^T|g(t)|_0^2dt$. In the same way one can show that if $\Gamma\in\mathscr{F}_t$ then

$$\mathbb{E}\left(e^{i\lambda\int_t^T\langle g(s),d\widehat{W}(s)\rangle_0}\mathbb{1}_\Gamma\right)=e^{\frac{-\lambda^2}{2}\int_t^T|g(s)|_0^2ds}\widehat{\mathbb{P}},\quad \lambda\in\mathbb{R}^1. \tag{10.29}$$

Consequently, the random variables $\int_t^T\langle g(s),d\widehat{W}(s)\rangle_0$ are also independent of $\mathscr{F}_t$. This way the proof of the theorem is complete under the condition that ψ is a bounded process. For a general process ψ satisfying (10.23) consider a sequence $\{\psi_N\}$ of bounded processes such that

$$\lim_{N\to\infty}\int_0^T|\psi_t-\psi_N(t)|_0^2dt=0,\quad \mathbb{P}\text{-a.s.}$$

and define processes

$$\widehat{W}_N(t)=W(t)-\int_0^t\psi_N(s)ds,\quad t\in[0,T],\ N\in\mathbb{N}.$$

It follows from (10.23) and (10.29) that for $\Gamma\in\mathscr{F}_t$

$$\begin{aligned}&\mathbb{E}(e^{\int_0^T\langle\psi_N(s),dW(s)\rangle_0-\frac{1}{2}\int_0^T|\psi_N(s)|_0^2ds}e^{i\int_t^T\langle g(s),dW(s)\rangle_0-i\int_t^T\langle g(s),\psi_N(s)\rangle_0ds}\mathbb{1}_\Gamma)\\&=e^{-\frac{1}{2}\int_t^T\|g(s)\|_0^2ds}\mathbb{E}(e^{\int_0^T\langle\psi_N(s),dW(s)\rangle_0-\frac{1}{2}\int_0^T\|\psi_N(s)\|_0^2ds}\mathbb{1}_\Gamma).\end{aligned} \tag{10.30}$$

One can pass however in (10.30) to the limit as $N\to\infty$ to obtain that (10.29) holds in the general case. The passage to the limit is made possible by the following lemma.

Lemma 10.16 *Let ξ, ξ_N, $N \in \mathbb{N}$ be nonnegative random variables such that $\int_\Omega \xi d\mathbb{P} = \int_\Omega \xi_N d\mathbb{P} = 1$ and such that $\{\xi_N\}$ converges to ξ as $N \to \infty$ in probability. Then*

$$\lim_{N\to\infty} \int_\Omega |\xi - \xi_N| d\mathbb{P} = 0.$$

Proof Fix $\varepsilon > 0$ and define

$$A_N = \{\omega \in \Omega : \ |\xi_N(\omega) - \xi(\omega)| < \varepsilon\}.$$

Then $\lim_{N\to\infty} \mathbb{P}(A_N) = 1$ and therefore

$$\lim_{N\to\infty} \int_{A_N} \xi d\mathbb{P} = \int_\Omega \xi d\mathbb{P} = 1.$$

Since

$$\left| \int_{A_N} \xi d\mathbb{P} - \int_{A_N} \xi_N d\mathbb{P} \right| \le \int_{A_N} |\xi - \xi_N| d\mathbb{P} < \varepsilon$$

therefore, for sufficiently large N,

$$\int_{A_N} \xi d\mathbb{P} \ge 1 - \varepsilon, \qquad \int_{A_N} \xi_N d\mathbb{P} \ge 1 - 2\varepsilon.$$

Finally

$$\begin{aligned} \int_{A_N} |\xi - \xi_N| d\mathbb{P} &= \int_{A_N^c} |\xi - \xi_N| d\mathbb{P} + \int_{A_N} |\xi - \xi_N| d\mathbb{P} \\ &\le \int_{A_N^c} \xi d\mathbb{P} + \int_{A_N^c} \xi_N d\mathbb{P} + \int_{A_N} |\xi - \xi_N| d\mathbb{P} \le 4\varepsilon. \end{aligned}$$

□

Proposition 10.17 *One of following conditions is sufficient in order for* (10.23) *to hold:*

(i) $\mathbb{E}\left(e^{\frac{1}{2}\int_0^T |\psi(t)|^2 dt} \right) < +\infty$,

(ii) there exists $\delta > 0$ *such that* $\sup_{t\in[0,T]} \mathbb{E}\left(e^{\delta |\psi(t)|^2} \right) < +\infty$.

Proof The result is true for real valued processes ψ and real valued Wiener processes W. By Lemma 10.15 it is therefore true in general. □

We will apply the Girsanov theorem to a special situation of two equations of the form

$$dX = AXdt + B(X)dW(t), \quad X(0) = x,$$

$$d\widetilde{X} = (A\widetilde{X} + \widetilde{F}(\widetilde{X}))dt + B(\widetilde{X})dW(t), \quad \widetilde{X}(0) = x$$

with the same diffusion coefficient B.

Theorem 10.18 *Assume that coefficients B and $\widetilde{F}$ satisfy the hypotheses of either Theorem 7.2 or Theorem 7.5. If for some $\delta > 0$*

$$\sup_{t\in[0,T]} \mathbb{E}\left(e^{\delta|B^{-1}(X(t))[\widetilde{F}(X(t))]|_0^2}\right) < +\infty,$$

then $\mathscr{L}(\widetilde{X}(\cdot))$ is absolutely continuous with respect to $\mathscr{L}(X(\cdot))$ (on $C([0,T];H)$).

Proof Assume that ψ is a process satisfying the conditions of Theorem 10.14. Then the process

$$\widehat{W}(t) = W(t) - \int_0^t \psi(s)ds$$

is a Q-Wiener process with respect to the measure $\widehat{\mathbb{P}}$ and the filtration $\{\mathscr{F}_t\}$. Note that

$$\begin{aligned} X(t) &= S(t)x + \int_0^t S(t-s)B(X(s))dW(s) \\ &= S(t)x + \int_0^t S(t-s)B(X(s))d\hat{\hat{W}}(s) \\ &\quad + \int_0^t S(t-s)B(X(s))\psi(s)ds. \end{aligned} \tag{10.31}$$

It follows from the assumptions and Proposition 10.17 that ψ given by

$$\psi(t) = B^{-1}(X(t))\widetilde{F}(X(t)), \quad t \in [0,T] \tag{10.32}$$

fulfills the hypothesis of Theorem 10.14. With such a choice of ψ one has that

$$X(t) = S(t)x + \int_0^t S(t-s)\widetilde{F}(X(s))ds + \int_0^t S(t-s)B(X(s))\widehat{W}(s).$$

Consequently the process X satisfies the same equation as $\widetilde{X}$. It follows from the uniqueness of the solutions that the law of X calculated with respect to the measure $\widetilde{\mathbb{P}}$ is the same as the law of $\widetilde{X}$ calculated with respect to $\mathbb{P}$. Therefore for an arbitrary Borel set $\Gamma \subset C([0,T];H)$ one has

$$\begin{aligned} \mathbb{P}(\widetilde{X}(\cdot) \in \Gamma) &= \widetilde{\mathbb{P}}(X(\cdot) \in \Gamma) \\ &= \mathbb{E}\left(e^{\int_0^T \langle \psi(s), dW(s)\rangle_0 - \frac{1}{2}\int_0^T |\psi(s)|_0^2 ds}; X(\cdot) \in \Gamma\right). \end{aligned} \tag{10.33}$$

Therefore if $\mathbb{P}(X)(\cdot) \in \Gamma) = 0$ then $\mathbb{P}(\widetilde{X}(\cdot) \in \Gamma) = 0$. This means that $\mathscr{L}(\widetilde{X}(\cdot))$ is absolutely continuous with respect to $\mathscr{L}(X(\cdot))$. □

Remark 10.19 Let $h : C([0,T];H) \to \mathbb{R}^1_+$ be a Borel function such that

$$\mathbb{E}\left(e^{\int_0^T \langle\psi(s),dW(s)\rangle_0 - \frac{1}{2}\int_0^T |\psi(s)|_0^2 ds} | \sigma(X(\cdot))\right) = h(X(\cdot)), \quad \mathbb{P}\text{-a.s.}, \tag{10.34}$$

then h is a density of $\mathscr{L}(\widetilde{X}(\cdot))$ with respect to $\mathscr{L}(X(\cdot))$. Formally, replacing dW in (10.34) by $B^{-1}(X)[dX - AXdt]$ and using (10.32) one arrives at the following formula for the density

$$h(X) = e^{\int_0^T \langle B^{-1}\widetilde{F}(X), B^{-1}[dX - AXdt]\rangle_0 - \frac{1}{2}\int_0^T |B^{-1}\widetilde{F}(X)|_0^2 dt}.$$

To justify the formula obtained in this way one needs additional conditions such as the existence of strong solutions. □

10.3 Application to weak solutions

We continue our considerations from Chapter 8 and use the second method of constructing martingale solutions of the problem

$$\begin{cases} dX = (AX + F(X))dt + BdW(t) \\ X(0) = x, \end{cases} \tag{10.35}$$

where $W(\cdot)$ is a cylindrical Wiener process on U with the covariance I, B is a linear operator from U into H and F is a Borel transformation from E into H. [(1)]

Assume that process $\widetilde{X}$ is a mild solution of a simpler equation

$$\begin{cases} d\widetilde{X} = A\widetilde{X}dt + BdW(t) \\ \widetilde{X}(0) = x. \end{cases} \tag{10.36}$$

We will show that, under appropriate conditions on F and $W(\cdot)$, the process $\widetilde{X}$ solves equation (10.35) on a probability space $(\Omega, \mathscr{F}, \mathbb{P})$ with a new measure $\widehat{\mathbb{P}}$ and a new Wiener process $\widehat{W}$.

We will assume the following.

Hypothesis 10.2 *There exists a Banach space E, continuously and as a Borel subset embedded into a Hilbert space H, such that the stochastic convolution $W_A(\cdot)$ has paths on $L^2(0,T;E)$, $\mathbb{P}$-a.s.*

The following result is due to [336], see also [461].

(1) The equation (10.35) was investigated in Chapter 5.

Theorem 10.20 *Let F be a Borel mapping from E into H such that for a constant $k > 0$*

$$|B^{-1}F(y)| \leq k(1 + \|y\|), \quad y \in E.^{(2)} \tag{10.37}$$

If $Z = W_A$ satisfies Hypothesis 10.2 *and*

$$\int_0^T \|S(t)x\|^2 dt < +\infty, \quad \forall\, x \in H,$$

then the equation (10.36) *has a martingale solution on the interval* $[0, T]$.

Proof The process

$$\psi(t) = B^{-1}F(Z(t) + S(t)x), \quad t \in [0, T]$$

is well defined and U-valued. We will show that

$$\mathbb{E}\left(e^{\int_0^T \langle\psi(s), dW(s)\rangle_0 - \frac{1}{2}\int_0^T |\psi(s)|^2 ds}\right) = 1. \tag{10.38}$$

By (10.37)

$$|\psi(t)| \leq k(1 + \|W_A(t)\| + \|S(t)x\|), \quad t \in [0, T].$$

Therefore, for arbitrary $0 \leq s \leq t \leq T$

$$\mathbb{E}\left(e^{\frac{1}{2}\int_s^t |\psi(r)|^2 dr}\right) \leq e^{\frac{k^2}{2}\int_s^t (1+\|S(r)x\|^2) dr}\, \mathbb{E}\left(e^{\frac{1}{2}\int_s^t \|W_A(r)\|^2 dr}\right). \tag{10.39}$$

We claim now that there exists a partition $0 = t_0 < t_1 < \cdots < t_m = T$ of $[0, T]$ such that

$$\mathbb{E}\left(e^{\frac{k^2}{2}\int_{t_j}^{t_{j+1}} \|W_A(s)\|^2 ds}\right) < +\infty, \quad j = 1, 2, \ldots, m-1. \tag{10.40}$$

To see this note that, for arbitrary $0 = t_0 < t_1 < \cdots < t_m = T$ and $s > 0$,

$$\sup_{j \leq m-1} \mathbb{P}\left(\int_{t_j}^{t_{j+1}} \|W_A(r)\|^2 dr > s\right) \leq \frac{1}{s} \sup_{j \leq m-1} \mathbb{E}\left(\int_{t_j}^{t_{j+1}} \|Z(r)\|^2 dr\right). \tag{10.41}$$

Since $\mathbb{E}\left(\int_0^T \|W_A(r)\|^2 dr\right) < +\infty$ then, by choosing a sufficiently fine partition of $[0, T]$, we can make the right hand side of (10.41) as small as we wish. Now, taking into account the Fernique theorem, Theorem 2.7, we prove (10.40). The estimates

(2) Implicitly we assume that $F(H) \subset B(H)$. B^{-1} stands for the pseudo-inverse of B, see Appendix B.

(10.39) and (10.40) and Proposition 10.17 imply

$$\mathbb{E}e^{\frac{1}{2}\int_{t_j}^{t_{j+1}}\langle\psi(s),dW(s)\rangle-\frac{1}{2}\int_{t_j}^{t_{j+1}}|\psi(s)|^2ds}=1,\quad j=0,\ldots,m-1.$$

Consequently the processes

$$M_t^j=e^{\int_{t_j}^{t}\langle\psi(s),dW(s)\rangle-\frac{1}{2}\int_{t_j}^{t}\|\psi(s)\|^2ds},\qquad t\in[t_j,t_{j+1}],\ j=0,\ldots,m-1,$$

are $\mathscr{F}_t$-martingales. In particular

$$\mathbb{E}\left(e^{\int_{t_j}^{t_{j+1}}\langle\psi(s),dW(s)\rangle-\frac{1}{2}\int_{t_j}^{t_{j+1}}\|\psi(s)\|^2ds}|\mathscr{F}_{t_j}\right)=1,\quad j=0,\ldots,m-1.$$

Taking into account that

$$e^{\int_0^T\langle\psi(s),dW(s)\rangle-\frac{1}{2}\int_0^T\|\psi(s)\|^2ds}=M_{t_1}^0M_{t_2}^1\cdots M_{t_m}^{m-1}$$

one finds, by an induction argument, that (10.38) holds.

Let $\widehat{\mathbb{P}}$ and $\widehat{W}(\cdot)$ be the probability measure and the Wiener process given by Theorem 10.14. Then, for $t\in[0,T]$,

$$\begin{aligned}W_A(t)+S(t)x&=S(t)x+\int_0^t S(t-s)BdW(s)\\&=S(t)x+\int_0^t S(t-s)BB^{-1}F(W_A(s)+S(s)x)ds\\&\quad+\int_0^t S(t-s)Bd\widehat{W}(s)\\&=S(t)x+\int_0^t S(t-s)F(W_A(s)+S(s)x)ds\\&\quad+\int_0^t S(t-s)Bd\widehat{W}(s),\quad \mathbb{P}\text{-a.s.}\end{aligned}$$

Consequently the process $\widetilde{X}(t)=W_A(t)+S(t)x,\ t\in[0,T]$, is a mild solution of the equation (10.36) on the probability space $(\Omega,\mathscr{F},\widehat{\mathbb{P}})$ and with respect to the Q-Wiener process $\widehat{W}(\cdot)$. Thus the equation (10.36) has a martingale solution. □

Remark 10.21 Let us consider the case $E=H$. As remarked in Section 6.4, even if F is a continuous and bounded mapping then the deterministic equation

$$\dot{X}=F(X),\quad X(0)=x\in H$$

may not have even a local solution. Therefore for existence of weak solutions, either F has to be more regular or a condition of type (10.37) must hold. □

As an immediate corollary of the theorem one obtains the following proposition.

Proposition 10.22 *If $E = H = U$ and F is a Borel mapping with at most linear growth and $\int_0^T \|S(t)\|^2_{L_2^0} dt < +\infty$ then the equation*

$$\begin{cases} dX = (AX + F(X))dt + dW(t) \\ X(0) = x \in H \end{cases} \tag{10.42}$$

has a martingale solution.

11

Large time behavior of solutions

This chapter is devoted to the existence and uniqueness of invariant measures for solutions of stochastic equations as well as to weak convergence of their transition probabilities. We give an almost complete answer to the problem for linear systems with additive and multiplicative noise. Existence and uniqueness of invariant measures for nonlinear equations are studied, under dissipativity or compactness conditions.

11.1 Basic concepts

In this chapter we study asymptotic properties of solutions of stochastic equations as time goes to infinity. To fix ideas, let us consider again equation (7.1) (with coefficients independent of t)

$$\begin{cases} dX = (AX + F(X))dt + B(X)dW(t), \\ X(0) = \xi, \end{cases} \tag{11.1}$$

and assume that the hypotheses of either Theorem 7.2 or Theorem 7.5 are fulfilled, and that coefficients F and B depend only on $x \in H$. If ξ is an H-valued random variable, $\mathscr{F}_0$ measurable, then the equation (11.1) has a unique mild solution $X(t,\xi)$, $t \geq 0$, and our main preoccupation in this chapter will be with the behavior of laws $\mathscr{L}(X(t,\xi))$ as $t \to +\infty$.

Let P_t and $P(t,x,\Gamma)$, $t \geq 0$, $x \in H$, $\Gamma \in \mathscr{B}(H)$, be the corresponding transition semigroup and transition function. Thus

$$P_t\varphi(x) = \mathbb{E}[\varphi(X(t,x))], \quad \varphi \in B_b(H),\ t \geq 0,\ x \in H, \tag{11.2}$$

and

$$P(t,x,\Gamma) = P_t 1\!\!1_\Gamma(x) = \mathscr{L}(X(t,\xi))(\Gamma), \quad t \geq 0,\ x \in H,\ \Gamma \in \mathscr{B}(H). \tag{11.3}$$

In the following we shall also write $P_t(x,\Gamma)$ instead of $P(t,x,\Gamma)$.

Let $\mathscr{M}(H)$ be the space of all bounded measures on $(H, \mathscr{B}(H))$, and $\mathscr{M}_1^+(H)$ be the subset of $\mathscr{M}(H)$ consisting of all probability measures. For any $\varphi \in B_b(H)$ and any $\mu \in \mathscr{M}(H)$, we set

$$\langle \varphi, \mu \rangle = \int_H \varphi(x)\mu(dx).$$

It is convenient to introduce at this point the family of operators P_t^*, $t \geq 0$, acting on $\mathscr{M}(H)$, which plays the role of the adjoint semigroup. Namely we set, for $t \geq 0$, $\mu \in \mathscr{M}(H)$,

$$P_t^*\mu(\Gamma) = \int_H P(t, x, \Gamma)\mu(dx), \quad \Gamma \in \mathscr{B}(H).$$

We can easily check that

$$\langle \varphi, P_t^*\mu \rangle = \langle P_t\varphi, \mu \rangle, \quad \forall\, \varphi \in B_b(H),\ \mu \in \mathscr{M}(H).$$

Proposition 11.1 *If $\mathscr{L}(\xi) = \nu$ then $P_t^*\nu = \mathscr{L}(X(t, \xi))$. In particular $P_t^*\delta_x = \mathscr{L}(X(t, x))$, $t \geq 0$, $x \in H$.*

Proof By the Markov property

$$\begin{aligned}\mathbb{E}[\varphi(X(t, \xi))] &= \mathbb{E}[\mathbb{E}(\varphi(X(t, \xi)) | \mathscr{F}_0))] = \mathbb{E}[P_t\varphi(\xi)]\\ &= \int_H P_t\varphi(x)\nu(dx) = \langle P_t\varphi, \nu \rangle = \langle \varphi, P_t^*\nu \rangle.\end{aligned}$$

□

A measure μ in $\mathscr{M}_1^+(H)$ is said to be an *invariant (stationary) measure* for (11.1) if

$$P_t^*\mu = \mu, \quad \forall\, t > 0.$$

The following conditions are equivalent for the invariance:

$$\int_H P(t, x, \Gamma)\mu(dx) = \mu(\Gamma), \quad \forall\, \Gamma \in \mathscr{B}(H),$$

and

$$\widehat{\mu}(\lambda) = \int_H (P_t e^{i\langle \cdot, \lambda \rangle})(x)\mu(dx), \quad \forall\, \lambda \in H,$$

where $\widehat{\mu}(\cdot)$ is the characteristic functional of μ,

$$\widehat{\mu}(\lambda) = \int_H e^{i\langle x, \lambda \rangle}\mu(dx), \quad \forall\, \lambda \in H.$$

We have the following elementary result.

Proposition 11.2 *If for some initial condition ξ,*

$$\mathscr{L}(X(t, \xi)) \to \widetilde{\mu} \quad \textit{weakly as } t \to +\infty,$$

equivalently if

$$\mathbb{E}[P_t\varphi(\xi)] \to \int_H \varphi(x)\widetilde{\mu}(dx), \quad \forall\, \varphi \in C_b(H) \text{ as } t \to +\infty,$$

then $\widetilde{\mu}$ *is an invariant measure for the equation* (11.1).

Proof Fix $\varphi \in C_b(H)$ and let $\nu = \mathscr{L}(\xi)$. Then $P_t^*\nu = \mathscr{L}(X(t,\xi))$ and for arbitrary $r > 0$

$$\langle \varphi, P_{t+r}^*\nu\rangle = \langle P_r\varphi, P_t^*\nu\rangle. \tag{11.4}$$

Since $P_r\varphi \in C_b(H)$, we can pass to limit for $t \to \infty$ in (11.4) to obtain

$$\langle \varphi, \widetilde{\mu}\rangle = \langle P_r\varphi, \widetilde{\mu}\rangle = \langle \varphi, P_r^*\widetilde{\mu}\rangle.$$

So $\widetilde{\mu} = P_r^*\widetilde{\mu}$ for $r \geq 0$ as required. □

This is why one of our aims will be to establish existence of invariant measures. Here is an important generalization of Proposition 11.2 due to [464].

Proposition 11.3 *If for some initial condition* ξ *and sequence* $t_n \uparrow +\infty$

$$\frac{1}{t_n}\int_0^{t_n} \mathscr{L}(X(s,\xi))ds \to \widetilde{\mu} \quad \text{weakly as } n \to +\infty,$$

then $\widetilde{\mu}$ *is an invariant measure for the equation* (11.1).

Proof Fix $\varphi \in C_b(H)$, denote $\nu = \mathscr{L}(\xi)$ and remark that by (11.3)

$$\int_0^t \mathscr{L}(X(s,\xi))ds = \int_0^t P_s^*\nu ds.$$

Write for $r \geq 0$

$$\begin{aligned}
\langle \varphi, P_r^*\widetilde{\mu}\rangle &= \langle P_r\varphi, \widetilde{\mu}\rangle = \left\langle P_r\varphi, \lim_{n\to\infty}\frac{1}{t_n}\int_0^{t_n} P_s^*\nu ds\right\rangle \\
&= \lim_{n\to\infty}\frac{1}{t_n}\left\langle P_r\varphi, \int_0^{t_n} P_s^*\nu ds\right\rangle = \lim_{n\to\infty}\frac{1}{t_n}\left\langle \varphi, \int_r^{t_n+r} P_s^*\nu ds\right\rangle \\
&= \lim_{n\to\infty}\left[\frac{1}{t_n}\left\langle \varphi, \int_0^{t_n} P_s^*\nu\right\rangle + \frac{1}{t_n}\left\langle \varphi, \int_{t_n}^{t_n+r} P_s^*\nu ds\right\rangle\right. \\
&\quad \left. - \frac{1}{t_n}\left\langle \varphi, \int_0^r P_s^*\nu ds\right\rangle\right] = \langle \varphi, \widetilde{\mu}\rangle.
\end{aligned}$$

□

It is also important to establish the uniqueness of the invariant measure. Only if μ is the unique invariant measure of $P_t,\ t \geq 0$, there is a chance that

$$\text{weak}\lim_{t\to+\infty} P_t^*\nu = \mu, \quad \forall\, \nu \in \mathscr{M}_1^+(H), \tag{11.5}$$

see also Remark 11.6 below.

Proposition 11.4 *Assume that there exists $\mu \in \mathscr{M}_1^+(H)$ such that*

$$P_t^*\delta_x \to \mu \quad \text{weakly as } t \to \infty\,, \forall\, x \in H. \tag{11.6}$$

Then (11.5) *holds.*

Proof For arbitrary $\varphi \in C_b(H)$ we have

$$\lim_{t\to+\infty} P_t\varphi(x) = \lim_{t\to+\infty} \langle \varphi, P_t^*\delta_x\rangle = \langle \varphi, \mu\rangle.$$

So

$$\lim_{t\to+\infty} \langle \varphi, P_t^*\nu\rangle = \lim_{t\to+\infty} \langle P_t\varphi, \nu\rangle = \langle \varphi, \mu\rangle.$$

□

One of the most important properties of invariant measures is formulated in the following.

Proposition 11.5 *If μ is an invariant measure for* (11.1) *and ξ is an H-valued random variable $\mathscr{F}_0$ measurable such that $\mathscr{L}(\xi) = \mu$, then the process $X(\cdot, \xi)$ is stationary.*

Proof Let $\varphi_1, \ldots, \varphi_n \in B_b(H)$, $0 \le u \le s$ and $0 \le t_1 \le t_2 \le \cdots \le t_n$. We have to show that the expectation

$$I := \mathbb{E}[\varphi_1(X(t_1 + h, \xi)) \cdots \varphi_n(X(t_n + h, \xi))]$$

is independent of $h \ge 0$. This is certainly true if $n = 1$, because

$$\begin{aligned}\mathbb{E}[\varphi_1(X(t_1 + h, \xi))] &= \mathbb{E}[P_{t_1+h}\varphi_1(\xi)] \\ &= \int_H P_{t_1+h}\varphi_1(x)\mu(dx) = \int_H \varphi_1(x)\mu(dx).\end{aligned}$$

Note that

$$\begin{aligned}&\mathbb{E}[\varphi_1(X(t_1 + h, \xi)) \cdots \varphi_n(X(t_n + h, \xi))] \\ &= \mathbb{E}\Big[\varphi_1(X(t_1 + h, \xi)) \cdots \varphi_{n-1}(X(t_{n-1} + h, \xi))\mathbb{E}[\varphi_n(X(t_n + h, \xi)|\mathscr{F}_{t_{n-1}+h}))\Big].\end{aligned}$$

By the Markov property

$$I = \mathbb{E}[\varphi_1(X(t_1 + h)) \cdots \varphi_n(X(t_{n-1} + h))P_{t_n - t_{n-1}}\varphi_n X(t_{n-1} + h)].$$

So if the result is true for $n - 1$ it is true for n, and consequently it holds by induction. □

Remark 11.6 It follows from Proposition 11.5 and the individual ergodic theorem of Birkhoff, see [272], that if μ is the unique invariant measure for (11.1) and $\mathscr{L}(\xi) = \mu$, or $\mathscr{L}(\xi)$ is absolutely continuous with respect to μ then, for arbitrary $\varphi \in B_b(H)$,

$$\lim_{T\to+\infty} \frac{1}{T}\int_0^T \varphi(X(t,\xi))dt = \int_H \varphi(x)\mu(dx), \quad \mathbb{P}\text{-a.e.},$$

see [630] for additional information. □

11.2 The Krylov–Bogoliubov existence theorem

Before we investigate various properties of invariant measures in the following sections, we prove here a general result on existence of invariant measures which will often be used.

Let $P_t,\ t \geq 0$, be the transition semigroup defined by (11.2). As noticed in Chapter 9, from Theorem 9.1 it follows that for arbitrary $\varphi \in C_b(H)$ the function

$$[0, +\infty) \times H, \quad (t, x) \mapsto P_t\varphi(x)$$

is continuous. This property is expressed by saying that $P_t,\ t \geq 0$, is a *Feller semigroup.*

For every $x \in H$ and $T > 0$ the formula

$$\frac{1}{T}\int_0^T P_t(x,\Gamma)dt = R_T(x,\Gamma),\ \Gamma \in \mathscr{H},$$

defines a probability measure. For any $\nu \in \mathscr{M}_1(H)$, $R_T^*\nu$ is defined in the obvious way:

$$R_T^*\nu(\Gamma) = \int_H R_T(x,\Gamma)\nu(dx), \quad \Gamma \in \mathscr{B}(H).$$

It is clear that for any $\varphi \in B_b(H)$

$$\langle R_T^*\nu, \varphi\rangle = \frac{1}{T}\int_0^T \langle P_t^*\nu, \varphi\rangle\, dt.$$

In this sense we write that

$$R_T^*\nu = \frac{1}{T}\int_0^T P_t^*\nu dt.$$

The method of constructing an invariant measure described in the following theorem is due to Krylov–Bogoliubov [464].

Theorem 11.7 *If for some $\nu \in \mathscr{M}_1(H)$ and some sequence $T_n \uparrow +\infty$, $R_{T_n}^*\nu \to \mu$ weakly as $n \to \infty$, then μ is an invariant measure for $P_t,\ t \geq 0$.*

Proof Fix $r > 0$ and $\varphi \in C_b(H)$. Then $P_r\varphi \in C_b(H)$ and

$$\begin{aligned}\langle \varphi, P_r^*\mu\rangle &= \langle P_r\varphi, \mu\rangle = \langle P_r\varphi, \lim_{n\to\infty} R_{T_n}^*\nu\rangle \\ &= \lim_{n\to\infty} \frac{1}{T_n}\left\langle P_r\varphi, \int_0^{T_n} P_s^*\nu ds\right\rangle \\ &= \lim_{n\to\infty} \frac{1}{T_n}\left\langle \varphi, \int_r^{T_n+r} P_s^*\nu ds\right\rangle \\ &= \lim_{n\to\infty} \left[\frac{1}{T_n}\left\langle \varphi, \int_0^{T_n} P_s^*\nu ds\right\rangle + \frac{1}{T_n}\left\langle \varphi, \int_{T_n}^{T_n+r} P_s^*\nu ds\right\rangle \right. \\ &\quad \left. - \frac{1}{T_n}\left\langle \varphi, \int_0^{r} P_s^*\nu ds\right\rangle\right] = \langle \varphi, \mu\rangle.\end{aligned}$$

Consequently $P_r^*\mu = \mu$ for arbitrary $r > 0$ and the result follows. □

Corollary 11.8 *If for some $\nu \in \mathscr{M}_1(E)$ and some sequence $T_n \uparrow +\infty$ the sequence $\{R_{T_n}^*\nu\}$ is tight, then there exists an invariant measure for $P_t,\ t \geq 0$.*

Proof For the proof it is enough to remark that any tight sequence of measures contains a weakly convergent subsequence and apply Theorem 11.7. □

Let μ be an invariant measure for P_t. We say that μ is *ergodic* if

$$\lim_{T\to\infty} \frac{1}{T}\int_0^T P_t\varphi dt = \overline{\varphi} \quad \text{for all } \varphi \in L^2(H, \mu), \tag{11.7}$$

where

$$\overline{\varphi} = \int_H \varphi(x)\mu(dx).$$

We now state the main result concerning ergodicity of an invariant measure $\mu \in \mathscr{M}_1(E)$, for a proof see for example [220].

Theorem 11.9 *Let μ be an invariant measure with respect to $P_t,\ t \geq 0$. Then the following conditions are equivalent.*

(i) *μ is ergodic.*

(ii) *If $\varphi \in L^2(H, \mu)$ and*

$$P_t\varphi = \varphi, \quad \mu\text{-a.s. for all } t > 0,$$

then φ is constant μ-a.s.

(iii) *If for a set $\Gamma \in \mathscr{B}(H)$ and all $t > 0$*

$$P_t 1\!\!1_\Gamma = 1\!\!1_\Gamma \ \mu\text{-a.s.}$$

then either $\mu(\Gamma) = 0$ or $\mu(\Gamma) = 1$.

(iv) For arbitrary $\varphi \in L^2(E, \mu)$

$$\lim_{T\to+\infty} \frac{1}{T}\int_0^T P_s\varphi ds = \langle \varphi, 1\rangle \ \text{ in } L^2(H,\mu)).$$

By Theorem 11.9 we deduce the following important corollary.

Proposition 11.10 *If μ and ν are ergodic measures with respect to $P_t,\ t \ge 0$, and if $\mu \ne \nu$, then μ and ν are singular.*

Proof Let $\Gamma \in \mathscr{B}(H)$ be such that

$$\nu(\Gamma) \ne \mu(\Gamma).$$

By Theorem 11.9 there exists a sequence $\{T_N \uparrow +\infty\}$ such that

$$\lim_{N\to\infty} \frac{1}{T_N}\int_0^{T_N} P_s 1\!\!1_\Gamma ds = \mu(\Gamma), \qquad \mu\text{-a.s.},$$

$$\lim_{N\to\infty} \frac{1}{T_N}\int_0^{T_N} P_s 1\!\!1_\Gamma ds = \nu(\Gamma), \qquad \nu\text{-a.s.}$$

Define

$$A = \left\{ x \in E : \lim_{N\to\infty} \frac{1}{T_N}\int_0^{T_N} P(s,x,\Gamma)ds = \mu(\Gamma)\right\},$$

$$B = \left\{ x \in E : \lim_{N\to\infty} \frac{1}{T_N}\int_0^{T_N} P(s,x,\Gamma)ds = \nu(\Gamma)\right\}.$$

It is clear that $A \cap B = \varnothing$ and that $\mu(A) = \nu(B) = 1$. Therefore μ and ν are singular. □

Theorem 11.11 *If $\mu \in \mathscr{M}_1(E)$ is the unique invariant measure for the semigroup $P_t,\ t > 0$, then it is ergodic.*

Proof Assume by contradiction that μ is the unique invariant measure for the semigroup $P_t,\ t > 0$, and that there exists a set $\Gamma \in \mathscr{B}(H)$ with $\mu(\Gamma) \in (0, 1)$, such that, for all $t > 0$

$$P_t 1\!\!1_\Gamma = 1\!\!1_\Gamma, \quad \mu\text{-a.s.} \tag{11.8}$$

We will check that the measure $\widetilde{\mu}$

$$\widetilde{\mu}(A) = \frac{1}{\mu(\Gamma)}\mu(A\cap\Gamma), \quad A \in \mathscr{E},$$

is also invariant for the semigroup $P_t,\ t > 0$. Note that for arbitrary $A \in \mathscr{B}(E)$ and $t \geq 0$

$$\begin{aligned} P_t^* \widetilde{\mu}(A) &= \int_H P_t(x, A)\widetilde{\mu}(dx) = \frac{1}{\mu(\Gamma)} \int_\Gamma P_t(x, A)\mu(dx) \\ &= \frac{1}{\mu(\Gamma)} \int_\Gamma P_t(x, A \cap \Gamma)\mu(dx) + \frac{1}{\mu(\Gamma)} \int_\Gamma P_t(x, A \cap \Gamma^c)\mu(dx). \end{aligned}$$

It follows from (11.8) that

$$P_t(x, A \cap \Gamma^c) \leq P_t(x, \Gamma^c) = 0, \quad \mu\text{-a.s. on } \Gamma,$$

and

$$P_t(x, A \cap \Gamma) \leq P_t(x, \Gamma) = 0, \quad \mu\text{-a.s. on } \Gamma^c.$$

Therefore, by the invariance of μ,

$$\begin{aligned} P_t^* \widetilde{\mu}(A) &= \frac{1}{\mu(\Gamma)} \int_E P_t(x, A \cap \Gamma)\mu(dx) \\ &= \frac{1}{\mu(\Gamma)} \mu(A \cap \Gamma) = \widetilde{\mu}(A), \end{aligned}$$

and so $\widetilde{\mu}$ is invariant for $P_t,\ t > 0$. □

Another characterization of ergodic measures is provided by the following result, see for example [220].

Proposition 11.12 *An invariant probability measure for the semigroup P_t, $t \geq 0$, is ergodic if and only if it is an extremal point of the set of all the invariant probability measures for the semigroup.*

11.2.1 Mixing and recurrence

Let μ be an invariant probability measure for $P_t,\ t \geq 0$. The measure μ is said to be *weakly mixing* (respectively *strongly mixing*) if for all $\varphi \in C_b(H)$ we have

$$\lim_{t \to +\infty} P_t \varphi = \int_H \varphi d\mu, \quad \text{weakly (respectively strongly) in } L^2(H, \mu).$$

11.2.2 Regular, strong Feller and irreducible semigroups

The transition semigroup $P_t,\ t \geq 0$, is said to be *regular* if all transition probabilities $P_t(x, \cdot),\ t > 0,\ x \in H$, are mutually equivalent.

If μ is an invariant measure for a regular semigroup $P_t,\ t \geq 0$, then all transition probability measures $P_t(x, \cdot)$, $t > 0,\ x \in H$, are equivalent to μ.

A sufficient condition for regularity can be phrased in terms of the so called strong Feller and irreducibility properties.

P_t, $t \geq 0$, is said to be *strong Feller* if for arbitrary $t > 0$ and $\varphi \in B_b(E)$ we have $P_{t_0}\varphi \in C_b(E)$.

P_t, $t \geq 0$, is said to be *irreducible* if for arbitrary $t > 0, x \in \Gamma$ and nonempty open set Γ

$$P_{t_0}(x,\Gamma) = P_{t_0}\mathbb{1}_\Gamma(x) > 0.$$

The following result is due to R. Z. Khas'minskii, for a proof see for example [220].

Proposition 11.13 *If the semigroup P_t, $t \geq 0$, is strongly Feller and irreducible then it is regular.*

The main result concerned with regular semigroups is due to Doob [264], for a proof see also [220].

Theorem 11.14 *Let μ be an invariant measure with respect to P_t, $t \geq 0$. If P_t, $t \geq 0$, is regular then*

(i) μ is strongly mixing and for arbitrary $x \in E$ and $\Gamma \in \mathscr{B}(H)$

$$\lim_{t\to+\infty} P_t(x,\Gamma) = \mu(\Gamma),$$

(ii) μ is the unique invariant probability measure for the semigroup P_t, $t \geq 0$,
(iii) μ is equivalent to all measures $P_t(x,\cdot)$, for all $x \in E$ and all $t > t_0$.

11.3 Linear equations with additive noise

We are concerned in this section with the linear equation

$$\begin{cases} dX(t) = AX(t)dt + BdW(t), \\ X(0) = x \in H \end{cases} \tag{11.9}$$

and assume as usual the following.

Hypothesis 11.1

(i) A is the infinitesimal generator of a strongly continuous semigroup $S(\cdot)$ in H.
(ii) We have

$$\mathrm{Tr}\, Q_t = \int_0^t \mathrm{Tr}\,[S(r)QS^*(r)]dr < +\infty$$

for all $t \geq 0$ where $Q = BB^*$ and W is a cylindrical Wiener process.

If Hypothesis 11.1 holds then $S(\cdot) \in L^2(0,T,L_2^0)$ and the equation (11.9) has a unique mild solution given by

$$X(t,x) = S(t)x + W_A(t), \quad t \geq 0,$$

where

$$W_A(t) = \int_0^t S(t-s)dW(s).$$

Therefore $P(t,x,\cdot) = \mathcal{N}(S(t)x, Q_t)$ and the transition semigroup $P_t,\ t \geq 0$, corresponding to $X(t,x)$ is given by the formula

$$P_t\varphi(x) = \mathbb{E}[\varphi(X(t,x))] = \int_H \varphi(y)\mathcal{N}(S(t)x, Q_t)(dy), \quad \forall\, t \geq 0,\ x \in H. \tag{11.10}$$

Proposition 11.15 *A probability measure μ on $(H, \mathscr{B}(H))$ is invariant for the semigroup $P_t,\ t \geq 0$, if and only if*

$$\widehat{\mu}(\lambda) = e^{-\frac{1}{2}\langle Q_t\lambda,\lambda\rangle}\,\widehat{\mu}(S^*(t)\lambda), \quad \forall\, t \geq 0,\ \lambda \in H. \tag{11.11}$$

Proof Let $\lambda \in H$ and $t \geq 0$. Then taking into account the invariance of μ we have

$$\widehat{\mu}(\lambda) = \int_H e^{i\langle\lambda,x\rangle}\mu(dx) = \int_H P_t(e^{i\langle\lambda,\cdot\rangle})(x)\mu(dx). \tag{11.12}$$

On the other hand, by (11.10) we have

$$P_t(e^{i\langle\lambda,\cdot\rangle})(x) = \int_H e^{i\langle\lambda,y\rangle}\mathcal{N}(S(t)x, Q_t)(dy) = e^{i\langle\lambda,S(t)x\rangle}e^{-\frac{1}{2}\langle Q_t\lambda,\lambda\rangle}.$$

Therefore from (11.12) it follows that

$$\widehat{\mu}(\lambda) = e^{-\frac{1}{2}\langle Q_t\lambda,\lambda\rangle}\int_H e^{i\langle S^*(t)\lambda,x\rangle}\mu(dx) = e^{-\frac{1}{2}\langle Q_t\lambda,\lambda\rangle}\,\widehat{\mu}(S^*(t)\lambda), \tag{11.13}$$

as required. □

Remark 11.16 We notice that the concept of invariant measure obviously applies to the deterministic system

$$X'(t) = AX(t), \quad X(0) = x \in H, \tag{11.14}$$

whose corresponding transition semigroup is given by

$$P_t\varphi(x) = \varphi(S(t)x), \quad \varphi \in C_b(H),\ x \in H. \tag{11.15}$$

In this case μ is invariant for $X(t),\ t \geq 0$, (or for $S(t),\ t \geq 0$) if and only if

$$\int_H \varphi(S(t)x)\mu(dx) = \int_H \varphi(x)\mu(dx), \quad \forall\, t \geq 0 \tag{11.16}$$

and if and only if

$$\widehat{\mu}(\lambda) = \widehat{\mu}(S^*(t)\lambda), \quad \forall\, t \geq 0,\ \lambda \in H. \tag{11.17}$$

□

11.3.1 Characterization theorem

We denote by $\Sigma(H)$ the set of all symmetric and by $\Sigma^+(H)$ the set of all symmetric and nonnegative operators from $L(H)$. Invariant measures of system (11.9) are characterized by the following result, see [727].

Theorem 11.17 *Assume Hypothesis* 11.1. *Then the following conditions are equivalent.*

(i) *There exists an invariant measure for problem* (11.9).
(ii) *There exists a trace class operator* $P \in \Sigma^+(H)$ *satisfying the equation*

$$2\langle PA^*x, x\rangle + \langle Qx, x\rangle = 0, \quad \forall\, x \in D(A^*). \tag{11.18}$$

(iii) *We have* $\sup_{t\geq 0}\,[\mathrm{Tr}\; Q_t] < +\infty$.

If any of conditions (i), (ii), (iii) holds, then an invariant measure for (11.9) *is of the form* $\nu * \mathcal{N}(0, \overline{P})$ *where* ν *is an invariant measure for the semigroup* $S(\cdot)$ *(see Remark* 11.16*) and* $\overline{P}$ *is given by*

$$\overline{P}x = \int_0^\infty S(r)QS^*(r)x\, dr, \quad \forall\, x \in H. \tag{11.19}$$

Proof (i) $\Rightarrow$ (iii) Let us assume that μ is an invariant measure for (11.9) and let $\widehat{\mu}$ denote its characteristic functional. Then $\widehat{\mu}$ fulfills (11.11), so that:

$$\langle Q_t\lambda, \lambda\rangle \leq 2\log\left(\frac{1}{|\mathrm{Re}\;\widehat{\mu}(\lambda)|}\right), \quad \forall\, \lambda \in H.$$

Moreover, by the Bochner theorem (Theorem 2.27), there exists a trace class operator $S_0 \in \Sigma^+(H)$ such that

$$\lambda \in H,\ \langle S_0\lambda, \lambda\rangle \leq 1 \Rightarrow \mathrm{Re}\;\widehat{\mu}(\lambda) \geq \frac{1}{2}.$$

Thus the following implication holds:

$$\lambda \in H,\ \langle S_0\lambda, \lambda\rangle \leq 1 \Rightarrow \langle Q_t\lambda, \lambda\rangle \leq 2\log 2,$$

which yields

$$0 \leq Q_t \leq 2\log 2 S_0, \quad \forall\, t > 0.$$

(ii) $\Rightarrow$ (iii) Let $P \in \Sigma^+(H)$ be a solution of (11.18) and let $x \in D(A^*)$. Then we have

$$\frac{d}{dt}\langle PS^*(t)x, S^*(t)x\rangle = -\langle QS^*(t)x, S^*(t)x\rangle.$$

By integrating this identity between 0 and t we get

$$\langle Px, x\rangle = \langle PS^*(s)x, S^*(s)x\rangle ds + \langle Q_tx, x\rangle, \quad \forall\, x \in H, \tag{11.20}$$

which implies $\mathrm{Tr}\; Q_t \leq \mathrm{Tr}\; P$.

(iii) ⇒ (ii) If (iii) holds then there exists $\overline{P} \in \Sigma^+(H)$ of trace class such that

$$\overline{P}x = \int_0^{+\infty} S(s)QS^*(s)x\,ds = \lim_{t\uparrow+\infty} \int_0^t S(s)QS^*(s)x\,ds, \quad \forall\, x \in H.$$

On the other hand, if $x \in D(A^*)$ we have

$$\begin{aligned} 2\langle Q_t A^* x, x\rangle &= \int_0^t \frac{d}{ds}\langle QS^*(s)x, S^*(s)x\rangle ds \\ &= \langle QS^*(t)x, S^*(t)x\rangle - \langle Qx, x\rangle. \end{aligned} \tag{11.21}$$

Since

$$\int_0^\infty \langle QS^*(s)x, S^*(s)x\rangle ds < +\infty,$$

there exists a sequence $t_n \uparrow +\infty$ such that

$$\lim_{n\to\infty} \langle QS^*(t_n)x, S^*(t_n)x\rangle = 0.$$

Thus, setting $t = t_n$ in (11.21), and letting n tend to infinity, we find

$$2\langle \overline{P}A^*x, x\rangle = -\langle Qx, x\rangle,$$

and (ii) is proved.

(iii) ⇒ (i) If (iii) holds then (11.19) defines a trace class operator $\overline{P}$. Let us prove that $\mu = \mathscr{N}(0, \overline{P})$ is an invariant measure. For this it suffices to show that (11.11) holds. We have in fact

$$\widehat{\mu}(\lambda) = e^{-\frac{1}{2}\langle \overline{P}\lambda, \lambda\rangle},$$

which implies

$$\widehat{\mu}(S^*(t)\lambda) = e^{-\frac{1}{2}\,\langle S(t)\overline{P}S^*(t)\lambda, \lambda\rangle} = e^{\frac{1}{2}\,\langle Q_t\lambda, \lambda\rangle} e^{-\frac{1}{2}\langle \overline{P}\lambda, \lambda\rangle},$$

so (i) is proved.

We want to show now the last part of the theorem. Let μ be an invariant measure for (11.9) and let

$$\overline{P}x = \int_0^\infty S(s)QS^*(s)x\,ds, \quad x \in H.$$

Then, letting t tend to $+\infty$ in (11.11) we have

$$\widehat{\mu}(\lambda) = e^{-\frac{1}{2}\langle \overline{P}\lambda, \lambda\rangle}\widehat{\psi}(\lambda), \quad \lambda \in H, \tag{11.22}$$

where

$$\widehat{\psi}(\lambda) = \lim_{t\uparrow+\infty} \widehat{\mu}(S^*(t)\lambda), \quad \lambda \in H.$$

It remains only to prove that $\widehat{\psi}(\lambda)$ is the characteristic function of a probability measure ν, which is invariant for $S(\cdot)$ since (recall (11.17))

$$\widehat{\nu}(S^*(s)\lambda) = \lim_{t\uparrow+\infty} \widehat{\mu}(S^*(t+s)\lambda) = \widehat{\nu}(\lambda), \quad \lambda \in H.$$

In fact by Bochner's theorem (Theorem 2.27), given $\varepsilon > 0$ there exists a trace class positive operator S such that

$$\operatorname{Re} \widehat{\mu}(\lambda)(x) \geq 1 - \varepsilon \quad \text{if } \langle Sx, x\rangle \leq 1.$$

Thus, if $x \in H$ is such that $\langle Sx, x\rangle \leq 1$, it follows that

$$\operatorname{Re} \widehat{\psi}(\lambda)(x) = \operatorname{Re} \widehat{\mu}(\lambda)(x) e^{\frac{1}{2}\langle \overline{P}\lambda, \lambda\rangle} \geq 1 - \varepsilon.$$

So, using once again Bochner's theorem, there exists a probability measure ν in $(H, \mathscr{B}(H))$ such that $\widehat{\psi}(\cdot) = \widehat{\nu}(\cdot)$ as required. In conclusion, by (11.22) we have $\mu = \mathscr{N}(0, \overline{P}) * \nu$. □

Remark 11.18 Assume that there exists an invariant measure for (11.9) and let $\overline{P}$ be given by (11.19). Then $\overline{P}$ is the *minimal solution* of (11.18). In fact let $P \in \Sigma^+(H)$ be a solution of (11.18), then, by (11.20),

$$\langle Px, x\rangle \geq \langle Q_t x, x\rangle, \quad \forall\, t > 0.$$

Now, letting t tend to infinity we get $P \geq \overline{P}$. □

Example 11.19 Let us consider the equation

$$dX = AXdt + bd\beta, \quad X(0) = x, \tag{11.23}$$

where $b \in H$ and β is a one dimensional Wiener process. Then equation (11.23) can be written in the form (11.1) defining the mapping

$$B : \mathbb{R} \to H, \quad \xi \mapsto B\xi = b.$$

In this case we have

$$B^* : H \to \mathbb{R}, \quad x \mapsto \langle x, b\rangle$$

and

$$Q = BB^*x = \langle x, b\rangle b = (b \otimes b)(x), \quad \forall\, x \in H.$$

Therefore

$$Q_t = \int_0^t (S(s)b) \otimes (S(s)b) ds, \quad t \geq 0,$$

so that $\operatorname{Tr} Q_t = \int_0^t |S(s)b|^2 ds$. It follows from Theorem 11.17 that an invariant measure for (11.23) exists if and only if

$$\int_0^\infty |S(t)b|^2 dt < +\infty. \tag{11.24}$$

□

11.3.2 Uniqueness of the invariant measure and asympotic behavior

We study now the uniqueness of the invariant measure and asymptotic behavior of $\mathscr{L}(X(t,x)) = P_t^*\delta_x$ as $t \to +\infty$.

Theorem 11.20 *The following statements hold.*

(i) If $\lim_{t\to+\infty} S(t)x = 0,\ \forall\, x \in H$, *then there exists at most one invariant measure for system* (11.9). *If an invariant measure* μ *exists we have*

$$\lim_{t\to+\infty} P_t^*\delta_x = \mu, \quad \textit{weakly for all } x \in H. \tag{11.25}$$

(ii) If $\lim_{t\to+\infty} \|S(t)\| = 0$, *then there exists a unique invariant measure for system* (11.9) *and* (11.25) *holds.*

Proof (i) Assume that there is an invariant measure for the system (11.9). Then by Theorem 11.17 it is of the form $\nu * \mathscr{N}(0,\overline{P})$ where ν is an invariant measure for the semigroup $S(\cdot)$ and $\overline{P}$ is given by (11.19). We have to prove that $\nu = \delta_x$. We have in fact, by the definition of invariant measure,

$$\int_H \frac{|S(t)x|}{1+|S(t)x|}\,\nu(dx) = \int_H \frac{|x|}{1+|x|}\,\nu(dx)$$

and the assertion follows letting $t \to \infty$.

Let us prove (11.25). We have to show that

$$\lim_{t\to+\infty} \mathscr{N}(S(t)x, Q_t) = \mathscr{N}(0,\overline{P}) \ \text{ weakly.}$$

For this it is enough to prove that

$$\lim_{t\to+\infty} \int_H e^{i\langle y,h\rangle}\mathscr{N}(S(t)x,Q_t)(dy) = \int_H e^{i\langle y,h\rangle}\mathscr{N}(0,\overline{P})(dy), \quad \forall\, h \in H.$$

We have in fact for any $h \in H$ as $t \to \infty$,

$$\int_H e^{i\langle y,h\rangle}\mathscr{N}(S(t)x,Q_t)(dy) = e^{i\langle S(t)x,h\rangle}e^{-\frac{1}{2}\langle Q_t h,h\rangle} \to e^{-\frac{1}{2}\langle \overline{P}h,h\rangle}.$$

Finally, let us prove (ii). It is enough to check condition (iii) of Theorem 11.17. Note that, for any $T > 0$

$$\begin{aligned}
\mathrm{Tr}[Q_{nT}] &= \sum_{k=0}^{n-1} \int_{kT}^{(k+1)T} \mathrm{Tr}[S(r)QS^*(r)]dr \\
&= \sum_{k=0}^{n-1} \mathrm{Tr}\left[S(kT)\int_0^T S(r)QS^*(r)dr\; S^*(kT)\right] \\
&\le \sum_{k=0}^{n-1} \|S(T)\|^{2k}\ \mathrm{Tr}[Q_T].
\end{aligned}$$

For sufficiently large T, $\|S(T)\| < 1$, so the result follows. □

11.3.3 Strong Feller case

Let us recall, see Section 11.2.2, that a transition semigroup $P_t,\ t \geq 0$, is *irreducible* if the supports of measures [(1)] $P(t, x, \cdot),\ t > 0,\ x \in H$, are identical with H. This is equivalent to saying that Ker $Q_t = \{0\},\ \forall\, t > 0$.

Assume that dim $H < +\infty$ and that the transition semigroup corresponding to equation (11.9) is irreducible. Then if there exists an invariant measure μ the operator $\overline{P}$, defined by (11.19), is nonsingular. Thus by (11.18) and Appendix B.1 it follows that $S(\cdot)$ is stable. [(2)] Now by Theorem 11.20(ii), μ is the unique invariant measure.

In the infinite dimensional case the situation is different as the following example shows.

Example 11.21 Take $H = L^2(0, +\infty)$ and define for $\lambda \geq 0$ and $\kappa > 0$ a semigroup $S(\cdot)$ and an element $b \in H$ as follows:

$$\begin{aligned} &(S(t)x)(\theta) = e^{\lambda t} x(\theta + t), \quad x \in H,\ t \geq 0, \\ &b(\theta) = e^{-\kappa\theta^2}, \quad \theta \geq 0. \end{aligned} \tag{11.26}$$

Then we have

$$(S(t)b)(\theta) = e^{\lambda t} e^{-\kappa(t+\theta)^2}, \quad \theta \geq 0.$$

It follows that (11.24) holds true and so there exists an invariant measure.

We now prove that the transition semigroup is irreducible. Assume by contradiction that there exists $x_0 \in L^2(0, \infty)$ not identically equal to 0, such that $\langle Q_t x_0, x_0\rangle = 0$. Since

$$\langle Q_t x, x\rangle = \int_0^t |\langle S(s)b, x\rangle|^2 ds, \quad t \geq 0,\ x \in H,$$

this implies

$$\langle S(t)b, x_0\rangle = \int_0^{+\infty} e^{\lambda t} e^{-\kappa(\theta+t)^2} x_0(\theta) d\theta = 0, \quad t > 0,$$

and so,

$$\int_0^{+\infty} e^{-2\kappa\theta t} [e^{-\kappa\theta^2} x_0(\theta)] d\theta = 0, \quad \forall\, t > 0.$$

This means that the Laplace transform of the function $\theta \to e^{-\kappa\theta^2} x_0(\theta)$ vanishes identically and this implies $x_0 = 0$, a contradiction.

Finally, if $\lambda = 0$ then $S(t)x \to 0$ as $t \to \infty$, whereas if $\lambda > 0$ then, except in the finite dimensional case, there exists more than one invariant measure. Indeed, setting $x_\lambda(\theta) = e^{-\lambda\theta}, \theta \geq 0$, one can easily check that the Dirac measure δ_{x_λ} is invariant for the semigroup $S(\cdot)$. □

(1) The support of a probability measure ν is the smallest closed set F such that $\nu(F) = 1$.
(2) That is $\lim_{t\to+\infty} \|S(t)\| = 0$.

With a stronger concept of nondegeneracy, the finite dimensional result mentioned before Example 11.21 has an exact infinite dimensional generalization.

We say that the transition semigroup $P_t,\ t \geq 0$, is *strongly Feller* at a moment $r > 0$ if, for any bounded Borel function $\varphi : H \to \mathbb{R}^1$, $P_t\varphi$ is a continuous function for any $t \geq r$, compare Section 11.2.2. This happens if and only if

$$Q_t^{1/2}(H) \supset S(t)(H), \quad \forall\, t \geq r, \tag{11.27}$$

see Section 9.4.1.

Moreover condition (11.27) is equivalent to the null controllability in time r of the the controlled system

$$y' = Ay + Q^{1/2}u, \quad y(0) = x, \tag{11.28}$$

see Appendix B. □

We have the following result, compare [727].

Theorem 11.22 *Assume that there exists an invariant measure μ for system* (11.9) *and that the semigroup $P_t,\ t \geq 0$, is strongly Feller at a moment $r > 0$.*

(i) *We have $\lim\limits_{t\to+\infty} \|S(t)\| = 0$, μ is the unique invariant measure for* (11.9) *and*

$$\lim_{t\to+\infty} P_t^*\delta_x = \mu, \quad \textit{weakly for all } x \in H.$$

(ii) *All transition probabilities $P_t^*\delta_x,\ t \geq r,\ x \in H$, are absolutely continuous with respect to μ.*

Proof (i) The fact that $\lim_{t\to+\infty} \|S(t)\| = 0$ follows from Appendix B.1. By Theorem 11.20(ii), part (i) follows.

(ii) We recall that the invariant measure μ is Gaussian, $\mu = \mathcal{N}(0, \overline{P})$, where

$$\langle \overline{P}x, x\rangle = \int_0^{+\infty} \langle S(s)QS^*(s)x, x\rangle dx, \quad x \in H.$$

We now prove that the probability measures $\mathcal{N}(0, \overline{P})$ and $\mathcal{N}(0, Q_t)$ are equivalent for all $t \geq r$. By recalling the Feldman–Hajek theorem, Theorem 2.25, it suffices to prove that

(i) $Q_t^{1/2}(H) \supset \overline{P}^{1/2}(H), \quad \forall\, t \geq r,$
(ii) the linear operator $C_tC_t^* - I$ is Hilbert–Schmidt and nonnegative, where $C_t = Q_t^{-1/2}\overline{P}^{1/2}$ (remark that the linear operator C_t is well defined in virtue of (i)).

Let us introduce linear operators acting from $L^2(0, \infty, U)$ into H:

$$L_tu = \int_0^t S(s)Q^{1/2}u(s)ds, \quad u \in L^2(0, \infty; U),\ t > 0,$$

and

$$L_\infty u = \int_0^\infty S(s)Q^{1/2}u(s)ds, \quad u \in L^2(0, \infty, U).$$

We remark that the definition of L_∞ is meaningful since $S(\cdot)$ is stable. Moreover the adjoint operator of L_∞ in $L^2(0, \infty; U)$ is given by

$$(L_\infty^* x)(t) = Q^{1/2} S^*(t)x, \quad t \geq 0, \ x \in H.$$

It follows that

$$|L_\infty^* x|^2 = \int_0^{+\infty} |Q^{1/2} S^*(t)x|^2 dt = |\overline{P}^{1/2} x|^2, \quad x \in H,$$

which by Proposition B.1 implies

$$\overline{P}^{1/2}(H) = L_\infty(H). \tag{11.29}$$

In a similar way

$$Q_t^{1/2}(H) = L_t(H), \quad t \geq r.$$

On the other hand, for any $t \geq r$ and any $u \in L^2(0, +\infty; U)$, we have

$$\begin{aligned} L_\infty u &= \int_0^t S(r)Q^{1/2}u(r)dr + S(t)\int_0^{+\infty} S(r)Q^{1/2}u(r+t)dr \\ &= L_t u + S(t)L_\infty u(\cdot + t), \end{aligned}$$

which implies, using the strong Feller property, $L_\infty(H) \subset S(t)(H)$; since obviously $L_\infty(H) \supset S(t)(H)$, therefore using the strong Feller property, we conclude that

$$L_\infty(H) = L_t(H) = Q_t^{1/2}(H), \quad \forall\, t \geq r,$$

which, along with (11.28) proves (i). Finally, we prove (ii). We have

$$P - Q_t = \int_t^{+\infty} S(s)QS^*(s)S(t)ds = S(t)PS^*(t), \quad t \geq 0. \tag{11.30}$$

Applying the operator $Q_t^{-1/2}$ to both sides of (11.30) one obtains for all $x \in H$

$$C_1 P^{1/2}x - Q_t^{-1/2}x = (Q_t^{-1/2}S(t))PS^*(t)x,$$

which implies

$$C_1 C_1^* - I = D\overline{P}D^*,$$

where $D = Q_t^{-1/2}S(t)$ is a bounded operator and D^* is its adjoint. Since $\overline{P}$ is trace class therefore the operator $D\overline{P}D^*$ is trace class as well and consequently Hilbert–Schmidt. This finishes the proof of (ii). □

11.4 Linear equations with multiplicative noise

11.4.1 Bounded diffusion operators

We consider here the equation

$$\begin{cases} dX = AXdt + B(X)dW(t), \\ X(0) = x \in H \end{cases} \tag{11.31}$$

and assume the following.

Hypothesis 11.2

(i) A generates a C_0-semigroup $S(\cdot)$.
(ii) $B \in L(H; L_2^0)$.

Then problem (11.31) has a unique mild solution $X(\cdot, x) \in \mathcal{N}_W^2(0, T; H)$, see Theorem 6.7. Note that for system (11.31) there exists always the trivial invariant measure $\mu = \delta_0$. It is therefore more natural in this case to investigate strong rather than weak convergence of $P_t^*\delta_x$ as $t \to \infty$. So we introduce the following concept.

The system (11.31) is said to be *mean square stable* if there exist constants $M > 0$ and $\omega > 0$ such that

$$\mathbb{E}|X(t, x)|^2 \le Me^{-\omega t}|x|^2, \quad \forall\, x \in H,\ t \ge 0. \tag{11.32}$$

In this subsection we characterize completely systems (11.31) which are mean square stable, and we prove a stochastic version of a classical theorem due to [223].

To formulate the main result we have to introduce the following stationary Liapunov type equation:

$$A^*R + RA + \Delta(R) + I = 0, \tag{11.33}$$

where, for any $R \in L(H)$, $\Delta(R)$ represents the linear bounded operator defined by

$$\langle \Delta(R)x, y\rangle = \mathrm{Tr}\,[R(B(x)Q^{1/2})(B(y)Q^{1/2})^*], \quad \forall\, x, y \in H. \tag{11.34}$$

An operator $R \in L(H)$ is said to be a solution of (11.33) if for any $x, y \in D(A)$,

$$\langle Rx, Ay\rangle + \langle RAx, y\rangle + \langle \Delta(R)x, y\rangle + \langle x, y\rangle = 0. \tag{11.35}$$

The following result is due to [398, 413, 727].

Theorem 11.23 *Assume Hypothesis 11.2, then the following statements are equivalent.*

(i) There exist $M > 0, \omega > 0$ such that

$$\mathbb{E}|X(t, x)|^2 \le Me^{-\omega t}|x|^2, \quad \forall\, x \in H,\ t \ge 0. \tag{11.36}$$

(ii) For any $x \in H$ we have

$$\mathbb{E}\int_0^\infty |X(t,x)|^2 dt < +\infty. \tag{11.37}$$

(iii) Equation (11.33) *has a solution $R \in \Sigma^+(H)$.*

Proof For the proof we need several analytical results on the equation (11.33) and on the evolutionary Lyapunov equation

$$\begin{cases} P' = A^* P + PA + \Delta(P) + I, \\ P(0) = P_0 \in L(H). \end{cases} \tag{11.38}$$

We will be looking for a solution P of equation (11.33) in the space $C_s([0,T]; L(H))$ of all strongly continuous mappings $P : [0,T] \to L(H)$ (that is such that $P(\cdot)x$ is continuous in $[0,T]$ for any $x \in H$). The space $C_s([0,T]; L(H))$, endowed with the norm

$$\|P\| = \sup_{t\in[0,T]} \|P(t)\|,$$

is a Banach space (note that the norm is finite in virtue of the uniform boundedness theorem).

We say that a function $P \in C_s([0,T]; L(H))$ is a *mild solution* of (11.38) if it fulfills the following integral equation:

$$P(t)x = S^*(t)P_0S(t)x + \int_0^t S^*(t-s)[I + \Delta(P(s))]S(t-s)x\,ds, \quad \forall\, x \in H. \tag{11.39}$$

Let us consider first the approximating equation

$$\begin{cases} P_n' = A_n^* P_n + P_n A_n + \Delta(P_n) + I, \\ P_n(0) = P_0 \in L(H), \end{cases} \tag{11.40}$$

which clearly has a unique solution $P_n \in C^1([0,T]; L(H))$. The following lemma gives an explicit expression for the solution of (11.40).

Lemma 11.24 *Assume Hypothesis* 11.2. *Then problem* (11.40) *has a unique solution given by:*

$$\langle P_n(t)x, x\rangle = \mathbb{E}\langle P_0 X_n(t,x), X_n(t,x)\rangle + \mathbb{E}\int_0^t |X_n(s,x)|^2 ds, \tag{11.41}$$

where $X_n(\cdot,x)$ is the strong solution to the problem [3]

$$dX_n(t,x) = A_n X_n(t,x)dt + B(X_n(t,x))dW(t), \quad X_n(0,x) = x.$$

[3] Operators A_n are the Yosida approximations of A (see Appendix A).

Proof By Itô's formula we have

$$
\begin{aligned}
&d_s\langle P_n(t-s)X_n(s,x), X_n(s,x)\rangle\\
&\quad = 2\langle A_nP_n(t-s)X_n(s,x), dX_n(s,x)\rangle + \langle \Delta(P_n(t-s))X_n(s,x), X_n(s,x)\rangle\\
&\quad = -|X_n(s,x)|^2ds + 2\langle P_n(t-s)X_n(s,x), B(X_n(s,x))dW(s)\rangle.
\end{aligned}
$$

By integrating this expression between 0 and t and by taking expectation, we find (11.41). □

Proposition 11.25 *Assume Hypothesis* 11.2. *Then for any* $P_0 \in \Sigma(H)$ *there exists a unique mild solution* P *of problem* (11.38) *given by*

$$
\langle P(t)x, x\rangle = \mathbb{E}\langle P_0X(t,x), X(t,x)\rangle + \mathbb{E}\int_0^t |X(s,x)|^2ds, \quad \forall\, x \in H. \tag{11.42}
$$

Proof By using successive approximations one can easily show that there exists a unique solution $P \in C_s([0,T]; L(H))$ of (11.39) and that $P_n(\cdot)x \to P(\cdot)$, $\forall\, x \in H$, where $P_n(\cdot)$ is the solution of (11.40). The conclusion follows now letting n tend to infinity in (11.41). □

Proposition 11.26 *Assume Hypothesis* 11.2, *then the following statements hold.*

(i) *If* $R \in \Sigma^+(H)$ *is a solution of* (11.33), *then we have*

$$
\mathbb{E}\int_0^{+\infty} |X(s,x)|^2ds < +\infty, \quad x \in H, \tag{11.43}
$$

and R *is given by*

$$
\langle Rx, x\rangle = \mathbb{E}\int_0^{+\infty} |X(s,x)|^2ds, \quad x \in H. \tag{11.44}
$$

(ii) *If* (11.43) *holds, then equation* (11.33) *has exactly one solution* $R \in \Sigma^+(H)$, *given by* (11.44). (4)

Proof (i) Let $R \in \Sigma^+(H)$ be a solution of (11.33). Then by Itô's formula we have

$$
\mathbb{E}\langle RX_n(t,x), X_n(t,x)\rangle = \langle Rx, x\rangle + \int_0^t [2\langle RX_n(s,x), A_nX_n(s,x)\rangle]ds,
$$

where $A_n = AJ_n = nA(nI-A)^{-1}$ are the Yosida approximations of A.

On the other hand,

$$
2\langle RX_n(s,x), A_nX_n(s,x)\rangle = -\langle \Delta(R)X_n(s,x), J_nX_n(s,x)\rangle - \langle X_n(s,x), J_nX_n(s,x)\rangle.
$$

(4) For the definition of $\Sigma^+(H)$ see the beginning of Section 11.3.1.

It follows that

$$\mathbb{E}\langle RX_n(t,x), X_n(t,x)\rangle = \langle Rx, x\rangle - \int_0^t \mathbb{E}\langle X_n(s,x), J_nX_n(s,x)\rangle$$
$$+ \int_0^t \mathbb{E}[\langle \Delta(R)X_n(s,x), X_n(s,x) - J_nX_n(s,x)\rangle]ds.$$

As $n \to \infty$ we find

$$\langle Rx, x\rangle = \mathbb{E}\langle RX(t,x), X(t,x)\rangle + \mathbb{E}\int_0^t |X(s,x)|^2 ds. \tag{11.45}$$

This implies (11.43). Consequently there exists a sequence $\{t_n\} \uparrow +\infty$ such that

$$\lim_{n\to\infty} \mathbb{E}\langle RX(t_n,x), X(t_n,x)\rangle = 0,$$

and R is given by (11.42).

(ii) Assume now that (11.43) holds and denote by $\overline{P}(\cdot)$ the mild solution to equation (11.38) when $P_0 = 0$. Clearly $\overline{P}(t)$ is increasing and bounded by a classical Baire category argument, so that, by a well known result, it is strongly convergent to a linear operator $R \in \Sigma^+(H)$, such that (11.43) holds. It remains to show that R is a solution to (11.33). To this end set $Q_n(t) = \overline{P}(t+n)$, $n \in N$. Then Q_n is the mild solution to (11.38) corresponding to $P_0 = \overline{P}(n)$ and so

$$\overline{P}(t+n)x = S^*(t)\overline{P}(n)S(t)x$$
$$+ \int_0^t S^*(t-s)(I + \Delta(\overline{P})(s+n))S(t-s)x\,ds, \quad x \in H.$$

Letting n tend to infinity, we find

$$Rx = S^*(t)RS(t)x + \int_0^t S^*(t-s)(I + \Delta(R)S(t-s)x\,ds, \quad x \in H, \tag{11.46}$$

which implies easily the conclusion. □

Proof of Theorem 11.14 We shall denote by $\overline{P}(\cdot)$ the solution of (11.38) corresponding to $P_0 = 0$. (i) $\Rightarrow$ (ii) and (ii) $\Rightarrow$ (iii) follow from Proposition 11.26. It remains to show that (iii) $\Rightarrow$ (i). Let R be a symmetric nonnegative solution of (11.34). Then

$$2\langle Rx, Ax\rangle = -\langle (I + \Delta(R))x, x\rangle, \quad \forall\, x \in D(A).$$

Since $\Delta(R) \geq 0$ then $(I + \Delta(R))$ is onto and has a bounded inverse. Thus, by Datko's theorem the semigroup $S(\cdot)$ is stable, that is, there exist positive constants, M_2 and ω_1 such that

$$\|S(t)\| \leq M_2 e^{-\omega_1 t}, \quad \forall\, t \geq 0. \tag{11.47}$$

Set now $\psi(t) = \mathbb{E}\langle RX(t,x), X(t,x)\rangle$. By (11.45) it follows that

$$\psi'(t) = -\mathbb{E}|X(t,x)|^2 \le -\frac{1}{\|R\|}\psi(t), \quad \psi(0) = \langle Rx, x\rangle,$$

and so

$$\psi(t) \le \langle Rx, x\rangle e^{-\frac{1}{\|R\|}t}, \quad t \ge 0. \tag{11.48}$$

Since

$$X(t,x) = S(t)x + \int_0^t S(t-s)B(X(s,x))dW(s),$$

by recalling Hypothesis 11.2 and (11.37), it follows that there exist two positive constants a, b such that

$$\mathbb{E}|X(t,x)|^2 \le ae^{-2\omega t}|x|^2 + b\int_0^t e^{-2\omega(t-s)}\mathbb{E}|X(s,x)|^2 ds.$$

Integrating by parts we get

$$\mathbb{E}|X(t,x)|^2 \le e^{-2\omega t}(a|x|^2 + \langle Rx, x\rangle) - b\psi(t) - 2b\omega\int_0^t e^{-2\omega(t-s)}\psi(s)ds.$$

Now, using (11.47) it follows that there exists a constant $c > 0$ such that

$$\mathbb{E}|X(t,x)|^2 \le ce^{-\frac{2}{\|R\|}t}|x|^2, \quad \forall\, x \in H,\ \forall\, t \ge 0.$$

The proof is complete. □

In general it is not easy to check the hypotheses of Theorem 11.23. We now present an example, due to [726], of a system whose stability can be completely characterized.

Example 11.27 Consider the system

$$dX = AXdt + \sum_{j=1}^N b_j\langle c_j, X\rangle d\beta_j, \quad X(0) = x, \tag{11.49}$$

where A is the generator of a stable C_0-semigroup $S(\cdot)$, $b_j, c_j, j = 1, \dots, N$ are nonzero given elements in H and $\beta_j, j = 1, \dots, N$ are independent real valued Wiener processes. We have here $U = \mathbb{R}^N$, $Q = I$, and $B(\cdot)$ is given by

$$B(x) = \sum_{j=1}^N b_j\langle c_j, x\rangle u_j, \quad (u_1, \dots, u_n) \in U. \tag{11.50}$$

Moreover

$$\Delta(R) = \sum_{j=1}^N \langle Rb_j, b_j\rangle c_j \otimes c_j.$$

We remark that, since $S(\cdot)$ is stable, equation (11.33) is equivalent to the following

$$
\begin{aligned}
R &= \int_0^{+\infty} S(t)\Delta(R)S^*(t)dt + \int_0^{+\infty} S(t)S^*(t)dt \\
&= \sum_{j=1}^N \langle Rb_j, b_j\rangle \int_0^{+\infty} (S(t)c_j)\otimes(S(t)c_j) + \int_0^{+\infty} S(t)S^*(t)dt. \quad (11.51)
\end{aligned}
$$

In order to solve equation (11.51) it suffices to find a vector $\xi \in \mathbb{R}^N$, $\xi = \{\langle Rb_j, b_j\rangle : j = 1, \dots, N\}$, with nonnegative components such that

$$\xi = M\xi + \eta, \tag{11.52}$$

where M is the $N \times N$ matrix defined as

$$M_{ij} = \int_0^{\infty} \langle S^*(t)b_i, c_j\rangle^2 dt, \quad i, j = 1, \dots, N,$$

and η is the vector

$$\eta_j = \int_0^{\infty} |S^*(t)b_j|^2 dt, \quad j = 1, \dots, N.$$

Since the entries of matrix M are all nonnegative and the components of η are strictly positive, equation (11.51) has a nonnegative solution, and so system (11.48) is stable if and only if the eigenvalues of M are all of modulus less than 1 (see [726] for details). □

11.4.2 Unbounded diffusion operators

We show here that the results of the previous section allow extension to the case when the operator B is unbounded. It is possible to do this under several hypotheses, as in Section 7.5. For brevity we shall consider the case when A is variational only (see Section A.4.2). Let V be a Hilbert space continuously and densely embedded in H and let a be a continuous coercive, bilinear form on $V \times V$ such that

$$\langle Ax, y\rangle = a(x, y), \quad \forall\, x, y \in D(A).$$

We shall endow V with the norm:

$$\|x\|_V^2 = -a(x, x), \quad \forall\, x \in V.$$

We write problem (11.31) as

$$
\begin{cases}
dX = AXdt + \sum_{j=1}^{\infty} \sqrt{\lambda_j} B_j X d\beta_j, \\
X(0) = x,
\end{cases} \tag{11.53}
$$

under the assumptions listed below.

Hypothesis 11.3

(i) A is variational.
(ii) $B_j \in L(V; H)$, $j \in \mathbb{N}$, and there exists $\eta \in (0,1)$ such that

$$\frac{1}{2}\sum_{j=1}^{\infty} \lambda_j |B_j x|^2 + \eta a(x,x) \leq 0, \quad \forall\, x \in V.$$

(iii) $\{\beta_j\}$ is a sequence of one dimensional standard Wiener processes, mutually independent.

By Theorem 7.20 we know that problem (11.53) has a unique solution $X \in \mathcal{N}_W^2(0,T;H)$. We have the following counterpart of Theorem 11.23.

Theorem 11.28 *Assume Hypothesis 11.3. Then the following statements are equivalent.*

(i) There exists $M > 0$, $\omega > 0$ such that

$$\mathbb{E}|X(t,x)|^2 \leq M e^{-\omega t}|x|^2, \quad t \geq 0,\ x \in H. \tag{11.54}$$

(ii) For any $x \in H$ we have

$$\mathbb{E}\int_0^{\infty} |X(t,x)|^2 dt < +\infty. \tag{11.55}$$

(iii) The stationary Liapunov equation

$$A^*P + PA + \sum_{k=1}^{\infty} \lambda_k B_k^* P B_k + I = 0 \tag{11.56}$$

has a solution $P \in \Sigma^+(H)$.

Proof We shall not go into all the details of the proof, because some of them are very similar to those of the proof of Theorem 11.23. We shall only discuss existence of solutions to the Liapunov equation, see Proposition 11.29 below, since it requires a new proof. For a complete discussion of the problem see [204].

The evolution Liapunov equations can be written formally as follows:

$$\begin{cases} P' = A^*P + PA + \sum_{k=1}^{\infty} \lambda_k B_k^* P B_k + I, \\ P(0) = P_0 \in L(H). \end{cases} \tag{11.57}$$

To give a precise meaning to (11.57) we introduce its mild form

$$P(t)x = S^*(t)P_0 S(t)x + (\gamma(P))(t)x + \int_0^t S^*(t-s)S(t-s)x\, ds, \tag{11.58}$$

where $x \in H$ and the mapping

$$\gamma : C_s([0,T];\Sigma(H)) \to C_s([0,T];\Sigma(H)),$$

is defined below. For any $t \in [0,T]$ and any $Q \in C_s([0,T];\Sigma(H))$, we define a bilinear symmetric form $\varphi_{t,Q}$ on $D(A) \times D(A)$ by setting

$$\varphi_{t,Q}(x,y) = \sum_{k=1}^{\infty} \lambda_k \int_0^t \langle Q(s)B_k S(t-s)x, B_k S(t-s)y\rangle ds,$$

for all $x, y \in D(A)$. From Proposition A.11, we have:

$$\begin{aligned} 0 \le \varphi_{t,Q}(x,y) &\le \|Q\|^2 \sum_{k=1}^{\infty} \lambda_k \int_0^t |B_k S(s)x|^2 ds \\ &\le -2\eta \|Q\|^2 \int_0^t a(S(s)x, S(s)x) ds \le \eta \|Q\|^2 |x|^2. \end{aligned}$$

Thus $\varphi_{t,Q}$ has a unique extension to $H \times H$, still denoted by $\varphi_{t,Q}$, and there exists $G(t,Q) \in \Sigma(H)$ such that

$$\varphi_{t,Q}(x,y) = \langle G(t,Q)x, y\rangle, \quad \forall\, x, y \in H.$$

Now we set

$$\gamma(Q)(t) = G(t,Q), \quad t \in [0,T]. \tag{11.59}$$

It is easy to check that $\gamma(Q) \in C_s([0,T];\Sigma(H))$ and

$$\|\gamma(Q)\| \le \eta \|Q\|, \quad \forall\, Q \in C_s([0,T];\Sigma(H)). \tag{11.60}$$

We can now prove the result,

Proposition 11.29 *Assume Hypothesis* 11.3, *then if* $P_0 \in \Sigma^+(H)$, *equation* (11.58) *has a unique strongly continuous solution* P *and* $P(t) \in \Sigma^+(H)$ *for all* $t \ge 0$.

Proof The conclusion follows immediately by the contractions principle, since problem (11.57) is equivalent to equation (11.58) and $\|\gamma\| \le \eta < 1$. □

11.5 General linear equations

We consider a more general linear system

$$\begin{cases} dX(t) = AX(t)dt + B(X(t))dW(t) + dW_1(t), \\ X(0) = x \in H, \end{cases} \tag{11.61}$$

and generalize an earlier result from [413]. Our approach, see [200], is different and does not require that equation (11.61) defines a stochastic flow. We assume the following.

Hypothesis 11.4

(i) Operators (A, B) satisfy either Hypothesis 11.2 or Hypothesis 11.3.
(ii) $W(\cdot)$ and $W_1(\cdot)$ are independent Wiener processes with values in U and H respectively.
(iii) $\mathrm{Cov}(W_1(1)) = Q_1$ is of trace class.

Theorem 11.30 *We assume Hypothesis 11.4 and that the system (11.31) is mean square stable. Then there exists a unique invariant measure ν for system (11.61) and for arbitrary $\nu_1 \in \mathscr{M}_1^+(H)$ we have*

$$P_t^* \nu \to \nu \quad \textit{weakly as } t \uparrow +\infty. \tag{11.62}$$

Proof We shall give the proof under the condition that B is bounded; the other case can be treated in a similar way. It will be convenient to consider equation (11.61) on the whole real line. Therefore we define processes $W(t)$ and $W_1(t)$ for $t < 0$ by choosing mutually independent Wiener processes $\widetilde{W}(\cdot)$ and $\widetilde{W}_1(\cdot)$, independent also on W and W_1, with the same laws as $W(\cdot)$ and $W(\cdot)$ respectively and setting

$$W(t) = \widetilde{W}(-t), \quad W_1(t) = \widetilde{W}(-t), \quad t \le 0. \tag{11.63}$$

Now, for any $\lambda > 0$, denote by $X_\lambda(t, x), t \ge -\lambda$, the unique mild solution of the equation

$$\begin{cases} dX = AXdt + B(X)dW + dW_1, \\ X(-\lambda) = x \in H. \end{cases} \tag{11.64}$$

It is easily seen that

$$\mathscr{L}(X_\lambda(0, x)) = \mathscr{L}(X(\lambda, x)), \quad \lambda \ge 0. \tag{11.65}$$

It is therefore enough, to prove the theorem, to show that

$$\lim_{\lambda\to\infty} \mathscr{L}(X_\lambda(0, x)) = \mu \quad \text{weakly for some } \mu \in M_1^+(H) \text{ and all } x \in H,$$

see also Proposition 11.3. In fact, we shall prove even more, that there exists a random variable η such that

$$\lim_{\lambda\to\infty} \mathbb{E}|X_\lambda(0, x) - \eta|^2 = 0, \quad \forall\, x \in H. \tag{11.66}$$

The measure $\mu = \mathscr{L}(\eta)$ is the required stationary distribution. We prove first that (11.66) is true for $x = 0$ and, to simplify the notation, we put $X_\lambda(t) = X_\lambda(t, 0)$. Remark that, because of Proposition 11.26, there exists a unique mild solution $P(\cdot)$ with values in $\Sigma^+(H)$ of the problem

$$P' = A^* P + PA + \Delta(P) + I, \quad P(0) = I. \tag{11.67}$$

By Itô's formula we have

$$d_s\langle P(t-s)X_\lambda(s), X_\lambda(s)\rangle = -|X_\lambda(s)|^2 ds + \text{ Tr }(Q_1 P(t-s))ds$$
$$+ 2\langle P(t-s)X_\lambda(s), B(X_\lambda(s))dW(s) + dW_1(s)\rangle.$$

By integrating from $-\lambda$ to t and taking expectation, we find

$$\mathbb{E}|X_\lambda(t)|^2 = -\int_{-\lambda}^{t} \mathbb{E}|X_\lambda(s)|^2 ds + \int_{-\lambda}^{t} \text{ Tr }[Q_1 P(t-s)]ds.$$

By solving the integral equation above, we have

$$\mathbb{E}|X_\lambda(t)|^2 = \int_{-\lambda}^{t} e^{-(t-s)} \text{ Tr }[Q_1 P(s+\lambda)]ds.$$

Since the system (11.31) is stable and by (11.42) $P(\cdot)$ is bounded, there exists a constant $C > 0$ such that

$$\mathbb{E}|X_\lambda(t)|^2 \le C, \quad \forall\, \lambda > 0,\ \forall\, t \in [-\lambda, +\infty). \tag{11.68}$$

We can now prove (11.66). Let $\mu > \lambda$ and set $Z_{\lambda\mu}(t) = X_\lambda(t) - X_\mu(t), \quad t \ge -\lambda$. Then $Z_{\lambda\mu}$ is a mild solution to the problem

$$\begin{cases} dZ = AZdt + B(Z)dW(t), \\ Z(-\lambda) = -X_\mu(-\lambda). \end{cases}$$

By recalling (11.32) it follows that

$$\mathbb{E}|Z_{\lambda\mu}(t)|^2 \le Me^{-\omega(t+\lambda)}\mathbb{E}|X_\mu(-\lambda)|^2, \quad t \ge -\lambda,$$

and so

$$\mathbb{E}|X_\lambda(0) - X_\mu(0)|^2 \le MCe^{-\omega\lambda}, \quad M \ge \lambda.$$

Thus there exists a random variable η such that $\mathbb{E}|X_\lambda(0) - \eta|^2 \to 0$ as $\lambda \to \infty$. Proceeding similarly we show that

$$\lim_{\lambda\to\infty} \mathbb{E}|X_\lambda(0,x) - X_\lambda(0)|^2 = 0, \quad \forall\, x \in H.$$

This ends the proof. □

11.6 Dissipative systems

The results of this section are based on [200] and [215].

11.6.1 Regular coefficients

We consider now a general semilinear system

$$\begin{cases} dX = (AX + F(X))dt + B(X)dW(t) \\ X(0) = x, \end{cases} \tag{11.69}$$

under Hypothesis 7.1 and the following dissipativity condition.

Hypothesis 11.5 *There exists $\omega > 0$ such that*

$$2\langle A_n(x-y), x-y\rangle + 2\langle F(x) - F(y), x-y\rangle + \|B(x) - B(y)\|_2^2 \le -\omega|x-y|^2,$$
$$\forall\, x, y \in H, n \in \mathbb{N},$$

where $A_n = nA(n-A)^{-1}$ are the Yosida approximations of A.

Theorem 11.31 *Assume that Hypotheses 7.1 and 11.5 hold. Then there exists exactly one invariant measure μ for (11.69) and, for arbitrary $\nu \in \mathscr{M}_1^+(H)$, $P_t^*\nu \to \mu$ weakly as $t \uparrow +\infty$.*

Proof The proof uses similar arguments to those of Theorem 11.30, so we will only sketch it. Let $X_\lambda(t)$ and $X_{\lambda,n}(t)$, $t \ge -\lambda$, be solutions of the following problems

$$\begin{cases} dX = (AX + F(X))dt + B(X)dW, \\ X(-\lambda) = 0, \end{cases} \tag{11.70}$$

and

$$\begin{cases} dX = (A_nX + F(X))dt + B(X)dW, \\ X(-\lambda) = 0, \end{cases} \tag{11.71}$$

respectively. Here $W(\cdot)$ is the Wiener process defined in $\mathbb{R}^1$ as in Section 11.4. Problems (11.70) and (11.71) have unique solutions by Theorem 7.5. Applying Itô's formula to $|X_{\lambda,n}(t)|^2$, $t \ge -\lambda$, and then taking the expectation, we arrive at

$$\mathbb{E}|X_{\lambda,n}(t)|^2 = \mathbb{E}\int_{-\lambda}^t G_n(X_{\lambda,n}(s))ds,$$

where

$$G_n(x) = 2\langle A_nx, x\rangle + 2\langle F(x), x\rangle + \|B(x)\|^2_{L_2^0}, \quad x \in H.$$

Consequently

$$\frac{d}{dt}\,\mathbb{E}|X_{\lambda,n}(t)|^2 = \mathbb{E}(G_n(X_{\lambda,n}(t))), \quad t \ge -\lambda.$$

It follows from Hypothesis 11.5 that for any $\varepsilon > 0$, there exists $C_\varepsilon > 0$ such that

$$G_n(x) \le -(\omega - \varepsilon)|x|^2 + C_\varepsilon, \quad x \in H,\ n \in \mathbb{N}$$

and so

$$\frac{d}{dt}\,\mathbb{E}|X_{\lambda,n}(t)|^2 \le \mathbb{E}|X_{\lambda,n}(t)|^2 + C_\varepsilon, \quad t \ge -\lambda.$$

Choosing $\varepsilon < \omega$ and letting n tend to infinity, we find the estimate

$$\mathbb{E}|X_\lambda(t)|^2 \le \frac{C_\varepsilon}{\omega - \varepsilon}, \quad \forall\,\lambda > 0,\ \forall\,t \ge -\lambda. \tag{11.72}$$

Now let $\mu > \lambda > 0$. In a similar way, but applying Itô's formula to $|X_{\lambda,n}(t) - X_{\mu,n}(t)|^2,\ t \ge -\lambda$, we arrive at the following estimate:

$$\begin{aligned}\mathbb{E}|X_\lambda(t) - X_\mu(t)|^2 &\le \mathbb{E}|X_\mu(-\lambda)|e^{-\omega(t+\lambda)}\\ &\le \frac{C_\varepsilon}{\omega - \varepsilon}e^{-\omega(t+\lambda)}, \quad \forall\,\lambda > 0,\ t \ge -\lambda.\end{aligned} \tag{11.73}$$

By the obtained estimate (11.73) it follows that the sequence $\{X_\lambda(0)\}$ is mean square convergent as $\lambda \to \infty$ to a random variable η. As in the proof of Theorem 11.30, we can show that the law μ of η is the required invariant measure. □

11.6.2 Discontinuous coefficients

We show here applicability of the method from the previous subsection to systems with irregular coefficients. We will limit our considerations to equations with additive noise and dissipative nonlinearities discussed in Sections 7.2.2 and 7.2.3, but it seems possible to obtain results by the same method for nonlinear equations with multiplicative noise. The different case of the so called *gradient* nonlinearities is studied in [315, 531, 732].

Let E be a separable Banach space continuously and as a Borel dense subset embedded into H. We consider the equation

$$\begin{cases} dX = (AX + F(X))dt + dW(t)\\ X(0) = x, \end{cases} \tag{11.74}$$

with initial condition x either in E or in H. Throughout this section we will assume the following.

Hypothesis 11.6

(i) A generates a C_0-semigroup $S(\cdot)$ on H.
(ii) The part A_E of A in E generates either a C_0-semigroup or an analytic semigroup $S_E(\cdot)$ on E.
(iii) $F : E \to E$ is uniformly continuous on bounded sets of E.
(iv) $W_A(\cdot)$ is an E-continuous process.

We will need dissipative and growth conditions of two types.

Hypothesis 11.7

(i) $A_E + \omega I$ *is dissipative in* E *for some* $\omega > 0$.
(ii) F *is dissipative in* E.
(iii) *For some* $c > 0, m > 0$ *we have*

$$\|F(x)\| \le c(1 + \|x\|^m), \quad x \in E.$$

Hypothesis 11.8

(i) $A + \omega I$ *is dissipative in* H *for some* $\omega > 0$.
(ii) F *is dissipative in* E *and in* H.
(iii) *For some* $c > 0, m > 0$ *we have*

$$|F(x)| \le c(1 + |x|^m), \quad x \in E.$$

Remark 11.32 Hypotheses 11.7 and 11.8 are similar, however, in important cases the operator A_E may be only dissipative with $\omega = 0$ but not with $\omega > 0$, whereas, at the same time, A may be dissipative with $\omega < 0$, see Example 11.36. □

Assume that Hypotheses 11.6 and 11.7 are satisfied. Then, by Theorem 7.11, equation (11.74) has a unique solution $X(\cdot, x)$; the associated transition semigroup P_t is Feller in E.

We now prove the following.

Theorem 11.33 *Assume that Hypotheses* 11.6 *and* 11.7 *hold and that* $W_A(\cdot)$ *is bounded on* E *in probability.* [(5)] *Then there exists exactly one invariant measure* $\mu \in \mathscr{M}_1^+(E)$ *for* (11.74), *and for arbitrary probability measure* ν *on* $\mathscr{M}_1^+(E)$, $P_t^*\mu \to \mu$ *weakly* [(6)] *as* $t \uparrow +\infty$.

Proof Proceeding as in the proof of Theorem 11.30, for any $\lambda > 0$, denote by $X_\lambda(\cdot, x)$ the solution to

$$\begin{cases} dX = (AX + F(X))dt + dW(t) \\ X(-\lambda) = x, \end{cases}$$

where $W(\cdot)$ is the Wiener process defined on $\mathbb{R}^1$ by (11.63). Set $X_\lambda(\cdot) = X_\lambda(\cdot, 0)$. Therefore

$$X_\lambda(t) = \int_{-\lambda}^t S_E(t-s)F(X_\lambda(s))ds + W_{A,\lambda}(t),$$

where

$$W_{A,\lambda}(t) = \int_{-\lambda}^t S_E(t-s)dW(s), \quad t \ge -\lambda.$$

(5) For all $\varepsilon > 0$ there exists $r > 0$ such that for all $t \ge 0$ we have $\mathbb{P}(\|W_A\| \ge r) \le \varepsilon$.
(6) In the topology of the Banach space E.

We show now that there exists $c_1 > 0$ such that

$$\mathbb{E}(\|X_\lambda(t)\|) \le c_1, \quad \forall\,\lambda > 0,\ \forall\, t > -\lambda. \tag{11.75}$$

We first remark that $Z_\lambda(t) = X_\lambda(t) - W_{A,\lambda}(t),\ t \ge -\lambda$, is the mild solution of the problem

$$\begin{cases} \dfrac{d}{dt} Z = AZ + F(Z + W_{A,\lambda}) \\ Z(-\lambda) = 0,\ t \ge -\lambda. \end{cases}$$

It follows, by denoting by $x^*_{\lambda,t}$ an element from the subdifferential of $\|Z_\lambda(t)\|$, that

$$\begin{aligned} \frac{d^-}{dt}\|Z_\lambda(t)\| &= \langle A_E Z_\lambda(t) + F(Z_\lambda(t) + W_{A,\lambda}(t)) - F(W_{A,\lambda}(t)), x^*_{\lambda,t}\rangle \\ &\quad + \langle F(W_{A,\lambda}(t)), x^*_{\lambda,t}\rangle \\ &\le -\omega\|Z_\lambda(t)\| + \|F(W_{A,\lambda}(t))\| \\ &\le -\omega\|Z_\lambda(t)\| + c(1 + \|W_{A,\lambda}(t))\|^m). \end{aligned} \tag{11.76}$$

Consequently

$$\|Z_\lambda(t)\| \le c\int_{-\lambda}^t e^{-\omega(t-s)}(1 + \|W_{A,\lambda}(s))\|^m)ds, \quad t > -\lambda. \tag{11.77}$$

Note that

$$\sup_{s\ge-\lambda\ge 0} \mathbb{E}[\|W_{A,\lambda}(s))\|^m] = \sup_{s\ge 0}\mathbb{E}[\|W_A(s))\|^m].$$

Since the process $W_A(\cdot)$ is bounded in probability, one can find $r > 0$ such that

$$\log\left(\frac{\mathbb{P}(\|W_A(s)\| > r}{\mathbb{P}(\|W_A(s)\| \le r}\right) \le -2,$$

so, for sufficiently small $\gamma > 0$

$$\log\left(\frac{\mathbb{P}(\|W_A(s)\| > r}{\mathbb{P}(\|W_A(s)\| \le r}\right) \le -2 + 32\gamma^2 r < -1.$$

Therefore, by Fernique's theorem,

$$\sup_{s\ge 0}\mathbb{E}\left(e^{\gamma}\|W_A(s))\|^2\right) < +\infty.$$

Consequently

$$\sup_{s\ge 0}\mathbb{E}[(\|W_{A,\lambda}(s))\|^m)] = c_1 < +\infty. \tag{11.78}$$

Taking into account (11.77) and (11.78) and

$$\sup_{t\ge 0}\mathbb{E}[\|X_\lambda(t)\|] \le \sup_{t\ge 0}\mathbb{E}[\|W_{A,\lambda}(t)\|] \le \sup_{t\ge 0}\mathbb{E}[\|Z_\lambda(t)\|], \quad t \ge -\lambda,$$

we arrive at (11.75). In a similar way we show that for a constant $c_2 > 0$ and all $\mu > \lambda > 0, t > -\lambda$

$$\mathbb{E}\|X_\lambda(t) - X_\mu(t)\| \le c_2 e^{-\omega(t+\lambda)}.$$

Therefore there exists an E-valued random variable η such that

$$\lim_{\lambda\to+\infty} \mathbb{E}\|X_\lambda(0) - X_\lambda(0,x)\| = 0, \quad \forall\, x \in E.$$

The law $\mathscr{L}(\eta)$ is the required invariant measure. □

Assume now that Hypotheses 11.6 and 11.8 hold. Then by Theorem 7.14 for arbitrary $x \in H$, there exists a generalized solution $X(\cdot, x)$ of (11.74). The corresponding transition semigroup $\widetilde{P}_t$ on $B_b(H)$ is an extension of the semigroup P_t from the previous theorem. It follows from the definition of generalized solution that it is Feller. We have the following analog of Theorem 11.33.

Theorem 11.34 *Assume that Hypotheses* 11.6 *and* 11.8 *hold. Then there exists exactly one invariant measure* $\mu \in \mathscr{M}_1^+(H)$ *for* (11.74)*, and for arbitrary probability measure* $\nu \in \mathscr{M}_1^+(H)$*,* $\widetilde{P}_t^*\nu \to \mu$ *weakly* [(7)] *as* $t \uparrow +\infty$.

Proof The proof is similar to that of the previous theorem. The only difference concerns the proof of boundedness in probability of $W_A(\cdot)$. This is an immediate consequence of the stability of $S(\cdot)$, see Theorem 11.19(ii). □

Remark 11.35 Hypotheses 11.6, 11.7(i)(ii) and 11.8(i)(ii) were discussed in Section 7.2. To have some idea for the remaining conditions of the theorem, take $H = L^2(\mathscr{O})$, $E = C(\overline{\mathscr{O}})$, $\mathscr{O}$ open bounded set of $\mathbb{R}^d$. The bounds on F from Hypotheses 11.7, 11.8 are satisfied for the Nemytskii operator $F(x)(\xi) = \varphi(x(\xi))$, $\xi \in \mathscr{O}$ if function φ is continuous and of polynomial growth.

Let A be a self-adjoint negative operator on $L^2(\mathscr{O})$ described in Example 5.24. Arguing as in the proof of Lemma 5.21 we see that, given $\gamma \in [0, 1)$, there exists $c_3 > 0$ such that

$$\mathbb{E}|W_A(t,\xi_1) - W_A(t,\xi_2)|^2 \le c_3|\xi_1 - \xi_2|^{2\gamma}, \quad \xi_1, \xi_2 \in \mathscr{O}.$$

It follows from the proof of the Kolmogorov test that, for any $\delta > 0$,

$$\sup_{t\ge 0} \mathbb{E}\|W_A(t)\|^\delta < +\infty.$$

This way we obtain boundedness in probability of W_A. □

(7) In the sense of the Hilbert space H.

Example 11.36 Let $\mathscr{O} = (0, \pi)$, $H = L^2([0, \pi])$, and $W(\cdot)$ be the white noise in $L^2([0, \pi])$

$$W(t, \xi) = \sqrt{\frac{2}{\pi}} \sum_{k=1}^{\infty} \sin(k\xi)\beta_k(t), \quad t \geq 0, \quad \xi \in [0, \pi].$$ [8]

Set $F(x) = -x^3$, $x \in E$ and let A be the linear operator [9]

$$\begin{cases} D(A) = \{x \in H^2(0, \pi) : \; x(0) = x(\pi), \; x'(0) = x'(\pi)\} \\ Ax = x''. \end{cases}$$

Then all hypotheses of Theorem 11.33 are fulfilled.

If in the definition of $D(A)$ we replace periodic conditions by Dirichlet boundary conditions, then there are no $\omega > 0$ such that $\|S_E(t)\| \leq e^{-\omega t}$, $t \geq 0$. So Theorem 11.33 is not applicable. It is easy to see, however, that Theorem 11.34 is. □

11.7 The compact case

We will prove now, following [200], existence of the invariant measure for the general system (11.1), starting from Proposition 11.3, more specifically, from its consequence which we formulate as a separate proposition.

Proposition 11.37 *If for some initial condition ξ and some $t > 0$ the family $\mathscr{L}(X(s, \xi))$, $s \geq t$, is tight, then there exists an invariant measure for* (11.1).

Proof It follows immediately from the assumption that the sequence of measures

$$\mu_n = \frac{1}{n} \int_0^n \mathscr{L}(X(s, \xi))ds, \quad n \in \mathbb{N},$$

is tight. By the Prokhorov theorem (Theorem 2.3) there exists $\widetilde{\mu} \in \mathscr{M}_1^+(H)$ and a subsequence $\{\mu_{n_m}\}$ such that

$$\lim_{m \to \infty} \mu_{n_m} = \widetilde{\mu}, \quad \text{weakly.}$$

Now by Proposition 11.3 measure $\widetilde{\mu}$ is invariant for (11.1). □

For proving tightness of the family $\mathscr{L}(X(s, \xi))$, $s \geq t$, we will apply the factorization method in a similar way as in Chapter 8. We will need the following basic assumption.

[8] As usual $\{\beta_k\}$ is a sequence of mutually independent real valued Brownian motions.
[9] The prime denotes derivation with respect to ξ.

Hypothesis 11.9

(i) *$S(t)$ is compact for $t > 0$.*
(ii) *$X(\cdot, \xi)$ is bounded in probability for some $\xi \in \mathscr{F}_0$.*

Note that Hypothesis 11.9 is necessary for the existence of an invariant measure. In fact, if μ is invariant for (11.1) and $\mathscr{L}(\xi) = \mu$ then $\mathscr{L}(X(s, \mu)) = \mathscr{L}(\mu),\ s \geq 0$ and boundedness in probability follows trivially.

We divide the section into two subsections, considering separately the case when $\mathrm{Tr}[\mathrm{Cov}\ W(1)] < +\infty$ or $\mathrm{Cov}\ W(1) = I$.

11.7.1 Finite trace Wiener processes

We prove the following result.

Theorem 11.38 *Assume that assumptions of Theorem 7.2 are satisfied and Hypothesis 11.9 holds. Then there exists an invariant measure for* (11.1).

Proof For the proof we define, as in Section 8.3, operators $G_\alpha : L^p(0, 1; H) \to L^p(0, 1; H)$ by

$$G_\alpha h(t) = \int_0^t (t - s)^{\alpha-1} S(t - s)h(s)ds.$$

It follows from Proposition 8.4 that if Hypothesis 11.9 holds and $\alpha \in [1/p, 1]$, then operators G_α are compact from $L^p(0, 1; H)$ into $C(0, 1; H)$. Consequently for arbitrary $r > 0$ the sets $K(r)$ defined by

$$K(r) := \{x \in H : x = S(1)z + G_1 g(1) + G_\alpha h(1), \quad |z| \leq r, |g|_p \leq r, |h|_p \leq r\}, \tag{11.79}$$

are relatively compact in H. [10] We now prove the following lemma.

Lemma 11.39 *Assume that Hypothesis 7.1 is satisfied. Then, for any $0 < p^{-1} < \alpha < \frac{1}{2}$ there exists a constant $c > 0$ such that for arbitrary $r > 0$ and all $x \in H$ such that $|x| \leq r$,*

$$\mathbb{P}(X(1, x) \in K(r)) \geq 1 - cr^{-p}(1 + |x|^p), \quad r > 0,\ |x| \leq r.$$

Proof Fix numbers α, p such that $0 < p^{-1} < \alpha < \frac{1}{2}$. Let

$$\gamma(x, s) = \mathbb{E}|X(s, x)|^p, \quad s \in [0, 1],\ x \in H,$$

[10] $|\cdot|_p$ stands for the norm in $L^p(0, 1; H)$.

for any $0 \le s \le 1$. By standard calculations, for appropriate constants $\{k_n\}_{n\in\mathbb{N}}$ and all $s \in [0, 1]$, $x \in H$,

$$\begin{aligned}\gamma(x,s) &\le k|x|^p + k\mathbb{E}\left|\int_0^s S(s-u)F(X(u,x))du\right|^p \\ &\quad + k\mathbb{E}\left|\int_0^s S(s-u)B(X(u,x))dW(u)\right|^p \\ &\le k|x|^p + k_1 + k_1\int_0^s \gamma(x,u)du + k_1\mathbb{E}\left(\int_0^s \|S(s-u)B(X(u,x))\|_2^2 du\right)^{p/2} \\ &\le k|x|^p + k_1 + k_1\int_0^s \gamma(x,u)du + k_2\mathbb{E}\left(\int_0^s \|B(X(u,x))\|_2^2 du\right)^{p/2} \\ &\le k|x|^p + k_3 + k_3\int_0^s \gamma(x,u)du.\end{aligned}$$

Hence, by Gronwall's lemma for a new constant k_4 we have

$$\gamma(x,s) \le k_4(|x|^p + 1), \quad s \in [0,1]\,, x \in H. \tag{11.80}$$

Define $F(s,x) = F(X(s,x))$ and

$$Y(s,x) = \int_0^s (s-u)^{-\alpha} S(s-u)B(X(u,x))dW(u), \quad s \in [0,1],\ x \in H.$$

Then for possibly new constants

$$\begin{aligned}\mathbb{E}\int_0^1 |Y(s,x)|^p ds &= \mathbb{E}\int_0^1 \left|\int_0^s (s-u)^{-\alpha} S(s-u)B(X(u,x))dW(u)\right|^p ds \\ &\le k_5\mathbb{E}\int_0^1 \left(\int_0^s (s-u)^{-2\alpha}\|B(X(u,x))\|_{L_2^0}^2 du\right)^{p/2} ds,\end{aligned}$$

and by applying Young's inequality,

$$\begin{aligned}\mathbb{E}\int_0^1 |Y(s,x)|^p ds &\le k_7 + k_7\int_0^1 \mathbb{E}|X(u,x)|^p du \\ &\le k_8(|x|^p + 1).\end{aligned} \tag{11.81}$$

Taking into account (11.80) and (11.81) we have

$$\mathbb{E}(|Y(\cdot,x)|_p^p + |F(\cdot,x)|_p^p) \le k_9(|x|^p + 1). \tag{11.82}$$

However, for $x \in H$, $\mathbb{P}$-a.s.

$$X(1,x) = S(1)x + G_1F(\cdot,x)(1) + \frac{\sin(\pi\alpha)}{\pi} G_\alpha Y(\cdot,x)(1), \tag{11.83}$$

see the proof of Proposition 7.3. Hence if $|x| \le r$, $|F(\cdot,x)|_p \le r$ and $|Y(\cdot,x)|_p \le \frac{\pi r}{\sin(\alpha\pi)}$ then $X(1,x) \in K(r)$. Assume that $\|x\| \le r$, then

$$\mathbb{P}(X(1,x) \notin K(r)) \le \mathbb{P}(|F(\cdot,x)|_p > r) + \mathbb{P}\left(|Y(\cdot,x)|_p > \frac{\pi r}{\sin(\alpha\pi)}\right)$$

and, by the Chebyshev inequality,

$$\mathbb{P}(X(1,x) \notin K(r)) \le r^{-p}\, \mathbb{E}(|F(\cdot,x)|_p^p) + r^{-p}\, \frac{\sin(\alpha\pi)}{\pi}^p \mathbb{E}(|Y(\cdot,x)|_p^p).$$

Taking into account (11.82) we have, for a constant $c > 0$,

$$\mathbb{P}(X(1,x) \notin K(r)) \le c r^{-p}(1+|x|^p), \quad r > 0,\ |x| \le r.$$

This finishes the proof of the lemma. □

We go back to the proof of Theorem 11.38. Fix $0 < 1/p, < \alpha < 1/2$. As we have noticed before, the set $K(r)$ is relatively compact. Let $X(\cdot) = X(\cdot,\xi)$ be a bounded in probability solution of (11.1). Then for any $t > 1$, by the Markov property

$$\begin{aligned}\mathbb{P}(X(t) \in K(r)) &= \mathbb{E}(\mathbb{P}(X(t) \in K(r))|\mathscr{F}_{t-1}))\\ &= \mathbb{E}(P(1, X(t-1), K(r))).\end{aligned}$$

By Lemma 11.39, for arbitrary numbers, $r > r_1 > 0$,

$$\begin{aligned}\mathbb{P}(X(t) \in K(r)) &= \mathbb{E}(P(1, X(t-1), K(r)))\\ &\ge \mathbb{E}\left(\left\{P(1, X(t-1), K(r))\mathbb{1}_{|X(t-1)|\le r_1}\right\}\right)\\ &\ge (1 - c(r^{-p}(1+r_1^p))\mathbb{P}(|X(t-1)|^p \le r_1).\end{aligned}$$

Since the process $X(\cdot)$ is bounded in probability, we can choose $r > r_1$ such that

$$\mathbb{P}(X(t) \in K(r)) > 1 - \varepsilon.$$

This finishes the proof of tightness of $\mathscr{L}(X(t))$, $t > 0$, and therefore the proof of the theorem is complete. □

Example 11.40 Let $\mathscr{O}$ be a bounded open set in $\mathbb{R}^d$ and let φ and b be real valued functions satisfying Lipschitz conditions and β a one dimensional standard Brownian motion. Consider the following stochastic equation:

$$\begin{cases} du(t,\xi) = [\Delta_\xi u(t,\xi) + \varphi(u(t,\xi))]dt + b(u(t,\xi))]d\beta(t),\\ u(t,\xi) = 0, \quad t \ge 0,\ \xi \in \partial\mathscr{O}\\ u(0,\xi) = x(\xi), \quad \xi \in \partial\mathscr{O},\end{cases} \tag{11.84}$$

where $x \in H = L^2(\mathscr{O})$. Defining

$$\begin{cases} D(A) = H^2(\mathscr{O}) \cap H_0^1(\mathscr{O})\\ Au = \Delta\xi u, \quad \forall\, u \in D(A)\end{cases}$$

and using Theorem 7.2 we see that the equation (11.84) has a unique mild solution. Assume in addition that

$$2\varphi(r) + b^2(r) \le k_1 - k_2 x^2, \quad \forall\, r \in \mathbb{R}^1. \tag{11.85}$$

Then, by (11.85) and a standard application of Ito's formula,

$$\frac{d}{dt}\,\mathbb{E}|u(t,\cdot)|^2 \le k_1 - k_2\mathbb{E}|u(t,\cdot)|^2.$$

Hence $\sup_{t\ge 0}\mathbb{E}|u(t,\cdot)|^2 < +\infty$. Consequently the solution of (11.84) is bounded in probability. Since the semigroup generated by A is compact, by Theorem 11.38, we get existence of an invariant measure for the equation (11.84). □

11.7.2 Cylindrical Wiener processes

We assume now that $W(\cdot)$ is a cylindrical Wiener process on U with the identity covariance operator. We have the following analog of Theorem 11.38.

Theorem 11.41 *Assume that Hypotheses* 7.2 *and* 11.9 *are satisfied. There exists a unique invariant measure for equation* (11.1).

Proof The proof is very similar to that of Theorem 11.38. It is based on a modified version of Lemma 11.39 in which Hypothesis 7.1 is replaced by Hypothesis 7.2, with the conclusion unchanged. Moreover, the proof of the modified version of the lemma is analogous to that of Lemma 11.39 with the exception of the estimate (11.81) which we describe now. Namely we have

$$\begin{aligned}\mathbb{E}\int_0^1 |Y(s,x)|^p ds &= \mathbb{E}\int_0^1 \left|\int_0^s (s-u)^{-\alpha} S(s-u)B(X(u,x))dW(u)\right|^p ds\\ &\le k\mathbb{E}\int_0^1 \left(\int_0^s (s-u)^{-2\alpha}\|S(s-u)\|^2_{L^2(H)}\|B(X(u,x)\|^2 du\right)^{p/2} ds,\end{aligned} \tag{11.86}$$

and, by Young's inequality,

$$\begin{aligned}\mathbb{E}\int_0^1 |Y(s,x)|^p ds &\le k_1\left(\int_0^1 t^{-2\alpha}\|S(t)\|^2_{L^2(H)}dt\right)^{p/2}\mathbb{E}\left(\int_0^1 \|B(X(s,x)\|^p ds\right)\\ &\le k_2 + k_2\int_0^1 \mathbb{E}|X(u,x)|^p du \le k_3(|x|^p+1).\end{aligned}$$

In the final estimate we have used inequality (7.30). We leave the remaining details of the proof to the reader. □

We will finish this section by giving a sufficient condition for boundedness in probability of solutions of (11.1) to hold.

Proposition 11.42 *Assume that*

(i) $\sup_{x\in H}\|B(x)\|_{L(U;H)} < +\infty$,

(ii) $\int_0^{+\infty} \|S(t)\|^2_{L_2(H)}dt < +\infty$,

(iii) for positive constants a, b we have

$$\langle Ax + F(x+y), x\rangle \le a(1+|y|^2) - 2b|x|^2.$$

Then $\sup_{t\ge 0} \mathbb{E}|X(t,x)|^2 < +\infty, \quad x \in H.$

Proof Denote $X(t,x)$ by $X(t)$ and set

$$Z(t) = \int_0^t S(t-s)B(X(s))dW(s), \quad Y(t) = X(t) - Z(t).$$

By assumption (ii),

$$\sup_{t\ge 0} \mathbb{E}|Z(t)|^2 \le \sup_{x\in H} \|B(x)\|^2 \int_0^\infty \|S(t)\|_2^2\, dt < +\infty. \tag{11.87}$$

Obviously

$$Y(t) = S(t)x + \int_0^t S(t-s)F(X(s))ds.$$

Let $Y_\lambda(t) = \lambda R_\lambda Y(t)$ and $F_\lambda(x) = \lambda R_\lambda F(x)$, where R_λ is the resolvent of the operator A. Since $Y_\lambda(t) \in D(A)$ for any $t \ge 0$ and

$$Y_\lambda(t) = \lambda R_\lambda x + \int_0^t S(t-s)F_\lambda(X(s))ds,$$

Y_λ satisfy the following equations

$$\frac{d}{dt}Y_\lambda(t) = AY_\lambda(t) + F_\lambda(X(s)) = AY_\lambda(t) + F(Y_\lambda(t) + Z(t)) + \delta_\lambda(t),$$

where $\delta_\lambda(t) = F_\lambda(X(t)) - F(Y_\lambda(t) + Z(t)) \to 0$.

Therefore, by assumption (iii),

$$\begin{aligned}\frac{d}{dt}|Y_\lambda(t)|^2 &= \langle AY_\lambda(t) + F(Y_\lambda(t) + Z(t)) + \delta_\lambda(t), Y_\lambda(t)\rangle \\ &\le a(1+|Z(t)|^2) - 2b|Y_\lambda(t)|^2||\delta_\lambda(t)||Y_\lambda(t)| \\ &\le k + k|Z(t)|^2 + k|\delta_\lambda(t)|^2 - b|Y_\lambda(t)|^2\end{aligned}$$

for some constant k. By the Gronwall lemma

$$|Y_\lambda(t)|^2 \le e^{-bt}|Y_\lambda(0)|^2 + k\int_0^t e^{-b(t-s)}(1+|Z(s)|^2 + |\delta_\lambda(s)|^2)ds.$$

Letting λ tend to infinity we obtain

$$|Y(t)|^2 \le e^{-bt}|x|^2 + k\int_0^t e^{-b(t-s)}(1+|Z(s)|^2)ds.$$

Therefore

$$\mathbb{E}|Y(t)|^2 \le e^{-bt}|x|^2 + k\int_0^t e^{-b(t-s)}(1+\mathbb{E}|Z(s)|^2)ds$$

and by (11.87)

$$\sup_{t\geq 0}\mathbb{E}|X(t)|^2 \leq k_1\left(1+\sup_{t\geq 0}\mathbb{E}|Z(t)|^2\right) < +\infty.$$

□

Example 11.43 Let $\varphi : \mathbb{R} \to \mathbb{R}$ be a Lipschitz continuous function such that $\varphi(0) = 0$ and

$$|\varphi(\xi) - \varphi(\eta)| \leq k_0|\xi - \eta|$$

for some $k_0 > 0$ and all $\xi, \eta \in \mathbb{R}^1$. Let $b : \mathbb{R}^1 \to \mathbb{R}^1$ be a bounded, Lipschitz continuous function and $W(t, \xi)$ a space-time white noise. For $k > k_0$ consider the following stochastic equation

$$\begin{cases} du(t,\xi) = \left(\dfrac{d^2}{d\xi^2} - k\right) u(t,\xi) + \varphi(u(t,\xi))dt \\ \qquad\qquad + b(u(t,\xi))dW(t,\xi), \quad t > 0,\ \xi \in (0,1) \\ u(t,0) \ = u(t,1) = 0, \quad t > 0 \\ u(0,\xi) \ = x_\xi, \quad x(\cdot) \in L^2(0,1) = H. \end{cases} \tag{11.88}$$

It is easy to see that assumptions for Proposition 11.42 and Theorem 11.41 are satisfied and therefore there exists an invariant measure for (11.88). This generalizes existence results in [315, 522, 653]. □

12

Small noise asymptotic behavior

This chapter is devoted to asymptotic properties of solutions of stochastic equations with small diffusion coefficients. We show first that the laws of solutions satisfy the large deviation principle with the rate function determined by an associated control system. Proofs are based on a large deviation result for a family of Gaussian measures on a Banach space. Then we study asymptotic behavior of the mean exit time and of the exit place of solutions from a given domain. Finally, explicit formulae are given for the so called gradient stochastic systems.

We only consider regular nonlinear coefficients. In the last few years several interesting results have been proved for equations with irregular coefficients, for example reaction diffusion, Burgers and Navier–Stokes equations. See the following chapter for bibliographic comments.

12.1 Large deviation principle

Let us consider the stochastic equation

$$\begin{cases} dX = (AX + F(X))dt + \sqrt{\varepsilon}B(X)dW(t), \\ X(0) = x, \end{cases} \tag{12.1}$$

which depends on a parameter $\varepsilon \geq 0$ and assume that the coefficients A, B, F fulfill Hypothesis 7.1 or 7.2. It follows from Theorem 7.2 or 7.5 that there exists a unique continuous solution $X^{x,\varepsilon}(\cdot)$ of equation (12.1) with arbitrary $x \in H$ and $\varepsilon \geq 0$.

Let $z^x(\cdot)$ be the solution to the deterministic equation

$$\begin{cases} z' = Az + F(z), \\ z(0) = x. \end{cases} \tag{12.2}$$

It is intuitively clear that if $\varepsilon \downarrow 0$ then processes $X^{x,\varepsilon}(\cdot)$ converge to $z^x(\cdot)$. We have in fact the following result.

Proposition 12.1 *Assume that Hypothesis* 7.1 *or* 7.2 *is satisfied. Then, for arbitrary* $T > 0$, *and* $r > 0$

$$\lim_{\varepsilon \downarrow 0} \mathbb{P}\left(\sup_{t \in [0,T]} |X^{x,\varepsilon}(t) - z^x(t)| \geq r \right) = 0. \tag{12.3}$$

Proof We prove the proposition under Hypothesis 7.1. The case when Hypothesis 7.2 is fulfilled can be treated in the same way.

Let us fix $p > 2$ and $T > 0$. It follows from Theorem 7.2 that there exists a constant $c_0 > 0$ such that

$$\sup_{t \in [0,T]} \mathbb{E}|X^{x,\varepsilon}(t)|^p \leq c_0(1 + |x|^p). \tag{12.4}$$

Note that

$$X^{x,\varepsilon}(t) - z^x(t) = \int_0^t S(t-s)\,[F(X^{x,\varepsilon}(s)) - F(z^x(s))]\,ds$$
$$+ \sqrt{\varepsilon} \int_0^t S(t-s)B(X^{x,\varepsilon}(s))dW(s), \quad t \in [0,T].$$

Using Proposition 7.3 one obtains that for some constants c_1, c_2,

$$\mathbb{E} \sup_{u \in [0,t]} |X^{x,\varepsilon}(u) - z^x(u)|^p \leq c_1 \int_0^t \mathbb{E} \sup_{u \in [0,s]} |X^{x,\varepsilon}(u) - z^x(u)|^p ds$$
$$+ c_2 \varepsilon^{p/2} \mathbb{E} \int_0^T (1 + |X^{x,\varepsilon}(s)|^p) ds.$$

By Gronwall's lemma and (12.4) one obtains that

$$\mathbb{E} \sup_{u \in [0,T]} |X^{x,\varepsilon}(u) - z^x(u)|^p \leq 2_0 c_1 e^{c_1 T} \varepsilon^{p/2}(1 + |x|^p), \quad x \in H,$$

and from Chebyshev's inequality (12.3) follows. □

It follows from Proposition 12.1 that if Γ is a Borel subset of $C_T := C([0,T];H)$ such that $d_{C_T}(z^x(\cdot), \Gamma)$ is positive, [(1)] then

$$\lim_{\varepsilon \downarrow 0} \mathbb{P}(X^{x,\varepsilon}(\cdot) \in \Gamma) = 0. \tag{12.5}$$

In the same way if Γ is a Borel subset of H, such that $d_H(z^x(T), \Gamma)$ is positive, then

$$\lim_{\varepsilon \downarrow 0} \mathbb{P}(X^{x,\varepsilon}(T) \in \Gamma) = 0. \tag{12.6}$$

It will be of prime interest in the first part of the chapter to give the exact rate of convergence in (12.5), (12.6). In fact we will show that the laws of $X^{x,\varepsilon}(\cdot)$ or $X^{x,\varepsilon}(T)$ satisfy the so called *large deviation principle* of Varadhan and we will determine the corresponding rate functions.

We will also show how the obtained convergence results can be applied to an analysis of the destabilizing effect of the noise term in (12.1) on trajectories of (12.2).

[(1)] $d_{C_T}(a, \Gamma)$ represents the distance in C_T between the element $a \in C_T$ and the subset Γ.

To simplify the exposition we will treat equation (12.1) with additive noise only. In the present section we follow [651] and [213] and use systematically control theoretic interpretation.

12.1.1 Formulation and basic properties

In this subsection we shall denote by (E, ρ) a complete separable metric space. We are given a family of probability measures $\{\mu_\varepsilon\}_{\varepsilon>0}$ on E and a lower semicontinuous function $I : E \to [0, +\infty]$, not identically equal to $+\infty$ and such that its level sets,

$$I_r := \{x \in E : I(x) \le r\}, \quad r > 0, \tag{12.7}$$

are compact for arbitrary $r \in [0, +\infty)$. The family $\{\mu_\varepsilon\}$ is said to satisfy the *large deviation principle* (LDP) or to have the *large deviation property* with respect to the *rate function* I if

(L) *for all closed sets* $\Gamma \subset E$ *we have*

$$\limsup_{\varepsilon \downarrow 0}[\varepsilon \log \mu_\varepsilon(\Gamma)] \le -\inf_{x\in\Gamma} I(x), \tag{12.8}$$

(U) *for all open sets* $G \subset E$ *we have*

$$\liminf_{\varepsilon \downarrow 0}[\varepsilon \log \mu_\varepsilon(G)] \ge -\inf_{x\in G} I(x). \tag{12.9}$$

Remark 12.2 Assume that $\{\mu_\varepsilon\}_{\varepsilon>0}$ fulfills the principle of large deviations with respect to I. Let $G \subset E$ be an open set such that

$$\inf_{x\in G} I(x) = \inf_{x\in \overline{G}} I(x),$$

where $\overline{G}$ is the closure of G. Then by (12.8) we have

$$\limsup_{\varepsilon \downarrow 0}[\varepsilon \log \mu_\varepsilon(G)] \le \limsup_{\varepsilon \downarrow 0}[\varepsilon \log \mu_\varepsilon(\overline{G})] \le -\inf_{x\in \overline{G}} I(x),$$

which compared with (12.9), yields

$$\lim_{\varepsilon \to 0}[\varepsilon \log(\mu_\varepsilon(G))] = -\inf_{x\in G} I(x). \tag{12.10}$$

□

It is often not easy to check estimates (L) and (U) for general open and closed subsets of E. For this reason it is useful to find equivalent conditions concerning suitable subsets of E.

12.1.2 Lower estimates

We introduce here the following condition:

(L_1) *for any* $x_0 \in E,\ \delta > 0,\ \gamma > 0$, *there exists* $\varepsilon_0 > 0$ *such that,*

$$\mu_\varepsilon(B(x_0, \delta)) \ge e^{-\frac{1}{\varepsilon}(I(x_0)+\gamma)}, \quad \forall\, \varepsilon \le \varepsilon_0. \tag{12.11}$$

Proposition 12.3 *Condition* (L) *is equivalent to condition* (L_1).

Proof (L) $\Rightarrow$ (L_1) Let $x_0 \in E$, $\delta > 0$, and $G = B(x_0, \delta)$. Then by (L) for any $\gamma > 0$ there exists $\varepsilon_0 > 0$ such that

$$\varepsilon \log\left(\mu_\varepsilon(B(x_0,\delta))\right) \geq -\inf_{x\in B(x_0,\delta)} I(x) - \gamma \geq -I(x_0) - \gamma,\ \varepsilon \leq \varepsilon_0.$$

Thus (12.10) follows.

(L_1) $\Rightarrow$ (L) Given G open, let $x_0 \in G$ and $\delta > 0$ such that $B(x_0, \delta) \subset G$. Then by (L_1) for any $\gamma > 0$ there exists $\varepsilon_0 > 0$ such that

$$\mu_\varepsilon(G) \geq \mu_\varepsilon(B(x_0,\delta)) \geq e^{-\frac{1}{\varepsilon}(I(x_0)+\gamma)},\ \varepsilon \leq \varepsilon_0.$$

Consequently

$$\liminf_{\varepsilon\to 0} \varepsilon\,[\log(\mu_\varepsilon(G)] \geq -I(x_0) - \gamma.$$

The conclusion follows now from the arbitrariness of x_0 and γ. □

12.1.3 Upper estimates

We now introduce the following condition:

(U_1) *for any* $r > 0,\ \delta > 0,\ \gamma > 0,\ \exists\, \varepsilon_0 > 0$ *such that,*

$$\mu_\varepsilon(B^c(I_r,\delta)) \leq e^{-\frac{1}{\varepsilon}(r-\gamma)}, \quad \forall\, \varepsilon \leq \varepsilon_0, \tag{12.12}$$

where

$$B^c(I_r,\delta) = \{x \in E:\ I(y) \leq r \Rightarrow |x-y| \geq \delta\}.$$

Proposition 12.4 *Condition* (U) *is equivalent to condition* (U_1).

Proof (U) $\rightarrow$ (U_1). Let $r > 0, \delta > 0, \gamma > 0$, and $\Gamma = B^c(I_r, \delta)$. Then by (U) there exists $\varepsilon_0 > 0$ such that

$$\mu_\varepsilon(B^c(I_r,\delta)) \leq e^{-\frac{1}{\varepsilon}(\inf_{x\in B^c(I_r,\delta)} - \gamma)}, \quad \varepsilon \leq \varepsilon_0.$$

But

$$x \in B^c(I_r,\delta) \Rightarrow I(x) \geq r,$$

which implies

$$\mu_\varepsilon(B^c(I_r,\delta)) \leq e^{-\frac{1}{\varepsilon}(r-\gamma)}, \quad \varepsilon \leq \varepsilon_0.$$

Thus (12.12) follows.

$(U_1) \to (U)$. Set $r = \inf_{x\in K} I(x)$. If $r = 0$ the conclusion is obvious, because

$$\limsup_{\varepsilon\to 0} [\varepsilon \log \mu_\varepsilon(\Gamma)] \le 0 = -\inf_{x\in\Gamma} I(x).$$

Let us assume that $r > 0$. Let $r_1 < r$ so that $K \cap I^{r_1} = \varnothing$. Let δ be the distance between K and I^{r_1}. We have $\delta > 0$ since I^{r_1} is compact. Therefore by (U_1), given $\gamma > 0$ there exists $\varepsilon_0 > 0$ such that

$$\mu_\varepsilon(K) \le \mu_\varepsilon(B^c(I^{r_1}, \delta)) \le e^{-\frac{1}{\varepsilon}(r_1-\gamma)}, \quad \varepsilon \le \varepsilon_0.$$

It follows that

$$\limsup_{\varepsilon\to 0} [\varepsilon \log \mu_\varepsilon(\Gamma)] \le -r_1 + \gamma.$$

By the arbitrariness of r_1 and γ we find

$$\limsup_{\varepsilon\to 0} [\varepsilon \log \mu_\varepsilon(\Gamma)] \le -r = -\inf_{x\in K} I(x).$$

□

The inequalities (12.11) and (12.12) are often referred to as *exponential estimates* of Freidlin–Wentzell.

12.1.4 Change of variables

Let E, F be Polish spaces and let Φ be a homeomorphism of E onto F. Let ν_ε be the image measure of μ_ε by Φ,

$$\nu_\varepsilon(A) = \mu_\varepsilon(\Phi^{-1}(A)), \quad \forall A \in \mathscr{B}(F).$$

Setting $J(y) = I(\Phi^{-1}(y))$, $y \in F$, we have $J_r := \{y \in F : J(y) \le r\} = \Phi^{-1}(I_r)$ for all $r > 0$. Consequently J fulfills the same properties as I, that is J is lower semicontinuous and its level sets J_r are compact for any $r > 0$.

Proposition 12.5 *Assume that $\{\mu_\varepsilon\}_{\varepsilon>0}$ fulfills the principle of large deviations with respect to I. Then $\{\nu_\varepsilon\}_{\varepsilon>0}$ fulfills the principle of large deviations with respect to J.*

Proof Let H be an open subset of F. Then $\Phi^{-1}(H)$ is open in E and so we have

$$\begin{aligned}\liminf_{\varepsilon\to 0}[\varepsilon \log \nu_\varepsilon(H)] &= \liminf_{\varepsilon\to 0}[\varepsilon \log \mu_\varepsilon(\Phi^{-1}(H))] \\ &\ge -\inf_{x\in\Phi^{-1}(H)} I(x) = -\inf_{y\in H} J(y).\end{aligned}$$

In a similar way we can prove that for any closed subset M of F we have

$$\limsup_{\varepsilon\to 0}[\varepsilon \log \nu_\varepsilon(M)] \le -\inf_{y\in M} J(y).$$

□

12.2 LDP for a family of Gaussian measures

Let E be a separable Banach space with norm $\|\cdot\|$, μ a symmetric Gaussian measure on E and H_μ its reproducing kernel with the norm $|\cdot|_\mu$ and inner product $\langle\cdot,\cdot\rangle_\mu$.

We recall, see Theorem 2.12, that if $\{e_n\}\subset H_\mu^0$ is a complete orthonormal basis in H_μ, then for arbitrary sequence $\{\xi_n\}$ of real valued independent variables with $\mathscr{L}(\xi_n)=\mathscr{N}(0,1)$, $n\in\mathbb{N}$, the series $\sum_{k=1}^\infty \xi_n e_n$ converges a.s. to a random variable S with law μ.

Define a family $\{\mu_\varepsilon\}$ by setting

$$\mu_\varepsilon(\Gamma)=\mu(\varepsilon^{-1/2}\Gamma)=\mathbb{P}(\varepsilon^{1/2}S\in\Gamma),\quad \forall\,\Gamma\in\mathscr{B}(E),\ \varepsilon>0. \tag{12.13}$$

Then $\mu_\varepsilon=\mathscr{L}(\sqrt{\varepsilon}S)$, $\quad\varepsilon>0$.

We are going now to prove that the family $\{\mu_\varepsilon\}$ satisfies the Friedlin–Ventzell estimates (L1), (U1) with the rate function

$$I(x)=\begin{cases}\dfrac{1}{2}\,|x|_\mu^2 & \text{if } x\in H_\mu,\\ +\infty & \text{otherwise,}\end{cases} \tag{12.14}$$

in a slightly stronger form, needed later. The theorem below is due to [434], the proof is partially based on [651].

For any $r>0$ we set

$$K(r)=\left\{x\in H_\mu:\frac{1}{2}\,|x|_\mu^2\le r\right\}.$$

We start with the upper estimate.

Proposition 12.6 *Let $\{\mu_\varepsilon\}$ be the family defined by* (12.13). *Then for all $r_0>0$, $\delta>0$, $\gamma>0$, there exists $\varepsilon_0>0$ such that for all $\varepsilon\in(0,\varepsilon_0)$ and $r\in(0,r_0)$,*

$$\mu_\varepsilon(B(K(r),\delta))\ge 1-e^{-\frac{1}{\varepsilon}(r-\gamma)}.$$

Proof Let $r_0>0,\delta>0,\gamma>0,\varepsilon\in(0,1)$ be fixed. Let us remark that for $r\in(0,r_0)$, $n\in\mathbb{N}$, $a\in(0,\frac{1}{2})$ and $b>0$ we have

$$\begin{aligned}\mu_\varepsilon((B(K(r),\delta)^c)&=\mathbb{P}(\varepsilon^{1/2}S\notin B(K(r),\delta))\\ &=\mathbb{P}\left(\varepsilon^{1/2}S\notin B(K(r),\delta)\text{ and }\varepsilon^{1/2}\sum_{k=1}^n\xi_ke_k\notin K(r)\right)\\ &\quad+\mathbb{P}\left(\varepsilon^{1/2}S\notin B(K(r),\delta)\text{ and }\varepsilon^{1/2}\sum_{k=1}^n\xi_ke_k\in K(r)\right).\end{aligned}$$

It follows that

$$
\begin{aligned}
\mu_\varepsilon((B(K(r),\delta)^c) &\le \mathbb{P}\left(\varepsilon^{1/2}\sum_{k=1}^n \xi_k e_k \notin K(r)\right) + \mathbb{P}\left(\left\|\varepsilon^{1/2}\sum_{k=n+1}^\infty \xi_k e_k\right\| \ge \delta\right) \\
&= \mathbb{P}\left(\sum_{k=1}^n \xi_k^2 > \frac{2r}{\varepsilon}\right) + \mathbb{P}\left(\left\|\sum_{k=n+1}^\infty \xi_k e_k\right\| \ge \frac{\delta}{\sqrt{\varepsilon}}\right) \\
&= \mathbb{P}\left(e^{a\sum_{k=1}^n \xi_k^2} > e^{2a\frac{r}{\varepsilon}}\right) + \mathbb{P}\left(e^{b\left\|\sum_{k=n+1}^\infty \xi_k e_k\right\|^2} \ge e^{b\frac{\delta^2}{\varepsilon}}\right) \\
&\le (1-2a)^{-n/2}e^{-2a\frac{r}{\varepsilon}} + e^{-\frac{b\delta^2}{\varepsilon}}\mathbb{E}\left(e^{b\left\|\sum_{k=n+1}^\infty \xi_k e_k\right\|^2}\right) \\
&= I_1 + I_2.
\end{aligned}
$$

We first choose b in such a way that

$$
e^{-\frac{b\delta^2}{\varepsilon}} \le \frac{e^2-1}{4e^2}\, e^{-\frac{1}{\varepsilon}(r-\gamma)}, \quad \forall\, r \in [0, r_0),\ \forall\, \varepsilon > 0.
$$

Then we choose n such that

$$
\mathbb{E}\left(e^{b\left\|\sum_{k=n+1}^\infty \xi_k e_k\right\|^2}\right) \le \frac{e^2}{e^2-1}.
$$

This is possible by Theorem 2.7 and because

$$
\lim_{n\to\infty}\sum_{k=n+1}^\infty \xi_k e_k = 0, \quad \text{in probability.}
$$

Finally we choose $a \in (0, 1/2)$ such that $(1-2a)r_0 < \frac{1}{2}\gamma$. It is easily seen now that for $\varepsilon > 0$ sufficiently small and all $r \in (0, r_0)$

$$
I_1 \le \frac{1}{2}\, e^{-\frac{1}{\varepsilon(r-\gamma)}}, \quad I_2 \le \frac{1}{2}\, e^{-\frac{1}{\varepsilon(r-\gamma)}}
$$

as required. □

To prove the second estimate we need a lemma of independent interest.

Lemma 12.7 *For arbitrary $r > 0$ and $h \in H_\mu$ we have*

$$
\mu(B(h,r)) \ge \mu(B(0,r))e^{-\frac{1}{2}|h|_\mu^2}.
$$

Proof By Theorem 2.23, taking into account the symmetry of μ we have

$$\begin{aligned}
\mu(\{x \in E: \|x - h\| < r\}) &= \mu^h(\{x \in E: \|x\| < r\}) \\
&= \int_{\{x \in E: \|x\| < r\}} e^{-\langle h,x\rangle_\mu - \frac{1}{2}|h|_\mu^2} \mu(dx) \\
&= e^{-\frac{1}{2}|h|_\mu^2} \int_{\{x \in E: \|x\| < r\}} e^{-\langle h,x\rangle_\mu} \mu(dx) \\
&= \frac{1}{2} e^{-\frac{1}{2}|h|_\mu^2} \int_{\{x \in E: \|x\| < r\}} \left(e^{-\langle h,x\rangle_\mu} + e^{\langle h,x\rangle_\mu}\right) \mu(dx) \\
&\geq e^{-\frac{1}{2}|h|_\mu^2} \int_{\{x \in E: \|x\| < r\}} \mu(dx) \\
&= e^{-\frac{1}{2}|h|_\mu^2} \mu(\{x \subset E: \|x\| < r\}).
\end{aligned}$$

□

The following proposition, which is a stronger version of the second estimate, is an immediate consequence of Lemma 12.7.

Proposition 12.8 *Let $\{\mu_\varepsilon\}$ be the family defined by* (12.13). *Then for all $r_0 > 0$, $\delta > 0$, $\gamma > 0$, there exists $\varepsilon_0 > 0$ such that for all $\varepsilon \in (0, \varepsilon_0)$ and x such that $|x|_\mu^2 \leq r_0$,*

$$\mu_\varepsilon(B(x, \delta)) \geq 1 - e^{-\frac{1}{\varepsilon}(\frac{1}{2}|x|_\mu^2 + \gamma)}.$$

As a corollary of Propositions 12.6 and 12.8 we have the following.

Theorem 12.9 *The family $\{\mu_\varepsilon\}$ given by* (12.13) *satisfies the LDP with the rate function* (12.14).

Here is an application of Theorem 12.9.

Proposition 12.10 *Assume that μ is a Gaussian measure $\mathscr{N}(0, Q)$ on a Hilbert space H. Then the family $\{\mu_\varepsilon\}$ given by* (12.13), *satisfies the LDP with the rate function*

$$I(x) = \begin{cases} \frac{1}{2}\,|Q^{-1/2}x|^2 & \text{if } x \in Q^{1/2}(H) \\ +\infty & \text{otherwise.} \end{cases}$$

Proof It is enough to notice that the space $\widehat{H} = Q^{1/2}(H)$ equipped with the norm $|x|_{\widehat{H}} = |Q^{-1/2}x|^2$ is the reproducing kernel for μ, see Section 2.2.2. □

12.3 LDP for Ornstein–Uhlenbeck processes

We now derive the large deviation estimates for solutions of linear equations

$$\begin{cases} dX = AXdt + \sqrt{\varepsilon}BdW(t) \\ X(0) = x \in H, \end{cases} \tag{12.15}$$

where A generates a C_0-semigroup $S(\cdot)$ on H, $W(\cdot)$ is a Q-Wiener process on a Hilbert space U and $B \in L(U; H)$. Let

$$Q_T = \int_0^T S(t)BQB^*S^*(t)dt. \tag{12.16}$$

We recall (see Section 5.2) that if Tr $Q_T < +\infty$ the problem (12.15) has a unique mild solution given by

$$X^{x,\varepsilon}(t) = S(t)x + \sqrt{\varepsilon}\int_0^t S(t-s)BdW(s), \quad t \in [0,T]. \tag{12.17}$$

Moreover the measure

$$\mathscr{M}_\varepsilon^x = \mathscr{L}(X^{x,\varepsilon}(\cdot))$$

is Gaussian and concentrated on $L^2(0,T;H) =: \mathscr{H}$.

Let us consider the following control system

$$g' = Ag + BQ^{1/2}u, \quad g(0) = x,$$

and the corresponding mild solution

$$g^{x,u}(t) = S(t)x + \int_0^t S(t-s)BQ^{1/2}u(s)ds, \quad t \in [0,T]. \tag{12.18}$$

Theorem 12.11 *For arbitrary $T > 0$ and $x \in H$ the family $\{\mathscr{M}_\varepsilon^x\}_{\varepsilon>0}$ satisfies on $\mathscr{H}$ the LDP with the rate functional*

$$I^x(h) = \inf\left\{\frac{1}{2}\int_0^T |u(s)|_U^2 ds : g^{x,u} = h\right\}, \quad h \in L^2(0,T;H).$$

Proof It is enough to prove the result for $x = 0$. By Theorem 5.1(iii) $\mathscr{M}_\varepsilon^0$ is a symmetric Gaussian measure on $\mathscr{H}$ with covariance operator $\mathscr{Q}$ given by

$$\mathscr{Q}\varphi(t) = \int_0^T G(t,s)\varphi(s)ds, \quad t \in [0,T],$$

where

$$G(t,s) = \int_0^{t\wedge s} S(t-r)BQB^*S^*(s-r)dr, \quad t,s \in [0,T]$$

and $t \wedge s = \min\{t, s\}$. Then by the large deviation principle for Gaussian measures it follows that the quasi-potential I is given by

$$I^x(h) = \frac{1}{2} \|\mathcal{Q}^{-1/2} h\|^2_{L^2(0,T;H)}.$$

So the conclusion follows from Proposition 12.10 and Remark B.7. □

Under additional conditions the rate functional can be written in a more explicit way:

$$I^x(h) = \frac{1}{2} \int_0^T |(BQ^{1/2})^{-1}(h'(t) - Ah(t))|^2 dt, \quad h \in L^2(0, T; H), \; h(0) = x.$$

We also have the following result.

Theorem 12.12 *For arbitrary $T > 0$ and $x \in H$ the laws $\{\mathcal{L}(X^{x,\varepsilon}(T))\}_{\varepsilon>0}$ satisfy the LDP with the rate functionals*

$$\begin{aligned} I_T^x(y) &= \inf\left\{\frac{1}{2}\int_0^T |u(s)|^2 ds : \; g^{x,u}(T) = y\right\} \\ &= \frac{1}{2}\left|Q_T^{-1/2}(y - S(T)x)\right|^2, \quad y \in H. \end{aligned}$$

Proof It is enough to remark that the covariance operator of $\mathcal{L}(X^{x,\varepsilon}(T))$ is given by εQ_T, and use Proposition 12.10 and Remark B.9. □

Example 12.13 Assume that A is self-adjoint with the spectrum contained in $(-\infty, -\alpha]$ for some $\alpha > 0$ and let $Q = I$. Then

$$Q_T = (-2A)^{-1}(I - e^{2AT}).$$

To simplify the notation assume that $x = 0$. Then

$$\frac{1}{2}|(-A)^{1/2}y|^2 \le I_T^0(y) \le \frac{1}{2}(1 - e^{2\alpha T})|(-A)^{1/2}y|^2, \quad y \in H. \tag{12.19}$$

So the rate functional I_T^0 is equivalent to the square of the fractional power norm $|\cdot|_{1/2}$:

$$|y|_{1/2} = \begin{cases} |(-A)^{1/2}y| & y \in D((-A)^{1/2}) \\ +\infty & \text{otherwise.} \end{cases}$$

□

More generally we have the following result, see Remark B.8.

Proposition 12.14 *Let $S(\cdot)$ be an analytic and exponentially stable semigroup. Then the rate functional I_T^0 is equivalent to $|\cdot|^2_{1/2,2}$.*

Remark 12.15 The results are even more explicit for the family of invariant measures $\{\mu_\varepsilon\}$ of system (12.17). We recall that conditions for existence and uniqueness of invariant measures have been given in Theorem 11.17. In particular, for the existence of invariant measures the operator

$$Q_\infty = \int_0^\infty S(r)BQS^*(r)dr,$$

has to be nuclear and in this case the measures

$$\mu_\varepsilon = \mathscr{N}(0, \varepsilon Q_\infty), \quad \varepsilon > 0,$$

are invariant for equation (12.15).[2] By Proposition 12.10 the rate functional for $\{\mu_\varepsilon\}_{\varepsilon>0}$ is exactly

$$I_\infty(y) = \frac{1}{2}\|Q_\infty^{-1/2}y\|^2, \quad y \in H.$$

Thus in the particular case when A is self-adjoint negative and $B = I$, we have

$$I_\infty(y) = \frac{1}{2}\|(-A)^{1/2}y\|^2, \quad y \in H.$$

□

So far we have considered laws on the Hilbert spaces $L^2(0, T; H)$ and H. Now let E be a Banach space continuously, and as a Borel set, embedded into H. Then also the spaces $C([0, T]; E)$ and $L^2(0, T; E)$ are continuously, and as a Borel set, embedded into $L^2(0, T; H)$. Taking into account Proposition 12.6 and Proposition 12.8, one obtains for instance the following result. In its formulation we denote by $K_T^x(r)$ the set of all mild solutions $g^{x,u}$ of (12.18) with u satisfying $\frac{1}{2}\int_0^T |u(s)|^2 ds \le r^2$.

Theorem 12.16 *Assume that the process $X(t)$, $t \in [0, T]$, is E-continuous. Then for all $r_0 > 0$, $\delta > 0$, $\gamma > 0$ there exists $\varepsilon_0 > 0$ such that for all $\varepsilon \in (0, \varepsilon_0)$, $r \in (0, r_0)$ and $x \in E$ we have*

$$P\left(d_{C([0,T];E)}(X^{x,\varepsilon}, K_T^x(r)) < \delta\right) \ge 1 - e^{-\frac{1}{\varepsilon}(r^2-\gamma)}.$$

Moreover, for all $r_0 > 0$, $\delta > 0$, $\gamma > 0$ there exists $\varepsilon_0 > 0$ such that for all $\varepsilon \in (0, \varepsilon_0)$, all u satisfying $\frac{1}{2}\int_0^T |u(s)|^2 ds \le r^2$ and all $x \in E$ we have

$$\mathbb{P}\left(\sup_{t\in[0,T]} \|X^{x,\varepsilon}(t) - g^{x,u}(t)\|_E < \delta\right) \ge e^{-\frac{1}{\varepsilon}\left(\frac{1}{2}\int_0^T |u(s)|^2 ds + \gamma\right)}.$$

Similar results can be formulated for the families of random variables $X^{x,\varepsilon}(T)$, $\varepsilon > 0$.

[2] But there might be other invariant measures for (12.15).

12.4 LDP for semilinear equations

We are here concerned with equations on a Banach space $E \subset H$:

$$\begin{cases} dX = (AX + F(X))dt + \sqrt{\varepsilon} B dW(t), t \geq 0 \\ X(0) = x \in E \end{cases} \tag{12.20}$$

studied in Section 7.2.1. We assume that conditions 7.3(i) and 7.4 hold. Instead of 7.4 it would be enough to assume that F is Lipschitz continuous on bounded sets. Formulation of the result would be longer due to the possible existence of exploding solutions. It follows from Theorem 7.7 that, for arbitrary $x \in E$ and $\varepsilon > 0$, equation (12.20) has a unique E-continuous mild solution.

Let us also consider the associated control system

$$f' = (Af + F(f)) + BQ^{1/2}u, \quad f(0) = x, \tag{12.21}$$

and denote by $f^{x,u}$ the solution of (12.21). Let finally

$$K_T^x(r) = \left\{ f \in C([0, T]; E) : \ f = f^{x,u}; \ \frac{1}{2}\int_0^T |u(s)|^2 ds \leq r \right\}.$$

Theorem 12.17 *Assume Hypotheses 7.3 and 7.4. Let $R_0 > 0, r_0 > 0$ and $T > 0$ be numbers such that all sets $K_T^x(r_0)$, $\|x\| \leq R_0$, are contained in a bounded subset of $C([0, T]; E)$.*

(i) *For all $\delta > 0$, $\gamma > 0$, there exists $\varepsilon_0 > 0$ such that for all $\varepsilon \in (0, \varepsilon_0)$, $x \in E$ with $\|x\| \leq R_0$ we have*

$$\mathbb{P}\left(d_{C([0,T];E)}(X^{x,\varepsilon}, K_T^x(r)) < \delta\right) \geq 1 - e^{-\frac{1}{\varepsilon}(r-\gamma)}.$$

(ii) *For all $\delta > 0$, $\gamma > 0$, there exists $\varepsilon_0 > 0$ such that for all $\varepsilon \in (0, \varepsilon_0)$, $u \in L^2(0, T; U)$ satisfying $\frac{1}{2}\int_0^T |u(s)|^2 ds \leq r_0$ and $x \in E$ with $\|x\| \leq R_0$ we have*

$$\mathbb{P}\left(\sup_{t\in[0,T]} \|X^{x,\varepsilon}(t) - f^{x,u}(t)\|_E < \delta\right) \geq e^{-\frac{1}{\varepsilon}\left(\frac{1}{2}\int_0^T |u(s)|^2 ds + \gamma\right)}.$$

Proof Let $R_1 > 0$ be a number such that for all $x \in E$, $\|x\| \leq R_0$ and all $f \in K_T^x(r_0)$, $\|f(t)\| \leq R_1$, $t \in [0, T]$. Since the basic estimates depend only on the behavior of the solutions $X^{x,\varepsilon}$ and $y^{x,u}$ in the ball of radius $R_1 + \delta$, without any loss of generality, we can assume that F is globally Lipschitz. For arbitrary $f \in C([0, T]; E)$ with $f(0) = 0$, denote by $\gamma(t, f)$, $t \in [0, T]$, the unique solution of the equation

$$\gamma(t) = \int_0^t S(t-s)F(\gamma(s))ds + f(t), \quad t \in [0, T].$$

Then there exists a constant $L > 0$ such that

$$\sup_{t\in[0,T]} \|\gamma(t, f) - \gamma(t, g)\| \le L \sup_{t\in[0,T]} \|f(t) - g(t)\|.$$

Since $X^{x,\varepsilon}(t) = \gamma(t, Z^{x,\varepsilon})$ and

$$f^{x,u}(t) = \gamma(t, g^{x,u}), \quad x \in E,\ t \in [0, T],$$

we have

$$\sup_{t\in[0,T]} \|X^{x,\varepsilon}(t) - f^{x,u}(t)\| \le L \sup_{t\in[0,T]} \|Z^{x,\varepsilon}(t) - g^{x,u}(t)\|.$$

Consequently the conclusion easily follows from Theorem 12.16. □

12.5 Exit problem

We consider here the framework of Section 12.1.4. In particular, we assume that conditions of Theorem 12.17 are satisfied, and denote by $X^{x,\varepsilon}$ the mild solution of (12.20) and by $f^{x,u}$, that of (12.21).

We assume moreover that $F(0) = 0$. Then 0 is an equilibrium point for the deterministic equation

$$z' = Az + F(z), \quad z(0) = x. \tag{12.22}$$

Let $z^x(\cdot)$ denote the solution of (12.22) and assume that there exists an open bounded neighborhood $D \subset E$ of 0 which is *uniformly attracted* to 0 by (12.22), more precisely, we have the following.

Hypothesis 12.1 *For all $r > 0$ there exists $T > 0$ such that*

$$\|z^x(t)\| \le r, \quad \forall\, t \ge T,\ x \in \overline{D}.$$

Note that $z^x = X^{x,0}$. So the assumption 12.1 means that $X^{x,0} \to 0$ as $t \to \infty$. However, for $\varepsilon > 0$ the behavior of $X^{x,\varepsilon}(\cdot)$ will be completely different. Under the influence of the additive noise the solution, starting from D, will eventually reach the boundary ∂D. To see this denote by $\tau^{x,\varepsilon}$ the exit time of the process $X^{x,\varepsilon}$ from D:

$$\tau^{x,\varepsilon} = \inf\{t \ge 0 :\ X^{x,\varepsilon}(t) \in D^c\}. \tag{12.23}$$

To study properties of $\tau^{x,\varepsilon}$ we need the following elementary lemma essential in the proofs of upper estimates for the exit times.

Lemma 12.18 *Let ξ be a nonnegative random variable and let $p \in (0, 1)$. If $P(\xi \ge k) \le p^k$ for all $k \in \mathbb{N}$ we have $\mathbb{E}(\xi) \le \frac{1}{1-p}$.*

Proof Write

$$\mathbb{E}(\xi) = \int_0^{+\infty} P(\xi \geq t)dt = \sum_{k=0}^{\infty} \int_k^{k+1} P(\xi \geq t)dt$$
$$\leq \sum_{k=0}^{\infty} P(\xi \geq k) \leq \frac{1}{1-p}.$$

□

Now we can prove the following result.

Proposition 12.19 *If the process $BW(\cdot)$ is not identically 0, then for arbitrary $x \in D$ and $\varepsilon > 0$, $\mathbb{E}[\tau^{x,\varepsilon}] < +\infty$.*

Proof We can assume that $D = \{x \in E : \|x\| < R\}$ for some $R > 0$ and that F is a bounded transformation. Since the process $BW(\cdot)$ is nondegenerate, there exists $\varphi \in E^*$ with $\|\varphi\| = 1$ such that the one dimensional Gaussian variable $\varphi(W_A(1))$ is nondegenerate as well. Define $q_\varepsilon(x) = \mathbb{P}(\tau^{x,\varepsilon} > 1)$. Since

$$\{\tau^{x,\varepsilon} > 1\} = \left\{ \sup_{s\in[0,1]} \|X^{x,\varepsilon}(s)\| \leq R \right\},$$

we have

$$q_\varepsilon(x) \leq \mathbb{P}(\|X^{x,\varepsilon}(1)\| \leq R) \leq \mathbb{P}(|\varphi(X^{x,\varepsilon}(1))| \leq R).$$

Since

$$X^{x,\varepsilon}(1) = S(1)x + \int_0^1 S(1-s)F(X^{x,\varepsilon}(s))ds + \sqrt{\varepsilon}\, W_A(1),$$

and F is bounded, there exists $R_1 > 0$ such that

$$|\varphi(X^{x,\varepsilon}(1))| \geq \sqrt{\varepsilon}|\varphi(W_A(1))| - R_1, \quad \mathbb{P}\text{-a.e.}$$

It follows that

$$q_\varepsilon(x) \leq \mathbb{P}\left(|\varphi(W_A(1))| \leq \frac{R + R_1}{\sqrt{\varepsilon}}\right) =: p_\varepsilon < 1, \quad \forall\, x \in D.$$

Moreover for arbitrary $k \in \mathbb{N}$

$$\mathbb{P}(\tau^{x,\varepsilon} > k+1) = \mathbb{P}(A_k \cap B_k),$$

where

$$A_k = \{\|X^{x,\varepsilon}(t)\| < R, \quad \forall\, t \in [0,k]\},$$
$$B_k = \{\|X^{x,\varepsilon}(k+s)\| < R, \quad \forall\, s \in [0,1]\}.$$

It follows that

$$\mathbb{P}(\tau^{x,\varepsilon} > k+1) = \mathbb{E}\left[\mathbb{P}(A_k \cap B_k|\mathscr{F}_k)\right] = \mathbb{E}\left[\mathbb{1}_{A_k}\mathbb{P}(B_k|\mathscr{F}_k)\right].$$

On the other hand,

$$\mathbb{P}(B_k|\mathscr{F}_k) = \mathbb{P}(X^{x,\varepsilon}(k+\cdot) \in \Gamma|\mathscr{F}_k),$$

where

$$\Gamma = \{f \in C([0,+\infty); E) : \ \|f(t)\| < R, \ \forall\, t \in [0,1]\}.$$

By the Markov property, see Corollary 9.13, $\mathbb{P}$-a.s.

$$\mathbb{P}(X^{x,\varepsilon}(k+\cdot) \in \Gamma|\mathscr{F}_k) = \mathbb{P}^{X^{x,\varepsilon}(k)}(\Gamma) = q_\varepsilon(X^{x,\varepsilon}(k)) \le p_\varepsilon.$$

Consequently, by induction

$$\mathbb{P}(\tau^{x,\varepsilon} > k) \le p_\varepsilon^k, \quad k \in \mathbb{N}.$$

Therefore $\mathbb{E}(\tau^{x,\varepsilon}) \le \frac{1}{1-p_\varepsilon}$ by Lemma 12.18. □

12.5.1 Exit rate estimates

It is intuitively clear that

$$\lim_{\varepsilon \downarrow 0} \mathbb{E}(\tau^{x,\varepsilon}) = +\infty. \tag{12.24}$$

In this subsection we will calculate the rate of divergence in (12.24). To achieve this we will use the exponential estimates of Theorem 12.17. We define

$$\overline{e} = \inf\left\{\frac{1}{2}\int_0^T |u(s)|^2 ds \ : \ f^{0,u}(T) \in (\overline{D})^c,\, T > 0\right\}. \tag{12.25}$$

Interpreting the integral $\int_0^T |u(s)|^2 ds$ as energy dissipated by the control u, we can say that $\overline{e}$ is the minimal energy required by the control system (12.21) to transfer the equilibrium state 0 outside $\overline{D}$. We will call $\overline{e}$ the *upper exit rate*. For any $r > 0$ let

$$e_r = \inf\left\{\frac{1}{2}\int_0^T |u(s)|^2 ds \ : \ f^{x,u}(T) \in (\overline{D})^c,\, T > 0,\, \|x\| \le r\right\}. \tag{12.26}$$

We call the number

$$\underline{e} = \lim_{r \downarrow 0} e_r \tag{12.27}$$

the *lower exit rate*. Note that always $\underline{e} \le \overline{e}$.

The main result of this section is the following theorem. In its formulation we set

$$D^0 = \{x \in D : \ z^x(t) \in D, \forall\, t \ge 0\}.$$

Theorem 12.20 *We assume the hypotheses of Theorems* 12.17 *and* 12.1. *Then we have*

$$\limsup_{\varepsilon\downarrow 0} \varepsilon \ \log \mathbb{E}(\tau^{x,\varepsilon}) \le \overline{e}, \quad \forall\, x \in D, \tag{12.28}$$

$$\liminf_{\varepsilon\downarrow 0} \varepsilon \ \log \mathbb{E}(\tau^{x,\varepsilon}) \ge \underline{e}, \quad \forall\, \in D^0. \tag{12.29}$$

Theorem 12.20 is a generalization of finite dimensional results due to Freidlin and Ventzell [316]. The presentation follows Zabczyk [729, 730]. However, probabilistic considerations are the same as in [316]. In a more special case the generalization was carried out in [315].

Proof Without any loss of generality, we can assume that F is globally Lipschitz, with constant L and that $\|S(t)\| \le M, t \ge 0$.

Part 1 Proof of the upper estimate (12.28). The proof is similar to that of Proposition (12.19).

Let us fix a control $\widehat{u}$ such that $f^{0,\widehat{u}}(T) \in (\overline{D})^c$. It is enough to show that

$$\limsup_{\varepsilon\downarrow 0} \varepsilon \ \log \mathbb{E}(\tau^{x,\varepsilon}) \le \frac{1}{2}\int_0^T |\widehat{u}(s)|^2 ds = \widehat{r}. \tag{12.30}$$

It follows from the continuous dependence on initial data of the solutions of (12.21), that one can find positive numbers $\delta_1 > 0, \delta_2 > 0$ such that

$$\|x\| < \delta_1 \Rightarrow \ \mathrm{dist}_E\left(f^{x,\widehat{u}}(T), \overline{D}\right) \ge \delta_2.$$

Since

$$\mathbb{P}(\tau^{x,\varepsilon} < T) \ge \mathbb{P}\left(\sup_{t\in[0,T]} \|X^{x,\varepsilon}(t) - f^{x,\widehat{u}}(T)\| < \delta_2\right), \quad \|x\| < \delta_1,$$

one obtains from part (ii) of Theorem 12.17 that for any $\gamma > 0$ there exists $\varepsilon_0 > 0$ such that for all $\varepsilon \in (0, \varepsilon_0)$, and $x \in E$ with $\|x\| < \delta_1$:

$$q_\varepsilon(x) = \mathbb{P}(\tau^{x,\varepsilon} < T) \ge e^{-\frac{1}{\varepsilon}(\widehat{r}+\gamma)}.$$

Taking into account 12.1 and the continuous dependence on initial data of the solutions of (12.20), one can find positive numbers T_1 and p_1 such that

$$\mathbb{P}(\|X^{x,\varepsilon}(T_1)\| \le \delta_1) \ge p_1,$$

for all $x \in D$ and all sufficiently small $\varepsilon > 0$. Consequently, by the Markov property

$$\begin{aligned}
&\mathbb{P}(\tau^{x,\varepsilon} < T + T_1)\\
&\quad \ge \mathbb{P}(\|X^{x,\varepsilon}(T_1)\| \le \delta_1 \ \text{ and } \ X^{x,\varepsilon}(T_1 + s) \in \overline{D}^c \ \text{ for some } s \in [0, T])\\
&\quad \ge \mathbb{E}(q_\varepsilon(X^{x,\varepsilon}(T_1))\mathbb{1}_{\{\|X^{x,\varepsilon}(T_1)\|\le\delta_1\}}) \ge e^{-\frac{1}{\varepsilon}(\widehat{r}+\gamma)}\mathbb{P}(\|X^{x,\varepsilon}(T_1)\| \le \delta_1))\\
&\quad \ge p_1 e^{-\frac{1}{\varepsilon}(\widehat{r}+\gamma)}, \quad x \in D,
\end{aligned}$$

or, equivalently,

$$P(\tau^{x,\varepsilon} \geq T + T_1) \leq p = 1 - p_1 e^{-\frac{1}{\varepsilon}(\widehat{r}+\gamma)}, \quad x \in D. \tag{12.31}$$

By a consecutive application of the Markov property

$$P(\tau^{x,\varepsilon} \geq k(T + T_1)) \leq p^k, \quad \forall\, k \in \mathbb{N}.$$

By Lemma 12.18

$$\mathbb{E}\left(\frac{\tau^{x,\varepsilon}}{T + T_1}\right) \leq \frac{1}{1-p} = \frac{1}{p_1}\; e^{\frac{1}{\varepsilon}(\widehat{r}+\gamma)}.$$

Thus

$$\limsup_{\varepsilon \downarrow 0} \varepsilon \;\; \log \mathbb{E}(\tau^{x,\varepsilon}) \leq \widehat{r} + \gamma.$$

Since $\gamma > 0$ was an arbitrary positive number, (12.28) follows.

Part 2 Proof of the estimate (12.29).

The proof will be divided into several steps. Fix $\gamma > 0$ and choose a number $r > 0$ such that $e_r \geq \underline{e} - \gamma$. We can assume that the closed ball $\overline{B}(0, r)$, denoted as $\overline{B}(r)$, is contained in D. Let r_0 be a positive number smaller than r.

Define a sequence of stopping times $\sigma_k^{x,\varepsilon}$, $k = 0, 1, \dots$
$\sigma_0^{x,\varepsilon} = 0$ and

$$\sigma_{k+1}^{x,\varepsilon} = \inf\{t > \sigma_k^{x,\varepsilon} : \|X^{x,\varepsilon}(t)\| = r,\ \exists\, t_1 \in [\sigma_k^{x,\varepsilon}, t],\ \|X^{x,\varepsilon}(t_1)\| = r_0\}.$$

Then, in particular,

$$\sigma_1^{x,\varepsilon} = \inf\left\{t \geq 0; \|X^{x,\varepsilon}(t)\| = r, \|X^{x,\varepsilon}(t_1)\| = r_0 \text{ for some } t_1 \in [0, t]\right\}.$$

Step 1 We will show first that there exists $\varepsilon_1 > 0$ such that if $\varepsilon \in (0, \varepsilon_1)$, $\|x\| = r$, then

$$p_1^\varepsilon(x) = \mathbb{P}(\sigma_1^{x,\varepsilon} < \tau^{x,\varepsilon}) \geq 1 - e^{-\frac{1}{\varepsilon}(e_r - \gamma)}. \tag{12.32}$$

Note that

$$p_1^\varepsilon(x) = \mathbb{P}(\|X^{x,\varepsilon}(t)\| = r_0, \quad \text{for some } \; t < \tau^{x,\varepsilon}),$$

$$q_1^\varepsilon(x) = 1 - p_1^\varepsilon(x) = \mathbb{P}(\|X^{x,\varepsilon}(t)\| > r_0, \quad \text{for all } \; t < \tau^{x,\varepsilon}).$$

For arbitrary $T > 0$ we have therefore

$$q_1^\varepsilon(x) \leq \mathbb{P}(X^{x,\varepsilon}(t) \in K, \forall\, t \in [0, T]) + \mathbb{P}(\tau^{x,\varepsilon} \leq T) = I_1 + I_2,$$

where $K = \overline{D} \backslash B(r_0)$. To estimate I_1 we need the following important lemma.

Lemma 12.21 *Assume that the condition* 12.1 *holds. Then for all* $r_0 > 0,\ L > 0,$ *there exist* $T > 0$ *and* $\varepsilon_0 > 0$ *such that for all* $x \in K = \overline{D} \backslash B(r_0)$ *and* $\varepsilon \in (0, \varepsilon_0)$ *we* *have*

$$\mathbb{P}(X^{x,\varepsilon}(s) \in K, s \in [0,T]) \le e^{-\frac{L}{\varepsilon}}. \tag{12.33}$$

Proof Let K_δ be a δ neighborhood of K, $\delta < r_0/3$. There exists $T_1 > 0, \delta > 0$ such that if $t \ge T_1$ and $x \in K_\delta$ then $\|z^x(t)\| \le r_0/3$. Let $x \in K$, $u \in L^2(0,T;U)$ be a control such that $f^{x,u}(s) \in K_\delta$ for all $s \in [0,T_1]$. If $M_1 = \|BQ^{1/2}\|$, then

$$\|z^x(t) - f^{x,u}(t)\| \le ML \int_0^t \|z^x(s) - f^{x,u}(s)\| ds + MM_1 \int_0^t |u(s)| ds.$$

Consequently

$$r_0/3 \le \|z^x(T_1) - f^{x,u}(T_1)\| \le e^{MLT_1} MM_1\ T_1 \int_0^{T_1} |u(s)|^2 ds)^{1/2},$$

and

$$\frac{1}{2}\int_0^{T_1} |u(s)|^2 ds \ge \left(\frac{r_0}{3MM_1} T_1^{-1/2} e^{-MLT_1} \right)^2 = M_2.$$

By a simple induction argument, one obtains that if $f^{x,u}(t) \in K_\delta$ for some $j \in \mathbb{N}$ and all $t \in [0, jT_1]$, then

$$\frac{1}{2}\int_0^{jT_1} |u(s)|^2 ds \ge jM_2.$$

Let us remark that if $T = jT_1 > 2L$ then

$$P(X^{x,\varepsilon}(s) \in K, \forall\, s \in [0,T]) \le P(d_{C([0,T];E)}(X^{x,\varepsilon}, K_T^x(2L)) \ge r_0/3).$$

Taking into account part (i) of Theorem 12.17 one can find $\varepsilon_0 > 0$ such that (12.33) holds for $\varepsilon \in (0, \varepsilon_0)$ and all $x \in K$. This finishes the proof of the lemma. □

We go back to the estimate of I_1. Applying Lemma 12.21, one can find $\varepsilon_0 > 0$ such that

$$I_1 < e^{-\frac{1}{\varepsilon}(e_r - \gamma)}, \quad \forall\, \varepsilon \in (0, \varepsilon_0).$$

It remains to estimate I_2. Let us remark that again by part (i) of Theorem 12.17 and by the definition of e_r,

$$P(\tau^{x,\varepsilon} \le T) \le P(d_{C([0,T];E)}(X^{x,\varepsilon}, K_T^x(e_r)) \ge r/2) \le e^{-\frac{1}{\varepsilon}(e_r - \gamma)},$$

for sufficiently small ε and all $x \in D$ such that $\|x\| = r$. This way the proof of (12.32) is complete.

Step 2 We will now generalize the estimate (12.32) and show

$$p_k^\varepsilon(x) = \mathbb{P}(\sigma_k^{x,\varepsilon} < \tau^{x,\varepsilon}) \ge (1 - e^{-\frac{1}{\varepsilon}(e_r - \gamma)})^k, \quad k \in \mathbb{N},\ \|x\| = r \tag{12.34}$$

first for if $\|x\| = r$ and then if $x \in D_0$.

By the strong Markov property,

$$p_{k+1}^{\varepsilon}(x) = \mathbb{E}(p_1^{\varepsilon}(X^{x,\varepsilon}(\sigma_k^{x,\varepsilon}))1\!\!1_{\sigma_k^{x,\varepsilon}<\tau^{x,\varepsilon}})$$
$$\geq (1 - e^{-\frac{1}{\varepsilon}(e_r-\gamma)})\mathbb{P}(\sigma_k^{x,\varepsilon} < \tau^{x,\varepsilon}),\ \|x\| = r.$$

So (12.34) holds by induction.

If now $x \in D^0$ we can assume that $\|x\| > r$. Then there exists $p \in (0,1)$ and $T_2 > 0$ such that

$$P(\|X^{x,\varepsilon}(T_2)\| = r) \geq p.$$

Using the strong Markov property one obtains immediately

$$p_k^{\varepsilon}(x) \geq p\left(1 - e^{-\frac{1}{\varepsilon}(e_r-\gamma)}\right)^k, \quad k \in \mathbb{N}. \tag{12.35}$$

Step 3 To complete the proof we will need the following lemma.

Lemma 12.22 *Assume that $\{C_k\}$ is a decreasing sequence of events and $\{K_k\}$ a sequence of nonnegative random variables, then*

$$\sum_{k=0}^{\infty} \mathbb{E}((K_0 + \cdots + K_k)1\!\!1_{C_k \setminus C_{k+1}}) + \mathbb{E}\left(\left(\sum_{k=0}^{\infty} K_k\right)1\!\!1_{\bigcap_{k=0}^{\infty} C_k}\right)$$
$$= \sum_{k=0}^{\infty} \mathbb{E}(K_k 1\!\!1_{C_k}).$$

Proof Define

$$\nu = \begin{cases} k \ \text{ on } \ C_k \setminus C_{k+1}, & \forall\, k \in \mathbb{N} \\ +\infty \ \text{ on } \ \bigcap_{k=0}^{\infty} C_k. \end{cases}$$

then

$$\sum_{k=0}^{\infty} \mathbb{E}((K_0 + \cdots + K_k)1\!\!1_{C_k \setminus C_{k+1}}) + \mathbb{E}\left(\left(\sum_{k=0}^{\infty} K_k\right)1\!\!1_{\bigcap_{k=0}^{\infty} C_k}\right)$$
$$= \mathbb{E}\sum_{k=0}^{\nu} K_k = \mathbb{E}\sum_{k=0}^{\infty} 1\!\!1_{\{k\leq\nu\}}K_k$$
$$= \sum_{k=0}^{\infty} \mathbb{E}(K_k . 1\!\!1_{k\leq\nu}) = \sum_{k=0}^{\infty} \mathbb{E}(K_k 1\!\!1_{C_k}).$$

□

Step 4 Final Define $C_0^{x,\varepsilon} = \Omega$ and

$$C_k^{x,\varepsilon} = \{\sigma_k^{x,\varepsilon} < \tau^{x,\varepsilon}\}, \quad k \in \mathbb{N}.$$

Let moreover

$$K_k^{x,\varepsilon} = \begin{cases} \inf\{s \geq 0; \|X^{x,\varepsilon}(\sigma_k^{x,\varepsilon} + s) - X^{x,\varepsilon}(\sigma_k^{x,\varepsilon})\| \geq \frac{r-r_0}{2}\} & \text{if } \sigma_k^{x,\varepsilon} < +\infty \\ +\infty \quad \text{otherwise}. \end{cases}$$

Taking into account (12.34) and (12.35) one has

$$\mathbb{P}(C_k^{x,\varepsilon}) \geq p(1 - e^{-\frac{1}{\varepsilon}(e_r - \gamma)})^k, \quad k \in \mathbb{N},\ x \in D^0. \tag{12.36}$$

We leave it to the reader to show that

$$\mathbb{P}\left(\bigcap_{k=0}^{\infty} C_k^{x,\varepsilon}\right) = 0.$$

It is clear that

$$\tau^{x,\varepsilon} \geq K_k^{x,\varepsilon} + \cdots + K_k^{x,\varepsilon} \ \text{ on } \ C_k^{x,\varepsilon} \backslash C_{k+1}^{x,\varepsilon}, \quad k \in \mathbb{N}$$

and therefore

$$\mathbb{E}(\tau^{x,\varepsilon}) \geq \sum_{k=0}^{\infty} \mathbb{E}(K_0^{x,\varepsilon} + \cdots + K_k^{x,\varepsilon}) \mathbb{1}_{C_k^{x,\varepsilon} \backslash C_{k+1}^{x,\varepsilon}}).$$

To complete the proof of (12.29) note that by the lemma, for a positive constant C

$$\mathbb{E}(\tau^{x,\varepsilon}) \geq \sum_{k=0}^{\infty} \mathbb{E}(K_k^{x,\varepsilon} \mathbb{1}_{C_k^{x,\varepsilon}}) \geq C \sum_{k=0}^{\infty} \mathbb{P}(C_k^{x,\varepsilon}).$$

Consequently by (12.36),

$$\mathbb{E}(\tau^{x,\varepsilon}) \geq \sum_{k=0}^{\infty} Cp(1 - e^{-\frac{1}{\varepsilon}(e_r - \gamma)})^k \geq Cpe^{\frac{1}{\varepsilon},(e_r - \gamma)}$$

and the estimate follows. □

12.5.2 Exit place determination

A closed set $\mathscr{E} \subset \partial D$ is called an *exit set* for equation (12.20) and set D if for arbitrary $\delta > 0$ and all x sufficiently close to 0:

$$\lim_{\varepsilon \downarrow 0} \mathbb{P}(d_E(X^{x,\varepsilon}(\tau^{x,\varepsilon}), \mathscr{E}) > \delta) = 0.$$

It turns out that for a large class of equations (12.20) and domains D one can often find an exit set occupying only a small portion of the boundary ∂D. We will introduce a family of exit sets which have a useful control theoretic interpretation and are sufficiently small.

Let us also recall that the trajectories of the control system (12.21)

$$f' = (Af + F(f)) + BQ^{1/2}u, \quad f(0) = x, \tag{12.37}$$

associated to the stochastic system (12.20), were denoted by $f^{x,u}$.

Define for $x \in E$, $T > 0$ and $r > 0$ the following reachable sets:

$$\gamma_T^x(r) = \text{ closure}\left\{y \in E : y = f^{x,u}(t), t \in [0, T], \frac{1}{2}\int_0^T |u(s)|^2 ds \leq r\right\},$$

and

$$\gamma^x(r) = \text{ closure}\left\{ \bigcup_{T\geq 0} \gamma_T^x(r) \right\}, \quad r \geq 0, x \in E.$$

Set

$$\mathscr{E}_r = \text{ closure}\left\{ y \in \partial D; d_E(y, \bigcup_{\|x\|\leq r} \gamma^x(\overline{e}+r)) \leq r \right\}.$$

Note that $\mathscr{E}_0$ is exactly the closure of the set of all elements of ∂D which can be reached from 0, by the system (12.21), with the minimal possible energy $\overline{e}$, see the definition (12.25). In several cases, one can show that $\bigcap_{r>0} \mathscr{E}_r = \mathscr{E}_0$.

Our aim is to prove that for positive $r > 0$ sets $\mathscr{E}_r$ are exits sets. We give an extension, see Zabczyk [729], of a finite dimensional result due to Friedlin and Ventzell [316]. Probabilistic arguments are the same as in Varadhan [692] and Freidlin and Ventzel [316], see also Freidlin [315] for an extension in a more special case. The control arguments, useful when treating systems with degenerate noise, are from Zabczyk [729].

Theorem 12.23 *Under the conditions of* (12.20) *and* (12.1), *for all* $r > 0$ *and* $x \in D_0$

$$\lim_{\varepsilon\downarrow 0} \mathbb{P}(X^{x,\varepsilon}(\tau^{x,\varepsilon}) \in \mathscr{E}_r) = 0.$$

Proof For the proof set $\pi^{x,\varepsilon} = X^{x,\varepsilon}(\tau^{x,\varepsilon})$. For a fixed number $\overline{r} > 0$, for arbitrary numbers $0 < r_0 < r < \overline{r}$, $T > 0$ and for arbitrary $x \in D$, $\|x\| = r$ define events

$$A^{x,\varepsilon,T} = A_1^{x,\varepsilon,T} \cup \tilde{A}_1^{x,\varepsilon,T}, \; B^{x,\varepsilon,T} = B_1^{x,\varepsilon,T} \cup \tilde{B}_1^{x,\varepsilon,T},$$

where

$$A_1^{x,\varepsilon,T} = \{\tau^{x,\varepsilon} \leq T \;\text{ and } \pi^{x,\varepsilon} \in \mathscr{E}_{\overline{r}}\},$$

$$\tilde{A}_1^{x,\varepsilon,T} = \{\tau^{x,\varepsilon} > T \;\text{ and } \pi^{x,\varepsilon} \in \mathscr{E}_{\overline{r}}\} \cap \{X^{x,\varepsilon}(t) \in \overline{D}\backslash B(r_0) \;\text{ for } t \in [T, \tau^{x,\varepsilon}]\},$$

$$B_1^{x,\varepsilon,T} = \{\tau^{x,\varepsilon} \leq T \;\text{ and } \pi^{x,\varepsilon} \in (\mathscr{E}_{\overline{r}})^c\},$$

$$\tilde{B}_1^{x,\varepsilon,T} = \{\tau^{x,\varepsilon} > T \;\text{ and } \pi^{x,\varepsilon} \in (\mathscr{E}_{\overline{r}})^c\} \cap \{X^{x,\varepsilon}(t)\overline{D}\backslash B(r_0) \;\text{ for } t \in [T, \tau^{x,\varepsilon}]\}\}.$$

Dependence of the introduced events on r_0 and r has been dropped to simplify the notation.

The proof of the theorem consists of two main parts.

Part 1 The first part consists in proving the following lemma.

Lemma 12.24 *For arbitrary* $\overline{r} > 0$ *there exists* $0 < r_0 < r < \overline{r}$ *and* $T > 0$ *such that*

$$\lim_{\varepsilon\downarrow 0} \frac{\sup\{\mathbb{P}(B^{x,\varepsilon,T}) : \; \|x\| = r\}}{\inf\{\mathbb{P}(A^{x,\varepsilon,T}) : \; \|x\| = r\}} = 0. \tag{12.38}$$

Proof Remark that

$$\mathbb{P}(A^{x,\varepsilon,T}) \geq \mathbb{P}(\tau^{x,\varepsilon} \leq T \text{ and } \pi^{x,\varepsilon} \in \mathscr{E}_{\overline{r}}).$$

Moreover the event $B^{x,\varepsilon,T}$ is the intersection of $\{\pi^{x,\varepsilon} \in (\mathscr{E}_{\overline{r}})^c\}$ and

$$(\{\tau^{x,\varepsilon} \leq T\}) \cup (\{\tau^{x,\varepsilon} > T\} \cap \{X^{x,\varepsilon}(t) \in \overline{D}\backslash B(r_0) \text{ for } t \in [T, \tau^{x,\varepsilon}]\}).$$

In addition, for every $T_1 > 0$,

$$\{\tau^{x,\varepsilon} > T\} = \{T < \tau^{x,\varepsilon} \leq T + T_1\} \cup \{\tau^{x,\varepsilon} > T + T_1\}.$$

Therefore,

$$\begin{aligned}\mathbb{P}(B^{x,\varepsilon,T}) &\leq \mathbb{P}(X^{x,\varepsilon}(t) \in \overline{D}\backslash B(r_0), \forall \in [T, T + T_1]) \\ &\quad + \mathbb{P}(X^{x,\varepsilon}(t) \in (\mathscr{E}_{\overline{r}})^c \text{ for some } t \leq T + T_1) \\ &\leq J_1 + J_2.\end{aligned}$$

We first estimate the denominator in (12.38). Let $T > 0$ and $u \in L^2(0, T; U)$ be chosen such that

$$f^{0,u}(T) \in (\overline{D})^c \quad \text{and} \quad \frac{1}{2}\int_0^T |u(s)|^2 ds \leq \overline{e} + \frac{\overline{r}}{4}.$$

From the continuous dependence of the solutions on initial data, there exists $r < \overline{r}/2$ such that

$$r < d_E(f^{x,u}(T), \overline{D}), \quad \text{for } x \in B(r).$$

Consequently, applying Theorem 12.17 there exists $\varepsilon_0 > 0$ such that, for all $\varepsilon \in (0, \varepsilon_0)$,

$$\begin{aligned}\mathbb{P}(A^{x,\varepsilon,T}) &\geq \mathbb{P}(\tau^{x,\varepsilon} \leq T \text{ and } \pi^{x,\varepsilon} \in \mathscr{E}_{\overline{r}}) \\ &\geq \mathbb{P}(\sup_{t\in[0,T]} |X^{x,\varepsilon}(t) - f^{x,u}(t)| < r) \\ &\geq e^{-\frac{1}{\varepsilon}(\frac{1}{2}\int_0^T |u(s)|^2 ds + \frac{\overline{r}}{4})} \\ &\geq e^{-\frac{1}{\varepsilon}(\overline{e} + \frac{\overline{r}}{2})}.\end{aligned}$$

Therefore, for $\varepsilon \in [0, \varepsilon_0)$,

$$\inf\{\mathbb{P}(A^{x,\varepsilon,T}); |x| = r\} \geq e^{-\frac{1}{\varepsilon}(\overline{e} + \frac{\overline{r}}{2})}.$$

To estimate the numerator in (12.38) we start from $r_0 \in (0, r)$ and we use Lemma 12.21 to find $T_1 > 0$ such that for small ε and $x \in \overline{D}\backslash B(r_0)$,

$$\begin{aligned}q_\varepsilon(x) &= \mathbb{P}(X^{x,\varepsilon}(s) \in \overline{D}\backslash B(r_0), \forall s \in [0, T_1]) \\ &\leq e^{-\frac{1}{\varepsilon}(\overline{e} + \frac{3}{4}\overline{r})}.\end{aligned}$$

By the Markov property

$$J_1 = \mathbb{P}(X^{x,\varepsilon}(t) \in \overline{D}\backslash B(r_0), \forall\, t \in [T, T+T_1])$$

$$= \mathbb{E}(q_\varepsilon(X^{x,\varepsilon}(T)); X^{x,\varepsilon}(T) \in \overline{D}\backslash B(r_0))$$

$$\leq e^{-\frac{1}{\varepsilon}(\overline{e}+\frac{3}{4}\overline{r})}.$$

Finally, recalling that

$$K_S^x(s) = \left\{ f \in C([0,S];E) :\ f = f^{x,u};\ \frac{1}{2}\int_0^S |u(v)|^2 dv \leq s \right\}, \qquad S, s > 0,$$

we have,

$$J_2 = \mathbb{P}(X^{x,\varepsilon}(t) \in (\mathscr{E}_{\overline{r}})^c \ \text{ for some } t \leq T+T_1)$$
$$\leq \mathbb{P}(d_{C([0,T+T_1];E)}(X^{x,\varepsilon}, K_{T+T_1}^x(\overline{e}+\overline{r}) > \overline{r})),$$

and, by Theorem 12.17(i)

$$J_2 \leq e^{-\frac{1}{\varepsilon}(\overline{e}+\frac{3}{4}\overline{r})}, \forall\, \varepsilon \in (0, \varepsilon_2),$$

where $\varepsilon_2 > 0$ is a properly choosen number. Combining the above estimates of J_1 and J_2 we see that for $\varepsilon \in (0, \varepsilon_1 \wedge \varepsilon_2)$:

$$\sup_{\|x\|=r} \mathbb{P}(B^{x,\varepsilon,T}) \leq J_1 + J_2 \leq 2e^{-\frac{1}{\varepsilon}(\overline{e}+\frac{3}{4}\overline{r})}.$$

Therefore, for $\varepsilon \in (0, \varepsilon_0 \wedge \varepsilon_1 \wedge \varepsilon_2)$

$$\frac{\sup_{\|x\|=r} \mathbb{P}(B^{x,\varepsilon,T})}{\inf_{\|x\|=r} \mathbb{P}(A^{x,\varepsilon,T})} \leq 2\frac{e^{-\frac{1}{\varepsilon}(\overline{e}+\frac{3}{4}\overline{r})}}{e^{-\frac{1}{\varepsilon}(\overline{e}+\frac{1}{2}\overline{r})}} \leq e^{-\frac{\overline{r}}{4\varepsilon}}.$$

This finishes the proof of the lemma. □

Part 2 The second part of the proof of Theorem 12.23 consists in showing that, for arbitrary $x \in \overline{D}$ with $\|x\| = r$

$$\lim_{\varepsilon \downarrow 0} \frac{\mathbb{P}(B^{x,\varepsilon})}{\mathbb{P}(A^{x,\varepsilon})} = 0 \tag{12.39}$$

where

$$A^{x,\varepsilon} = \{\tau^{x,\varepsilon} < +\infty,\ \pi^{x,\varepsilon} \in \mathscr{E}_{\overline{r}}\},$$

$$B^{x,\varepsilon} = \{\tau^{x,\varepsilon} < +\infty,\ \pi^{x,\varepsilon} \in (\mathscr{E}_{\overline{r}})^c\}.$$

Note that this will end the proof of the theorem for $x \in D$, such that $\|x\| = r$. In fact if (12.39) holds then $\lim_{\varepsilon \downarrow 0} \mathbb{P}(A^{x,\varepsilon}) = 1$ because $\mathbb{P}(A^{x,\varepsilon} \cup B^{x,\varepsilon}) = 1$. The case of general $x \in D^0$ follows easily as well.

We pass to the proof of (12.39) and divide the time interval $[0, \tau^{x,\varepsilon}]$ into random, possibly an infinite number, subintervals $[0, \kappa_1^{x,\varepsilon}]$, $]\kappa_1^{x,\varepsilon}, \kappa_2^{x,\varepsilon}]$, ... and consider the events

$$A_n^{x,\varepsilon,T} = \{\kappa_{n-1}^{x,\varepsilon} < \tau^{x,\varepsilon} \le \kappa_n^{x,\varepsilon} \text{ and } \pi^{x,\varepsilon} \in \mathscr{E}_{\bar{r}}\},$$

$$B_n^{x,\varepsilon,T} = \{\kappa_{n-1}^{x,\varepsilon} < \tau^{x,\varepsilon} \le \kappa_n^{x,\varepsilon} \text{ and } \pi^{x,\varepsilon} \in (\mathscr{E}_{\bar{r}})^c\}$$

where $n \in \mathbb{N}$ and $\kappa_0^{x,\varepsilon} = 0$. Then

$$A^{x,\varepsilon,T} = \bigcup_{n=1}^{\infty} A_n^{x,\varepsilon,T}, \quad B^{x,\varepsilon,T} = \bigcup_{n=1}^{\infty} B_n^{x,\varepsilon,T},$$

and

$$\mathbb{P}\left(A^{x,\varepsilon,T}\right) = \bigcup_{n=1}^{\infty} \mathbb{P}\left(A_n^{x,\varepsilon,T}\right), \quad \mathbb{P}\left(B^{x,\varepsilon,T}\right) = \bigcup_{n=1}^{\infty} \mathbb{P}\left(B_n^{x,\varepsilon,T}\right).$$

To estimate the probabilities $\mathbb{P}(A_n^{x,\varepsilon,T})$ and $\mathbb{P}(B_n^{x,\varepsilon,T})$ we first specify stopping times $\kappa_n^{x,\varepsilon}$ in the way suggested by the first part of the proof. To simplify the notation we first define stopping times τ, σ, κ on the canonical probability space $C([0, +\infty); E)$ setting for $f \in C([0, +\infty); E)$:

$$\tau(f) = \inf\{t \ge 0 : \ f(t) \in \partial D\},$$

$$\sigma(f) = \inf\{t \ge T : \ \|f(t)\| = r \text{ and } \|f(t_1)\| = r_0 \text{ for some } t_1 \in [\tau, T]\},$$

$$\kappa(f) = \min(\tau(f), \sigma(f)).$$

Obviously $\tau^{x,\varepsilon} = \tau(X^{x,\varepsilon})$. Moreover

$$\sigma_1^{x,\varepsilon,T} = \sigma(X^{x,\varepsilon}), \quad \kappa_1^{x,\varepsilon,T} = \kappa(X^{x,\varepsilon}),$$

are well defined stopping times. In an inductive way

$$\sigma_{n+1}^{x,\varepsilon,T} = \sigma_n^{x,\varepsilon,T} + \sigma\left(X^{x,\varepsilon}(\cdot + \kappa_n^{x,\varepsilon,T})\right),$$

$$\kappa_{n+1}^{x,\varepsilon,T} = \kappa_n^{x,\varepsilon,T} + \sigma\left(X^{x,\varepsilon}(\cdot + \kappa_n^{x,\varepsilon,T})\right),$$

$n \in \mathbb{N}$. Note that if

$$\Gamma^{x,\varepsilon,T} = \{f \in C([0, +\infty); E); \tau(f) \le \kappa(f), f(\tau(f)) \in \mathscr{E}_{\bar{r}}\},$$

$$\Delta^{x,\varepsilon,T} = \{f \in C([0, +\infty); E); \tau(f) \le \kappa(f), f(\tau(f)) \in (\mathscr{E}_{\bar{r}})^c\},$$

then

$$A^{x,\varepsilon,T} = \{X^{x,\varepsilon}(\cdot) \in \Gamma^{x,\varepsilon,T}\},$$

$$B^{x,\varepsilon,T} = \{X^{x,\varepsilon}(\cdot) \in \Delta^{x,\varepsilon,T}\}.$$

In addition for $n \in \mathbb{N}$, $x \in D$, $\|x\| = r$

$$\begin{aligned} A_n^{x,\varepsilon,T} &= \{\tau^{x,\varepsilon} \le \kappa_n^{x,\varepsilon}, \pi^{x,\varepsilon} \in \mathscr{E}_{\overline{r}}, \kappa_{n-1}^{x,\varepsilon} < \tau^{x,\varepsilon}\} \\ &= \{X^{x,\varepsilon}(\cdot + \kappa_{n-1}^{x,\varepsilon}) \in \Gamma^{x,\varepsilon,T} \text{ and } \kappa_{n-1}^{x,\varepsilon} < \tau^{x,\varepsilon}\}, \\ B_n^{x,\varepsilon,T} &= \{\tau^{x,\varepsilon} \le \kappa_n^{x,\varepsilon}, \pi^{x,\varepsilon} \in (\mathscr{E}_{\overline{r}})^c, \kappa_{n-1}^{x,\varepsilon} < \tau^{x,\varepsilon}\} \\ &= \{X^{x,\varepsilon}(\cdot + \kappa_{n-1}^{x,\varepsilon}) \in \Delta^{x,\varepsilon,T} \text{ and } \kappa_{n-1}^{x,\varepsilon} < \tau^{x,\varepsilon}\}. \end{aligned}$$

Define for $x \in D$, $\|x\| = r$:

$$q_{\varepsilon,T}(x) = \mathbb{P}(A^{x,\varepsilon,T}), \ r_{\varepsilon,T}(x) = \mathbb{P}(B^{x,\varepsilon,T}),$$

then, by the strong Markov property of the solutions of (12.1),

$$\begin{aligned} \mathbb{P}(A_n^{x,\varepsilon,T}) &= \mathbb{E}(q_{\varepsilon,T}(X^{x,\varepsilon}(\kappa_{n-1}^{x,\varepsilon})); \mathbb{1}_{\kappa_{n-1}^{x,\varepsilon} < \tau^{x,\varepsilon}}) \\ &\ge \inf\{q_{\varepsilon,T}(y); \|y\| = r\}\mathbb{P}(\kappa_{n-1}^{x,\varepsilon} < \tau^{x,\varepsilon}) \end{aligned}$$

$$\begin{aligned} \mathbb{P}(B_n^{x,\varepsilon,T}) &= \mathbb{E}(r_{\varepsilon,T}(X^{x,\varepsilon}(\kappa_{n-1}^{x,\varepsilon})); \kappa_{n-1}^{x,\varepsilon} < \tau^{x,\varepsilon}) \\ &\le \sup\{r_{\varepsilon,T}(y); \|y\| = r\}\mathbb{P}(\kappa_{n-1}^{x,\varepsilon} < \tau^{x,\varepsilon}). \end{aligned}$$

Consequently

$$\begin{aligned} \frac{\mathbb{P}(B^{x,\varepsilon})}{\mathbb{P}(A^{x,\varepsilon})} &\le \lim_{N\to\infty} \frac{\sum_{k=1}^N \mathbb{P}(A_n^{x,\varepsilon,T})}{\sum_{k=1}^N \mathbb{P}(B_n^{x,\varepsilon,T})} \\ &\le \lim_{N\to\infty} \frac{\sum_{k=1}^N \sup_{\|y\|=r} r_{\varepsilon,T}(y)\mathbb{P}(\kappa_{n-1}^{x,\varepsilon} < \tau^{x,\varepsilon})}{\sum_{k=1}^N \inf_{\|y\|=r} q_{\varepsilon,T}(y)\mathbb{P}(\kappa_{n-1}^{x,\varepsilon} < \tau^{x,\varepsilon})} \\ &\le \frac{\sup_{\|y\|=r} r_{\varepsilon,T}(y)}{\inf_{\|y\|=r} q_{\varepsilon,T}(y)}, \end{aligned}$$

and, by the first part of the proof (12.39) holds. □

12.5.3 Explicit formulae for gradient systems

Problems of calculating exit rates $\underline{e}$ and $\overline{e}$ and determining the exit set can often be reduced to an optimization problem. For any $x, y \in E$, let

$$I(x, y) = \inf\left\{\frac{1}{2}\int_0^T |u(s)|^2 ds : \ f^{x,u}(T) = y, T > 0\right\}. \tag{12.40}$$

Then

$$\overline{e} = \inf\left\{I(0, y) : \ y \in (\overline{D})^c\right\},$$

and, under fairly general conditions

$$\overline{e} = \inf\{I(0, y) : \; y \in \partial D\}. \tag{12.41}$$

Moreover

$$\mathscr{E}_0 = \{y \in E : \; I(0, y) = \overline{e}\}. \tag{12.42}$$

The problems of finding minimal values in (12.40) and (12.41) are called the *first* and *second minimum energy problem.* The functional $I(\cdot, \cdot)$ is called *quasi-potential*. If T is fixed in the right hand side of (12.40) then the infimum will be denoted by $I_T(\cdot, \cdot)$.

In this subsection we show that explicit formulae for $I(0, \cdot)$ are available for two important classes of stochastic systems

$$dX = [AX - U'(X)]dt + dW(t) \tag{12.43}$$

and

$$\begin{cases} dX = Y dt \\ dY = [AX - U'(X) - \kappa Y]dt + dW(t). \end{cases} \tag{12.44}$$

Systems (12.43) and (12.44) are called *gradient systems* respectively of the first and second order.

We make the following assumptions.

Hypothesis 12.2

(i) *A is a negative definite operator on H.*
(ii) *$U \in C^1(V; [0, +\infty))$, where $V = D((-A)^{1/2})$, $U(0) = 0$, $DU(0) = 0$.*
(iii) *There exists a mapping $U' : V \to H$, Lipschitz on bounded sets and such that $DU(x; h) = \langle U'(x), h\rangle$ for all $x, h \in V$, where $DU(x; h)$ denotes the derivative of u at x in the direction h.*
(iv) *κ is a positive constant.*

To discuss the wave type equation (12.44) we proceed as in Example A.5.4 and introduce the Hilbert space $\mathscr{H} = D((-A)^{1/2}) \oplus X_2$ endowed with the inner product

$$\left\langle \begin{pmatrix} y \\ z \end{pmatrix}, \begin{pmatrix} y_1 \\ z_1 \end{pmatrix} \right\rangle = \left\langle \sqrt{\Lambda} y, \sqrt{-A} y_1 \right\rangle + \langle z, z_1 \rangle.$$

We define in $\mathscr{H}$ the linear operator:

$$\begin{cases} D(\mathscr{A}) = D(-A) \oplus D((-A)^{1/2}) \\ \mathscr{A} \begin{pmatrix} y \\ z \end{pmatrix} = \begin{pmatrix} 0 & 1 \\ -\Lambda & 0 \end{pmatrix} \begin{pmatrix} y \\ z \end{pmatrix}, \quad \forall \begin{pmatrix} y \\ z \end{pmatrix} \in D(\mathscr{A}). \end{cases}$$

The associated control systems are the following

$$\dot{f} = Af - U'(f) + u, \quad f(0) = a \in H, \tag{12.45}$$

$$\ddot{f} = Af - U'(f) - \kappa \dot{f} + u, \quad f(0) = a \in V, \quad \dot{f}(0) = b \in H. \tag{12.46}$$

The quasi-potential for system (12.44) will be denoted as

$$I\left(\begin{pmatrix} a \\ b \end{pmatrix}, \begin{pmatrix} a_1 \\ b_1 \end{pmatrix}\right), \begin{pmatrix} a \\ b \end{pmatrix}, \begin{pmatrix} a_1 \\ b_1 \end{pmatrix} \in \mathscr{H}.$$

Finally we denote by $z(a)$, $z(a, b)$ the solutions of the deterministic systems

$$\begin{cases} \dot{z} = Az - U'(z), \\ z(0) = a \in H, \end{cases} \tag{12.47}$$

and

$$\begin{cases} \ddot{z} = Az - U'(z) - \kappa \dot{z}, \\ z(0) = a \in V, \quad \dot{z}(0) = b \in H, \end{cases} \tag{12.48}$$

respectively. The following result is taken from [207]. It extends an earlier result from [313].

Theorem 12.25 *Assume that* (12.48) *is fulfilled.*

(i) If $a \notin D((-A)^{1/2})$ *then* $I(0, a) = +\infty$.
(ii) If $a \in D((-A)^{1/2})$ *and* $(-A)^{1/2} z^a(t) \to 0$ *as* $t \to \infty$ *in* H, *then*

$$I(0, a) = |(-A)^{1/2} a|^2 + 2U(a). \tag{12.49}$$

(iii) If $\begin{pmatrix} a \\ b \end{pmatrix} \in \mathscr{H}$ *and* $z(a, b) \to 0$ *as* $t \to \infty$ *in* $\mathscr{H}$, *then*

$$I\left(\begin{pmatrix} 0 \\ 0 \end{pmatrix}, \begin{pmatrix} a \\ b \end{pmatrix}\right) = \kappa[|(-A)^{1/2} a|^2 + 2U(a) + |b|^2]. \tag{12.50}$$

Proof The proof is based on the following identities. For the system (12.45) with $a \in V$

$$\begin{aligned} \frac{1}{2} \int_0^T |u(s)|^2 ds = \frac{1}{2} \int_0^T & |u(s) + 2Af(s) - 2U'(f(s))|^2 ds \\ & + [|(-A)^{1/2} f(T)|^2 + 2U(f(T)) \\ & - |(-A)^{1/2} f(0)|^2 - 2U(f(0))]. \end{aligned} \tag{12.51}$$

For the system (12.46) with $a \in V, b \in H$

$$\begin{aligned} \frac{1}{2} \int_0^T |u(s)|^2 ds = \frac{1}{2} \int_0^T & |u(s) - 2\kappa \dot{f}(s)|^2 ds \\ & + \kappa[|(-A)^{1/2} f(T)|^2 + 2U(f(T)) - |\dot{f}(t)|^2 \\ & - (-A)^{1/2} f(0)|^2 - 2U(f(0)) - |\dot{f}(0)|^2]. \end{aligned} \tag{12.52}$$

To show that (12.51) holds let us use the fact that the mild solution of (12.45) is in fact a strong solution. Elementary calculations give

$$\frac{1}{2}\int_0^T |u(s)|^2 ds = \frac{1}{2}\int_0^T |u(s) + 2Af(s) - 2U'(f(s))|^2 ds$$
$$-2\int_0^T \langle \dot{f}(s), Af(s) - U'(f(s)) \rangle \, ds. \tag{12.53}$$

It remains to show that

$$\int_0^T \langle \dot{f}(s), U'(f(s)) \rangle \, ds = U(f(T)) - U(f(0)) \tag{12.54}$$

and

$$-2\int_0^T \langle \dot{f}(s), Af(s) \rangle \, ds = |(-A)^{1/2} f(T)|^2 - (-A)^{1/2} f(0)|^2. \tag{12.55}$$

In fact the identities (12.54) and (12.55) are true for arbitrary functions f from $W^{1,2}(0, T; H) \cap L^2(0, T; D(A))$. To see this consider a sequence $\{f_n\} \subset C^1([0, T]; D(A))$ converging to f both in $W^{1,2}(0, T; H)$ and in $L^2(0, T; D(A))$ topologies. Such a sequence exists since $D(A)$ is dense in H. For each n and all $t \in [0, T]$

$$\frac{d}{dt} U(f_n(t)) + DU(f_n(t); \dot{y}_n(t)) = \langle U'(f_n(t)), \dot{y}_n(t) \rangle$$

and

$$\frac{d}{dt}\left|(-A)^{1/2} f_n(t)\right|^2 = \frac{d}{dt}\langle Af_n(t), f_n(t)\rangle = 2\langle Af_n(t), \dot{y}_n(t)\rangle.$$

So the identities (12.54) and (12.55) hold for each y_n, $n \in \mathbb{N}$, and eventually for general f, letting n tend to ∞. To prove (12.52), note that if the control and the initial conditions are smooth, we have

$$\frac{1}{2}\int_0^T |u(s)|^2 ds = \frac{1}{2}\int_0^T |u(s) - 2\kappa \dot{f}(s) + 2\kappa \dot{f}(s)|^2 ds$$
$$= \frac{1}{2}\int_0^T |u(s) - 2\kappa \dot{f}(s)|^2 ds$$
$$+2\int_0^T \langle \dot{y}(s) - Ay(s) + U'(y(s)), \dot{y} \rangle \, ds$$

and consequently (12.52) holds in this case. The general case is obtained by approximation.

Note that if $\dot{z} = Az(t) - U'(z(t))$, $t \in [0, T]$, then for $f(t) = z(T - t)$ we have

$$\dot{f}(t) = Af(t) - U'(f(t)) = 0, \quad f(0) = z(T), \quad f(T) = z(0), \quad t \in (0, T).$$

Moreover the function f is a solution of (12.45) when $u = -2Ay + 2U'(y)$ and for this control the first term on the right hand side of (12.51) vanishes. Hence

$u \in L^2(0, T; H)$ and

$$\begin{aligned} I_T(0, a) &\geq |(-A)^{1/2}a|^2 + 2U(a) \\ I_T(z^a(T), a) &\geq |(-A)^{1/2}a|^2 + 2U(a) - |(-A)^{1/2}z^a(T)|^2 + 2U(z^a(T)). \end{aligned} \tag{12.56}$$

But

$$I_T(0, a) = I_T(0, z^a(T)) + I_T(z^a(T), a).$$

Since $|(-A)^{1/2}z^a(T)| \to 0$ as $T \to \infty$, it follows from the results in Section B.3.1 that $I(0, z^a(T)) \to 0$ as $T \to \infty$. Thus

$$\lim_{T\to\infty} I_T(z^a(T), a) = |(-A)^{1/2}a|^2 + 2U(a),$$

$$\inf_{T>0} I_T(0, a) \leq |(-A)^{1/2}a|^2 + 2U(a),$$

and so formula (12.49) holds. Formula (12.50) can be proved in a similar way. □

Remark 12.26 With the explicit formulae given by Theorem 12.25 at hand one can attempt to find $\overline{e}$ and $\mathscr{E}_0$ solving some minimization energy problems. Some results in this direction are given in [207]. □

13

Survey of specific equations

In this chapter we present a list of stochastic partial differential equations which have been intensively studied. We restrict the discussion to equations with Gaussian noise. We do not claim that our list is complete, but we hope it will be helpful. Moreover, additional comments on specific equations and problems can also be found in the next chapter on recent developments. We start from an infinite sequence of ordinary equations and delay equations which form a natural link between ordinary and partial differential equations. Then we treat equations with growing orders of time and space derivatives involved. We end up with nonlinear Schrödinger equations, which posses many properties of equations with second time derivatives.

13.1 Countable systems of stochastic differential equations

An important class of stochastic equations in infinite dimensions, which is close to ordinary differential equations, consists of systems of one dimensional equations like

$$dX_\gamma(t) = F_\gamma(X(t))dt + \sum_{k\in\Gamma} G_{\gamma k}(X(t))dW_k, \quad X_\gamma(0) = x_\gamma, \ \gamma \in \Gamma, \ t > 0,$$

where Γ is a countable set, W_γ, $\gamma \in \Gamma$, is a family of standard Wiener processes, usually independent. Moreover F_γ, $G_{\gamma k}$, are real valued functions defined on the Cartesian product of copies $\mathbb{R}_\gamma$ of the real line $\mathbb{R}$, or on its subsets, to which $x = (x_\gamma)$ belongs. To such equations one can, sometimes, reduce stochastic PDEs where $X_\gamma(t)$, $\gamma \in \Gamma$, are the coefficients of the expansions of the solution $X(t)$ with respect to a Schauder basis $(e_\gamma,\ \gamma \in \Gamma)$, see for example [736]. A large number of papers is devoted to countable systems modeling the so called spin systems on d-dimensional lattices $\mathbb{Z}^d$, interpreted as rigid body consisting of atoms. A configuration x could be basically any real function defined on $\mathbb{Z}^d$ with x_γ denoting the state of the atom γ. The main questions studied are those of existence and uniqueness of solutions and existence of invariant measures, which are often interpreted as Gibbs measures. Early contributions are those of Doss and Royer see [265, 631]; see also Fritz [319] and Leha and Ritter [489]. The theory of stochastic dissipative systems was employed in Da Prato and Zabczyk [219]. For more recent contributions see the paper by Albeverio, Kondratiev, Röckner and Tsikalenko [11] and the book by Albeverio, Kondratiev, Kozitsky and Röckner [12]. Ergodic problems with discontinuous background noise were discussed recently by Xu and Zegarlinski [713] and Priola and Zabczyk [608].

13.2 Delay equations

Delay equations were introduced in the Introduction and can be treated as stochastic equations in infinite dimensional spaces. In this way, and using abstract theorems for stochastic evolution equations, one can obtain new results on existence and asymptotics of solutions, see for example Da Prato and Zabczyk [220]. For a modern introduction to this approach see the monograph by Engel and Nagel [282]. Various state spaces are possible. The case of continuous functions is discussed in a recent paper by van Neerven and Riedle [687]. For related results see Riedle [614, 615] and the paper by Bierkens, van Gaans and Lunel [71].

Existence and uniqueness of invariant measures, using coupling methods are discussed in the paper by Hairer, Mattingly and Scheutzow [391].

13.3 First order equations

As was noticed in the Introduction, the first order equations of the following type:

$$dr(t)(\xi) = \left(\frac{\partial}{\partial \xi} r(t)(\xi) + F(r(t))(\xi)\right) + G(r(t))(\xi) dW(t)$$

appear in the theory of bond markets. Here W is a real Wiener process, transformations F and G act on functions of the $\xi > 0$ variable and F is of the form:

$$F(r)(\xi) = \frac{1}{2} \frac{\partial}{\partial \xi} \left| \int_0^{\xi} G(r)(\eta) d\eta \right|^2 .$$

For some natural G the drift term is of quadratic form and the equation may not have solutions, see Morton [552]. Financial questions lead to the invariance problem, see Björk, Christensen and Bent [75], investigated also by Filipović [293] and Zabczyk [735]. For more recent studies of the equation we refer to Peszat and Zabczyk [599], Rusinek [635], Filipović, Tappe and Teichmann [294], Marinelli [525] and Barski and Zabczyk [49] . These authors study the equation with the noise process of Lévy type.

13.4 Reaction-diffusion equations

This is a very large class of equations describing physical, chemical as well as ecological situations. Usually the solution is a vector valued process $u(t,\xi) = (u_1(t,\xi),\ldots,u_m(t,\xi))$ with the space variable ξ in an open subset G of $\mathbb{R}^d$, with the components describing, for instance, the densities of chemical reactants. The equations can be written compactly as

$$\frac{\partial u}{\partial t}(t,\xi) = \mathscr{A}u(t,\xi) + f(u(t,\xi)) + g(u(t,\xi))\frac{\partial W}{\partial t}(t,\xi), \tag{13.1}$$

where $\mathscr{A}$ is a matrix of second order partial differential operators and f and g are nonlinear mappings, often polynomials, from $\mathbb{R}^m$ into $\mathbb{R}^m$. Their forms depend on the chemical reactions. Stochastic perturbations take into account random influences of the environment.

If $m = 1$, $\mathscr{A} = \Delta$, we arrive at a stochastic heat equation. There is a huge literature concerning the equation, we quote early papers by Faris and Jona-Lasinio [288], Gyöngy and Pardoux [383] and Gyöngy [373]. The problem was also studied using the theory of dissipative operators both in Hilbert and in Banach spaces by Da Prato and Zabczyk [216, 220], see also the paper by Peszat [593] for the Banach space case.

For the case $m > 1$ see Cerrai's book [149] and references therein and also the recent result by Cerrai [150]. For some results about exponential ergodicity for stochastic reaction-diffusion equations see Goldys and Maslovski [356].

Similar equations appear in the theory of infinite particle systems. One starts here from a system of particles on a lattice. The particles change locations according to some random mechanism and the approximate densities of the particle distributions, after proper normalization and in proper time scale, converge in distribution to continuous space-time densities. These densities satisfy a partial differential equation driven by a space-time white noise. In the paper by Mueller and Tribe [553], in this way the authors derived the following stochastic reaction-diffusion equation

$$\frac{\partial u}{\partial t}(t,\xi) = \frac{1}{6}\Delta u(t,\xi) + \theta_c u - u^2 + |2u|^{\frac{1}{2}}\frac{\partial W}{\partial t}(t,\xi). \tag{13.2}$$

13.4.1 Spatially homogeneous noise

A large amount of papers has been devoted to heat equations perturbed by spatially homogeneous noise. Formally they are written as follows

$$\frac{\partial X}{\partial t}(t,\xi) = \Delta X(t,\xi) + f(X(t,\xi)) + g(X(t,\xi))\frac{\partial W_\Gamma}{\partial t}(t,\xi), \tag{13.3}$$

plus appropriate initial conditions. Here W_Γ is a Wiener process with the covariance kernel Γ and spectral measure μ, see Section 4.1.4. The subject was initiated by Dawson and Salehi [228]. For more recent contributions we refer to Peszat and Zabczyk [597, 598], to Tessitore and Zabczyk [673], to Sanz-Solé [637] and to Karczewska and Zabczyk [440]. A study in Banach spaces was performed by Brzezniak and Peszat [120] and Brzeźniak and van Neerven [125].

The spatial variable ξ may vary not only over $\mathbb{R}^d$ but, more generally, over a Lie group, in particular over a torus. The Lie group setting is developed in the papers by Tindel and Viens [676] and Peszat and Tindel [595].

13.4.2 Skorohod equations in infinite dimensions

Let us consider a stochastic differential inclusion in a Hilbert space H,

$$\begin{cases} dX(t) + (AX(t) + N_K(X(t)))dt \ni dW(t), \\ X(0) = x. \end{cases} \tag{13.4}$$

Here $A\colon D(A) \subset H \to H$ is a self-adjoint operator, K is a convex closed set in H, $N_K(x)$ is the normal cone to K at x (1) and W is a cylindrical Wiener process in H.

When H is finite dimensional, a solution to (13.4) is a pair of continuous adapted processes (X, η) such that X is K-valued, η is of bounded variation with $d\eta$ concentrated on the set of times where $X(t) \in \partial K$ (the boundary of K) and

$$X(t) + \int_0^t AX(s)ds + \eta(t) = x + W(t), \quad t \geq 0, \ \mathbb{P}\text{-a.s.},$$

$$\int_0^T (d\eta(t), X(t) - z(t)) \geq 0, \quad \mathbb{P}\text{-a.s.},$$

for all $z \in C([0, T]; K)$. Equation (13.4) is a generalization of the equation, introduced by Skorohod [649] in finite dimensions, see also Tanaka [669], to construct solutions reflected at the boundary. The existence and uniqueness of a solution (X, η) of (13.4) (also with multiplicative noise) was first proven by Cépa [148].

One can construct a transition semigroup corresponding to X in $C(K)$, whose infinitesimal generator is the Kolmogorov operator

$$L\varphi = \frac{1}{2}\,\Delta\varphi + \langle Ax, D\varphi\rangle$$

equipped with a Neumann condition at the boundary ∂K of K.

With H infinite dimensional, (13.4) was studied by Nualart and Pardoux [562] when $H = L^2(0, 1)$, A is the Laplace operator with Dirichlet or Neumann boundary conditions and K is the convex set of all nonnegative functions of $L^2(0, 1)$, see also Haussman and Pardoux [399] and Donati-Martin and Pardoux [263].

The reflection problem arises in modeling the fluctuations of random interfaces, see Funaki and Olla [334]. A detailed study of the measure of the contact set was done by Zambotti [738, 739] and Dalang, Mueller and Zambotti [181]. For a discussion of asymptotic behavior see the end of Chapter 12.

(1) $N_K(x) = \{y \in H : \langle y, z - x\rangle \leq 0, \ \forall\, z \in K\}$ if $x \in K$ and $N_K(x) = \varnothing$ if $x \notin K$.

The Kolmogorov equation corresponding to problem (13.4) is naturally equipped with Neumann type boundary conditions. In the infinite dimensional case it was studied by Barbu, Da Prato and Tubaro see [45, 46].

13.5 Equations for manifold valued processes

Some physical problems lead to equations whose solutions take values in a manifold. For an early work on diffusions on manifolds see Funaki [332]. A typical example is provided by the Landau–Lifshitz–Gilbert equation, see recent work by Brzeźniak, Goldys and Jegaraj [109, 110]. The case of the stochastic wave equation is studied in detail by Brzeźniak and Ondreját [118] and Brzeźniak, Goldys and Ondreját [112]. General theory is developed in Albeverio, Brzezniak and Dalteski [8] and Brzeźniak, Goldys and Ondreját [111].

13.6 Equations with random boundary conditions

In several applications external forces enter the system through the boundary of a region where the system evolves. The same is true if the forces have a stochastic character. Consider for instance the nonlinear heat equation in the region $G = (0, \pi)^d \subset \mathbb{R}^d$, with the noise affecting only a portion Γ_0

$$\Gamma_0 = \{\xi = (\xi_1, \dots, \xi_d) \in \partial G : \xi_d = 0\},$$

of the boundary Γ of G. Thus the state equation is of the form

$$\begin{cases} \dfrac{\partial X}{\partial t}(t,\xi) = (\Delta - m)X(t,\xi) + f(X(t,\xi)), \quad t > 0, \\ X(0,\xi) = x(\xi),\ \xi \in G, \quad \dfrac{\partial X}{\partial \nu}(t,\xi) = 0,\ \xi \in \Gamma\backslash\Gamma_0,\ t > 0, \\ \dfrac{\partial X}{\partial \nu}(t,\xi) = \dfrac{\partial V}{\partial t}(t,\xi), \quad \xi \in \Gamma_0,\ t > 0, \end{cases} \tag{13.5}$$

where ν is the inner normal vector to the boundary at those points $\xi \in \Gamma$ where it is well defined. The process $V(t),\ t \geq 0$, is a Wiener process on $U = L^2(\Gamma_0)$ with the covariance operator Q not necessarily of trace class. The operator Q could be of integral type,

$$Qu(\xi) = \int_{\Gamma_0} g(\xi,\eta)u(\eta)d\eta, \quad u \in U,\ \xi \in \Gamma_0, \tag{13.6}$$

with the kernel g being the correlation function of the random field $V(1,\xi),\ \xi \in \Gamma_0$. Of special interest is the case when the field $V(1,\xi),\ \xi \in \Gamma_0$, is the restriction of a stationary random field on $\mathbb{R}^{d-1}$. Then there exists an integrable function

$q_0 : \mathbb{R}^{d-1} \to \mathbb{R}$ such that

$$q(\xi, \eta) = q_0(\xi - \eta), \quad \xi, \eta \in \Gamma_0. \tag{13.7}$$

Moreover q_0 is the Fourier transform of the spectral density of the random field.

The semi-group treatment of such problems was initiated by Da Prato and Zabczyk [218] for the one dimensional problem. The multidimensional case was treated by Sowers [654]. Some related results were recently obtained by Alòs and Bonaccorsi [17, 18] and by Brzeźniak, Goldys, Peszat and Russo [113], see also a discussion by Peszat and Zabczyk [599]. The case of hyperbolic equations is treated by Brzeźniak and Peszat in [123].

13.7 Equation of stochastic quantization

We continue the discussion of these equations started in the Introduction. Let us consider the reaction diffusion equation on $L^2([0, 2\pi]^d)$, $d \in \mathbb{N}$,

$$dX = (AX - X^k)dt + dW(t), \quad X(0) = x, \tag{13.8}$$

where $A = \frac{1}{2}(\Delta_\xi - X)$, Δ_ξ is the Laplacian on $[0, 2\pi]^d$ with periodic boundary conditions, k is an odd integer greater than 1 and W is a cylindrical Wiener process on $L^2([0, 2\pi]^d)$. It is easy to check that the corresponding stochastic convolution evolves in $L^2([0, 2\pi]^d)$ only for $d = 1$. However, for $d = 2$ it takes values in every negative Sobolev space $H^{-\varepsilon}((0, 2\pi)^2)$ with $\varepsilon > 0$; but we cannot solve the equation in these spaces because the power x^k is not defined in a distributional space.

For this reason the function x^k is replaced by the Wick power $:x^k:$, described in the Introduction and the original problem becomes

$$dX = (AX - :X^k:)dt + dW(t), \quad X(0) = x. \tag{13.9}$$

This procedure is called *renormalization* and is physically justified in quantum field theory. It has a long history, also in connection with the constructive field theory in the Euclidean framework, see Glimm and Jaffe [351], the monograph of Simon [646] and references therein. See also Parisi and Wu [585, Chapter V] where this problem was set in order to construct a dynamical system with an invariant measure ν of the form:

$$\nu(dx) = ce^{\frac{1}{k+1}\int_{[0,2\pi]^2} :x^{k+1}: d\xi}\mu(dx). \tag{13.10}$$

Here c is a normalizing constant and μ the invariant measure of the free system

$$dX = AXdt + dW(t), \quad X(0) = x.$$

The measure ν is well defined thanks to the important Nelson estimate, see Simon [646, Chapter V2]. This problem was first considered by Jona Lasinio and Mitter

[432], who were applying Girsanov transform, and then also by Borkar, Chari and Mitter [92] and Gątarek and Goldys [338].

Using the Dirichlet form approach, this problem was studied by Albeverio and Röckner [14] and by Liskevich and Röckner [502], proving strong uniqueness for a class of infinite dimensional Dirichlet operators, see also Da Prato and Tubaro for similar results [212]. Moreover Mikulevicius and Rozovskii applied their theory of martingale solutions for stochastic PDEs [545].

In all these papers (with the exception of [14], [545] and [191] below) the *renormalized* equation was of the form:

$$\begin{cases} dX = ((-A)^{1-\varepsilon}X + (-A)^{-\varepsilon} : X^k :)dt + (-A)^{-\varepsilon/2}dW(t) \\ X(0) = x, \end{cases} \tag{13.11}$$

where ε is a positive number subject to some restrictions. Notice that the invariant measure corresponding to (13.11) is still ν. This modification allows smoothing of the nonlinear term and, in some cases, use of the Girsanov transform.

The problem (13.9), in its original form, was also considered by Mikulevicius and Rozovskii [545]. In particular they showed, by a compactness method, that (13.9) has a stationary weak solution and if $\varepsilon = 0$, then the law of the stationary solution is singular with respect to the law of the stationary solution of the linear equation.

In Da Prato and Debussche [191] the existence of a strong solution (in the probability sense) was proved for the original problem (13.9) for μ-almost every initial data x. The main argument consists in splitting the unknown process: $X = Y + Z$, where $Z(t)$ is the stochastic convolution in $(-\infty, t]$

$$Z(t) = \int_{-\infty}^{t} e^{(t-s)A} dW(s).$$

Then Z is a stationary solution to the linear version of (13.9) and Y is smoother than X and therefore one can define $: X^k := \sum_{l=0}^{k} C_k^l Y^l : Z^{k-l} :$. Consequently the problem (13.9) becomes

$$\begin{cases} \dfrac{dY}{dt} = AY + \displaystyle\sum_{k=0}^{n} a_k \sum_{l=0}^{k} C_k^l Y^l : Z^{k-l} :, \\ Y(0) = x - z(0). \end{cases} \tag{13.12}$$

Since the law of $Z(t)$ is equal to μ for any $t \in \mathbb{R}$, we can define $: Z^n :$ in the classical way by the formula

$$\mathbb{E}[g(: Z^n :)] = \int g(: x^n :)\mu(dx),$$

where g is any Borel bounded real function. The main advantage of considering (13.12) is that now the nonlinear term is a continuous function with respect to the unknown. Then (13.12) can be solved by a fixed point on a suitable Besov space.

13.8 Filtering equations

The classical nonlinear filtering equation was deduced by Fujisaki, Kallianpur and Kunita (FKK equation) see [330]. In the equation the diffusion operator is both nonlinear and nonlocal and causes some problems for the direct SPDE methods. For an early contribution, see Levieux [494].

The FKK equation, in the simplest situation, when the state and the observation equations are one dimensional, is of the form:

$$dp_t(x) = L^* p_t(x)dt + p_t(x)(G(x) - \int_{-\infty}^{+\infty} G(y)p_t(y)dy)dV, \quad t > 0,\ x \in \mathbb{R}^1. \tag{13.13}$$

Here L^* denotes the adjoint of the characteristic operator of a one dimensional Markov process X observed through the observation equation

$$dy(t) = G(X(t))dt + dW(t), \tag{13.14}$$

where W is the Wiener process, independent of X, describing the noise affecting the process of observation. Moreover V is, under an equivalent probability measure, a Wiener process. As was noticed in the Introduction, Zakaï [737] derived a linear equation for unnormalized conditional density. It is of the form

$$dq_t(x) = L^* q_t(x) + q_t(x)G(x)dy(t), \quad t > 0,\ x \in \mathbb{R}^1. \tag{13.15}$$

Due to the equivalence of these equations, established under rather mild conditions, the FKK equation was not intensively studied.

The case when the unobserved process is infinite dimensional was investigated by Ahmed, Fuhrman and Zabczyk [6].

Recent emphasis in filtering theory is on numerical solutions of the filtering equations, see for example Hu, Kallianpur and Xiong [408], Gobet, Pagés, Pham and Printem [352] and Crisan and Lyons [174].

13.9 Burgers equations

An important stochastic variant of a model of turbulence was introduced in 1939 by Burgers. Let $U = U(t)$ be the primary velocity of a fluid, parallel to the walls of the channel, let $v = v(t, \xi)$, $t \geq 0$, $\xi \in (0, 1)$, be the secondary velocity, of turbulent motion, and let P be the exterior force. Denote the density of the fluid by ρ and its viscosity by μ and write $\nu := \frac{\mu}{\rho}$. The stochastic version of the Burgers system looks as follows:

$$\frac{dU(t)}{dt} = P - \nu U(t) - \int_0^1 v^2(t, \xi)\, d\xi, \quad \text{for } t > 0 \tag{13.16}$$

and, for $\{(t,\xi)\colon t>0,\ \xi\in(0,1)\}$,

$$\begin{cases} \frac{\partial v}{\partial t}(t,\xi) = \nu\, \frac{\partial^2 v}{\partial \xi^2}(t,\xi) + U(t)v(t,\xi) \\ \qquad\qquad - \frac{\partial}{\partial \xi}\big(v^2(t,\xi)\big) + g\big(v(t,\xi)\big)\frac{\partial^2 W(t,\xi)}{\partial t \partial \xi}. \end{cases} \tag{13.17}$$

The Burgers system is considered with appropriate initial and Dirichlet boundary conditions. In (13.17), $W(t,\xi)$ denotes a Brownian sheet. Note that the system does not satisfy the conditions of the general existence results, due to the nonlinear term which depends on the first derivative. Some existence results for the system (13.16)–(13.17) can be found in the papers by Twardowska and Zabczyk [682, 683].
The most studied problem is equation (13.17) with U identically equal to zero. Here one should mention papers by Bertini, Cancrini and Jona-Lasinio [64], by Da Prato, Debussche and Temam [194] and by Da Prato and Gątarek [199]. The equation in the real line was studied by Gyöngy and Nualart [382] and Kim [448]. The paper by E, Khanin, Mazel, and Sinai [275] is a comprehensive study of invariant measures for the stochastic Burgers equation. Goldys and Maslowski [354] studied exponential ergodicity. For the study of large deviations see Cardon-Weber [142].

The existence and uniqueness of a stationary distribution for the random forced inviscid Burgers equation on the torus is proved in dimension one by E, Khanin, Mazel and Sinai [275] and in arbitrary dimensions by Iturriaga and Khanin [426]. Moreover, Tessitore and Zabczyk [673] proved existence of an invariant measure for a generalization of the Burgers equation in the real line.

The stochastic fractional Burgers equation is discussed by Brzeźniak and Debbi [105].

A new approach for studying infinite dimensional Kolmogorov equations corresponding to generalized Burgers equations was presented by Röckner and Sobol [617, 618].

13.10 Kardar, Parisi and Zhang equation

This equation was proposed by Kardar, Parisi and Zhang [442] to model the behavior of the surface, called the interface, separating two phases of a substance in $\mathbb{R}^d$. The fluctuations are described by the interface height X satisfying

$$\frac{\partial X}{\partial t}(t,\xi) = \Delta X(t,\xi) + \sum_{i=1}^{d} \left|\frac{\partial X}{\partial \xi_i}(t,\xi)\right|^2 + \frac{\partial W}{\partial t}(t,\xi).$$

For a physically motivated study in dimension $d=1$ we refer to Bertini and Giacomin [65]. A modified Kardar, Parisi and Zhang model, see [442], is studied in Da Prato, Debussche and Tubaro [197].

A comprehensive analysis of the equation, with suitable initial and boundary conditions, was recently provided by Hairer [387].

13.11 Navier–Stokes equations and hydrodynamics

We are here concerned with the following stochastic Navier–Stokes equation,

$$\begin{cases} dX - \nu\Delta X dt + (X\cdot\nabla)X dt = \nabla p dt + \sqrt{Q} dW, & \text{in } D\times\mathbb{R}^+,\\ \operatorname{div} X = 0, & \text{in } D\times\mathbb{R}^+,\\ X(t,x) = 0, & \text{on } \mathbb{R}^+\times\partial D,\\ X(0,x) = x, & \text{in } D, \end{cases} \tag{13.18}$$

where D is an open bounded domain of $\mathbb{R}^d$, $d = 2, 3$, with smooth boundary ∂D and ν is the viscosity. $X(t,\xi)$ and $p(t,\xi)$ represent the velocity and the pressure of a fluid for $t \geq 0$ at the point ξ of D.

We consider Dirichlet boundary conditions but the majority of the results also hold for other boundary conditions, such as Neumann or periodic. We set

$$H = \{x \in (L^2(D))^2 : \ \operatorname{div} x = 0 \text{ in } D,\ x\cdot n = 0 \text{ on } \partial D\},$$

where n is the outward normal to ∂D. Moreover, W is a cylindrical Wiener process on H associated with a stochastic basis $(\Omega, \mathscr{F}, \mathbb{P}, \{\mathscr{F}_t\}_{t\geq 0})$. The operator $Q \in L(H)$ is nonnegative, symmetric and of trace class.

We denote by A the Stokes operator, see for example Temam [671]:

$$A = P\Delta, \quad D(A) = (H^2(D))^2 \cap (H_0^1(D))^2 \cap H,$$

where P is the orthogonal projection of $(L^2(D))^2$ on the space of divergence free vectors. We set

$$V = (H_0^1(D))^2 \cap H$$

and define a linear operator $B : V \to V$ setting

$$(B(y), z) = b(y, y, z), \quad y, z \in V,$$

where

$$b(y,\theta,z) = \sum_{i,j=1}^{d} \int_D y_i\ D_i\theta_j\ z_j\ d\xi, \quad y,\theta,z \in V.$$

Then we may rewrite problem (13.18) as

$$\begin{cases} dX(t) = (\nu AX(t) - B(X(t)))dt + \sqrt{Q}\, dW(t)\\ X(0) = x. \end{cases} \tag{13.19}$$

13.11.1 Existence and uniqueness for $d = 2$

The well posedness of the Navier–Stokes equation (13.19), when $d = 2$, has been investigated in many articles. We quote in particular Bensoussan and Temam [58], Albeverio and Cruzeiro [9], Brzezniak, Capinski and Flandoli [103], Capinski and

Gątarek [141], Flandoli [301], Flandoli and Gątarek [304], Kuksin and Shirikyan [474].

In Da Prato and Zabczyk [220, Chapter 13], a fixed point argument is used instead of the more common Galerkin–Faedo approximation scheme.

Kuksin, see [470, 471], studied the two dimensional Navier–Stokes equation with small viscosity proving, under suitable assumptions, that when the viscosity tends to 0, the distribution of the stationary solution of the Navier–Stokes equation converges to an invariant measure of the Euler equation. Brzeźniak and Li proved in [115] that two dimensional stochastic Navier–Stokes equations in unbounded domains are asymptotically compact. This result implies existence of invariant measures.

Existence and uniqueness of invariant measures was proved, under different assumptions, by Flandoli and Maslowski [306], Bricmont, Kupiainen and Lefevere [99], E, Mattingly and Sinai [276], Hairer and Mattingly [388, 389], Kuksin, Piatnitski and Shirikyan [472], Kuksin and Shirikyan [474]. In [306] the authors proved that the transition semigroup corresponding to the two dimensional Navier–Stokes equation is irreducible and strong Feller. Moreover, Hairer and Mattingly, [388, 389], proved the uniqueness of the invariant measure for the equation with a very degenerate noise using the asymptotic strong Feller property of the transition semigroup, see also Section 11.15. The coupling argument was used in [472] and [474].

For large deviations results see Chang [161], Sritharan and Sundar [656], Bessaih and Millet [69] and Chueshov and Millet [171]. For the Boussinesq equation see Duan and Millet [268].

13.11.2 Existence and uniqueness for $d = 3$

In the deterministic case (that is when $Q = 0$), there exists a global weak solution (in the PDE sense) in H but uniqueness of such a solution is an open problem. On the other hand, when considering smoother initial data, there exists a unique solution but it is not known whether it is globally defined. After the seminal paper by Leray [493], we limit ourselves to quoting, from the huge existing literature on this topic, Kato [444], Ladyzhenskaya [482], Temam [671], Constantin and Foias [173] and von Wahl [701].

The stochastic equation (13.19) has also been studied by many authors who proved the existence of a martingale solution in H but uniqueness of such a solution is again an open problem. See Bensoussan and Temam [58], Viot [695], Vishik and Fursikov [700], Brzezniak, Capinski and Flandoli [104], Capinski and Cutland [139], Capinski and Gątarek [141], Flandoli and Gątarek [304], Flandoli and Romito [307], Mikulevicius and Rozovskii [546], and Flandoli [302].

Three dimensional Navier–Stokes equations in $\mathbb{R}^3$ with additive, spatially homogeneous noise, were studied by Basson [51].

The problem of pathwise uniqueness of the martingale solution (13.19) seems to be as difficult as in the deterministic case. However, it is possible to construct a

Markovian selection by solving directly the corresponding Kolmogorov equation, see Da Prato and Debussche [190, 193], Debussche and Odasso [251], or by proving a multi-valued version of the Markov property for sets of solutions and then applying a selection principle (due to Krylov [465], see also Stroock and Varadhan [665, Chapter 12]) see Flandoli and Romito [308]. In this direction see also Goldys, Röckner and Zhang [357]. Finally, we quote an interesting result by Romito [629] proving that all invariant measures corresponding to different Markov selections are equivalent. If one could show that they coincide this would imply uniqueness of the weak solution of the three dimensional and Navier–Stokes equation.

13.11.3 Stochastic magneto-hydrodynamics equations

Stochastic magneto-hydrodynamics (SMHD) equations were invented to model interactions between a conducting fluid or plasma and a magnetic field, see Landau and Lifshitz [483], in the presence of random perturbations. It is a system of equations, for the velocity field of the fluid $X = (X_1, X_2)$ and the magnetic field $B = (B_1, B_2)$, of the form:

$$\begin{cases} dX = (\nu\Delta X - (X\cdot\nabla)X + S(B\cdot\nabla)B - \nabla(P + \frac{1}{2}\,S|B|^2)dt + \sqrt{Q_1}\,dW_1(t) \\ dB = (\nu_1\Delta B - (X\cdot\nabla)B + (B\cdot\nabla)X)dt + \sqrt{Q_2}\,dW_2(t) \\ \operatorname{div} X = 0, \quad \nabla\cdot B = 0, \ B\cdot n = 0 \quad \text{in } (0,+\infty)\times\mathscr{O} \\ X = 0, \quad \operatorname{curl} B = 0 \quad \text{on } (0,+\infty)\times\partial\mathscr{O} \\ X(0,\xi) = x_0(\xi), \quad B(0,\xi) = b_0(\xi) \quad \text{in } \mathscr{O}. \end{cases} \tag{13.20}$$

Here $\mathscr{O}\subset\mathbb{R}^2$ is a bounded, open and simply connected domain and W_1, W_2 are independent cylindrical Wiener processes, defined in filtered probability space $(\Omega, \mathscr{F}, \mathscr{F}_t, \mathbb{P})$ and taking values in a space of divergence free functions:

$$H = \{y \in (L^2(\mathscr{O}))^2 : \ \nabla\cdot y = 0 \ \text{ in } \mathscr{O}, \ \ y\cdot n = 0 \text{ on } \partial\mathscr{O}\}.$$

Moreover $\nabla\cdot y = D_i y_i$, $y = \{y_1, y_2\}$, $D_i = \frac{\partial}{\partial\xi}$, n is the outward normal to $\partial\mathscr{O}$, $B = (B_1, B_2)$ and curl $B = D_2B_1 - D_1B_2$. In addition $\frac{1}{\nu}$ is the Reynold number, ν_1 the magnetic resistivity and $S = \nu_1 M^2$ where M is the Hartman number. The fluid pressure is denoted by P while the expression

$$S((B\cdot\nabla)B) - \nabla\left(P + \frac{1}{2}\,S|B|^2\right)$$

represents the Lorentz force, see Landau and Lifshitz [483]. Boundary conditions on B express the physical requirement that the boundary is perfectly conductive. The operators Q_1 and Q_2 are trace class defined on H.

The deterministic version of system (13.20) was extensively studied in the literature, see for example Sermange and Temam [644] and references therein.

Existence and uniqueness of a solution (X, B) of (13.20) was proved by Barbu and Da Prato in the paper [39], see also a recent paper by Chueshov and Millet [171].

13.11.4 The tamed Navier–Stokes equation

The three dimensional stochastic tamed Navier–Stokes equation, studied by Röckner and Zhang [623, 624] is obtained by adding to equation (13.19) a term $g_N(|u|^2)$ where the function $g_N : [0, +\infty) \to [0, +\infty)$, vanishes on $[0, N]$ and grows linearly at ∞. So, this equation looks like

$$\begin{cases} du(t) = (\nu Au(t) - B(u(t)) + g_N(|u|^2))dt + \sqrt{Q}\, dW(t) \\ u(0) = x. \end{cases} \tag{13.21}$$

The presence of the additional term allows us to find an a priori estimate of $|u(t)|_{L^4}$ which yields existence and uniqueness of a global solution.

It is worth noticing that if X is a global solution of (13.19), then it coincides with u, with probability close to 1, for N sufficiently large.

13.11.5 Renormalization of the Navier–Stokes equation

The renormalization of the two dimensional Navier–Stokes equation with white noise perturbation, was considered by Albeverio and Cruzeiro [9], Albeverio and Ferrario [10] and by Da Prato and Debussche [189].

13.11.6 Euler equations

This equation looks like the Navier–Stokes equation but the second order part is missing. Intensive studies have been carried out for the two dimensional equation, see Capinski and Cutland [140] and Biryuk [72].

The existence of martingale solutions was studied in the Ph.D. Thesis by Bessaih [66] and the subsequent papers by Bessaih and Flandoli [68] and Bessaih [67]. In the paper by Brzezniak and Peszat [122] the solution was constructed as the limit of a sequence of solutions of the Navier–Stokes equations as the viscosity converges to zero. For more recent work see Stannat [657] and Cruzeiro, Flandoli and Malliavin [178]. See also the survey article by Albeverio and Ferrario [10].

13.12 Stochastic climate models

The primitive equations are a set of nonlinear differential equations that are used to approximate global atmospheric flow in several models. They consist of three main sets of equations.

(1) Conservation of momentum which gives rise to a Navier–Stokes equation describing the hydrodynamical flow on the surface of a sphere under the assumption that vertical motion is much smaller than horizontal motion.

(2) A thermal energy equation relating the overall temperature of the system to heat sources and sinks.
(3) A continuity equation, representing the conservation of mass.

The case of dimension two was studied first. In a paper by Ewald, Petcu and Temam [286] the primitive equations are studied with additive noise. The case of multiplicative noise was studied by Glatt-Holtz and Ziane [350]. The existence of pathwise, strong solutions of the primitive equations was proved by Glatt-Holtz and Temam [349].

In the case of dimension three, local existence theory, for a class of abstract stochastic evolution equations which include the primitive equation of the oceans, was studied by Debussche, Glatt-Holtz and Temam [247]. Global existence was obtained later by Debussche, Glatt-Holtz, Temam and Ziane [248]. The exponential behavior of the stochastic primitive equations with multiplicative noise was studied by Medjo [541].

13.13 Quasi-geostrophic equation

Consider the equation in two dimensions:

$$\begin{cases} d\theta = -k(-\Delta)^{\alpha}\theta dt - u\nabla\theta dt + G(\theta)dW(t), \\ \theta(0) = \theta_0, \end{cases} \tag{13.22}$$

with $\alpha \in (0, 1)$ and periodic boundary conditions. Here θ is the temperature potential related to the speed of a fluid u through the relations:

$$u = \text{rot}\,\psi, \quad (-\Delta)^{1/2}\psi = -\theta,$$

where ψ is the so called stream function. Finally W is a cylindrical Wiener process.

The equation (13.22), called a *stochastic quasi-geostrophic equation*, is an important model in the theory of geophysical fluid dynamics. For the critical value $\alpha = 1/2$ it exhibits similar properties to the three dimensional Navier–Stokes equation: existence but not uniqueness of a weak solutions.

Equation (13.22) was studied by Röckner, R. Zou and X. Zou [627]. Existence and uniqueness of a strong solution is proved when $\alpha > 1/2$ (subcritical case) and ergodicity and exponential convergence to equilibrium when $\alpha > 2/3$ and the noise is nondegenerate. In the other cases existence of a weak solution (in the sense of martingales) and also of a Markov selection are proved. Large deviations were studied by Liu, Röckner and Zou [504].

13.14 A growth of surface equation

We are concerned with a model arising in the theory of growth of surfaces, where an amorphous material is deposited in high vacuum on an initially flat surface. Details on this model can be found in Raible, Linz and Hänggi [610]. After rescaling, the equation reads as follows

$$\dot{h} = -h_{xxxx} - h_{xx} + (h_x^2)_{xx} + \dot{W}, \tag{13.23}$$

with periodic boundary conditions on an interval, where the noise W is white in both space and time.

In the deterministic case the existence of a local solution of (13.23) was proved by Stein and Winkler [659]. Concerning the stochastic case, the existence of a unique local solution in $L^p([0,\tau), H^1) \cap C((0,\tau)), H^1)$, for initial conditions in H^γ with $\gamma > 1 - \frac{1}{p}$ and $p > 8$, was proved by Blömker and Gugg [78]. For the existence of a weak martingale solution see Blömker and Gugg [79] and Blömker, Gugg and Raible [80].

The main difficulty here, which is shared by both the deterministic and the stochastic models, is the lack of uniqueness for weak solutions. This is again very similar to the three dimensional Navier–Stokes equation. The existence of a Markov selection is presented in a paper by Blömker, Flandoli and Romito [77].

13.15 Geometric SPDEs

These are equations for time evolution of geometric objects, like surfaces, subject to some physically motivated rules.

Yip [717] proved the existence of hypersurfaces in $\mathbb{R}^n$, with normal velocity given by mean curvature and perturbed by noise. Uniqueness, under suitable assumptions, was proved by Souganidis and Yip [652].

A description of the short time behavior of solutions of the Allen–Cahn equation with a smoothened additive noise was studied by Weber [709]. The key result is that in the sharp interface limit solutions move according to motion by mean curvature with an additional stochastic forcing. This extends a similar result of Funaki [333] in spatial dimension two, to arbitrary dimensions. In a subsequent paper, Weber [710] proved the existence of an invariant measure for a one dimensional Allen–Cahn equation with an additive space-time white noise. In particular, an exponentially fast convergence of the pushforward of the invariant measure to the set of minimizers is proved.

Es-Sarhir, von Renesse and Stannat [284] proved moment estimates for the invariant measure of an SPDE describing motion by mean curvature.

In the paper by Es-Sarhir and von Renesse [283] the ergodicity is proved under suitable assumptions.

13.16 Kuramoto–Sivashinsky equation

This equation arises in the study of various pattern formation phenomena involving some kind of phase turbulence or phase transition, see for example Sell and

You [643, page 320] and references therein. After some scaling the equation looks like:

$$u_t + u_{xxxx} + u_{xx} + uu_x = 0.$$

Some stochastic models are provided by Cuerno *et al.* [175] and Lauritsen, Cuerno and Makse [487].

The stochastic version:

$$\begin{cases} du + (u_{xxxx} + u_{xx} + uu_x) = dW(t), & a < x < b, \\ u(t,a) = u(t,b) = 0, & t > 0, \end{cases} \tag{13.24}$$

was studied by Duan and Ervin [267] and Ferrario [292].

Existence and uniqueness of an invariant mesure was proved by Yang [715], who also studied the corresponding Kolmogorov operator.

13.17 Cahn–Hilliard equations

Cahn and Hilliard [133] introduced an equation to describe the evolution of the phase separation of a binary alloy when the temperature has been quenched from a value T_0, above the critical temperature T_c, to a temperature less than T_c. The unknown concentration u satisfies the equation:

$$\frac{\partial u}{\partial t} + \Delta^2 u + \Delta f(u) = 0, \tag{13.25}$$

where f is the derivative of the homogeneous free energy F. The energy F contains a logarithmic term often approximated by a polynomial. For existence and uniqueness results we refer to Debussche and Dettori [244] and references therein. In the stochastic version of (13.25):

$$du + (\Delta^2 u + \Delta f(u)) = \sqrt{Q}\, dW, \tag{13.26}$$

f is a polynomial of odd degree, Q is a symmetric, positive operator and W is a cylindrical Wiener process. Equation (13.26) is supplemented with Dirichlet or Neumann boundary conditions on a bounded open set $\mathscr{O} \subset \mathbb{R}^d$.

Existence and uniqueness of (13.26), when Q is a trace class operator, was proved by Elezoviċ and Mikeliċ [277]. The case when $Q = I$ was studied by Da Prato and Debussche [190] for $d = 1, 2, 3$, taking advantage of the strong regularizing power of the linear part. When $Q = (-\Delta)^{1/2}$ and $d = 1$, (13.26) defines a gradient system. In this case, some further properties, also in a more general situation, were studied by Da Prato and Debussche [188].

We pass now to the equation (13.26) on the interval [0, 1] and with reflecting boundary 0. This problem is motivated again by fluctuations of interfaces but with the requirement of the conservation of the area between the interface and the wall, see for instance Spohn [655].

The analysis of the equation (13.26) with reflection is much more difficult than that of equation (13.4). This is due to the presence of a fourth order elliptic operator and thus to the lack of the maximum principle, see existence and uniqueness results in Debussche and Zambotti [255], see also Goudenège [359]. For a discussion of the asymptotic behavior see the end of Chapter 12.

We quote finally an interesting generalization to a Cahn–Hilliard equation on $[-1, 1]$, with nonlinearity of logarithmic type and two reflections on -1 and 1, by Debussche and Goudenège [249].

13.18 Porous media equations

These equations describe the filtration of liquid through a porous medium and are of the following form

$$\begin{cases} dX(t,\xi) = \Delta\beta(X(t,\xi))dt + \sigma(X)dW(t,\xi), & \text{in } [0,+\infty)\times\mathscr{O}, \\ \beta(X(t,\xi)) = 0, \quad \text{on } [0,+\infty)\times\partial\mathscr{O}, \\ X(0,\xi) = x(\xi), \quad \text{in } \mathscr{O}, \end{cases} \tag{13.27}$$

where $\mathscr{O}$ is a bounded domain in $\mathbb{R}^d$ with regular boundary $\partial\mathscr{O}$, $\beta\colon \mathbb{R}\to\mathbb{R}$ is an increasing function (possibly multi-valued) and $\sigma(X)dW$ is the noise term.

For some results on deterministic theory ($\sigma = 0$) see Evans [285, pages 170, 180, 182].

A natural space for studying this problem is the negative Sobolev space $H^{-1}(\mathscr{O})$ because, as is easily seen, the nonlinear operator

$$\begin{cases} F(x) := \Delta\beta(x), \quad \forall\, x\in D(F), \\ D(F) = \{x\in H^{-1}(\mathscr{O})\cap L^1(\mathscr{O}) :\ \beta(x)\in H_0^1(\mathscr{O})\}, \end{cases}$$

is dissipative in $H^{-1}(\mathscr{O})$. For the definition of dissipative operators see Appendix D.

Let us write problem (13.27) as an abstract problem on $H^{-1}(\mathscr{O})$

$$\begin{cases} dX(t) = \Delta\beta(X(t))dt + \sigma(X)dW(t), \\ X(0) = x\in H^{-1}(\mathscr{O}), \end{cases} \tag{13.28}$$

with the Laplace operator regarded as a mapping:

$$\Delta : H_0^1(\mathscr{O}) \to H^{-1}(\mathscr{O}).$$

Let us describe some related results in the literature, starting with the *Stefan problem*. In this case β is given by

$$\beta(r) = \begin{cases} \alpha_1 r \quad \text{for } r<0, \\ 0 \quad \text{for } 0\le r\le\rho, \\ \alpha_2(r-\rho) \quad \text{for } r>\rho, \end{cases} \tag{13.29}$$

where $\alpha_1, \alpha_2, \rho > 0$. Equation (13.27) models the Stefan two phase heat transfer (melting solidification) in the presence of distributed stochastic Gaussian perturbation. Here β represents the inverse of the enthalpy function associated with the phase transition and X is related to the temperature θ by the transformation $\theta = \beta(X)$.

This free boundary problem was extensively studied in the deterministic case, see for example Elliot and Ockendon [279]. In the stochastic case, well posedness with additive noise and existence of an invariant measure, were proved in Barbu and Da Prato [38].

We pass now to *slow diffusions*. Here a typical assumption is that:

$$\beta(r) = |r|^m r, \quad r \in \mathbb{R}, \ m \in \mathbb{N}, \ m > 1. \tag{13.30}$$

Under this condition the equation (13.28) covers many important models describing the dynamics of an ideal gas in a porous medium, see for example. Aronson [28].

In the additive noise case, equation (13.27) was first studied by Da Prato, Röckner, Rozovskii and Wang [209]. In particular they proved existence and uniqueness of an invariant measure. In the multiplicative case Barbu, Da Prato and Röckner [41] proved the positivity of the solution X when $\sigma(X)$ depends linearly on X and $X(0)$ is positive.

We notice that assumption (13.30) excludes other significant physical models such as *plasma fast diffusion*, see for example Berryman and Holland [63], which arises for $\beta(s) = \sqrt{s}$, and phase transition or dynamics of saturated underground water flows (the *Richards* equation). In the latter case multi-valued monotone graphs might appear, as in Marinoschi [527]. These cases were studied by Ren, Röckner and Wang [613] using variational methods and (in a more general situation) by Barbu, Da Prato and Röckner [42] using nonlinear semigroup methods. They also established in [43] and [44] that, with high probability, the extinction of solutions in finite time may occur. Existence of an invariant measure for fast diffusion equations was proved by Barbu and Da Prato in [40] and the uniqueness was proved by Liu in [503].

The case when $\beta(r) = \operatorname{sign} r$ was also investigated in [43] and applied to *self-organized behavior* of stochastic nonlinear diffusion equations with critical states, see also Barbu *et al.* [37].

For more general results about extinction, see Röckner and Wang [621]. Harnack estimates were proved for special equations by Wang [706].

For the study of random attractors see Gess, Liu and Röckner [346] and Beyn, Gess, Lescot and Röckner [70].

We note finally that a probabilistic formula for the solution of some porous media equations can be found in Blanchard, Röckner and Russo [76] and in Barbu, Röckner and Russo [48].

13.19 Korteweg–de Vries equation

In short, the Korteweg–de Vries equation is a model for the propagation of long, unidirectional, weakly nonlinear waves. The random forcing is taken into account when modeling a noisy plasma or the surface of a shallow fluid, when the bottom is rough or when the fluid is submitted to a random exterior pressure field. It has then the form

$$\begin{cases} du + (u_{xxx} + uu_x) = G(u)dW(t), & (t,x) \in \mathbb{R}_+ \times \mathbb{R}, \\ u(0) = u_0. \end{cases} \tag{13.31}$$

13.19.1 Existence and uniqueness

To study (13.31), as well as the nonlinear Schrödinger equation below, special techniques, coming mainly from harmonic analysis, are needed. A first result was obtained for an additive noise, $G(u) \equiv G$, with a Hilbert–Schmidt covariance operator acting from $L^2(\mathbb{R})$ to $H^1(\mathbb{R})$. Generalizing to the stochastic context the method used by Kenig, Ponce and Vega [446] and by de Bouard and Debussche [230] proved existence and uniqueness of a global strong solution in $H^1(\mathbb{R})$. Then Printems [603] also showed that choosing properly an additive noise, strong solutions exist in $L^2(\mathbb{R})$.

New ideas introduced by Bourgain [93] have allowed the resolution of the deterministic Korteweg–de Vries equation in Sobolev spaces of negative order. The techniques of Bourgain have been generalized to the stochastic equation by de Bouard, Debussche and Tsutsumi [241, 242]. They obtained a local in time existence result under the assumption that the covariance operator is Hilbert–Schmidt from $H^{-s}(R)$ to $L^2(R)$, for some $s > -5/8$. In particular, this covers the case of a "localized" space-time white noise, i.e. the space-time white noise multiplied by a function decaying at infinity. Multiplicative noise was considered by de Bouard and Debussche [235]. In [235], using similar arguments as in [241], solutions were constructed in $H^1(\mathbb{R})$.

13.19.2 Soliton dynamic

The deterministic Korteweg–de Vries equation has special solutions called solitons. They have the form $u(x,t) = \varphi(x - ct)$ and exist for all $t \geq 0$. In [238, 239] a small additive or multiplicative noise was added to the equation and the exit time of the perturbed solution from a neighborhood of a modulated soliton was studied.

13.20 Stochastic conservation laws

A scalar conservation law on $\mathbb{R}^N$ is described by an equation of the form

$$\frac{d}{dt} u(t) + \operatorname{div} A(u(t)) = 0, \quad u(0) = u_0, \tag{13.32}$$

where $A \in C^2(\mathbb{R};\mathbb{R}^N)$ is called the *flux function*. For irregular A the equation does not have classical solutions and it may have many solutions in the space of Schwartz distributions.

The equation was studied by Kruẑkov [463], who introduced the notion of entropic solution motivated by physical considerations. In the papers by Kim [447] and Vallet and Wittbold [685], the stochastic problem with additive noise

$$\frac{d}{dt}\,u(t) + \operatorname{div}\, A(u(t)) = \Phi dW(t), \quad u(0) = u_0, \tag{13.33}$$

was studied and existence and uniqueness of an entropic solution was proved. When the noise Φ is multiplicative, existence was proved by Feng and Nualart [290] in any dimension N and uniqueness when $N = 1$. Debussche and Vovelle [254] proved existence and uniqueness for the problem with multiplicative noise in any dimension, showing also that its solution is the limit of a suitable parabolic approximation.

Mariani [524] proved some large deviation results for conservation laws.

13.21 Wave equations

The wave equation,

$$\frac{\partial^2 u}{\partial t^2}(t,\xi) = \frac{\partial^2 u}{\partial \xi^2}(t,\xi), \qquad t \geq 0,\ \xi \in \mathbb{R}^1,$$

was the first partial differential equation to be formulated and studied. It appeared around 1740 in the works of J. R. d'Alembert, D. Bernoulli and L. Euler. Its various generalizations, also nonlinear versions in domain $D \subset \mathbb{R}^d$, like

$$\frac{\partial^2 u}{\partial t^2}(t,\xi) = \sum_{i,j=1}^{d} a_{i,j}(\xi)\frac{\partial^2 u}{\partial \xi_i \partial \xi_j}(t,\xi) - \left|\frac{\partial u(t,\xi}{\partial t}\right|^{p-2}\frac{\partial u(t,\xi)}{\partial t} + f(u(t,\xi)),$$

$$u(t,\xi) = 0,\ \xi \in \partial D,\ t > 0, \quad u(0,\xi) = x_0(\xi), \quad \frac{\partial u}{\partial t}(0,\xi) = x_1(\xi),\ \ \xi \in D,$$

were the object of an enormous number of investigations. First results on stochastic wave equations were published around 1970. They were devoted to linear equations formally written as

$$\frac{\partial^2 u}{\partial t^2}(t,\xi) = \Delta u(t,\xi) + \frac{\partial}{\partial t}W(t,\xi), \quad t > 0, \quad \xi \in D,$$

where W is a Wiener process taking values in $L^2(D)$ and Δ is the Laplace operator, see Cabana [132]. General nonlinear, stochastic wave equations (of monotone type) were an object of the second part of Pardoux's Ph.D. thesis [575]. The results formed a stochastic version of the deterministic theory developed by Lions and Strauss [500, 662]. For recent results we refer to a series of papers by Ondreját [571, 572, 573].

For large deviation results see Sanz-Solé and Ortiz-López [574], Swiech [666] and Swiech and Zabczyk [667].

13.21.1 Spatially homogeneous noise

Recently a large amount of papers has been devoted to wave equations on the whole $\mathbb{R}^d$, perturbed by spatially homogeneous noise, formally written as

$$\frac{\partial^2 X}{\partial t^2}(t,\xi) = \Delta X(t,\xi) + f(X(t,\xi)) + g(X(t,\xi))\frac{\partial W_\Gamma}{\partial t}(t,\xi), \tag{13.34}$$

plus appropriate initial conditions. Here W_Γ is a Wiener process with the covariance kernel Γ and spectral measure μ, see Section 4.1.4. As usual the equation can be written as a system of two equations. If,

$$X := \begin{pmatrix} u \\ \frac{\partial u}{\partial t} \end{pmatrix} = \begin{pmatrix} u \\ v \end{pmatrix},$$

then

$$\frac{\partial X}{\partial t} = \begin{pmatrix} \frac{\partial u}{\partial t} \\ \frac{\partial v}{\partial t} \end{pmatrix} = \begin{pmatrix} v \\ \frac{\partial^2 u}{\partial t^2} \end{pmatrix}.$$

One can treat (13.34) as a stochastic evolution

$$dX = (AX + F(X))\,dt + G(X)dW_\Gamma$$

on a Hilbert space H and the operator A given by

$$H := \begin{pmatrix} L^2_\rho(\mathbb{R}^d) \\ H^{-2}_\rho(\mathbb{R}^d) \end{pmatrix},$$

$$A := \begin{pmatrix} 0 & I \\ \Delta & 0 \end{pmatrix}, \quad D(A) := \begin{pmatrix} H^1_\rho(\mathbb{R}^d) \\ L^2_\rho(\mathbb{R}^d) \end{pmatrix}. \tag{13.35}$$

Here ρ is a properly chosen weight, see Peszat [594]. Moreover,

$$F\begin{pmatrix} u \\ v \end{pmatrix} = \begin{pmatrix} 0 \\ f(u) \end{pmatrix}$$

and

$$G\begin{pmatrix} u \\ v \end{pmatrix}\varphi[\xi] = \begin{pmatrix} 0 \\ g(u(\xi))\varphi(\xi) \end{pmatrix},$$

and φ belongs to the RKHS of W_Γ.

For general existence results we refer to Peszat and Zabczyk [597, 598]. The paper by Peszat [594] provides the most general existence result in any space dimension. Several existence results can also be found in a recent book by Sanz-Solé [637], lecture notes by Dalang *et al.* [180] and in Sanz-Solé and Sarrá [638]. An extensive study of the regularity of the solutions is presented in Dalang and Sanz-Solé [182]. Since the Wiener process is, in general, a distribution valued process, it is not clear, even in the linear case with additive noise, when the solution is function valued. The question of the existence of function valued solutions for linear wave equations was raised in the paper by Dalang and Frangos [179] and solved by them in the case of two dimensional space variable. Their result for linear equations was then extended to all dimensions in papers by Karczewska and Zabczyk [440, 441] which also treated the case of the heat equation.

The support theorem for stochastic hyperbolic equations was proved by Millet and Sanz-Solé [547, 548].

13.21.2 Symmetric hyperbolic systems

A quasi-linear symmetric hyperbolic system in $\mathbb{R}^d$ perturbed by noise was investigated by Kim [449]. He proved global uniqueness and existence of a local solution. He also showed that the probability of global existence can be made arbitrarily close to 1 if the noise satisfies certain nondegeneracy conditions and the initial data are small enough.

13.21.3 Wave equations in Riemannian manifolds

These equations belong to a larger class of equations for manifold valued processes, see Section 13.5. In particular one can find several results in this direction in the paper by Brzeźniak and Ondreját [118].

13.22 Beam equations

Brzeźniak, Maslowski and Seidler [117] considered a generalized stochastic beam equation of the form

$$\frac{\partial^2 u}{\partial t^2} + A^2 u + f(u, \frac{\partial u}{\partial t}) + g(||B^{1/2}u||) = \sigma(u, \frac{\partial u}{\partial t})\frac{\partial}{\partial t}W$$

in a Hilbert space H, where A and B are positive self-adjoint operators, W is an infinite dimensional Wiener process, g is a nonnegative function, and f and σ satisfy appropriate conditions. They established, in particular, global existence and uniqueness of the solutions. Existence of an invariant measure for a stochastic extensible beam equation and for a stochastic damped wave equation with polynomial nonlinearities was proved by Brzeźniak, Ondreját and Seidler [119].

13.23 Nonlinear Schrödinger equations

The *focusing nonlinear Schrödinger equation* is one of the basic models for nonlinear waves. It arises in various areas of physics such as hydrodynamics, nonlinear optics and plasma physics. The equation has the form

$$\begin{cases} i\frac{du}{dt} = \Delta u + |u|^{2\sigma} u \\ u(0) = u_0, \end{cases} \tag{13.36}$$

where u is a complex valued process defined on $\mathbb{R}^n$ and $\sigma > 0$.

Equation (13.36) has been extensively studied, see Cazenave [147] and references therein. The natural space for the solutions is $C([0,T]; L^2(\mathbb{R}^n)) \cap L^{\frac{4(\sigma+1)}{n\sigma}}(0,T; L^{2\sigma+2}(\mathbb{R}^n))$. Using *Strichartz estimates*, local existence is proved if $\sigma < \frac{2}{n}$ and $u_0 \in L^2(\mathbb{R}^n)$ or if $\sigma < \frac{2}{n-2}$ and $u_0 \in H^1(\mathbb{R}^n)$. These solutions are global if $\sigma < \frac{2}{n}$.

13.23.1 Existence and uniqueness

In some circumstances, randomness has to be taken into account and a model was proposed in a paper by Bang, Christiansen and Rasmussen [35] in the context of molecular aggregates with thermal fluctuations. In this case the equation reads as follows

$$\begin{cases} i\frac{dz}{dt} = \Delta z + |z|^{2\sigma} z + \dot{\eta} z, \\ z(0) = z_0, \end{cases} \tag{13.37}$$

where $\dot{\eta}$ is a noise. Two extreme cases of noise are considered: time independent, spatially white noise, simply corresponding to disorder in the arrangement of the molecules, and pure white noise.

It is important that the $L^2(\mathbb{R}^n)$ norm of W is a conserved quantity and the multiplication $\dot{\eta} z$ has to be interpreted as the Stratonovitch product. Then the equation has the form:

$$\begin{cases} idz = (\Delta z + |z|^{2\sigma} z)dt + z \circ \Phi dW, \\ z(0) = z_0. \end{cases} \tag{13.38}$$

The stochastic nonlinear Schrödinger equation is also used to model propagation in optical fibers. Then the noise can be either multiplicative as above or additive.

In the case of a random dispersion, the noise is a time white noise multiplying the Laplace operator:

$$\begin{cases} idz = \Delta z \circ d\beta + |z|^{2\sigma} z dt, \\ z(0) = z_0, \end{cases} \tag{13.39}$$

where β is a Brownian motion.

Under suitable assumptions on the covariance operator Q of β, existence and uniqueness of a solution in L^2 of (13.39) for $z_0 \in L^2(\mathbb{R}^n)$ and $\sigma < \min(\frac{2}{n}, \frac{1}{n-1})$ was proved by de Bouard and Debussche [231]. The same authors studied the case when $u_0 \in H^1(\mathbb{R}^n)$ [233] for additive or multiplicative noise.

Equation (13.39) has been studied by de Bouard and Debussche [240] and by Debussche and Tsutsumi [253]. Global existence and uniqueness is proved in $L^2(\mathbb{R}^n)$ for $\sigma < \frac{2}{n}$ when $n \geq 2$ and for $\sigma = 2$ when $n = 1$.

For large deviation properties see Gautier [339] and Debussche and Gautier [246].

13.23.2 Blow-up

The influence of the noise on finite time blow-up of the solutions was studied by de Bouard and Debussche [232, 235]. They showed that, in the supercritical case, all solutions blow up in finite time. Note that in the corresponding deterministic case only some solutions may blow up.

This was confirmed numerically by Barton-Smith, Debussche and Di Menza [50, 245]. They also observed that a multiplicative space-time white noise, which had not been treated theoretically, seems to eliminate blow-up.

14
Some recent developments

As was mentioned in the Introduction, there have been many new developments in the theory of SPDEs since the first edition of this book in 1992. Only a few of them are presented in the new edition, due to the character of the book and to the space limitation. In this chapter we briefly indicate some of the omitted topics, providing references but not aiming for completeness. The material is organized as follows: Section 14.1 concerns solutions of equations, Section 14.2 deals with laws of solutions, and Section 14.3 concerns asymptotics of solutions.

14.1 Complements on solutions of equations

14.1.1 Stochastic PDEs in Banach spaces

In this book some SPDEs are discussed in the framework of Banach spaces, however they were, as a rule, with additive noise. Some equations with multiplicative noise were discussed by Peszat [593] using the factorization method. As was mentioned in the Introduction, Banach spaces play an important role when stochastic equations are treated in the framework of the so called Lion's triple $V \subset H \subset V^*$ with V and V^* a Banach space and its adjoint, and H a Hilbert space. The best regularity results in L^p spaces, $p \geq 2$, were obtained by Krylov [466, 467].

Another possibility is to develop first a stochastic integration theory in Banach spaces and then apply it to specific equations. This approach was initiated in the Ph.D. thesis by Neidhardt in 1978 [558], developed by Dettweiler in 1989 [259] and has been intensively studied recently. Not all Banach spaces are proper for stochastic integration and stochastic calculus. For instance the stochastic integral

$$\int_0^t f(s)dW(s), \quad t \geq 0$$

with respect to a one dimensional Wiener process W of a deterministic, continuous function with values in the space $C[0, 1]$ may not have values in $C[0, 1]$ although it has a well defined meaning and is a continuous process in the space $L^2(0, 1)$. For even more striking examples we refer to the paper by Brzeźniak, Peszat and Zabczyk [124] and for the first example, with less regular f, see Yor [720]. There exist important classes of Banach spaces in which a satisfactory stochastic integration theory can be developed and applied to a study of SPDEs. The case of the so called M-p spaces was investigated by Dettweiler and continued by Brzeźniak, see for example [100]. A refined theory for a more special class of the so called UMD spaces was built in recent papers by van Neerven, Veraar and Weiss [689, 690]. For Itô's formula in UMD spaces see the paper by Brzeźniak, van Neerven, Veraar and Weis [126]. Related results are obtained in the paper by van Neerven and Veraar [688]. For a nice exposition of the theory and for a wealth of new results we refer to the Ph.D. thesis of Veraar [693]. Stochastic convolutions were discussed in this framework by Brzeźniak [101], Brzeźniak and Peszat [121] and very recently by Seidler [641]. Specific results for stochastic equations were studied by Brzeźniak, Long and Simao [116] and by Brzeźniak and Gątarek [107]. Relations between various concepts of uniqueness of the solutions were discussed by Ondreját [569]. See also the very recent paper on SPDEs in Banach spaces by Zhang [740].

14.1.2 Backward stochastic differential equations

A backward stochastic differential equation (BSDE for short) on a bounded interval $[0, T]$ is an equation of the form

$$\begin{cases} dY_t = Z_t\, dW_t - BY_t\, dt - f(t, Y_t, Z_t)\, dt, \\ Y_T = \xi. \end{cases} \tag{14.1}$$

Here W is a cylindrical Wiener process in a Hilbert space Ξ, with completed natural filtration denoted $(\mathscr{F}_t)$. The unknown process is an $(\mathscr{F}_t)$-progressive pair (Y, Z), where Y takes values in another Hilbert space K and Z in the space $L_2(\Xi, K)$ of Hilbert–Schmidt operators from Ξ to K. An $\mathscr{F}_T$-measurable terminal condition ξ is given for the process Y and the equation is solved backwards in time. The coefficient f is called the generator, and for given $y \in K, z \in L_2(\Xi, K)$, the process $t \mapsto f(t, y, z)$ is assumed to be $(\mathscr{F}_t)$-progressive. B denotes the infinitesimal generator of a strongly continuous semigroup in K. Note that no progressive solution Y exists with $Z = 0$, even in the simplest cases. The occurrence of the stochastic differential $Z_t\, dW_t$ and the addition of another unknown process Z makes the problem well posed in the class of progressive processes, under appropriate conditions.

In the finite dimensional case $K = R^n$, $\Xi = R^d$, existence and uniqueness for this class of equations with generators f Lipschitz in (y, z), was first established by Pardoux and Peng [583]. The linear case was addressed earlier by J.-M. Bismut [73]. Since then, the literature has grown considerably. Systematic expositions of the finite dimensional theory can be found in El Karooui, Peng and Quenez [278], Pardoux [581, 582] and Ma and Yong [517]. General solvability results have been proved when

f is Lipschitz with respect to z and continuous, dissipative with respect to y (Briand *et al.* [97]), or in the case of quadratic growth with respect to z (Kobilanski [450], Briand and Hu [98]) .

The first extension to the infinite dimensional case is due to Hu and Peng [411], who used the concept of mild solution to (14.1). Further results can be found in Tessitore [672], Confortola [172], Pardoux and Rascanu [584], Fuhrman and Hu [323] under various assumptions on f and B, including cases where B is the generator of an analytic semigroup or the subdifferential of a lower-semicontinuous convex functional. A lot of results are known for partial differential equations of backward type: see for instance Ma and Yong [517], Hu, Ma and Yong [409], Du, Qiu and Tang [266]. One of the main motivations for the study of BSDEs is the fact that they provide probabilistic representation formulae for the solutions of some classes of nonlinear partial differential equations on finite and infinite dimensional spaces. More precisely, suppose that X is a process in a Hilbert space H, solving, on an interval $s \in [t, T] \subset [0, T]$, an Itô equation of the usual form

$$\begin{cases} dX_s = AX_s\,ds + F(s, X_s)\,ds + G(s, X_s)\,dW_s, \\ X_t = x \in H. \end{cases} \tag{14.2}$$

Associate with equation (14.2), on the same interval $[t, T]$, the backward equation

$$\begin{cases} dY_s = Z_s\,dW_s + \psi(s, X_s, Y_s, Z_s)\,ds, \\ Y_T = \phi(X_T), \end{cases} \tag{14.3}$$

where now Y is a scalar process. Thus $K = \mathbb{R}$, $L_2(\Xi, K) = \Xi^*$), $\phi : H \to \mathbb{R}$ and $\psi : [0, T] \times H \times \mathbb{R} \times \Xi^* \to \mathbb{R}$ are given deterministic functions. Since the solutions (X, Y, Z) depend on $t \in [0, T]$ and $x \in H$, they will be denoted as $(X_s^{t,x}, Y_s^{t,x}, Z_s^{t,x})_{s\in[t,T]}$. It turns out that the process $v(t, x) := Y_t^{t,x}$ is deterministic and is a solution of the equation

$$\begin{cases} \partial_t v(t, x) + \mathscr{L}_t v(t, x) = \psi(t, x, u(t, x), Du(t, x)G(t, x)), \\ v(T, x) = \phi(x), \qquad t \in [0, T],\ x \in H, \end{cases} \tag{14.4}$$

where $\mathscr{L}_t$ is the Kolmogorov operator of the process X defined by (14.2). This connection between BSDEs and PDEs was first investigated by Peng [588] in the finite dimensional case.

Another important motivation for BSDEs is a connection with stochastic control. In the case when (14.4) is the Hamilton–Jacobi–Bellman equation for the value function $v(t, x)$ of a stochastic control problem, the corresponding BSDE provides an alternative probabilistic formula for $v(t, x)$. In several cases, a direct connection can be established between the control problem and an appropriate BSDE, without introducing the Hamilton–Jacobi–Bellman equation explicitly. Among the earliest results of this kind in the finite dimensional case we cite Peng [589]. A related topic is the so called stochastic maximum principle of Pontryagin type, which historically marks the first occurrence of BSDEs of linear type. For a complete exposition the reader can consult the treatise by Yong and Zhou [718].

The connection of BSDEs with nonlinear PDEs and with optimal control problems, when H is an infinite dimensional Hilbert space, was first established by Fuhrman and Tessitore [325, 326, 328]. Extensions were given by Briand and Confortola [94, 95] and Masiero [530]. Some generalizations, including the case when X takes values in a Banach space, are presented in Masiero [529], Fuhrman, Masiero and Tessitore [324], Guatteri [364], Zhou and Liu [741], Zhou and Zang [742]. Some results on the stochastic maximum principle on Hilbert spaces are in Bensoussan [55], Hu and Peng [410], Li and Tang [496], Zou [743] and Guatteri [365].

Finally, we note that so far we have considered BSDEs on a bounded interval $[0, T]$. Most of the results generalize to the case of the infinite interval $[0, \infty)$, or even to the random time interval. This allows us to represent probabilistically solutions of elliptic equations of the form

$$\mathscr{L}v(x) = \psi\Big(x, v(x), Dv(x)G(x)\Big), \qquad x \in H. \tag{14.5}$$

Compare the parabolic version (14.4). The Kolmogorov operator $\mathscr{L}$ corresponds to the process X given by (14.2). In this way one can treat infinite horizon discounted control problems, or even ergodic control problems: see for instance Pardoux [581, 582] and Royer [632]. Extensions to the infinite dimensional case are given in Fuhrman and Tessitore [327], Briand and Confortola [96], Fuhrman, Tessitore and Hu [329] and Debussche, Hu and Tessitore [250].

14.1.3 Wiener chaos expansions

A new approach to stochastic evolution equations, based on the concept of Wiener chaos, was proposed and extensively developed by Lototsky and Rozovskii [506, 507, 508, 509], see also Kalpinelli, Frangos and Yannacopoulos [439], and by Yannacopoulos, Frangos and Karatzas [716]. By considering an expansion of the solution, with respect to an orthonormal basis in the Wiener space, one arrives at deterministic equations on the coefficients of the expansion. Those are rather complex systems but important information can be deduced from them. They also lead to numerical schemes.

14.1.4 Hida's white noise approach

To study stochastic models it was also proposed to use Hida's white noise theory. Fundamentals of the theory are presented in the monograph by Hida, Kuo, Potthoff and Streit [402]. We also refer to the monograph by Holden, Øksendal, Ubøe and Zhang [407] where the theory is applied to stochastic PDEs.

14.1.5 Rough paths approach

The theory of *rough paths integration* was invented by Terry Lyons. Its exposition and applications to stochastic differential equations can be found in the original paper by Lyons [513]. The theory is also presented in the monograph by Lyons and Qian [515], in a more recent presentation by Friz and Victoir [322] as well as in the St. Flour lecture notes by Lyons, Caruana and Lévy [514]. An alternative formulation of the basic results in terms of the notion of controlled path is provided by Gubinelli [366, 367].

The rough path theory starts from the concept of a rough path integral with respect to a nondifferentiable stochastic integrator X:

$$\int_0^t \varphi(X)dX, \quad t \geq 0,$$

where $\varphi : \mathbb{R}^d \to \mathbb{R}^d$ is a smooth function. To fix the ideas take X to be a d-dimensional Brownian motion.Then Itô's theory ensures that this integral can be understood as a suitable limit in probability of the (forward) Riemman sums

$$\sum_{i=1}^{n}\sum_{i=1}^{d} \varphi_i(X_{t_i})(X^i_{t_{i+1}} - X^i_{t_i})$$

where $\{[t_i, t_{i+1})\}_{i=1,\dots,n}$ is a finite partition of the interval $[0, t)$. A careful analysis of these discrete sums reveals that, after adding suitable correction terms, these sums converge almost surely. These corrections are defined in terms of *iterated integrals* of X. The simplest one is the second order iterated integral $(\mathbb{X}^{i,j}_{st} : i, j = 1, \dots, d; t, s \in [0, T])$:

$$\mathbb{X}^{i,j}_{s,t} = \int_s^t \int_s^{r_2} dX^i_{r_1} dX^j_{r_2} \tag{14.6}$$

where the integrals are of Itô type. It is possible to show that the corrected sums

$$\sum_{i=1}^{n}\sum_{i=1}^{d} \varphi_i(X_{t_i})(X^i_{t_{i+1}} - X^i_{t_i}) + \sum_{i,j=1}^{d} \nabla_j \varphi_i(X_{t_i})\mathbb{X}^{i,j}_{t_i,t_{i+1}}$$

converge almost surely for all $\varphi \in C^2(\mathbb{R}^d; \mathbb{R}^d)$ and the limits can be used to define the integral $\int_0^t \varphi(X_s)dX_s$. An important fact is that the exceptional set where convergence does not take place does not depend on the function φ. So it is possible to integrate random functions φ which depend on the paths of $(X_t)_{t\in[0,1]}$. The rough path machinery can be used to *define* integrals over processes which are not semimartingale, like fractional Brownian motions. A key element which is crucial for the rough path analysis to work is the possibility of identifying suitable iterated integrals like those in (14.6). The number of integrals needed depends on the regularity of the trajectory.

The theory can then be extended to solve and analyze differential equations driven by rough paths of the form

$$Y_t = Y_0 + \int_0^t \varphi(Y_s)dX_s, \quad t \geq 0$$

where the integral is understood as the rough path integral of the type described above. The theory works quite well for rough differential equations driven by finite dimensional or even infinite dimensional signals. Its application to PDEs driven by rough paths (RPDEs) is still a subject of lively research. Indeed a satisfactory understanding of the appropriate analytic framework to study RPDEs is still missing. Let us mention special cases for which some partial results are known.

(i) Rough evolution equations were considered by Gubinelli and Tindel [369] and Teichmann [670]. Formally they can be written in the mild form

$$Y_t = S_t Y_0 + \int_0^t S_{t-s}\,\sigma(Y_s)dX_s$$

where Y is a path in some Hilbert or Banach space of functions V. Here $(S_t)_{t\geq 0}$ is an analytic semigroup on V and $\sigma : V \to V$ is nonlinear operator of the form:

$$\sigma(Y)(x) = f(Y(x)),$$

with f a smooth function on a Euclidean space. The process X can be Gaussian with values in a space of distributions and with singular spatial covariance. A suitable analysis of the rough convolution:

$$\int_0^t S_{t-s}\,\sigma(Y_s)dX_s, \quad t \geq 0,$$

allows us to prove existence and uniqueness of local solutions in appropriate spaces of continuous paths on V. The alternative approach developed by Teichmann [670] is based on the fact that a time-dependent transformation of the state space maps the RPDE into a rough differential equation which can be dealt with using the standard rough path machinery.

(ii) Fully nonlinear equations were studied by Friz and coworkers [144, 145, 320, 321]. They also studied equations of the form

$$\partial_t Y_t(x) - F(t, x, DY_t(x), D^2 Y_t(x)) = (V(t, x)DY_t(x) + G(t, x)Y_t(x))\frac{dX_t}{dt}.$$

By a change of variables related to the characteristics for the linear transport equation

$$\partial_t u_t(x) = V(t, x)Du_t(x)\frac{dX_t}{dt},$$

together with a "Doss transformation" to reabsorb the term $GYdX$, the equation can be reduced to a classical fully nonlinear PDE of the form

$$\partial_t \tilde{Y}_t(x) - \tilde{F}(t, x, D\tilde{Y}_t(x), D^2\tilde{Y}_t(x)) = 0$$

which under appropriate conditions on $\tilde{F}$ fall in the domain of the theory of viscosity solutions.

(iii) SPDEs with strong nonlinearities were studied by Gubinelli [368] and Hairer [386]. Rough path theory can also be used to analyze semilinear PDEs where the nonlinear term is a priori not well defined in usual functional spaces. Two examples are the deterministic or stochastic Korteweg–de Vries equation and the vector Burgers equation with additive white noise perturbation.

14.1.6 Equations with fractional Brownian motion

Let us recall that a real valued, Gaussian, path continuous process (β_t) is called a fractional Brownian motion with Hurst parameter $H > 0$ if its mean is zero and its covariance is of the form:

$$\mathbb{E}[\beta_t^H \beta_s^H] = \frac{1}{2}(t^{2H} + s^{2H} - |t-s|^{2H}).$$

Let (e_i) be an orthonormal basis in a Hilbert space V and λ_i positive numbers such that $\sum_{i=1}^{\infty} \lambda_i < +\infty$. The process

$$B_t^H = \sum_{i=1}^{\infty} \sqrt{\lambda_i} e_i \beta_i^H(t),$$

where β_i^H are independent fractional Brownian motions with the Hurst parameter H, is called a V-valued fractional Brownian motion with the Hurst parameter H.

Equations in which the Hilbert space valued Wiener process is replaced by a fractional Brownian motion have been intensively studied. For a good start we recommend papers by Duncan, Pasik-Duncan and Maslowski [270], Maslowski and Nualart [534] and Duncan, Maslowski and Pasik-Duncan [269]. The studied equation is of the form:

$$dX_t = (AX_t + F(X_t))dt + G(X_t)dB_t^H$$

where A is the infinitesimal generator of an analytic semigroup on V.

Under some regularity and growth conditions on the coefficients F and G and for some values of H the existence and uniqueness of a mild solution is established. The results are applied to stochastic parabolic PDEs on bounded domains $D \subset \mathbb{R}^d$. In the proofs the authors combine techniques of fractional calculus with semigroup estimates.

14.1.7 Equations with Lévy noise

The first published paper on SPDEs with Lévy noise is that by Chojnowska-Michalik [165]. Around a decade later more works started to appear in great numbers, beginning with Albeverio, Wu and Zhang [15], Applebaum and Wu [24] and Mueller [554]. For more recent publications we refer to the book by Peszat and Zabczyk [599] which concentrates mainly on existence and uniqueness results and contains references to other works.

Some structural properties of the solutions are discussed in Brzeźniak and Zabczyk [127] and Priola and Zabczyk [608]. Unexpected lack of time regularity of the solutions was reported in a paper by Brzeźniak *et al.* [108], see also [599] and [127]. Maximal inequalities for stochastic integrals were discussed recently by Brzeźniak and Hausenblas [114] and Marinelli, Prevot and Röckner [526]. Large deviations were discussed by Röckner and Zhang [622], Swiech and Zabczyk [667] and Högele [406].

14.1.8 Equations with irregular coefficients

Let us consider, in a separable Hilbert space H, the following equation

$$dX_t = (AX_t + B(X_t))dt + dW_t, \qquad X_0 = x \in H$$

where $A : D(A) \subset H \to H$ is self-adjoint, negative and such that A^{-1} is of trace class, $B : H \to H$ and $W = (W_t)$ is a cylindrical Wiener process. We do not assume that B is regular. In the case of parabolic equation with space-time white noise in space dimension one, pathwise uniqueness has been proved, under various assumptions on the drift, by Gyöngy and coworkers, in a series of papers, see [7, 374, 378, 382, 383] and references therein. A general pathwise uniqueness result in the abstract case when B is Hölder continuous was obtained by Da Prato and Flandoli [198]; the proof uses some ideas from the finite dimensional case of Krylov and Röckner [468] and of Flandoli, Gubinelli and Priola [305]. Pathwise uniqueness for perturbations of reaction-diffusion equations was proved by Cerrai, Da Prato and Flandoli [153].

14.1.9 Yamada–Watanabe theory in infinite dimensions

A generalization to infinite dimensions of the Yamada–Watanabe theory was provided by Ondrejàt [569]. For the variational approach see Röckner, Schmuland and Zhang [616].

14.1.10 Numerical methods for SPDEs

The discretization of SPDEs has been the subject of many articles. At the end of the 1990s, Allen, Novosel and Zhang [16] proposed a finite element discretization for SPDEs. Also Gyöngy and Nualart [381] and Gyöngy [375, 376] proposed the analysis of finite difference schemes. Several authors have proposed similar works in various situations. Printems [604] has considered an implicit Euler scheme of a stochastic parabolic equation giving an application to Burger's equation. Hausenblas has studied approximation in Banach spaces [395, 396]. Yubin [722, 723, 724] and Walsh [703] have analyzed finite element methods for parabolic stochastic PDEs.

Gyöngy and Millet [379, 379] have analyzed evolution SPDEs driven by monotone operators. De Bouard and Debussche [234] have studied the stochastic nonlinear Schrödinger equation. Lord and Shardlow [505] have investigated the numerical approximation of a stochastic reaction-diffusion equation with colored noise. Baňas, Brzeźniak and Prohl [34] have studied finite elements discretization for stochastic Landau–Lifshitz–Gilbert equations and Carelli, Hausenblas and Prohl [143] have studied a time-splitting method for the stochastic incompressible time-dependent Stokes equation. Larsson and co-authors have analyzed the finite element method for the wave equation [459] and for parabolic equations [343, 485].

Jentzen and Kloeden [431] considered SPDEs with additive noise for which in general Itô's formula is not available and stochastic Taylor expansion of solutions cannot be derived by an iterated application of Itô's formula. They proposed an alternative method by using the mild formulation of the SPDE and introducing a suitable recursion technique. Jentzen considered in [430] the multiplicative noise case and in [429] the case of nonglobally Lipschitz coefficients.

All these articles study the error in the strong sense and pathwise error estimates are given. It is also important to understand the weak error. Indeed, the rate of convergence in the weak sense is expected to be better than the strong error rate. In many situations, it is important to simulate only the law of the solution. This has been studied, in the linear case by Debussche and Printems [252], for a semilinear parabolic SPDE by Debussche [243], and for the nonlinear Schrödinger equation by de Bouard and Debussche [236].

Approximations of the Wong–Zakai type for stochastic evolution equations were studied by Tessitore and Zabczyk [675]

Finally, Dunst, Hausenblas and Prohl [273] presented a discretization method for parabolic SPDEs with jumps.

14.2 Some results on laws of solutions

14.2.1 Applications of Malliavin calculus

Malliavin calculus is a tool which often allows us to prove existence of densities of random variables associated with solutions of stochastic differential equations in both finite and infinite dimensional spaces. A central role here is played by an integration by parts formula which involves Malliavin derivatives. It was used in the book, in a disguised form, in the derivation of the Bismut–Elworthy–Li formula. The monograph by Sanz-Solé [637] is well suited as an introduction to the subject as well as for applications to SPDEs. Let $X(t, \xi)$ denote the value of a solution of a SPDE at the moment $t > 0$ and the point ξ of a domain $G \subset \mathbb{R}^d$. It is of some interest to know whether the vector random variable $(X(t, \xi_1), \ldots, X(t, \xi_m))$, for a given choice of different points $\xi_1, \ldots, \xi_m$ has a density and how smooth the density is. It is shown in the book by Sanz-Solé [637], as well as in several papers, that this is really the case for solutions of various specific SPDEs. The Malliavin calculus in infinite dimensions turned out to be of great importance to prove the so called asymptotic strong Feller property which implies uniqueness of invariant measures. This was done for the two dimensional Navier–Stokes equation by Hairer, Mattingly and Pardoux [390] and Hairer and Mattingly [388, 389].

14.2.2 Fokker–Planck and mass transport equations

Consider a stochastic differential equation in $\mathbb{R}^d$

$$\begin{cases} dX = b(t,X)dt + \sigma(t,X)dW(t), \\ X(0) = x \in H, \end{cases} \tag{14.7}$$

where $b : [0,T] \times H \to H$ and $\sigma : [0,T] \times H \to L(H)$ are Lipschitz and regular and W is an $\mathbb{R}^d$ dimensional Brownian motion. Setting $P_{0,t}\varphi(x) := \mathbb{E}[\varphi(X(t,x))]$, $\forall\, \varphi \in C_b^2,(H)$, one has by Itô's formula

$$D_t P_{0,t}\varphi(x) = P_{0,t}\mathscr{L}(t)\varphi(x),$$

where $\mathscr{L}(t)$ is the Kolmogorov operator

$$\mathscr{L}(t)\varphi(x) := \frac{1}{2}Tr\,[\sigma(t,x)\sigma^*(t,x)D_x^2\varphi(x)] + \langle b(t,x), D_x\varphi(x)\rangle.$$

Now for any probability measure ζ and any $t \ge 0$ set $\mu_t = P_{0,t}^*\zeta$, where $P_{0,t}^*$ is the adjoint of $P_{0,t}$. Then we have

$$\frac{d}{dt}\int_H \varphi(x)\mu_t(dx) = \int_H \mathscr{L}(t)\varphi(x)\mu_t(dx), \tag{14.8}$$

for all $\varphi \in C_b^2(H)$. Equation (14.8) or the unknown, measure valued function $(\mu_t)_{t\ge 0}$ is called the *Fokker–Planck* equation.

The first basic reference in finite dimensions is Il'in and Hasminskii [419]. For the case when b and σ are only continuous and $\det(\sigma\sigma^*) > 0$, see Bogachev, Da Prato and Röckner [85, 86, 90] and references therein. In the deterministic case $\sigma = 0$ we quote important papers by Di Perna and Lions [262] and Ambrosio [19].

Now let H be an infinite dimensional separable Hilbert space and consider the stochastic differential equation

$$\begin{cases} dX = (AX + b(t,X))dt + BdW(t), \\ X(0) = x \in H, \end{cases} \tag{14.9}$$

where $A : D(A) \subset H \to H$ is self-adjoint negative, A^{-1} is of trace class, $B \in L(H)$, $b : [0,T] \times H \to H$, and $W(t)$ is a cylindrical Wiener process in H.

Given a probability measure ζ, one looks for measures $(\mu_t)_{t\ge 0}$ such that $\mu_0 = \zeta$ and equation (14.8) holds for any φ in suitable space of *test functions*, $\mathscr{E}$. We notice that in infinite dimensions the choice of $\mathscr{E}$ is not obvious. The space $C_b^2(H)$ is not suitable because $\mathscr{L}(t)\varphi$ is not well defined for all $\varphi \in C_b^2(H)$. One possibility is to take $\mathscr{E} = \mathscr{E}_A(H)$, the linear span of all the real parts of functions of the form $\varphi(x) = e^{i\langle x,h\rangle}$, $h \in D(A)$. Existence and uniqueness of the solution to (14.8) when $b(t,\cdot)$ is dissipative was studied by Bogachev, Da Prato and Röckner [87, 88] and when the noise is white or Q^{-1} is bounded, by [89].

When $B = 0$ the Di Perna–Lions theory has been generalized in the setting of abstract Wiener spaces by Ambrosio and Figalli [20], see also Fang and Luo [287].

14.2.3 Ultraboundedness and Harnack inequalities

Let P_t, $t \geq 0$, be a Markov semigroup in a Hilbert space H and let μ be an invariant measure for P_t, $t \geq 0$. The semigroup is called *ultrabounded* if it maps continuously $L^2(H,\mu)$ into $L^\infty(H,\mu)$. The key tool to show ultraboundedness is a Harnack estimate of the form

$$(P_t\varphi(x))^2 \leq P_t(\varphi^2)(y)e^{C_t|x-y|^2}, \quad \forall\, x, y \in H, \quad \forall\, \varphi \in B_b(H), \qquad (14.10)$$

where $C_t > 0$ is a suitable constant. It is well known that an Ornstein–Uhlenbeck semigroup is hypercontractive but not ultrabounded.

Harnack estimates for transition semigroups, independent of the dimension of the state space, were introduced and studied intensively by Wang [704, 705], see also Röckner and Wang [619]. In the paper by Da Prato, Röckner and Wang [210], ultraboundedness is proved for some reaction-diffusion equations with a nonlinearity growing faster than linearly.

14.2.4 Gradient flows in Wasserstein spaces and Dirichlet forms

Dirichlet forms are a powerful tool in the construction and study of stochastic processes and play an important role in infinite dimensional stochastic analysis.

In order to study the Dirichlet form:

$$\mathscr{E}(u,v) := \int_H \langle \nabla u, \nabla v\rangle\, d\gamma,$$

Ambrosio, Savaré and Zambotti [21] used the theory of gradient flows in Wasserstein spaces. In the definition of the form $\mathscr{E}$, the integral is over a separable Hilbert space H and γ is a *log-concave* probability measure on H. That is

$$\log\gamma\,((1-t)B + tC) \geq (1-t)\log\gamma(B) + t\log\gamma(C),$$

for all $t \in (0,1)$ and for all pairs of open sets B, $C \subset H$. It turns out that log-concavity implies closability of $\mathscr{E}$ without any requirement of the validity of the integration by parts formulae. Moreover, if γ_n is a sequence of log-concave probability measures converging to γ, then the stochastic processes associated to the respective Dirichlet forms converge in law.

This problem was also studied by Röckner, R. Zhu and X. Zhu [626, 628] using functions of bounded variation (BV) in Gelfand triples.

14.3 Asymptotic properties of the solutions

14.3.1 More on invariant measures

Extensive references to the literature on large time behavior of solutions to SPDEs up to year 1996 can be found in Da Prato and Zabczyk [220]. The book exploited dissipativity of the coefficients when analyzing the solutions as time decreased to $-\infty$. Uniqueness of the invariant measure as well as fast convergence of the laws of the solutions to the equilibrium were based on the strong Feller and irreducibility properties of the transition semigroup, see Chapter 11. An important tool to establish the strong Feller property is the so called BEL (Bismut–Elowrthy–Li) formula for the gradient of the transition semigroup, see Bismut [74], Li [495] and Elworthy [281], see Section 9.4.2. Its infinite dimensional extension can be found in Peszat and Zabczyk [596]. For an extension to Lévy noise see Priola and Zabczyk [607] and Marinelli, Prévôt and Röckner [532]. General studies on various concepts of strong Feller property can be found in recent papers by Maslowski and Seidler [537, 538]. Important results on exponential ergodicity were recently obtained by Goldys and Maslowski [354, 355].

A new method to study uniqueness and convergence to the equilibrium was recently developed by Hairer and Mattingly [388, 389]. It is based on the so called asymptotic strong Feller property and uses in an essential way infinite dimensional Malliavin calculus [390]. Some comments about the application of the method to delay equations is given in Section 13.2 and application to the two dimensional Navier–Stokes equations is discussed in Section 10.10.1 of this survey.

Another important approach, developed by Komorowski, Peszat and Szarek [451], is based on the *e-chain property* of the corresponding transition semigroup.

A powerful method to study invariant measures is based on the method of *coupling*, see Lindvall [497]. For an exposition of the method in the context of SPDEs we refer to the review paper by Hairer [384] and to the book by Kuksin and Shirikyan [474]. Its connection with Harris theorem is discussed in Hairer [385]. In particular you find in [385] a short proof of the theorem based on Liapunov's technique. For recent applications of the method to hydrodynamic equations see Kuksin and Shirikyan [473, 474] and Kuksin, Piatnitski and Shirikyan [472].

Several new results have been obtained for *equations with non-Gaussian noise*. For an early study of invariant measures for Ornstein–Uhlenbeck processes with Lévy noise see Chojnowska-Michalik [165].

Some recent results on limiting behavior of the solutions to SPDEs with Lévy noise can be found in the book by Peszat and Zabczyk [599]. Dissipativity type of results, extending those from [220], have been obtained by Rusinek [635], see also

[599]. Gradient estimates were used by Priola, Xu and Zabczyk [606] to establish exponential mixing. The coupling method is treated in Priola, Shirikyan, Xu and Zabczyk [605].

For the case of the noise being fractional Brownian motion see Maslowski and Pospišil [535].

14.3.2 More on large deviations

The large deviation principle for infinite dimensional Itô's equations has been studied in several papers, see Faris and Jona-Lasinio [288], Freidlin [315], Chow [168], Smolenski, Sztencel and Zabczyk [651], Peszat [591], Kallianpur and Xiong [438], de Acosta [229], Chenal and Millet [162], Swiech [666] and Feng and Kurtz [289].

Concerning equations with Lèvy noise. see Swiech and Zabczyk [667].

Large deviations for *systems* of reaction-diffusion equations have been considered by Cerrai and Röckner [157, 158]; moreover in [158] they proved large deviations for the corresponding invariant measure.

In a paper by Budhiraja, Dupuis and Maroulas [129] a new approach is presented. It is based on suitable variational representations of infinite dimensional Brownian motions. Proofs of large deviation properties are reduced to demonstrating basic qualitative properties of certain perturbations of the original process.

14.3.3 Stochastic resonance

Stochastic resonance in climate models and in the Landau–Ginzburg equation, were studied by Benzi, Parisi, Sutera and Vulpiani in the papers [61] and [62] respectively.

Blömker and Hairer derived an ϵ-expansion for the invariant measure of some semilinear SPDEs near a change of stability [81]. In the paper [83], a rigorous derivation of amplitude equations for SPDEs with quadratic nonlinearities (as Burgers equations) is presented. They studied moreover the stochastic bifurcation for the stochastic Swift–Hohenberg equation [82].

14.3.4 Averaging

The motion of a particle of a mass m in the field $b(q) + \sigma(q)\dot{W}_t$ with the damping proportional to the speed is described, according to the Newton law, by the equation

$$\begin{cases} m\ddot{q}(t,m) = b(q(t,m)) + \sigma((t,m))\dot{W}(t) - \dot{q}(t,m), \\ q(0,m) = q \in \mathbb{R}^n, \ \dot{q}(m,0)m = p \in \mathbb{R}^n. \end{cases} \tag{14.11}$$

Here $b(q)$ is the deterministic component of the force and $\sigma(q)\dot{W}(t)$ is the stochastic part. It is well known that, for $0 < m \ll 1$, $q(t,m)$ can be approximated by the solution of the first order equation

$$\begin{cases} \dot{q}(t) = b(q(t)) + \sigma((t))\dot{W}(t), \\ q(0) = q \in \mathbb{R}^n, \end{cases} \tag{14.12}$$

in the sense that

$$\lim_{m\to 0} \mathbb{P}\left(\max_{0\le t\le T} |q(t,m) - q(t)| \ge \delta\right) > 0. \tag{14.13}$$

Statement (14.13) is called the *Smoluchowski–Kramers* approximation of $q(t, m)$ by $q(t)$. This statement justifies the description of the motion of a small particle by the first order equation (1.2) instead of the second order equation (1.1).

The Smoluchowski–Kramers approximation for SPDEs was studied by Cerrai and Freidlin [154, 155].

An averaging principle for a class of stochastic reaction-diffusion equations was studied by Cerrai and Freidlin [156] and Cerrai [151, 152].

14.3.5 Short time asymptotic

Given a symmetric Markovian semigroup T_t, it is of interest to identify the behavior of

$$\int_A T_t \mathbb{1}_B d\mu,$$

as $t \downarrow 0$,where A, B are Borel sets and μ is the invariant measure for T_t . The limit

$$\lim_{t\downarrow 0} t \log \int_A T_t \mathbb{1}_B d\mu,$$

is studied in papers by Ramìrez [612] and by Hino and Ramìrez [404, 405].

Appendix A

Linear deterministic equations

A.1 Cauchy problems and semigroups

Linear evolution equations, as parabolic, hyperbolic or delay equations, can often be formulated as an evolution equation in a Banach space E:

$$\begin{cases} u'(t) = A_0 u(t), & t \geq 0, \\ u(0) = x \in E, \end{cases} \tag{A.1}$$

with A_0 being a linear operator, in general unbounded, defined in a dense linear subspace $D(A_0)$ of E. In (A.1), $u'(t)$ stands for the strong derivative of $u(t)$

$$u'(t) = \lim_{h \to 0} \frac{u(t+h) - u(t)}{h},$$

the limit being taken in the topology of E.

Problem (A.1) is the *initial value problem* or the *Cauchy problem* relative to the operator A_0.

Definition A.1 *We say that the Cauchy problem* (A.1) *is* well posed *if:*

(i) for arbitrary $x \in D(A_0)$ there exists exactly one strongly differentiable function $u(t, x)$, $t \in [0, +\infty)$, satisfying (A.1) *for all $t \in [0, +\infty)$;*
(ii) if $\{x_n\} \in D(A_0)$ and $\lim_{n \to \infty} x_n = 0$, then for all $t \in [0, +\infty)$ we have

$$\lim_{n \to \infty} u(t, x_n) = 0. \tag{A.2}$$

If the limit in (A.2) *is uniform in t on compact subsets of $[0, +\infty)$ we say that the Cauchy problem* (A.1) *is* uniformly well posed.

From now on we shall assume that the Cauchy problem (A.2) is uniformly well posed and define operators $S(t) : D(A_0) \to E$ by the formula:

$$S(t)x = u(t, x), \quad \forall\, x \in D(A_0), \quad \forall\, t \geq 0. \tag{A.3}$$

For all $t \geq 0$ the linear operator $S(t)$ can be uniquely extended to a linear bounded operator on the whole E, which we still denote by $S(t)$. We have clearly

$$S(0) = I, \tag{A.4}$$

moreover, by the uniqueness

$$S(t+s) = S(t)S(s), \quad \forall\, t, s \geq 0. \tag{A.5}$$

Finally, by the uniform boundedness theorem, it follows that:

$$S(\cdot)x \text{ is continuous in } [0, +\infty), \quad \forall\, x \in E. \tag{A.6}$$

In this way we are led directly from the study of the uniformly well posed problem to the family $S(t)$, $t \geq 0$, of linear bounded operators in E satisfying (A.4), (A.5) and (A.6).

Any family $S(\cdot)$ of bounded linear operators on E satisfying (A.4), (A.5) and (A.6) is called a C_0-semigroup of linear operators. So the concept of C_0-semigroup is in a sense equivalent to that of uniformly well posed Cauchy problem.

The *infinitesimal generator* A of $S(\cdot)$ is a linear operator defined as follows

$$\begin{cases} D(A) = \left\{ x \in H : \exists \lim\limits_{h \to 0^+} \dfrac{S(t)x - x}{h} \right\} \\ Ax = \lim\limits_{h \to 0^+} \dfrac{S(t)x - x}{h}, \quad \forall\, x \in D(A). \end{cases} \tag{A.7}$$

It is easy to see that A is an extension of A_0 and moreover that the problem

$$\begin{cases} u'(t) = Au(t), \quad t \geq 0, \\ u(0) = x \in E, \end{cases} \tag{A.8}$$

is also uniformly well posed with the same associated semigroup $S(\cdot)$. This is why in our following considerations we will consider the Cauchy problem (A.8) with the operator A being the infinitesimal generator of a C_0-semigroup.

A.2 Basic properties of C_0-semigroups

To formulate basic properties of C_0-semigroups, we introduce some notation. Let E be a real or complex Banach space. We say that a linear operator $L : D(L) \subset E \to E$ is *closed* if its graph:

$$\mathscr{G}_L = \{(x, y) \in E \times E : x \in D(L),\ Lx = y\},$$

is closed in $E \times E$ (endowed with the product topology). If L is closed we always endow the domain $D(L)$ with the *graph* norm

$$\|x\|_{D(L)} = \|x\| + \|Lx\|, \quad x \in D(L).$$

By definition, a complex number λ belongs to the *resolvent set* $\rho(L)$ of L if $\lambda I - L$ is one-to-one and onto. If $\lambda \in \rho(L)$, we set

$$R(\lambda, L) = (\lambda I - L)^{-1},$$

and call $R(\lambda, L)$ the *resolvent operator* of L. By the closed graph theorem, $R(\lambda, L)$ is bounded. The complement of $\rho(L)$ in $\mathbb{C}$ is called the *spectrum* of L.

For general C_0-semigroups we have the following basic properties (for proofs see for example [587, Theorems 1.2.4. and 1.3.1]).

Proposition A.2 *Let $S(\cdot)$ be a C_0-semigroup in E and let A be its infinitesimal generator. Then A is closed and the domain $D(A)$ is dense in E. Moreover, if $x \in D(A)$, then*

$$S(\cdot)x \in C^1([0, +\infty); E) \cap C([0, +\infty), D(A))$$

and

$$\frac{d}{dt}S(t)x = AS(t)x = S(t)Ax, \quad t \geq 0.$$

Theorem A.3 (Hille–Yosida) *Let $A : D(A) \subset E \to E$ be a linear closed operator on E. Then the following statements are equivalent.*

(i) A is the infinitesimal generator of a C_0-semigroup $S(\cdot)$ such that

$$\|S(t)\| \leq Me^{\omega t}, \quad \forall\, t \geq 0. \tag{A.9}$$

(ii) $D(A)$ is dense in E, the resolvent set $\rho(A)$ contains the interval $(\omega, +\infty)$ and the following estimates hold

$$\|R^k(\lambda, A)\| \leq \frac{M}{(\lambda - \omega)^k}, \quad \forall\, k \in \mathbb{N}. \tag{A.10}$$

Moreover if either (i) or (ii) holds then

$$R(\lambda, A)x = \int_0^\infty e^{-\lambda t} S(t)x\, dt, \quad \forall\, x \in X,\ \lambda > \omega. \tag{A.11}$$

Finally

$$S(t)x = \lim_{n\to\infty} e^{tA_n}x, \quad \forall\, x \in E, \tag{A.12}$$

where $A_n = nAR(n, A)$ and the following estimate holds

$$\|e^{tA_n}\| \leq Me^{\frac{\omega n t}{n-\omega}}, \quad \forall\, t \geq 0,\ n > \omega. \tag{A.13}$$

The operators $A_n = AJ_n$ where $J_n = nR(n, A)$, $n > \omega$, are called the *Yosida approximations* of A. The following properties of Yosida approximations will be frequently used.

Proposition A.4 *Let $A : D(A) \subset E \to E$ be the infinitesimal generator of a C_0-semigroup. Then*

$$\begin{cases} \lim_{n\to\infty} nR(n,A)x = x, & \forall\, x \in E, \\ \lim_{n\to\infty} A_n x = Ax, & \forall\, x \in D(A). \end{cases}$$

If the constant M in (A.9) is equal to 1, then the semigroup $S(\cdot)$ is called a *pseudo-contraction* C_0-semigroup; if in addition $\omega \le 0$ it is called a *contraction* C_0-semigroup. The number

$$\omega_0 = \liminf_{t\to+\infty} \frac{1}{t} \log \|S(t)\|, \tag{A.14}$$

is called the *type* of the semigroup $S(\cdot)$. If ω_0 is the type of $S(\cdot)$ then, for any $\varepsilon > 0$ there exists $M_\varepsilon \ge 1$ such that

$$\|S(t)\| \le M_\varepsilon e^{(\omega_0+\varepsilon)t}, \quad t \ge 0.$$

A.3 Cauchy problem for nonhomogeneous equations

We are here concerned with the initial value problem in E

$$\begin{cases} u'(t) = Au(t) + f(t), & t \in [0,T], \\ u(0) = x \in E, \end{cases} \tag{A.15}$$

where A is the infinitesimal generator of a C_0-semigroup $S(\cdot)$ in E and $f \in L^p(0,T;E)$, $p \ge 1$.

Definition A.5

(i) *A* strict solution *of problem* (A.15) *in $L^p(0,T;E)$, $p \in [1,\infty]$, is a function u that belongs to $W^{1,p}(0,T;E) \cap L^p([0,T];D(A))$ and fulfils* (A.15). [1]

(ii) *A* strict solution *of problem* (A.15) *in $C([0,T];E)$, is a function u that belongs to $C^1([0,T];E) \cap C([0,T];D(A))$ and fulfils* (A.15).

(iii) *A* weak solution *of problem* (A.15) *is a function $u \in C([0,T];E)$ such that*

$$\varphi(u(t)) = \varphi(x) + \int_0^t (A^*\varphi)(u(s))ds + \int_0^t f(s)ds, \quad \forall\, \varphi \in D(A^*). \tag{A.16}$$

Obviously, a strict solution is also a weak solution, but not conversely.

The following result is proved in [33].

[1] $W^{1,p}([0,T];E)$ is the set of all functions $u : [0.T] \to E$ such that there exists a sequence $\{u_n\}$ in $C^1([0,T];E)$ and an element $v \in L^p([0,T];E)$ such that $u_n \to u$ and $u_n' \to v$ in $L^p([0,T];E)$. The function v is unique and we set $v = u'$. If $u \in W^{1,p}([0,T];E)$ then there exists an absolute continuous function $\widetilde{u} \in C([0,T];E)$ which is equal to u a.e. in $[0,T]$. For general properties of vector valued functions see for instance [3].

Proposition A.6 *Let A be the infinitesimal generator of a C_0-semigroup $S(\cdot)$ in E and $f \in L^1(0,T;E)$. Then there exists a unique weak solution u of equation* (A.15) *and it is given by the variation of constants formula*

$$u(t) = S(t)x + \int_0^t S(t-s)f(s)ds, \quad t \in [0,T]. \tag{A.17}$$

The function $u(\cdot)$ defined by (A.17) is called the *mild* solution of problem (A.15).

Before proving a sufficient condition for the existence of strict solutions, it is convenient to introduce the approximating problem

$$\begin{cases} u_n'(t) = A_n u_n(t) + f(t), & t \in [0,T], \\ u_n(0) = x \in X, \end{cases} \tag{A.18}$$

where A_n are the Yosida approximations of A. Clearly problem (A.18) has a unique solution $u_n \in W^{1,1}(0,T;E)$, given by the variation of constants formula

$$u_n(t) = S_n(t)x + \int_0^t S_n(t-s)f(s)ds, \quad t \in [0,T], \tag{A.19}$$

where $S_n(t) = e^{tA_n}$, $t > 0$, and moreover

$$\lim_{n\to\infty} u_n = u \quad \text{in } C([0,T];E). \tag{A.20}$$

We can now prove the following.

Proposition A.7 *Let A be the infinitesimal generator of a C_0-semigroup $S(\cdot)$ in E.*

(i) If $x \in D(A)$ and $f \in W^{1,p}(0,T;E)$ with $p \geq 1$, then problem (A.15) *has a unique strict solution u in $C([0,T];E)$, given by formula* (A.19) *and moreover $u \in C^1([0,T];E) \cap C([0,T];D(A))$.*

(ii) If $x \in D(A)$ and $f \in L^p(0,T;D(A))$, then problem (A.15) *has a unique strict solution u in $L^p(0,T;E)$, given by formula* (A.17) *and moreover $u \in W^{1,p}(0,T;E) \cap C([0,T];D(A))$.*

Proof We sketch only the proof of (i), the proof of (ii) being similar. Let us consider the mild solution v of problem:

$$\begin{cases} v'(t) = Av(t) + f'(t), & t \in [0,T], \\ v(0) = Ax + f(0), \end{cases}$$

and the solution of

$$\begin{cases} v_n'(t) = A_n v_n(t) + f'(t), & t \in [0,T], \\ v_n(0) = A_n x + f(0). \end{cases}$$

Clearly $v_n(t) = u_n'(t)$. Since

$$u_n \to u, \quad v_n \to v \quad \text{in } C([0,T];E),$$

we have $u \in C^1([0,T];E)$ and $u' = v$. Moreover

$$A_n u_n = v_n - f \to u' - f \quad \text{in } C([0,T];E).$$

Since A is closed this implies that $u(t) \in D(A)\ \forall\, t \in [0,T]$ and $Au(t) = u'(t) + f(t)$. Consequently $Au \in C([0,T];E)$. □

We end this subsection by giving an existence result of a strict solution for a special form of problem (A.15). We consider two Hilbert spaces H and U (inner product $\langle \cdot, \cdot \rangle$) and the problem

$$\begin{cases} u'(t) = Au(t) + Bg(t), & t \in [0,T], \\ u(0) = 0, \end{cases} \tag{A.21}$$

where A is as before the infinitesimal generator of a C_0-semigroup $S(\cdot)$ in H, $g \in C([0,T];U)$ and $B \in L(U;H)$. We shall prove that problem (A.21) has a strict solution in $C([0,T];H)$ under the following assumption.

Hypothesis A.1 *There exists $K > 0$ such that*

$$\int_0^T |B^* A^* e^{sA^*} x|^2 ds \le K^2 |x|^2, \quad \forall\, x \in D(A^*). \tag{A.22}$$

We have in fact the following result, due to [486].

Proposition A.8 *Assume that the Hypothesis* A.1 *is fulfilled. Let $g \in L^2(0,T;U)$. Then a weak solution u of problem* (A.21) *is a strict solution in $C([0,T];E)$ and there exists $K > 0$ such that*

$$|Au(t)| \le K |g|_{L^2(0,T;U)}, \quad t \in [0,T]. \tag{A.23}$$

Proof We first prove that $u(t) \in D(A)$ for all $t \in [0,T]$. In fact, if $x \in D(A^*)$ we have

$$\langle u(t), A^* x \rangle = \int_0^t \langle g(s), D^* A^* e^{(t-s)A^*} x \rangle ds.$$

It follows that

$$\begin{aligned} |(u(t), A^* x)|^2 &\le |g|^2_{L^2(0,T;U)} \int_0^t |D^* A^* e^{(t-s)A^*} x|^2 ds \\ &\le K^2 |g|^2_{L^2(0,T;U)} |x|^2. \end{aligned}$$

This inequality implies that the linear form on H: $x \to \langle u(t), A^* x \rangle$, is continuous and so that $u(t) \in D(A)$ and (A.23) holds. This proves that $u \in C([0,T];H)$ by a standard approximation argument. □

A.4 Cauchy problem for analytic semigroups

A.4.1 Analytic generators

We proceed now to study an important class of semigroups, called *analytic semigroups*, and the associated Cauchy problems.

For any $\omega \in \mathbb{R}^1$ and $\theta \in (0, \pi)$ we denote by $S_{\omega,\theta}$ the sector in $\mathbb{C}$

$$S_{\omega,\theta} = \{\lambda \in \mathbb{C}\backslash\{\omega\} : \ |\arg(\lambda - \omega)| \leq \theta\}.$$

Assume that A is a linear closed operator such that the following holds.

Hypothesis A.2

(i) *There exist $\omega \in \mathbb{R}^1$ and $\theta_0 \in (\frac{\pi}{2}, \pi)$ such that $\rho(A) \supset S_{\omega,\theta_0}$.*
(ii) *There exists $M > 0$ such that*

$$\|R(\lambda, A)\| \leq \frac{M}{|\lambda - \omega|}, \qquad \forall\, \lambda \in S_{\omega,\theta_0}. \tag{A.24}$$

Then we can define a semigroup $S(\cdot)$ of bounded linear operators in E by setting $S(0) = I$ and:

$$S(t) = \frac{1}{2\pi i}\int_{\gamma_{\varepsilon,\theta}} e^{\lambda t} R(\lambda, A)d\lambda, \qquad t > 0. \tag{A.25}$$

In definition (A.25) θ belongs to the sector $(\frac{\pi}{2}, \theta_0)$ and $\gamma_{\varepsilon,\theta}$ is the following, oriented counterclockwise, path in $\mathbb{C}$

$$\gamma_{\varepsilon,\theta} = \gamma^+_{\varepsilon,\theta} \cup \gamma^-_{\varepsilon,\theta} \cup \gamma^0_{\varepsilon,\theta}.$$

$$\gamma^{\pm}_{\varepsilon,\theta} = \{z \in \mathbb{C};\ z = \omega + re^{\pm i\theta},\ r \geq \varepsilon\}.$$

$$\gamma^0_{\varepsilon,\theta} = \{z \in \mathbb{C};\ z = \omega + \varepsilon e^{i\eta},\ |\eta| \leq \theta\}.$$

Remark that the integral in (A.25) is well defined since $\theta > \frac{\pi}{2}$; moreover, by the Cauchy theorem for holomorphic functions, $S(t)$ does not depend on the choice of ε and θ. We say that $S(\cdot)$ is the semigroup generated by A, although it may not satisfy the continuity property (A.6) and the domain $D(A)$ may not be dense in E.

Theorem A.9 *Assume that A fulfils* (A.24) *and let $S(\cdot)$ be defined by* (A.25). *Then the following statements hold.*

(i) *The mapping $S : (0, +\infty) \to L(E)$, $t \mapsto S(t)$ is analytic. Moreover for any $x \in E$, $t > 0$ and $n \in \mathbb{N}$, $S(t)x \in D(A^n)$ and*

$$S^n(t)x = A^n S(t)x. \tag{A.26}$$

(ii) *We have $S(t + s) = S(t)S(s)$ for all $t, s \geq 0$.*
(iii) *$S(\cdot)x$ is continuous at 0 if and only if $x \in \overline{D(A)}$.*

(iv) There exists $M, N > 0$ such that

$$\|S(t)\| \leq Me^{\omega t}, \quad \|AS(t)\| \leq e^{\omega t}(N/t + \omega M), \quad \forall\, t \geq 0.$$

(v) $S(\cdot)$ can be extended to an analytic $L(E)$-valued function in $S_{0,\theta_0-\frac{\pi}{2}}$.

Because of property (v), we say that $S(\cdot)$ is an *analytic semigroup*. For the proof of Theorem A.9 see for example [648, Propositions 1.1 and 1.2].

Since the solutions of the Cauchy problem (A.8) with A satisfying Hypothesis A.2 have many properties of the solutions of classical parabolic equations, therefore in this case we say that the Cauchy problem (A.8) is *parabolic*.

Assume that A fulfils Hypothesis A.2. It follows from [648] that $S(\cdot)$ is a C_0-semigroup if and only if $D(A)$ is dense in E. In the general case let us consider the part A_F of A in $F = \overline{D(A)}$. We recall that if F is a subspace of E, then the *part* A_F of A in F is defined by

$$\begin{cases} D(A_F) = \{y \in D(A) \cap F : \; Ay \in F\}, \\ A_F y = Ay, \quad \forall\, y \in D(A_F). \end{cases} \tag{A.27}$$

It is easily seen that $D(A_F)$ is dense in F and the restriction $S_F(\cdot)$ of $S(\cdot)$ to F is a C_0-semigroup.

Remark A.10 Assume that A fulfils Hypothesis A.2 and set

$$\delta_0 = \sup_{\lambda \in \sigma(A)} \mathfrak{Re}\, \lambda.$$

Then, by using the representation formula (A.25) it is easy to check that, for any $\varepsilon > 0$, there exists $M_\varepsilon \geq 1$ such that

$$\|S(t)\| \leq M_\varepsilon e^{(\delta_0+\varepsilon)t}, \quad t \geq 0. \tag{A.28}$$

Thus, if $D(A)$ is dense in E, then δ_0 coincides with the type of the C_0-semigroup $S(\cdot)$. □

Remark A.11 Assume that A fulfils Hypothesis A.2 and that

$$\sigma(A) = \{\lambda_0\} \cup \sigma_1(A),$$

where λ_0 is a semisimple eigenvalue of A and that there exists $\eta > 0$ such that

$$\mathfrak{Re}\, \lambda \leq \lambda_0 - \eta, \quad \forall\, \lambda \in \sigma_1.$$

Then (A.28) holds with $\varepsilon = 0$ (see for instance [401]). □

A.4.2 Variational generators

We consider here an important subclass of analytic semigroups. We are given a Hilbert space H, norm $|\cdot|$, inner product $\langle\cdot,\cdot\rangle$. We say that a linear operator $A : D(A) \subset H \to H$ is *variational* if we have the following.

(i) There exists a Hilbert space V densely embedded in H and a continuous bilinear form $a : V \times V \to \mathbb{R}^1$ such that

$$-a(v, v) \geq \alpha \|v\|_V^2 - \lambda_0 |v|^2, \quad \forall\, v \in V, \tag{A.29}$$

for suitable constants $\alpha > 0$, $\lambda_0 \geq 0$.

(ii) $D(A) = \{u \in V : a(u, \cdot)$ is continuous in the topology of $H\}$.

(iii) $a(u, v) = \langle Au, v\rangle, \quad \forall\, u \in D(A),\ \forall\, v \in V.$

Proposition A.12 *Let A be a variational operator in H such that* (A.29) *holds. Then A generates an analytic semigroup $S(\cdot)$ such that $\|S(t)\| \leq e^{\lambda_0 t}$, $t \geq 0$. Moreover, if A is symmetric then A is self-adjoint.*

For the proof see [668].

Let us end this subsection by giving an abstract energy identity, which will be useful in what follows.

Proposition A.13 *Assume that A is a variational operator in H and let $S(\cdot)$ be the semigroup generated by A in H. Then the following identity holds*

$$\int_0^t a(S(s)x, S(s)x)ds = \frac{1}{2}[|S(t)x|^2 - |x|^2], \quad \forall\, t \geq 0,\ \forall\, x \in V. \tag{A.30}$$

Proof The conclusion follows by integrating both sides of the identity

$$a(S(s)x, S(s)x) = \frac{1}{2}\frac{d}{ds}|S(s)x|^2, \quad s \geq 0,\ x \in V.$$

□

A.4.3 Fractional powers and interpolation spaces

To study regularity properties of solutions to Cauchy problems it is convenient to introduce several scales of subspaces of E. To simplify the notation we assume, in addition to Hypothesis A.2, that the semigroup $S(\cdot)$ is of negative type, which means that $\omega < 0$. For any $\alpha \in (0, 1)$ we set

$$(-A)^{-\alpha}x = \frac{1}{2\pi i}\int_{\gamma_{\varepsilon,\theta}} (-\lambda)^{\alpha} R(\lambda, A)x\, d\lambda, \quad t > 0,\ x \in X,$$

where we have used the symbols of Section A.4.1.

As is easily checked, $(-A)^{-\alpha}$ is one-to-one. We shall denote by $(-A)^{\alpha}$ the inverse of $(-A)^{-\alpha}$ and by $D((-A)^{\alpha})$ its domain. It is not difficult to prove (see [587]) that

$$(-A)^{\alpha}(-A)^{\beta} = (-A)^{\alpha+\beta}, \quad \forall\, \alpha, \beta \in (0, 1),\ \alpha + \beta \leq 1.$$

The operators $(-A)^{\alpha}$ are called *fractional powers* of $-A$. The first scale of subspaces of E is provided by

$$D((-A)^{\alpha}), \quad \alpha \in (0, 1).$$

By (A.25) we obtain a representation formula for $(-A)^{\alpha}S(\cdot)$ which will be often used.

Proposition A.14 *Let A be a linear operator satisfying Hypothesis* A.2 *with $\omega < 0$. Then for any $\alpha \in (0,1)$ and $t > 0$ we have $S(t)x \in D((-A)^\alpha)$, $\forall\, x \in X$ and*

$$(-A)^\alpha S(t)x = \frac{1}{2\pi i}\int_{\gamma_{\varepsilon,\theta}} e^{\lambda t}(-\lambda)^\alpha R(\lambda, A)x\, d\lambda. \tag{A.31}$$

Moreover for any $\varepsilon > 0$, there exists $N_{\alpha,\varepsilon} > 0$ such that

$$\|(-A)^\alpha S(t)\| \le N_{\alpha,\varepsilon} t^{-\alpha} e^{(\omega+\varepsilon)t}, \quad t > 0. \tag{A.32}$$

We will need two other scales of spaces related to the so called *interpolation* theory. For any $\alpha \in (0,1)$ we set

$$\|x\|_{\alpha,\infty} = \sup_{t>0} t^{-\alpha}\, \|S(t)x - x\|, \quad x \in E,$$

and denote by $D_A(\alpha,\infty)$ the Banach space of all $x \in E$ such that $\|x\|_{\alpha,\infty} < +\infty$, endowed with the norm $\|\cdot\| + \|\cdot\|_{\alpha,\infty}$. We set

$$D_A(\alpha+1,\infty) = \{x \in D(A) :\ Ax \in D_A(\alpha,\infty)\}.$$

Moreover $D_A(\alpha,\infty)$ is an invariant subspace of $S(t)$, $t > 0$, and the restriction of $S(t)$ to $D_A(\alpha,\infty)$ generates a C_0 semigroup in $D_A(\alpha,\infty)$. Since $S(\cdot)$ is an analytic semigroup then an equivalent norm is the following, see [131], $\|\cdot\| + \widehat{\|\cdot\|}_{\alpha,\infty}$, where

$$\widehat{\|x\|}_{\alpha,\infty} = \sup_{t>0} t^{1-\alpha}\|AS(t)x\|, \quad x \in E.$$

Finally we define the spaces $D_A(\alpha,2)$, for any $\alpha \in (0,1/2]$

$$D_A(\alpha,2) = \left\{x \in E :\ |x|_\alpha^2 = \int_0^\infty \xi^{1-2\alpha}|AS(\xi)x|^2 d\xi < \infty\right\}. \tag{A.33}$$

For $\alpha \in (1/2,1)$

$$D_A(\alpha,2) = \left\{x \in E :\ |x|_\alpha^2 = \int_0^\infty \xi^{3-2\alpha}|A^2S(\xi)x|^2 d\xi < \infty\right\}. \tag{A.34}$$

It follows from definitions (A.33), (A.34) that *the restriction of $S(\cdot)$ to $D_A(\alpha,2)$ is a contractions semigroup in $D_A(\alpha,2)$* for all $\alpha \in (0,1)$. In fact, if $\alpha \in (0,\frac{1}{2})$, we have

$$|S(t)x|_\alpha^2 = \int_t^\infty (\eta - t)^{1-2\theta}|AS(\eta)x|^2 d\eta \le |x|_\alpha^2,$$

whereas if $\alpha \in (\frac{1}{2},1)$, we have

$$|S(t)x|_\alpha^2 = \int_t^\infty (\eta - t)^{3-2\theta}|A^2S(\eta)x|^2 d\eta \le |x|_\alpha^2.$$

We have the following relationship between some of the introduced spaces.

Proposition A.15 *Let A be a linear operator that fulfils Hypothesis* A.2 *with $\omega < 0$, and let $\theta \in (0,1)$. Then the following inclusions hold.*

(i) $D((-A)^\theta) \subset D_A(\theta, \infty)$.
(ii) $D_A(\theta, \infty) \subset D((-A)^{\theta-\varepsilon})$, for $0 < \varepsilon < \theta$.

Proof Let $x \in D((-A)^\theta)$, then by (A.32) there exists $C > 0$ such that

$$\|AS(t)x\| = \|(-A)^{1-\theta} S(t)(-A)^\theta x\| \le Ct^{\theta-1}\|(-A)^\theta x\|.$$

Thus $\|t^{1-\theta} AS(t)x\|$ is bounded and $x \in D_A(\theta, \infty)$.

Assume, conversely, that $x \in D_A(\theta, \infty)$. Then by recalling (A.11) it is not difficult to show that there exists a constant $C_1(x)$ such that

$$\|\lambda^\theta AR(\lambda, A)x\| \le C_1(x), \quad \lambda > 0.$$

Now, let $\varepsilon \in (0, \theta)$; then it is easy to show that $x \in D((-A)^{\theta-\varepsilon})$ and

$$(-A)^{\theta-\varepsilon}x = \frac{1}{2\pi i}\int_{\gamma_{\varepsilon,\theta}} \lambda^{\theta-\varepsilon-1} AR(\lambda, A)x \, d\lambda.$$

□

Remark A.16 We will need the following inclusion result, see [511, Lemma 1.1]:

$$C^{1,\alpha}([0, T]; E) \cap C^\alpha([0, T]; D(A)) \subset C^{1+\alpha-\beta}([0, T]; D_A(\beta, \infty)) \quad \text{(A.35)}$$

for all $\alpha \in (0, 1)$, $\beta \in (0, \alpha]$. □

Proposition A.17 *Let H be a Hilbert space and let A be the infinitesimal generator of an analytic C_0-semigroup $S(\cdot)$ of negative type. Then for any $\theta \in (0, 1)$ and any $\varepsilon \in (0, \theta)$ the following inclusion holds*

$$D((-A)^\theta) \subset D_A(\theta - \varepsilon, 2). \quad \text{(A.36)}$$

Proof We first remark that by (A.32) there exist two constants $C > 0$ and $\alpha > 0$ such that

$$\|(-A)^{1-\theta} S(t)\| \le Ct^{\theta-1}e^{-\alpha t}, \quad t \ge 0.$$

Now let $x \in D((-A)^\theta)$, then we have

$$\|AS(t)x\| = \|(-A)^{1-\theta} S(t)(-A)^\theta x\| \le Ct^{\theta-1}e^{-\alpha t}\|(-A)^\theta x\|.$$

But this implies the conclusion since

$$\int_0^\infty \xi^{1-2\theta-2\varepsilon}\|AS(\xi)x\|^2 d\xi \le C^2\|(-A)^\theta x\|^2 \int_0^\infty \xi^{2\varepsilon-1}e^{-\alpha\xi} d\xi < \infty.$$

□

Remark A.18 In [498] the following inclusion is proved

$$W^{1,2}(0, T; H) \cap L^2(0, T; D(A)) \subset C([0, T]; D_A(1/2, 2)).$$

□

The following result is due to [714].

Proposition A.19 *Assume that there exists $\gamma \in (0, 1)$ such that $D((-A)^\theta)$ is isomorphic to $D((-A^*)^\theta)$, $\forall\, \theta \in (0, \gamma)$, where A^* denotes the adjoint of A. Then $D_A(\theta, 2)$ is isomorphic to $D((-A)^\theta)$ for all $\theta \in (0, 1)$.*

Remark A.20 The hypothesis of Proposition A.19 is fulfilled if $S(\cdot)$ is a contraction semigroup (see [443]), and in particular if A is self-adjoint negative. □

The following result is due to [299].

Proposition A.21 *Let H be a Hilbert space and let A be the infinitesimal generator of an analytic semigroup $S(\cdot)$ in H of negative type. Assume that $D((-A)^{1/2})$ is isomorphic to $D_A(1/2, 2)$. Then there exists a constant $C > 0$ such that*

$$\int_0^t |(-A)^{1/2} S(s)x|^2 ds \le C|x|^2, \quad \forall\, x \in D(A). \tag{A.37}$$

Proof If $x \in D(A)$ we have

$$\begin{aligned}\int_0^t |(-A)^{1/2} S(s)x|^2 ds &= \int_0^t |AS(s)(-A)^{-1/2} x|^2 ds \\ &< |(-A)^{-1/2} x|^2_{D_A(1/2,2)},\end{aligned}$$

and the conclusion follows since $D((-A)^{1/2})$ is isomorphic to $D_A(1/2, 2)$. □

A.4.4 Regularity of solutions for the homogeneous Cauchy problem

To study regularity properties of the mild solution u of problem (A.15) we consider separately homogeneous and nonhomogeneous cases. Denote

$$u_1(t) = S(t)s, \quad u_2(t) = \int_0^t S(t-s) f(s) ds$$

and notice that $u = u_1 + u_2$. In this subsection we investigate regularity of u_1. First of all we remark that by formula (A.25), $u_1(\cdot)$ is analytic for $t > 0$; moreover $u_1(t) \in D(A^k)$, $k \in \mathbb{N}$, $t > 0$. In addition $u_1(\cdot)$ is continuous at 0 if and only if $x \in F = \overline{D(A)}$ and it is differentiable at 0 if and only if $x \in D(A_F)$ where A_F is the part of A in F, see [648].

In terms of interpolation spaces $D_A(\alpha, \infty)$ we have the following result.

Proposition A.22 *Let A be a linear operator satisfying Hypothesis* A.2 *with $\omega < 0$ and let $\alpha \in (0, 1)$. Then $S(\cdot)x$ is α-Hölder continuous if and only if $x \in D_A(\alpha, \infty)$.*

Finally, concerning $D_A(\alpha, 2)$ we have the result when $E = H$ is a Hilbert space.

Proposition A.23 *The following statements hold.*

(i) If $x \in D_A(\theta, 2))$ with $\theta \in (0, 1/2)$, then $u_1 \in L^2([0, \infty); D_A(\theta + 1/2, 2))$.
(ii) If $x \in D_A(1/2, 2))$ then $u_1 \in L^2([0, \infty); D(A))$.

Proof We first prove (i). Set $Z = L^2([0,\infty); D_A(\theta + 1/2, 2))$; then the conclusion follows from the estimates

$$\begin{aligned}
\|v\|_Z^2 &= \int_0^\infty dt \int_0^\infty \xi^{-2\theta} |AS(t+\xi)x|^2 d\xi \\
&= \int_0^\infty \xi^{-2\theta} d\xi \int_\xi^\infty |AS(\tau)x|^2 d\tau = \int_0^\infty |AS(\tau)x|^2 d\tau \\
&= \int_0^\tau \xi^{-2\theta} d\xi = \frac{1}{1-2\theta} |x|_\theta^2 < +\infty.
\end{aligned}$$

So (i) is proved. (ii) is clear because

$$\int_0^\infty |AS(t)x|^2 dt = |x|_{1/2}^2.$$

□

A.4.5 Regularity for the nonhomogeneous problem

We now pass to the solution $u_2(\cdot)$ of the nonhomogeneous problem

$$u_2'(t) = Au_2(t) + f(t), \quad u_2(0) = 0.$$

We start with the following result (see [587]).

Proposition A.24 *Assume that $f \in C^\alpha([0,T]; E)$ for some $\alpha \in (0,1)$. Then $u_2 \in C^1([0,T]; E) \cap C([0,T]; D(A))$.*

Now we state a maximal regularity result proved in [648] (Theorem 4.1)(part (i)) and in [201] (part (ii)).

Proposition A.25 *Let A be a linear operator which fulfils Hypothesis* A.2. *Then the following statements hold.*

(i) If $f \in C^\alpha([0,T]; E)$ for some $\alpha \in (0,1)$ and $T > 0$ and if $f(0) \in D_A(\alpha,\infty)$, then

$$u_2 \in C^{1,\alpha}([0,T]; E) \cap C^\alpha([0,T]; D(A)).$$

(ii) If $f \in C([0,T]; D_A(\alpha,\infty))$ for some $\alpha \in (0,1)$ and $T > 0$, then

$$u_2 \in C^1([0,T]; D_A(\alpha,\infty)) \cap C([0,T]; D_A(\alpha+1,\infty)).$$

We conclude with the following result.

Proposition A.26 *Let H be a Hilbert space and let A be the infinitesimal generator of an analytic C_0-semigroup $S(\cdot)$ of negative type. Let moreover $f \in L^2(0,\infty; H)$. Then $u_2 \in W^{1,2}(0,\infty : H) \cap L^2(0,\infty; D(A))$.*

Proof Set

$$\widetilde{f}(\lambda) = \int_0^{\infty} e^{i\lambda t} f(t)dt,$$

$$\widetilde{u_2}(\lambda) = \int_0^{\infty} e^{i\lambda t} u_2(t)dt,$$

then we have

$$\widetilde{u_2}(\lambda) = R(ik, A)\widetilde{f}(\lambda), \quad \lambda \in \mathbb{R}^1.$$

It follows that $\widetilde{u}_2(\lambda) \in D(A)$ and

$$\|A\widetilde{u}_2(\lambda)\| \leq 2M\|\widetilde{f}_2(\lambda)\|,$$

which implies

$$\int_{-\infty}^{+\infty} |A\widetilde{u_2}(\lambda)|^2 dk \leq 4M^2 \int_{-\infty}^{+\infty} |\widetilde{f}(\lambda)|^2 dk.$$

By the Parseval identity we get $Au_2 \in L^2(0, \infty, H)$ and then the conclusion follows. □

A.5 Example of deterministic systems

A.5.1 Delay systems

We are concerned here with the problem

$$\begin{cases} z'(t) = \int_{-r}^{0} a(d\theta)z(t+\theta) + f(t), & t \geq 0, \\ z(0) = h_0, \\ z(\theta) = h_1(\theta), & \theta \in [-r, 0], \ a.e. \end{cases} \tag{A.38}$$

where $a(\cdot)$ is an $N \times N$ matrix valued finite measure on $[-r, 0]$, $f : [0, +\infty) \to \mathbb{R}^N$ is a locally integrable function, $h_0 \in \mathbb{R}^N$, $h_1 \in L^2([-r, 0]; \mathbb{R}^N)$ and r is a positive number equal to the maximal *delay*. It is possible to prove that this problem has a unique absolute continuous solution z defined in $[-r, +\infty)$.

Assume for the moment $f \equiv 0$ and consider the Hilbert space $H = \mathbb{R}^N \oplus L^2([-r, 0]; \mathbb{R}^N)$. It turns out that the following formula

$$S(t)\begin{pmatrix} h_0 \\ h_1 \end{pmatrix} = \begin{pmatrix} z(t) \\ z_t \end{pmatrix}, \quad \forall \begin{pmatrix} h_0 \\ h_1 \end{pmatrix} \in H, \tag{A.39}$$

where z is the solution to (A.38), with $f \equiv 0$ and $z_t(\theta) = z(t+\theta), \theta \in [-r, 0]$, defines a C_0-semigroup on H. The following result is well known, see for instance [258].

Proposition A.27 *$S(\cdot)$ is a C_0-semigroup in H which moreover is differentiable for any $t \geq r$. The infinitesimal generator A of $S(\cdot)$ is given by:*

$$\begin{cases} D(A) = \left\{ \begin{pmatrix} h_0 \\ h_1 \end{pmatrix} \in H : \ h_0 \in \mathbb{R}^N, \ h_1 \in W^{1,2}(-r, 0, \mathbb{R}^N), \ h_1(0) = h_0 \right\} \\ Ah = \begin{pmatrix} \int_{-r}^{0} a(d\theta) h_1(\theta) \\ \frac{dh_1}{d\theta} \end{pmatrix} \end{cases} \tag{A.40}$$

and we have

$$\sigma(A) = \left\{ \lambda \in \mathbb{C} : \ \det \left(\lambda - \int_{-r}^{0} e^{\lambda\theta} a(d\theta) \right) = 0 \right\}. \tag{A.41}$$

We now consider the general problem (A.38). We have the following formula

$$(z(t), z_t) = S(t)(h_0, h_1) + \int_0^t S(t-s)(f(s), 0) ds. \tag{A.42}$$

So the problem (A.38) is equivalent to the Cauchy problem

$$u' = Au + Bf, \quad u(0) \in H$$

where A is given by (A.40) and $B = \begin{pmatrix} I \\ 0 \end{pmatrix}$.

A.5.2 Heat equation

From now on we consider an open set $\mathcal{O}$ (not necessarily bounded) of $\mathbb{R}^N$ with a regular boundary $\partial\mathcal{O}$. We shall denote by $\nu(\xi)$ the outward normal to $\partial\mathcal{O}$ at the point $\xi \in \partial\mathcal{O}$. We shall consider the spaces

$$X_p = \begin{cases} L^p(\mathcal{O}) & \text{if } p \in (1, \infty), \\ C(\overline{\mathcal{O}}) & \text{if } p = \infty, \end{cases}$$

as well as the Sobolev spaces $W^{k,p}(\mathcal{O})$, $k \in \mathbb{N}$, $p \in (1, \infty)$, and the spaces of Hölder continuous functions $C^\alpha(\overline{\mathcal{O}})$ and $C^{1,\alpha}(\overline{\mathcal{O}})$, $\alpha \in (0, 1)$.

We are given an elliptic operator

$$A_0 u = \sum_{i,j=1}^{N} a_{ij}(\xi) \frac{\partial^2 u}{\partial\xi_i \partial\xi_j} + \sum_{i=1}^{N} b_i(\xi) \frac{\partial u}{\partial\xi} + c(\xi) u, \tag{A.43}$$

where the coefficients a_{ij}, b_i, c , $i, j = 1, \ldots, N$, are continuous in $\overline{\mathcal{O}}$, the closure of $\mathcal{O}$, and there exists $\alpha_0 > 0$ such that

$$\sum_{i,j=1}^{N} a_{ij}(\xi) \alpha_i \alpha_j \geq \alpha_0 |\alpha|^2, \quad \forall\, \alpha = (\alpha_1, \ldots, \alpha_N) \in \mathbb{R}^N.$$

We denote by A_p the realization in X_p of A_0 under the Dirichlet boundary conditions, that is

$$\begin{cases} D(A_p) = W^{2,p}(\mathscr{O}) \cap W_0^{1,p}(\mathscr{O}) \\ Au = A_0 u, \quad \forall\, u \in D(A_p), \end{cases} \tag{A.44}$$

if $p \in [1, \infty)$ and

$$\begin{cases} D(A_\infty) = \{u \in C(\overline{\mathscr{O}}) : \ A_0 u \in C(\overline{\mathscr{O}}), \quad u = 0 \text{ on } \partial\mathscr{O}\} \\ A_\infty u = A_0 u, \quad \forall\, u \in D(A_\infty). \end{cases} \tag{A.45}$$

In the above definition $A_0 u$ is meant in the sense of distributions.

If $p \in [1, \infty)$ then $D(A_p)$ is dense in X_p and A_p is the infinitesimal generator of an analytic semigroup in $L^p(\mathscr{O})$ (see [5]).

We have moreover, see [362],

$$D_{A_p}(\theta, p) = \begin{cases} W^{2\theta,p}(\mathscr{O}) & \text{if } \theta \in (0, 1/(2p)) \\ \{u \in W^{2\theta,p}(\mathscr{O}) : \ u = 0 \quad \text{on } \partial\mathscr{O}\} & \text{if } \theta \in (1/(2p), 1). \end{cases} \tag{A.46}$$

If $p = \infty$ then $D(A_\infty)$ is not dense in X_∞, since the closure of $D(A_\infty)$ is $C_0(\overline{O})$, however A_∞ generates an analytic semigroup in $C(\overline{O})$ (see [660]).

Also, if $p = \infty$, we have, see [510],

$$D_{A_\infty}(\theta, \infty) = \begin{cases} C_0^{2\theta}(\mathscr{O}) & \text{if } \theta \in (0, 1/2), \\ C_0^{1,2\theta-1}(\mathscr{O}) & \text{if } \theta \in (1/2, 1). \end{cases} \tag{A.47}$$

We are now going to consider other realizations of the operator A_0. Let B_0 be the first order differential operator

$$B_0 u = \sum_{i=1}^{N} \beta_i(\xi) \frac{\partial u}{\partial \xi_i} + \gamma(\xi) u,$$

where $\beta_i, \gamma \in C^1(\overline{\mathscr{O}})$. We assume that B_0 is *nontangential*, that is

$$\sum_{i=1}^{N} \beta_i(\xi) \nu_i(\xi) \neq 0, \quad \forall\, \xi \in \partial\mathscr{O}.$$

We denote by C_p the realization in X_p of A_0 under mixed conditions, that is

$$\begin{cases} D(C_p) = \{u \in W^{2,p}(\mathscr{O}) : \ B_0 u = 0 \quad \text{on } \partial\mathscr{O}\} \\ C_p u = A_0 u, \quad \forall\, u \in D(C_p), \end{cases} \tag{A.48}$$

if $p \in [1, \infty)$ and

$$\begin{cases} D(C_\infty) = \{u \in C(\overline{\mathscr{O}}) : \ B_0 u = 0 \quad \text{on } \partial\mathscr{O}\} \\ C_\infty u = A_0 u, \quad \forall\, u \in D(C_\infty). \end{cases} \tag{A.49}$$

In this case $D(C_p)$ is dense in X_p for all $p \in (1, \infty]$ and C_p is the infinitesimal generator of an analytic semigroup in X_p (see [5] for $p \in (1, \infty)$ and [661]) for $p = \infty$.

From [362] we have

$$D_{C_p}(\theta, p) = \begin{cases} W^{2\theta,p}(\mathscr{O}) & \text{if } \theta \in (0, 1/(2p)) \\ \{u \in W^{2\theta,p}(\mathscr{O}); B_0 u = 0 \quad \text{on } \partial\mathscr{O}\} & \text{if } \theta \in (1/(2p), 1). \end{cases} \tag{A.50}$$

Also, if $p = \infty$, we have, see [2],

$$D_{C_\infty}(\theta, \infty) = \begin{cases} C^{2\theta}(\overline{\mathscr{O}}) & \text{if } \theta \in (0, 1/2), \\ \{u \in C^{1,2\theta-1}(\overline{\mathscr{O}}) : \ B_0 u = 0\} & \text{if } \theta \in (1/2, 1). \end{cases} \tag{A.51}$$

Remark A.28 Assume that $\mathscr{O}$ is bounded. Then by the Sobolev embedding theorem the embeddings

$$D(A_p) \subset X_p, \quad D(C_p) \subset X_p,$$

are compact. Consequently the resolvent operators of A_p and C_p, $p \in (1, \infty]$, are compact and the spectra of A_p and C_p consist of a sequence of eigenvalues. Moreover $\sigma(A_p)$ and $\sigma(C_p)$ are independent of p, see [5]. □

Remark A.29 Assume that $\mathscr{O}$ is bounded and that A_0 coincides with the Laplace operator Δ. In this case the operator A_2 is self-adjoint in X_2. We denote by $S_p(\cdot)$ the semigroup generated by A_p. $S_p(\cdot)$ is a C_0-semigroup if $p < \infty$.

By the Poincaré inequality (see [5]), A_2 is strictly negative. Thus the spectrum of A_2 (and consequently the spectrum of A_p, $p \in (1, \infty]$) consists of a sequence of real negative semisimple eigenvalues $-\lambda_1 \geq -\lambda_2 \geq \dots$. By recalling Remark A.11, there exists constants M_p, $p \in (1, \infty]$, such that

$$\|S_p(t)\| \leq M_p e^{-\lambda_1 t}, \quad t \geq 0, \ p \in (1, \infty]. \tag{A.52}$$

Moreover, by the maximum principle the estimate follows

$$\|S_\infty(t)\| \leq 1, \quad t \geq 0. \tag{A.53}$$

□

A.5.3 Heat equation in variational form

We are given an elliptic operator

$$\widehat{A}u = \sum_{i,j=1}^{N} \frac{\partial}{\partial \xi_j} a_{ij}(\xi) \frac{\partial u}{\partial \xi_i} + \sum_{i=1}^{N} \frac{\partial u}{\partial \xi_i}(b_i(\xi)u) + c(\xi)u, \tag{A.54}$$

where the coefficients a_{ij}, b_i, c , $i, j = 1, \dots, N$, are continuous in $\overline{\mathscr{O}}$ and there exists $\alpha_0 > 0$ such that

$$\sum_{i,j=1}^{N} a_{ij}(\xi)\alpha_i\alpha_j \geq \alpha_0|\alpha|^2, \quad \forall\, \alpha = (\alpha_1, \dots, \alpha_N) \in \mathbb{R}^N.$$

Let $\widehat{a}$ be the bilinear form on $H^1(\mathscr{O})$ defined by

$$\widehat{a}(u, v) = \int_{\mathscr{O}} \left(\sum_{i,j=1}^{N} a_{ij} \frac{\partial u}{\partial \xi_i} \frac{\partial v}{\partial \xi_j} + \sum_{i=1}^{N} b_i u \frac{\partial v}{\partial \xi_i} + c(\xi)uv \right) d\xi. \tag{A.55}$$

Let $a_D(\cdot, \cdot)$ (respectively $a_N(\cdot, \cdot)$) be the restriction of $\widehat{a}$ to $H_0^1(\mathscr{O})$ (respectively $H^1(\mathscr{O})$). Then it is easy to check that (A.29) holds with $V = H_0^1(\mathscr{O})$ (respectively $H^1(\mathscr{O})$). Thus there exists a variational operator A_D (respectively A_N) corresponding to the bilinear form a_D (respectively a_N). A_D (respectively A_N) is called the Dirichlet (respectively Neumann) realization of the elliptic operator $\widehat{A}$.

Assume now that the coefficients a_{ij}, b_i, c belong to $C^1(\overline{\mathscr{O}})$. Then the operators A_D and A_N are given respectively by

$$\begin{cases} D(A_D) = H^2(\mathscr{O}) \cap H_0^1(\mathscr{O}) \\ A_D u = \widehat{A}u, \quad \forall\, u \in D(A_D), \end{cases} \tag{A.56}$$

$$\begin{cases} D(A_N) = \left\{ u \in H^2(\mathscr{O}) : \ \displaystyle\sum_{i,j=1}^{N} a_{ij}\nu_j \frac{\partial u}{\partial \nu_i} = 0 \right\} \\ Au = \widehat{A}u, \quad \forall\, u \in D(A_N). \end{cases} \tag{A.57}$$

From [362] we have

$$D_{A_D}(\theta, 2) = \begin{cases} H^{2\theta}(\mathscr{O}) \quad \text{if } \theta \in (0, 1/4) \\ \{u \in H^{2\theta}(\mathscr{O}) : \ u = 0 \text{ on } \partial\mathscr{O}\} \quad \text{if } \theta \in (1/4, 1) \end{cases} \tag{A.58}$$

and

$$D_{A_N}(\theta, 2) = \begin{cases} H^{2\theta}(\mathscr{O}) \quad \text{if } \theta \in (0, \frac{3}{4}), \\ \left\{ u \in H^{2\theta}(\mathscr{O}) : \ \displaystyle\sum_{i,j=1}^{N} a_{ij}\nu_j \frac{\partial u}{\partial \nu_i} = 0 \quad \text{on } \partial\mathscr{O} \right\} \quad \text{if } \theta \in (3/4, 1). \end{cases} \tag{A.59}$$

Remark A.30 If the matrix $\{a_{ij}\}$ is symmetric then we have $A_N = A_N^*$. In the general case we have

$$D_{A_N}(\theta, 2) = D_{A_N^*}(\theta, 2), \quad \forall\, \theta \in (0, 3/4). \tag{A.60}$$

Moreover

$$D_{A_N}(\theta, 2) = D((\lambda_0 - A_N)^\theta), \quad \forall\, \theta \in (0, 1), \tag{A.61}$$

for any positive number λ_0 large enough. □

A.5.4 Wave and plate equations

In this section we assume for simplicity that the set $\mathscr{O}$ is bounded and we consider the wave equation with Dirichlet boundary conditions

$$\begin{cases} y_{tt}(t, \xi) = \Delta_\xi y(t, \xi), & t \in \mathbb{R}^1, \quad \xi \in \mathscr{O} \\ y(t, \xi) = 0, & t > 0,\ \xi \in \partial\mathscr{O} \\ y(0, \xi) = x_0(\xi), & y_t(0, \xi) = x_1(\xi),\ \xi \in \mathscr{O}. \end{cases} \tag{A.62}$$

Let us write this problem in abstract form; to this aim denote by Λ the positive self-adjoint operator

$$\begin{cases} D(\Lambda) = H^2(\mathscr{O}) \cap H_0^1(\mathscr{O}), \\ \Lambda y = -\Delta_\xi y, \quad \forall\, y \in D(\Lambda), \end{cases} \tag{A.63}$$

and introduce the Hilbert space $H = D(\Lambda^{1/2}) \oplus L^2(\mathscr{O})$ endowed with the inner product

$$\left\langle \begin{pmatrix} y \\ z \end{pmatrix}, \begin{pmatrix} y_1 \\ z_1 \end{pmatrix} \right\rangle = \langle \Lambda^{1/2} y, \Lambda^{1/2} y_1 \rangle + \langle z, z_1 \rangle.$$

Notice that by (A.44) and by Remark A.29, we have $D(\sqrt{\Lambda}) = H_0^1(\mathscr{O})$. In the formulae above we have set $H^2(\mathscr{O}) = W^{2,2}(\mathscr{O})$ and $H_0^1(\mathscr{O}) = W_0^{1,2}(\mathscr{O})$.

We define in H the linear operator

$$\begin{cases} D(A) = D(\Lambda) \oplus D(\Lambda^{1/2}) \\ A \begin{pmatrix} y \\ z \end{pmatrix} = \begin{pmatrix} 0 & 1 \\ -\Lambda & 0 \end{pmatrix} \begin{pmatrix} y \\ z \end{pmatrix}, \quad \forall \begin{pmatrix} y \\ z \end{pmatrix} \in D(A). \end{cases} \tag{A.64}$$

It is easy to check that

$$(\lambda - A)^{-1} = \begin{pmatrix} \lambda & 1 \\ -\Lambda & \lambda \end{pmatrix} (\lambda^2 + \Lambda)^{-1}, \quad \forall\, \lambda > 0,$$

and

$$\left\langle A \begin{pmatrix} y \\ z \end{pmatrix}, \begin{pmatrix} y \\ z \end{pmatrix} \right\rangle = 0, \quad \forall \begin{pmatrix} y \\ z \end{pmatrix} \in D(A).$$

This implies that

$$\|R(\lambda, A)\| \le \frac{1}{\lambda}, \quad \forall\, \lambda > 0,$$

and so A generates a contraction semigroup $S(\cdot)$ in H. We have

$$S(t)\begin{pmatrix} y \\ z \end{pmatrix} = \begin{pmatrix} \cos(\Lambda^{1/2}t) & \Lambda^{-1/2}\sin(\Lambda^{1/2}t) \\ -\Lambda^{1/2}\sin(\Lambda^{1/2}t) & \cos(\Lambda^{1/2}t) \end{pmatrix}\begin{pmatrix} y \\ z \end{pmatrix}. \tag{A.65}$$

We now consider the *plate equation*

$$\begin{cases} y_{tt}(t,\xi) = -(\Delta_\xi)^2 y(t,\xi), & t \in \mathbb{R}^1,\ \xi \in \mathscr{O}, \\ y(t,\xi) = 0, \quad \Delta y(t,\xi) = 0, & t > 0,\ \xi \in \partial\mathscr{O}, \\ y(0,\xi) = x_0(\xi), \quad y_t(0,\xi) = x_1(\xi), & \xi \in \mathscr{O}. \end{cases} \tag{A.66}$$

We consider the Hilbert space $H = D(\Lambda) \oplus L^2(\mathscr{O})$ endowed with the inner product

$$\left\langle \begin{pmatrix} y \\ z \end{pmatrix}, \begin{pmatrix} y_1 \\ z_1 \end{pmatrix} \right\rangle = \langle \Lambda y, \Lambda y_1 \rangle + \langle z, z_1 \rangle,$$

and define in H the linear operator:

$$\begin{cases} D(A) = D(\Lambda^2) \oplus D(\Lambda) \\ A\begin{pmatrix} y \\ z \end{pmatrix} = \begin{pmatrix} 0 & 1 \\ -\Lambda^2 & 0 \end{pmatrix}\begin{pmatrix} y \\ z \end{pmatrix}, \quad \forall \begin{pmatrix} y \\ z \end{pmatrix} \in D(A). \end{cases} \tag{A.67}$$

It is easy to check that

$$(\lambda - A)^{-1} = \begin{pmatrix} \lambda & 1 \\ \Lambda^2 & \lambda \end{pmatrix}(\lambda^2 + \Lambda^2)^{-1}, \quad \forall\, \lambda > 0,$$

and

$$\left\langle A\begin{pmatrix} y \\ z \end{pmatrix}, \begin{pmatrix} y \\ z \end{pmatrix} \right\rangle = 0, \quad \forall \begin{pmatrix} y \\ z \end{pmatrix} \in D(A).$$

This implies as before that A generates a contraction semigroup in H. □

A.5.5 Wave and plate equation with strong damping

As in the previous example we shall consider a bounded open set $\mathscr{O}$ in $\mathbb{R}^N$ and the linear operator Λ in X_2 defined by (A.63). As remarked, the spectrum of Δ consists of a sequence

$$\lambda_1 \le \lambda_2 \le \dots$$

of positive numbers. We consider here the strongly damped wave equation:

$$\begin{cases} y_{tt}(t,\xi) = \Delta_\xi y(t,\xi) + \rho\Delta_\xi y_t(t,\xi), & t \ge 0,\ \xi \in \mathscr{O} \\ y(t,\xi) = 0, \quad t > 0,\ \xi \in \partial\mathscr{O}, \\ y(0,\xi) = x_0(\xi), \quad y_t(0,\xi) = x_1(\xi),\ \xi \in \mathscr{O}, \end{cases} \tag{A.68}$$

where ρ is a positive constant. We still write the problem in abstract form in the Hilbert space $H = D(\Lambda^{1/2}) \oplus X_2$. Let A be the linear operator

$$\begin{cases} D(A) = \left\{ \begin{pmatrix} y \\ z \end{pmatrix} \in H; y + \rho z \in D(\Lambda) \right\} \\ A \begin{pmatrix} y \\ z \end{pmatrix} = \begin{pmatrix} 0 & 1 \\ -\Lambda & -\rho\Lambda \end{pmatrix} \begin{pmatrix} y \\ z \end{pmatrix}, \quad \forall \begin{pmatrix} y \\ z \end{pmatrix} \in D(A). \end{cases} \tag{A.69}$$

As far as the spectrum and the resolvent of A are concerned, we have:

$$\sigma(A) = \{-1/\rho\} \cup \left\{ \mu = -\frac{1}{2}(\rho\lambda_k \pm \sqrt{(\rho^2\lambda_k^2 - 4\lambda_k)} : k \in \mathbb{N} \right\}$$

and

$$(\lambda - A)^{-1} = \begin{pmatrix} \lambda + \rho\Lambda & 1 \\ -\Lambda & \lambda \end{pmatrix} (\lambda^2 + \rho\Lambda\lambda + \Lambda)^{-1}, \quad \forall \lambda \in \sigma(A).$$

From the formula above, it is not difficult to prove, using the spectral decomposition formula of Λ, that A is the infinitesimal generator of a C_0-semigroup of contractions in H and it is analytic.

Remark A.31 It is not difficult to find the interpolation spaces $D_A(\theta, 2)$, $\theta \in (0, 1)$. In particular we have

$$D_A(1/2, 2) = D(\Lambda^{1/2}) \oplus D(\Lambda^{1/2}). \tag{A.70}$$

□

We consider now the plate equation

$$\begin{cases} y_{tt}(t,\xi) = -(\Delta_\xi)^2 y(t,\xi) + \rho\Delta_\xi y_t(t,\xi), & t \in R^1, \ \xi \in \mathscr{O}, \\ y(t,\xi) = 0, \quad \Delta y(t,\xi) = 0, & t > 0, \ \xi \in \partial\mathscr{O}, \\ y(0,\xi) = x_0(\xi), \quad y_t(0,\xi) = x_1(\xi), & \xi \in \mathscr{O}, \end{cases} \tag{A.71}$$

where ρ is a positive constant. We write the problem in abstract form in the Hilbert space $H = D(\Lambda) \oplus E_2$. Let A be the linear operator

$$\begin{cases} D(A) = D(\Lambda^2) \oplus D(\Lambda) \\ A \begin{pmatrix} y \\ z \end{pmatrix} = \begin{pmatrix} 0 & 1 \\ -\Lambda^2 & -\rho\Lambda \end{pmatrix} \begin{pmatrix} y \\ z \end{pmatrix}, \quad \forall \begin{pmatrix} y \\ z \end{pmatrix} \in D(A). \end{cases} \tag{A.72}$$

The spectrum and the resolvent of A are given by

$$\sigma(A) = \left\{ \mu = -\frac{\lambda_k}{2} \left(\rho \pm \sqrt{\rho^2 - 4} \right) : k \in \mathbb{N} \right\},$$

and

$$(\lambda - A)^{-1} = \begin{pmatrix} \lambda + \rho\Lambda & 1 \\ -\Lambda^2 & \lambda \end{pmatrix} (\lambda^2 + \lambda\rho\Lambda + \Lambda^2)^{-1}, \quad \forall \lambda \in \sigma(A)$$

respectively. From the formula above, one can show as before that A is the infinitesimal generator of a C_0-semigroup and it is analytic.

Remark A.32 We have

$$D_A(1/2, 2) = D(\Lambda^{3/2}) \oplus D(\Lambda^{1/2}). \tag{A.73}$$

□

Appendix B
Some results on control theory

B.1 Controllability and stabilizability

We are given two Hilbert spaces H and U and a dynamical system governed by the differential equation

$$\begin{cases} y'(t) = Ay(t) + Bu(t), & t \in [0, T], \\ y(0) = x \in H \end{cases} \tag{B.1}$$

where $T > 0$ is fixed, $u \in L^2(0, T; U)$, A is the infinitesimal generator of a C_0-semigroup $S(\cdot)$ in H and B is a bounded operator from U into H. H and U represent the spaces of *states* and *controls* of the system respectively. We set $\mathscr{H} = L^2(0, T; H)$ and $\mathscr{U} = L^2(0, T; U)$. We know, by Section A.3, that Problem (B.1) has a unique mild solution $y = y^{x,u} \in C([0, T]; H)$ given by

$$y^{x,u}(t) = S(t)x + \int_0^t S(t - s)Bu(s)ds, \quad t \in [0, T]. \tag{B.2}$$

If $a = y^{x,u}(t)$ for some $u \in \mathscr{U}$, then the state a is said to be *reachable* from x in time t. If states reachable from 0 in time T form a dense set in H then (B.1) is said to be *approximately controllable* in time T. If 0 is a reachable state from arbitrary $x \in H$ in time T, then (B.1) is said to be *null controllable* in time T.

Semigroup $S(\cdot)$ is called *exponentially stable,* if for some positive numbers ω and M, we have

$$\|S(t)\| \le Me^{-\omega t}, \quad \forall\, t \ge 0. \tag{B.3}$$

The infinitesimal generator of an exponentially stable semigroup is called an *exponentially stable generator*. System (B.1) is said to be *stabilizable* if there exists an operator $K \in L(H; U)$ such that the generator $A + BK$ is exponentially stable.

It is well known that if the system (B.1) is null controllable in some time T then it is stabilizable [177].

The following equation for a nonnegative bounded operator P is called the *Liapunov equation*

$$2\langle PA^*x, x\rangle + \langle BB^*x, x\rangle = 0, \quad \forall\, x \in D(A^*). \tag{B.4}$$

It is well known, see [177, 725], that if the system (B.1) is stabilizable and equation (B.4) has a solution, then A is exponentially stable. In particular if the system (B.1), with $B = Q^{1/2}$, Q a nonnegative bounded operator, is null controllable in time T and there exists a solution P to the equation

$$2\langle PA^*x, x\rangle + \langle Qx, x\rangle = 0, \quad \forall\, x \in D(A^*), \tag{B.5}$$

then A is exponentially stable.

B.2 Comparison of images of linear operators

We recall here and prove some results on linear operators and their images which play an important role in control theory.

We are given three Hilbert spaces E_1, E_2 and E and two linear and bounded operators $A_1 : E_1 \to E$, $A_2 : E_2 \to E$. We shall denote by $A_1^* : E \to E_1$ and by $A_2^* : E \to E_2$ the corresponding adjoint operators.

Let us recall the notion of *pseudo-inverse* of A_1. For any $x \in A_1(E_1)$ consider the set

$$A_1^{-1}(x) := \{x_1 \in E_1 : \ A_1x_1 = x\}.$$

Then we set

$$A_1^{-1}x = \operatorname{argmin}\,\{|A_1^{-1}(\{x\})|_{E_1}\}.$$

Notice that $A_1^{-1}(E) := \{A^{-1}x_1 : \ x_1 = A_1(E)\}$ is the orthogonal subspace to $\operatorname{Ker}(A_1) := A_1^{-1}(0)$ in E_1.

We prove the following result.

Proposition B.1 *The following statements hold.*

(i) $A_1(E_1) \subset A_2(E_2)$ *if and only if there exists a constant* $k > 0$ *such that* $|A_1^*h| \le k|A_2^*h|$ *for all* $h \in E$.
(ii) If $|A_1^*h| = |A_2^*h|$ *for all* $h \in E$, *then* $A_1(E_1) = A_2(E_2)$ *and* $|A_1^{-1}h| = |A_2^{-1}h|$ *for all* $h \in A_1(E_1)$.

Proof We first show that $A_1(E_1) \subset A_2(E_2)$ if and only if there exists a constant $k > 0$ such that $\{A_1u : \ |u| \le 1\} \subset \{A_2v : \ |v| \le k\}$. Remark that the two sets above are closed since E is reflexive. The part "only if" is trivial, let us prove the part "if." Assume first $\operatorname{Ker}(A_2) = 0$, then $A_2^{-1}A_1$ is bounded and there exists a constant $k > 0$

such that

$$|A_2^{-1}A_1u| \le k \quad \text{if } |u| \le 1$$

so the conclusion follows. If $\mathrm{Ker}(A_2) \neq \{0\}$ we apply the previous argument to the operators A_1 and $\widehat{A}_2$ where $\widehat{A}_2$ is the restriction of A_2 to the orthogonal complement of Ker A_2.

Secondly we prove that $\{A_1u : |u| \le 1\} \subset \{A_2v : |v| \le k\}$ if and only if $|A_1^*h| \le k|A_2^*h|, \ \forall\, h \in E$. In fact if $\{A_1u : |u| \le 1\} \subset \{A_2v : |v| \le k\}$, then

$$\begin{aligned} |A_1^*h| &= \sup_{|u|\le 1} |\langle h, A_1u\rangle| \\ &\le \sup_{|v|\le k} |\langle h, A_2\rangle| = k|A_2^*h|. \end{aligned}$$

Conversely assume $|A_1^*h| \le k|A_2^*h|, \ \forall\, h \in E$ and, by contradiction, that there exists $u_0 \in E_1$ such that

$$|u_0| \le 1, \quad A_1u_0 \notin \{A_2v : |v| \le k\}.$$

Since the set $\{A_2v : |v| \le k\}$ is closed and convex then, by the separation Hahn–Banach theorem, there exists $h \neq 0 \in E$ such that

$$\langle h, A_1u_0\rangle > 1 \text{ and } \langle h, A_2v\rangle \le 1, \text{ if } |v| \le k.$$

Thus $|A_1^*h| > 1$ and $|A_2^*h| \le 1$, a contradiction. This complete the proof of (i).

(ii) The first statement follows from part (i), let us prove the last one. One can assume that A_1 and A_2 are invertible operators (otherwise one can take restrictions on orthogonal complements of respective kernels). We have to show that if $e \in E$ is such that $e = A_1h_1 = A_2h_2$, then $|h_1| = |h_2|$. Assume, by contradiction, that $|h_1| > |h_2| = 1$. Then

$$\frac{e}{|h_2|} = A_2\left(\frac{h_2}{|h_2|}\right) \in \{A_2v : |v| \le 1\} = \{A_1u : |u| \le 1\}.$$

But $\frac{e}{|h_2|} = A_1\left(\frac{h_1}{|h_2|}\right)$ and $\left|\frac{h_1}{|h_2|}\right| > 1$, therefore

$$\frac{e}{|h_2|} \notin \{A_1u : |u| \le 1\},$$

a contradiction. The proof of the proposition is complete. □

Remark B.2 The condition $|A_1^*h| = |A_2^*h|, \ \forall\, h \in E$, is equivalent to $A_1A_1^* = A_2A_2^*$. □

Corollary B.3 *Set $Q_1 = A_1A_1^*$; then we have $Q_1^{1/2}(E_2) = A_1(E_1)$ and*

$$|Q_1^{-1/2}x| = |A_1^{-1}x|, \quad \forall\, x \in A_1(E_1), \tag{B.6}$$

where $Q_1^{-1/2}$ denotes the pseudo-inverse of $Q_1^{1/2}$.

Proof For any $x \in E$ we have:

$$|Q_1^{1/2}x|^2 = \langle Q_1 x, x\rangle = |A_1^* x|^2.$$

This implies the conclusion by Proposition B.1. □

Remark B.4 Let

$$\widehat{E}_1 = (\mathrm{Ker}(A_1))^{\perp}, \quad \widehat{E}_2 = (\mathrm{Ker}(A_2))^{\perp}$$

be the orthogonal complements of $\mathrm{Ker}(A_1)$ and $\mathrm{Ker}(A_2)$. By Proposition B.1(ii) the transformations

$$\widehat{I}_{21} : \widehat{E}_1 \to \widehat{E}_2, \quad h_1 \mapsto A_2^{-1} A_1 h_1$$

and

$$\widehat{I}_{12} : \widehat{E}_2 \to \widehat{E}_1, \quad h_2 \mapsto A_1^{-1} A_2 h_2$$

are isometries onto $\widehat{E}_2$ and

$$\widehat{I}_{21}\widehat{I}_{12} h_2 = h_2,\ \forall\, h_2 \in \widehat{E}_2, \quad \widehat{I}_{12}\widehat{I}_{21} h_1 = h_1,\ \forall\, h_1 \in \widehat{E}_1.$$

Let Π_1 and Π_2 be the orthogonal projectors onto $\widehat{E}_1$ and $\widehat{E}_2$ respectively. Define

$$I_{21} = \widehat{I}_{21}\Pi_1 : E_1 \to E_2, \quad I_{12} = \widehat{I}_{12}\Pi_2 : E_1 \to E_1. \tag{B.7}$$

The following factorization formulae are now straightforward:

$$\begin{aligned} &A_1 = A_2 I_{21}, \quad A_2 = A_1 I_{12}, \\ &A_1^{-1} = I_{12} A_2^{-1}, \quad A_2^{-1} = I_{21} A_1^{-1}. \end{aligned} \tag{B.8}$$

The transformations I_{21} and I_{12} are called *partial isometries* determined by A_1, A_2. Note that $I_{21}^* = I_{12}$, $I_{12}^* = I_{21}$ and the operators $I_{12}I_{12}^* = I_{12}I_{21}$, $I_{21}I_{21}^* = I_{21}I_{12}$ are orthogonal projectors from E_1 onto $\widehat{E}_1$ and E_2 onto $\widehat{E}_2$ respectively. □

B.3 Operators associated with control systems

Assume that a control system is given and define the following two operators (recall that $\mathscr{H} = L^2(0, T; H)$ and $\mathscr{U} = L^2(0, T; U)$),

$$\mathscr{L} : \mathscr{U} \to \mathscr{H}, \quad (\mathscr{L}u)(t) = \int_0^t S(t-s)Bu(s)ds, \quad t \in [0, T] \tag{B.9}$$

and

$$L : \mathscr{U} \to H, \quad Lu = \int_0^T S(T-s)Bu(s)ds. \tag{B.10}$$

Note that for any $x \in H, u \in \mathscr{U}$, we have

$$y^{x,u}(t) = S(t)x + (\mathscr{L}u)(t), \quad t \in [0, T]$$

and

$$y^{x,u}(T) = S(T)x + Lu.$$

Therefore $\mathscr{L}(\mathscr{U})$ consists of all trajectories which start from 0 and are generated by (B.1) and $L(\mathscr{U})$ consists of all states reachable in time T from 0. Consequently system (B.1) is approximately controllable in time T if and only if

$$\overline{L(\mathscr{U})} = H.$$

Moreover system (B.1) is null controllable in time T if and only if

$$S(T)(H) \subset L(\mathscr{U}). \tag{B.11}$$

We will give now some useful characterizations of $\mathscr{L}(\mathscr{U})$ and $L(\mathscr{U})$.

B.3.1 Characterization of $\mathscr{L}(\mathscr{U})$

Let us consider the adjoint $\mathscr{L}^* : \mathscr{H} \to \mathscr{U}$ of the linear operator $\mathscr{L}$ defined by (B.9) and the linear operator $\mathscr{R} := \mathscr{L}\mathscr{L}^*$. They are given respectively by:

$$(\mathscr{L}^* y)(t) = \int_t^T B^* S^*(s-t) y(s) ds, \quad y \in \mathscr{H}, t \in [0, T] \tag{B.12}$$

and

$$(\mathscr{R} y)(t) = \int_0^T R(t,s) y(s) ds, \quad y \in \mathscr{H}, t \in [0, T], \tag{B.13}$$

where

$$R(t,s)x = \int_0^{t\wedge s} S(t-r) B B^* S^*(s-r) x \, ds, \quad x \in H, \ t, s \in [0, T]. \tag{B.14}$$

Corollary B.5 *We have:*

$$\mathscr{L}(\mathscr{U}) = \mathscr{R}^{1/2}(\mathscr{H}) \tag{B.15}$$

and

$$|\mathscr{R}^{-1/2} z|^2_{\mathscr{H}} = \min\left\{ \int_0^T |u(s)|^2 ds : \ y^{0,u} = z \right\}. \tag{B.16}$$

Proof From Corollary B.3 it follows that

$$\mathscr{R}^{-1/2}(\mathscr{H}) = L(\mathscr{U}),$$

and for any $y \in \mathscr{H}$ we have

$$|\mathscr{R}^{-1/2}z|_{\mathscr{H}} = |\mathscr{L}^{-1}y|_{\mathscr{H}}.$$

The conclusion follows. □

Remark B.6 It is not easy in general to characterize the images of $\mathscr{L}$ and $\mathscr{R}^{1/2}$. We list some cases in which it is possible.

(i) If H is finite dimensional and B is invertible then

$$\mathscr{L}(C_0^1([0,T];H)) = \{u \in C_0^1([0,T];H) : u(0) = 0\}. \tag{B.17}$$

(ii) Assume that $S(\cdot)$ is an analytic semigroup and that $B^{-1} \in L(H;U)$, then

$$\mathscr{L}(U) = W_0^{1,2}([0,T];H) \cap L^2([0,T];D(A)) \tag{B.18}$$

where

$$W_0^{1,2}([0,T];H) = \{y \in W^{1,2}([0,T];H) : y(0) = 0\}.$$

In fact the inclusion $\mathscr{L}(U) \subset W_0^{1,2}([0,T];H) \cap L^2([0,T];D(A))$ follows from Proposition A.24; conversely, if

$$y \in W_0^{1,2}([0,T];H) \cap L^2([0,T];D(A)),$$

setting $u = B^{-1}(y' - Ay)$ we have $y = \mathscr{L}u$. □

B.3.2 Characterization of the image of L

Let us consider the adjoint L^* of the linear operator L defined by (B.10) and the linear operator $R = LL^*$. They are given respectively by

$$(L^*x)(s) = B^*S^*(T-s)x, \quad x \in H,\ s \in [0,T] \tag{B.19}$$

and

$$Rx = \int_0^T S(t)BB^*S^*(t)x\,dt, \quad x \in H. \tag{B.20}$$

From Corollary B.3 we have the following result.

Corollary B.7 *We have*

$$L(U) = R^{1/2}(H) \tag{B.21}$$

and

$$|R^{-1/2}x|^2 = \min\left\{\int_0^T |u(s)|^2 ds \,:\, y^{0,u}(T) = x\right\}. \tag{B.22}$$

Remark B.8 Assume that $S(\cdot)$ is an analytic semigroup and that $B^{-1} \subset L(H;U)$. Then

$$L(U) = D_A(1/2, 2). \tag{B.23}$$

In fact, the inclusion

$$L(U) \subset D_A(1/2, 2)$$

follows by Remark A.16. Moreover if $x \in D_A(1/2, 2)$ then, by the definition of $D_A(1/2, 2)$ as a space of traces (see [499]), there exists $z \in W^{1,2}([0, T]; H) \cap L^2(0, T; D(A))$ such that $x = z(0)$. Let φ be a scalar function of class C^∞ such that $\varphi(0) = 0$ and $\varphi(1) = 1$. Set

$$y(t) = \begin{cases} \varphi(t)z(T-t) & \text{if } t \in [0, T/2] \\ z(T-t) & \text{if } t \in [T/2, T] \end{cases}$$

and

$$u(t) = B^{-1}(y'(t) - Ay(t)).$$

Then, as easily seen, $Lu = x$. □

Remark B.9 (i) Note that system (B.1) is null controllable in time T if

$$R^{1/2}(H) \supset S(T)(H). \tag{B.24}$$

It follows directly from (B.22) that

$$|R^{-1/2}S(T)x|^2 = \min\left\{\int_0^T |u(s)|^2 ds; y^{x,u}(T) = 0\right\}. \tag{B.25}$$

Interpreting the integral as energy needed to implement control u, we can say that quantities

$$|R^{-1/2}x|^2 \text{ and } |R^{-1/2}S(T)x|^2$$

are equal to the *minimal energy* needed to transfer 0 to x and x to 0 in time T respectively.

(ii) Assume that B is invertible and that $B^{-1} \in L(H; U)$; then $S(T)(H) \subset L(U)$, that is the system is null controllable. In fact, given $x \in H$ and setting

$$\widetilde{u}(t) = -\frac{1}{T} B^{-1} S(t)x, \quad t \in (0, T]$$

we have $y^{x,\widetilde{u}}(T) = 0$.

(iii) Assume that $B(U) \supset S(t)(H),\ t \in [0, T]$, and that

$$\int_0^T |B^{-1} S(t)x|^2 dt < +\infty, \quad x \in H.$$

Then the system is null controllable by the argument in (ii). □

Appendix C

Nuclear and Hilbert–Schmidt operators

Let E, G be Banach spaces and let $L(E, G)$ be the Banach spaces of all linear bounded operators from E into G endowed with the usual supremum norm. We write $L_1(E)$ instead of $L_1(E, E)$. We denote by E^* and $G^\star$ the dual space of E and G respectively. An element $T \in L(E, G)$ is said to be a *nuclear* or *trace class* operator if there exist two sequences $\{a_j\} \subset G$, $\{\varphi_j\} \subset E^\star$ such that

$$\sum_{j=1}^{\infty} \|a_j\| \, \|\varphi_j\| < +\infty \tag{C.1}$$

and T has the representation

$$Tx = \sum_{j=1}^{\infty} a_j \varphi_j(x), \quad x \in E.$$

The spaces of all nuclear operators from E into G, endowed with the norm

$$\|T\|_1 = \inf \left\{ \sum_{j=1}^{\infty} \|a_j\| \, \|\varphi_j\| : Tx = \sum_{j=1}^{\infty} a_j \varphi_j(x) \right\}$$

is a Banach space (see [271]), and will be denoted as $L_1(E, G)$. Let K be another Banach space; it is clear that if $T \in L_1(E, G)$ and $S \in L(G, K)$ then $TS \in L_1(E, K)$ and $\|TS\|_1 \le \|T\| \|S\|_1$. Let H be a separable Hilbert space and let $\{e_k\}$ be a complete orthonormal system in H. If $T \in L_1(H, H)$ then we define

$$\text{Tr}\, T = \sum_{j=1}^{\infty} \langle Te_j, e_j \rangle.$$

Proposition C.1 *If* $T \in L_1(H)$ *then* Tr T *is a well defined number independent of the choice of the orthonormal basis* $\{e_k\}$.

Proof Let $\{a_j\} \subset H$ and $\{\varphi_j\} \subset H^*$ be two sequences such that

$$Th = \sum_{j=1}^{\infty} a_j \varphi_j(h), \quad h \in H$$

and (C.1) holds. Let $b_j \in H$ such that $\varphi_j(h) = \langle h, b_j \rangle$. Then

$$\langle Te_k, e_k \rangle = \sum_{j=1}^{\infty} \langle e_k, a_j \rangle \langle e_k, b_j \rangle.$$

Moreover

$$\begin{aligned}\sum_{k=1}^{\infty} |\langle Te_k, e_k \rangle| &\leq \sum_{j=1}^{\infty} \sum_{k=1}^{\infty} |\langle e_k, a_j \rangle \langle e_k, b_j \rangle| \\ &\leq \sum_{j=1}^{\infty} \left(\sum_{k=1}^{\infty} |\langle e_k, a_j \rangle|^2 \right)^{1/2} \left(\sum_{k=1}^{\infty} |\langle e_k, b_j \rangle|^2 \right)^{1/2} \\ &\leq \sum_{j=1}^{\infty} |a_j||b_j| < +\infty.\end{aligned}$$

Since

$$\sum_{k=1}^{\infty} \langle Te_k, e_k \rangle = \sum_{j=1}^{\infty} \sum_{k-1}^{\infty} \langle e_k, a_j \rangle \langle e_k, b_j \rangle = \sum_{j=1}^{\infty} \langle a_j, b_j \rangle,$$

the definition of Tr T is independent of the basis $\{e_k\}$. □

Note also that

$$|\text{Tr } T| \leq \|T\|_1, \quad \forall\, T \in L_1(H). \tag{C.2}$$

Corollary C.2 *If $T \in L_1(H)$ and $S \in L(H)$, then $TS \in L_1(H)$ and*

$$\text{Tr } TS = \text{ Tr } ST \leq \|T\|_1 \|S\|. \tag{C.3}$$

Proposition C.3 *A nonnegative operator $T \in L(H)$ is of trace class if and only if for an orthonormal basis $\{e_k\}$ on H*

$$\sum_{j=1}^{\infty} \langle Te_j, e_j \rangle < +\infty.$$

Moreover in this case Tr $T = \|T\|_1$.

Proof We will show first that T is compact. Let $T^{1/2}$ denote the nonnegative square root of T. Then $T^{1/2}x = \sum_{j=1}^{\infty} \langle T^{1/2}x, e_j \rangle e_j$, and

$$\begin{aligned}\Big| T^{1/2}x - \sum_{j=1}^{N} \langle T^{1/2}x, e_j \rangle e_j \Big|^2 &= \sum_{k=N+1}^{\infty} |\langle T^{1/2}x, e_j \rangle|^2 \\ &\leq |x| \sum_{k=N+1}^{\infty} |T^{1/2}e_j|^2 \leq |x| \sum_{k=N+1}^{\infty} \langle Te_k, e_k \rangle, \quad x \in H.\end{aligned}$$

So the operator $T^{1/2}$ is a limit, in the operator norm, of finite rank operators. Therefore $T^{1/2}$ is compact and $T = T^{1/2}T^{1/2}$ is a compact nonnegative operator as well. Let $\{f_j\}$ be a sequence of all eigenvectors of T and let $\{\lambda_j\}$ be the corresponding sequence of eigenvalues. Then

$$Tx = \sum_{k=1}^{\infty} \lambda_k \langle x, f_k\rangle f_k, \quad x \in H. \tag{C.4}$$

Since

$$\langle Te_j, e_j\rangle = \sum_{k=1}^{\infty} \lambda_k |\langle f_j, e_k\rangle|^2,$$

wc have

$$\sum_{j=1}^{\infty} \langle Te_j, e_j\rangle = \sum_{j=1}^{\infty}\sum_{k=1}^{\infty} \lambda_k |\langle f_j, e_k\rangle|^2 = \sum_{k=1}^{\infty} \lambda_k < +\infty.$$

From this and the expansion (C.4) one has that T is trace class and Tr $T = \sum_{k=1}^{\infty} \lambda_k$. From (C.2) and (C.4) the identity Tr $T = \|T\|_1$ follows. □

Let E and F be two separable Hilbert spaces with complete orthonormal bases $\{e_k\} \subset H$, $\{f_j\} \subset F$, respectively. A linear bounded operator $T : H \to E$ is said to be *Hilbert–Schmidt* if

$$\sum_{k=1}^{\infty} |Te_k|^2 < \infty.$$

Since

$$\sum_{k=1}^{\infty} |Te_k|^2 = \sum_{k=1}^{\infty}\sum_{j=1}^{\infty} |\langle Te_k, f_j\rangle|^2 = \sum_{k=1}^{\infty} |T^* f_j|^2,$$

the definition of Hilbert–Schmidt operator, and the number $\|T\|_2 = \left(\sum_{k=1}^{\infty} |Te_k|^2\right)^{1/2}$, is independent of the choice of the basis $\{e_k\}$. Moreover $\|T\|_2 = \|T^*\|_2$.

One can check easily that the set $L_2(E, F)$ of all Hilbert–Schmidt operators from E into F, equipped with the norm

$$\|T\|_2 = \left(\sum_{k=1}^{\infty} |Te_k|^2\right)^{1/2}$$

is a separable Hilbert space, with the scalar product

$$\langle S, T\rangle_2 = \sum_{k=1}^{\infty} \langle Se_k, Te_k\rangle.$$

The double sequence of operators $\{f_j \otimes e_k\}_{j,k\in N}$ is a complete orthonormal basis in $L_2(E, F)$. [1]

Proposition C.4 *Let E, F, G be separable Hilbert spaces. If $T \in L_2(E, F)$ and $S \in L_2(F, G)$ then $ST \in L_1(E, G)$ and*

$$\|ST\|_1 \le \|S\|_2 \|T\|_2. \tag{C.5}$$

Proof Note that

$$STx = \sum_{j=1}^{\infty} \langle Tx, f_j\rangle S f_j, \quad x \in E.$$

Then it follows, from the definition of trace class operator, that

$$\begin{aligned}\|ST\|_1 &\le \sum_{j=1}^{\infty} |T^* f_j|\,|S f_j| \\ &\le \left(\sum_{j=1}^{\infty} |T^* f_j|^2\right)^{1/2} \left(\sum_{j=1}^{\infty} |S f_j|^2\right)^{1/2}.\end{aligned}$$

□

[1] For arbitrary $b \in E$, $a \in F$ we denote by $b \otimes a$ the linear operator defined by $(b \otimes a) \cdot h = \langle a, h\rangle b$, $h \in F$.

Appendix D

Dissipative mappings

In this appendix we recall basic facts on the subdifferential of the norm in a Banach space E and on general properties of dissipative mappings in E. For more details see for instance [47] or [185].

D.1 Subdifferential of the norm

Let $x, y \in E$, then the mapping

$$\varphi : \mathbb{R} \to \mathbb{R}, \quad h \mapsto \varphi(h) = \|x + hy\|$$

is convex, and so the mapping

$$\mathbb{R}^1 \backslash \{0\} \to \mathbb{R}^1, \quad h \to \frac{\|x + hy\| - \|x\|}{h}$$

is nondecreasing and the following limits:

$$D_+\|x\| \cdot y = \lim_{h \to 0^+} \frac{\|x + hy\| - \|x\|}{h}$$
$$D_-\|x\| \cdot y = \lim_{h \to 0^-} \frac{\|x + hy\| - \|x\|}{h}$$

do exist. The *subdifferential* $\partial\|x\|$ of $\|x\|$ is defined as:

$$\partial\|x\| = \{x^* \in E^* : \ D_-\|x\| \cdot y \le \langle y, x^* \rangle \le D_+\|x\| \cdot y, \ \forall y \in E\}$$

where E^* is the dual of E. One can prove easily that the set $\partial\|x\|$ is convex, closed nonempty and is given by:

$$\partial\|x\| = \{x^* \in X^* : \ \langle x, x^* \rangle = \|x\|, \ \|x^*\| = 1\}. \tag{D.1}$$

Moreover:

$$D_+\|x\| \cdot y = \max\{\langle y, x^*\rangle : x^* \in \partial\|x\|\}, \tag{D.2}$$

$$D_-\|x\| \cdot y = \min\{\langle y, x^*\rangle : x^* \in \partial\|x\|\}. \tag{D.3}$$

Example D.1 Let $E = H$ be a Hilbert space and $x \neq 0$, then $\partial\|x\|$ consists of the unique element [(1)]

$$\partial\|x\| = \{\tfrac{x}{\|x\|}\}.$$

□

Example D.2 Let $E = L^p(\mathscr{O}),\ p > 1$, where $\mathscr{O}$ is an open set of $\mathbb{R}^d$. If $x \neq 0$, then $\partial\|x\|$ consists of the unique element x^* of $L^q(\mathscr{O}),\ q = \frac{p}{p-1}$,

$$x^* = \|x\|^{1-p}_{L^p(\mathscr{O})}\,|x|^{p-2}x.$$

□

Example D.3 Let $E = C(\overline{\mathscr{O}})$ and x be a nonzero element on E. For the following characterization of the subdifferential of $\|x\|$ see [647].

Given $x \in E$, set

$$M_x = \{\xi \in \mathscr{O} : |x(\xi)| = \|x\|\}.$$

Then $\mu \in \partial\|x\|$ if and only if

(i) μ is a Radon measure on $\overline{\mathscr{O}}$ with $\|\mu\| = 1$,
(ii) the support of μ is included in M_x,
(iii) $\int_\Gamma \operatorname{sgn} x(\xi)\mu(d\xi) \geq 0, \quad \forall\, \Gamma \in \mathscr{B}(\overline{\mathscr{O}})$. [(2)]

If in particular M_x consists of a single element $\xi_0 \in \mathscr{O}$, then

$$\partial\|x\| = \begin{cases} \{\delta_{\xi_0}\} & \text{if } x(\xi_0) = \|x\| \\ \{-\delta_{\xi_0}\} & \text{if } x(\xi_0) = -\|x\|, \end{cases} \tag{D.4}$$

where δ_{ξ_0} is the Dirac mass concentrated at ξ_0. It is easy to see that the set

$$\{x \in E : \#(M_x) = 1\}$$

is dense in E. □

We end this section by giving a generalization of the chain rule.

Proposition D.4 *Let $u : [0, T] \to E,\ t \mapsto u(t)$ differentiable in $t_0 \in [0, T]$. Then the function $\gamma = \|u(\cdot)\|$ is differentiable on the right at t_0. Moreover if $t_0 > 0$ then γ*

(1) Here we identify H with its dual space.
(2) $\operatorname{sgn} x = 1$ if $x \geq 0$, $\operatorname{sgn} x = 0$ if $x < 0$.

is differentiable on the left at t_0. We have

$$\begin{aligned}\frac{\partial^+\gamma}{\partial t}(t_0) &= D_+\|u(t_0)\| \cdot u'(t_0)\\ &= \max\{\langle u'(t_0), x^*\rangle : \ x^* \in \partial\|u(t_0)\|\},\end{aligned} \tag{D.5}$$

$$\begin{aligned}\frac{\partial^-\gamma}{\partial t}(t_0) &= D_-\|u(t_0)\| \cdot u'(t_0)\\ &= \min\{\langle u'(t_0), x^*\rangle; \ x^* \in \partial\|u(t_0)\|\}.\end{aligned} \tag{D.6}$$

Proof Let $h > 0$, $t_0 + h \in [0, T[$, then

$$\begin{aligned}&\left|\gamma(t_0 + h) - \gamma(t_0) - (\|u(t_0) + hu'(t_0)\| - \|u(t_0)\|)\right|\\ &= \left|\ \|u(t_0 + h)\| - \|u(t_0) + hu'(t_0)\|\ \right|\\ &\le \|u(t_0 + h) - u(t_0) - hu'(t_0)\|.\end{aligned}$$

Dividing by h and letting h tend to 0 the conclusion follows. □

D.2 Dissipative mappings

We will need the following simple result.

Proposition D.5 *Let $x, y \in E$. The following assertions are equivalent:*

(i) $\|x\| \le \|x + \alpha y\|, \quad \forall\, \alpha \ge 0,$
(ii) there exists $x^* \in \partial\|x\|$ such that $\langle y, x^*\rangle \ge 0$.

Proof If (i) holds, we have

$$\frac{\|x + hy\| - \|x\|}{h} \ge 0, \quad \forall\, h > 0,$$

from which

$$D_+\|x\| \cdot y = \max\{\langle y, x^*\rangle : \ x^* \in \partial\|x\|\} \ge 0$$

which yields (ii). Conversely, if (ii) holds, from (D.2) it follows that $D_+\|x\| \cdot y \ge 0$ and so

$$\frac{\|x + hy\| - \|x\|}{h} \ge D_+\|x\| \cdot y \ge 0.$$

□

A mapping $f : D(f) \subset E \to E$ is said to be *dissipative* if

$$\|x - y\| \le \|x - y - \alpha(f(x) - f(y))\|, \quad \forall\, x, y \in D(f). \tag{D.7}$$

From Proposition D.5, $f : D(f) \subset E \to E$ is dissipative if and only if for any $x, y \in D(f)$ there exists $z^* \in \partial\|x - y\|$ such that

$$\langle f(x) - f(y), z^* \rangle \leq 0. \tag{D.8}$$

Example D.6 Let $E = H$ be a Hilbert space, with inner product $\langle \cdot, \cdot \rangle$. Then $f : D(f) \subset H \to H$ is dissipative if and only if

$$\langle f(x) - f(y), x - y \rangle \leq 0, \quad \forall\, x, y \in D(f). \tag{D.9}$$

□

Example D.7 Let $E = C(\overline{\mathscr{O}})$ as in Example D.3 and let $\varphi : \mathbb{R}^1 \to \mathbb{R}^1$ be a real nondecreasing function. Consider the Nemytskii operator $f : E \to E$,

$$f(x)(\xi) = \varphi(x(\xi)), \quad \xi \in \mathscr{O}. \tag{D.10}$$

Then f is dissipative. In fact, let $x, y \in E$ and $\alpha > 0$. Set

$$u = x - \alpha f(x), \quad v = y - \alpha f(y),$$

then

$$x(\xi) - y(\xi) - \alpha(\varphi(x(\xi) - \varphi(y(\xi))) = u(\xi) - v(\xi), \quad \xi \in \overline{\mathscr{O}}. \tag{D.11}$$

Let $\xi_0 \in \overline{\mathscr{O}}$ such that $\|x - y\| = |x(\xi_0) - y(\xi_0)|$. By possibly exchanging x with y we can assume $\|x - y\| = x(\xi_0) - y(\xi_0)$. Set now $\xi = \xi_0$ in (D.11) and multiply both sides of (D.11) by $x(\xi_0) - y(\xi_0)$. Since

$$(\varphi(x(\xi_0) - \varphi(y(\xi_0)))x(\xi_0) - y(\xi_0) \leq 0,$$

it follows that

$$\|x - y\| \leq u(\xi_0) - v(\xi_0) \leq \|u - v\|$$

as required. □

Example D.8 Let A be the infinitesimal generator of a contraction C_0-semigroup $S(\cdot)$ in E. Then A is dissipative, moreover

$$\langle Ax, x^* \rangle \leq 0, \quad \forall\, x^* \in \partial\|x\|.$$

In fact, if $x \in D(A)$ and $x^* \in \partial\|x\|$ we have

$$\langle Ax, x^* \rangle = \lim_{h \to 0} \frac{1}{h} (\langle S(h)x, x^* \rangle - \|x\|) \leq 0.$$

□

D.3 Continuous dissipative mappings

We consider the particular important case of a continuous dissipative mapping $f : E \to E$. The following result is due to [528].

Theorem D.9 *Let $f : E \to E$ be a continuous dissipative mapping in E. Then for any $x \in E$ there exists a unique $u : [0, +\infty) \to E$, continuously differentiable and such that*

$$u'(t) = f(u(t)), \quad t \geq 0, \quad u(0) = x. \tag{D.12}$$

Moreover, if there exists $\omega > 0$ such that $f + \omega I$ is dissipative, then there exists a unique $z \in E$ such that $f(z) = 0$ and $\lim_{t\to+\infty} u(t) = z$.

We shall need also the following result.

Corollary D.10 *Let $f : E \to E$ be a continuous dissipative mapping in E. Then for any $\alpha > 0$ and any $y \in E$ there exists a unique $x = J_\alpha(y)$ such that*

$$x - \alpha f(x) = y. \tag{D.13}$$

Proof Set $g(x) = \alpha f(x) - x - y,\ x \in E$, then $g + I$ is dissipative and, by the second part of Theorem D.9 there exists a unique $x \in E$ such that $g(x) = 0$, which is equivalent to (D.13). □

We define now the *Yosida approximations* $f_\alpha,\ \alpha > 0$, of f, by setting

$$f_\alpha(x) = f(J_\alpha(x)) = \frac{1}{\alpha}(J_\alpha(x) - x), \quad x \in E, \tag{D.14}$$

where

$$J_\alpha(x) = (I - \alpha f)^{-1}(x), \quad x \in E. \tag{D.15}$$

We finish this section by listing some useful properties of J_α and f_α.

Proposition D.11 *Let $f : E \to E$ be a continuous dissipative mapping in E, and let J_α and f_α be defined by* (D.14) *and* (D.15) *respectively.*

(i) For any $\alpha > 0$ we have

$$\|J_\alpha x - J_\alpha y\| \leq \|x - y\| \quad \forall\, x, y \in E. \tag{D.16}$$

(ii) For any $\alpha > 0$ f_α is dissipative and Lipschitz continuous:

$$\|f_\alpha(x) - f_\alpha(y)\| \leq \frac{2}{\alpha}\|x - y\|, \quad \forall\, x, y \in E \tag{D.17}$$

and

$$\|f_\alpha(x)\| \leq \|f(x)\|, \quad \forall\, x \in E. \tag{D.18}$$

(iii) We have

$$\lim_{\alpha \to 0} J_\alpha(x) = x, \quad \forall\, x \in E. \tag{D.19}$$

Proof (i) follows from the dissipativity of f and then (D.17) is clear. We prove now dissipativity of f_α. Let $x, y \in E$ and $\beta > 0$, we have

$$\begin{aligned} \|x - y - \beta(f_\alpha(x) - f_\alpha(y))\| &= \left\| \left(1 + \frac{\beta}{\alpha}\right)(x - y) - \frac{\beta}{\alpha}(J_\alpha(x) - J_\alpha(y)) \right\| \\ &\geq \|x - y\|, \end{aligned} \tag{D.20}$$

which implies dissipativity of f_α. Finally since

$$f_\alpha(x) = \frac{1}{\alpha}(J_\alpha(x) - J_\alpha(x - \alpha f(x))),$$

therefore, from (D.17), (D.18) follows. Finally we have

$$\|J_\alpha(x) - x\| = \alpha \|f_\alpha(x)\| \leq \alpha \|f(x)\|, \quad \forall\, x \in E,$$

which implies (D.19). □

Bibliography

[1] ACQUISTAPACE P. & TERRENI B. (1984) An approach to Ito linear equations in Hilbert spaces by approximation of white noise with coloured noise, *Stochastic Anal. Appl.*, **2**, no. 2, 131–186.

[2] ACQUISTAPACE P. & TERRENI B. (1987) Hölder classes with boundary conditions as interpolation spaces, *Math. Z.*, **195**, no. 4, 451–471.

[3] ADAMS R. A. (1975) *Sobolev Spaces*, Pure and Applied Mathematics, 65, Academic Press.

[4] AGMON S. A. (1962) On the eigenfunctions and on the eigenvalues of general elliptic boundary value problems, *Commun. Pure Appl. Math.*, **15**, 119–147.

[5] AGMON S. A. (1965) *Lectures on Elliptic Boundary Value Problems*, Van Nostrand.

[6] AHMED N., FUHRMAN M. & ZABCZYK J. (1997) On filtering equations in infinite dimensions, *J. Funct. Anal.*, **143**, no. 1, 180–204.

[7] ALABERT A. & GYÖNGY I. (2001) On stochastic reaction-diffusion equations with singular force term, *Bernoulli*, **7**, no. 1, 145–164.

[8] ALBEVERIO S., BRZEŻNIAK Z. & DALETSKI A. (2003) Stochastic differential equations on product loop manifolds, *Bull. Sci. Math.*, **127**, no. 7, 649–667.

[9] ALBEVERIO S. & CRUZEIRO A. B. (1990) Global flows with invariant (Gibbs) measures for Euler and Navier–Stokes two dimensional fluids, *Commun. Math. Phys.*, **129**, no. 3, 431–444.

[10] ALBEVERIO S. & FERRARIO B. (2008) Some methods of infinite dimensional analysis in hydrodynamics: an introduction. In *SPDE in Hydrodynamic: Recent Progress and Prospects*, G. Da Prato and M. Röckner (eds.), Lecture Notes in Mathematics, no. 1942, Springer-Verlag, 1–50.

[11] ALBEVERIO S., KONDRATIEV Yu., RÖCKNER M. & TSIKALENKO T. V. (2001) Glauber dynamics for quantum lattice systems, *Rev. Math. Phys.*, **13**, no. 1, 51–124.

[12] ALBEVERIO S., KONDRATIEV Y., KOZITSKY Y. & RÖCKNER M. (2009) *The Statistical Mechanics of Quantum Lattice Systems. A Path Integral Approach,* EMS Tracts in Mathematics, 8.

[13] ALBEVERIO S. & RÖCKNER M. (1989) Classical Dirichlet forms on topological vector spaces – the construction of the associated diffusion process, *Probab. Theory Relat. Fields*, **83**, no. 3, 405–434.

[14] ALBEVERIO S. & RÖCKNER M. (1991) Stochastic differential equations in infinite dimensions: solutions via Dirichlet forms, *Probab. Theory Relat. Fields*, **89**, no. 3, 347–386.

[15] ALBEVERIO S., WU J. L. & ZHANG T. S. (1998) Parabolic SPDEs driven by Poissonian noise, *Stochastic Process. Appl.*, **74**, no. 1, 21–36.

[16] ALLEN E. J., NOVOSEL S. J. & ZHANG Z. (1998) Finite element and difference approximation of some linear stochastic partial differential equations, *Stochastics Stochastics Rep.*, **64**, no. 1–2, 117–142.

[17] ALÒS E. & BONACCORSI S. (2002) Stochastic partial differential equations with Dirichlet white-noise boundary conditions, *Ann. Inst. H. Poincaré Probab. Stat.*, **38**, no. 2, 125–154.

[18] ALÒS E. & BONACCORSI S. (2002) Stability for stochastic partial differential equations with Dirichlet white-noise boundary conditions, *Infin. Dimens. Anal. Quantum Probab. Relat. Topics*, **5**, no. 4, 465–481.

[19] AMBROSIO L. (2004) Transport equation and Cauchy problem for BV vector fields, *Invent. Math.*, **158**, no. 2, 227–260.

[20] AMBROSIO L. & FIGALLI A. (2009) On flows associated to Sobolev vector fields in Wiener spaces: an approach à la DiPerna-Lions, *J. Funct. Anal.*, **256**, no. 1, 179–214.

[21] AMBROSIO L., SAVARÉ G. & ZAMBOTTI L. (2009) Existence and stability for Fokker–Planck equations with log-concave reference measure, *Probab. Theory Relat. Fields*, **145**, no. 3–4, 517–564.

[22] ANTONIADIS A. & CARMONA R. (1987) Eigenfunctions expansions for infinite dimensional Ornstein–Uhlenbeck processes, *Probab. Theory Relat. Fields*, **74**, no. 1, 31–54.

[23] APPLEBAUM D. (2004) *Lévy Processes and Stochastic Calculus*, Cambridge University Press.

[24] APPLEBAUM D. & WU J. L. (2000) Stochastic partial differential equations driven by Lévy space-time white noise, *Random Oper. Stochastic Equations*, **8**, no. 3, 245–261.

[25] ARNOLD L. (1974) *Stochastic Differential Equations,* Wiley Interscience.

[26] ARNOLD L. (1981) *Mathematical Models of Chemical Reactions, Stochastic Systems,* M. Hazewinkel and J. Willems (eds.), Dordrecht.

[27] ARNOLD L., CURTAIN R. F. & KOTELENEZ P. (1980) Nonlinear evolution equations in Hilbert spaces, Forschungsschwerpunkt Dynamische Systeme, Universität Bremen, Report no. 17.

[28] ARONSON D. G. (1986) The porous medium equation. In *Nonlinear Diffusion Problems (Montecatini Terme 1985)*, Lecture Notes in Mathematics, no. 1224, Springer-Verlag, 1–46.

[29] AUBIN J. P. (1979) *Applied Functional Analysis*, Wiley Interscience.

[30] AUBIN J. P. & DA PRATO G. (1990) Stochastic viability and invariance, *Ann. Scuola Norm. Sup. Pisa*, **17**, no. 4, 595–613.

[31] AUBIN J. P. & FRANKOWSKA H. (1990) *Set Valued Analysis, Systems and Control: Foundations and Applications,* Birkhäuser.

[32] BALAKRISHNAN A. V. (1971) *Introduction to Optimization Theory in a Hilbert Space,* Springer-Verlag.

[33] BALL J. M. (1977) Strongly continuous semigroups, weak solutions and the variation of constants formula, *Proc. Am. Math. Soc.,* **63**, no. 2, 370–373.

[34] BAŇAS L., BRZEŹNIAK Z. & PROHL A. (2012) Convergent finite element based discretization of the stochastic Landau–Lifshitz–Gilbert equations, Preprint, Universität Tübingen, Numerical Analysis Group.

[35] BANG O., CHRISTIANSEN P. L. & RASMUSSEN K. O. (1994) Temperature effects in a nonlinear model of monolayer Scheibe aggregates, *Phys. Rev. E*, **49**, 4627–4636.

[36] BARAS J. S., BLANKENSHIP G. L. & HOPKINS W. E. (1983) Existence, uniqueness and asymptotic behaviour of solutions to a class of Zakai equations with unbounded coefficients, *IEEE Trans. Automat. Control*, **28**, no. 2, 203–214.

[37] BARBU B., BLANCHARD P., DA PRATO G. & RÖCKNER M. (2009) Self-organized criticality via stochastic partial differential equations. In *Potential Theory and Stochastics*, *Albac, Aurel Cornea Memorial Volume*, Theta Series in Advanced Mathematics, Bucharest, 11–19.

[38] BARBU V. & DA PRATO G. (2002) The two phase stochastic Stefan problem, *Probab. Theory Relat. Fields*, **124**, no. 4, 544–560.

[39] BARBU V. & DA PRATO G. (2007) Existence and ergodicity for the two-dimensional stochastic magneto-hydrodynamics equations, *Appl. Math. Optim.*, **56**, no. 2, 145–168.

[40] BARBU V. & DA PRATO G. (2011) Ergodicity for the phase-field equations perturbed by Gaussian noise, *Infin. Dimens. Anal. Quantum Probab. Relat. Topics*, **14**, no. 1, 35–55.

[41] BARBU V., DA PRATO G. & RÖCKNER M. (2008) Existence and uniqueness of nonnegative solutions to the stochastic porous media equation, *Indiana Univ. Math. J.*, **57**, no. 1, 187–212.

[42] BARBU V., DA PRATO G. & RÖCKNER M. (2008) Existence of strong solutions for stochastic porous media equation under general monotonicity conditions, *Ann. Probab.*, **37**, no. 2, 428–452.

[43] BARBU V., DA PRATO G. & RÖCKNER M. (2009) Stochastic porous media equation and self-organized criticality, *Commun. Math. Phys.*, **285**, no. 3, 901–923.

[44] BARBU V., DA PRATO G. & RÖCKNER M. (2009) Finite time extinction for solutions to fast diffusion stochastic porous media equations, *C. R. Acad. Sci. Paris, Ser. I*, **347**, no. 1–2, 81–84.

[45] BARBU V., DA PRATO G. & TUBARO L. (2009) Kolmogorov equation associated to the stochastic reflection problem on a smooth convex set of a Hilbert space, *Ann. Probab.*, **37**, no. 4, 1427–1458.

[46] BARBU V., DA PRATO G. & TUBARO L. (2011) Kolmogorov equation associated to the stochastic reflection problem on a smooth convex set of a Hilbert space II, *Ann. Inst. H. Poincaré Probab. Stat.*, **47**, no. 3, 699–724.

[47] BARBU V. & PRECUPANU Th. (1978) *Convexity and Optimization in Banach Spaces,* Sijthoff & Noordhoff.

[48] BARBU V., RÖCKNER M. & RUSSO F. (2011) Probabilistic representation for solutions of an irregular porous media type equation: the degenerate case, *Probab. Theory Relat. Fields*, **151**, no. 1–2, 1–43.

[49] BARSKI M. & ZABCZYK J. (2012) Heath–Jarrow–Morton–Musiela equation with Lévy perturbation, *J. Differential Equations*, **253**, no. 9, 2657–2697.

[50] BARTON SMITH M., DEBUSSCHE A. & DI MENZA L. (2005) Numerical study of two-dimensional stochastic NLS equations, *Numer. Methods Partial Differential Equations*, **21**, no. 4, 810–842.

[51] BASSON A. (2008) Spatially homogeneous solutions of 3D stochastic Navier–Stokes equations and local energy inequality, *Stochastic Process. Appl.*, **118**, no. 3, 417–451.

[52] BELOPOLSKAYA T. & DALECKIJ Yu. L. (1990) *Stochastic Equations on Manifolds*, Kluwer.

[53] BENSOUSSAN A. (1971) *Filtrage Optimal des Systèmes Linéaires,* Dunod.

[54] BENSOUSSAN A. (1973) Généralisation du Théorème de Girsanov, Univ. Naz. Rosario, Separato de Math. Notae, no. 23.

[55] BENSOUSSAN A. (1983) Stochastic maximum principle for distributed parameter systems, *J. Franklin Inst.*, **315**, no. 5–6, 387–406.

[56] BENSOUSSAN A. (1992) Some existence results for stochastic partial differential equations, *Pitman Res. Notes Math.*, no. 268, G. Da Prato and L. Tubaro (eds.), 37–53.

[57] BENSOUSSAN A., GLOWINSKI R. & RASCANU A. (1989) Approximation of Zakai equation by the splitting up method, Research report U.H./M.D. no. 55, Department of Mathematics, University of Houston.

[58] BENSOUSSAN A. & TEMAM R. (1972) Équations aux dérivées partielles stochastiques nonlinéaires, *Isr. J. Math.*, **11**, 95–121.

[59] BENSOUSSAN A. & TEMAM R. (1973) Équations stochastiques du type Navier–Stokes, *J. Funct. Anal.,* **13**, 195–222.

[60] BENSOUSSAN A. & VIOT M. (1975) Optimal control of stochastic linear distributed parameter systems, *SIAM J. Control Optim.*, **13**, 904–926.

[61] BENZI R., PARISI G., SUTERA A. & VULPIANI A. (1982) Stochastic resonance in climatic change. *Tellus 34*, (1): 10-6. doi:10.1111/j.2153-3490.1982.

[62] BENZI R., SUTERA A. & VULPIANI A. (1985) Stochastic resonance in the Landau–Ginzburg equation. *J. Phys. A*, **18**, no. 12, 2239–2245.

[63] BERRYMAN J.G. & HOLLAND C. J. (1980) Stability of the separable solution for fast diffusion, *Arch. Rational Mech. Anal.*, **74**, no. 4, 379–388.

[64] BERTINI L., CANCRINI N. & JONA-LASINIO G. (1994) The stochastic Burgers equation, *Commun. Math. Phys.*, **165**, no. 2, 211–232.

[65] BERTINI L. & GIACOMIN G. (1997) Stochastic Burgers and KPZ equations from particle systems, *Commun. Math. Phys.*, **183**, no. 3, 571–607.

[66] BESSAIH H. (1999) *Stochastic PDEs of Euler type*, PhD Thesis, Scuola Normale Superiore, Pisa.

[67] BESSAIH H. (1999) Martingale solutions for stochastic Euler equations, *Stochastic Anal. Appl.*, **17**, no. 5, 713–725.

[68] BESSAIH H. & FLANDOLI F. (1999) 2D Euler equation perturbed by noise, *Nonlinear Differential Equations Appl.,* **6**, no. 1, 35–54.

[69] BESSAIH H. & MILLET A. (2009) Large deviation principle and inviscid shell models, *Electron. J. Probab.*, **14**, no. 89, 2551–2579.

[70] BEYN W., GESS B., LESCOT P. & RÖCKNER M. The global random attractor for a class of stochastic porous media equations, arXiv:1010.0551.

[71] BIERKENS J., VAN GAANS O. & LUNEL S. V. (2009) Existence of an invariant measure for stochastic evolutions driven by an eventually compact semigroup, *J. Evol. Equations*, **9**, 771–786.

[72] BIRYUK A. (2006) On invariant measures of the 2D Euler equation, *J. Stat. Phys.*, **122**, no. 4, 597–616.

[73] BISMUT J. M. (1973) Conjugate convex functions in optimal stochastic control, *J. Math. Anal. Appl.*, **44**, 384–404.

[74] BISMUT J. M. (1984) *Large Deviations and the Malliavin Calculus*, Birkhäuser.

[75] BJÖRK T., CHRISTENSEN B. J. & BENT J. (1999) Interest dynamics and consistent forward rate curves, *Math. Finance*, **9**, no. 4, 323–348.

[76] BLANCHARD P., RÖCKNER M. & RUSSO F. (2010) Probabilistic representation for solutions of an irregular porous media type equation, *Ann. Probab.*, **38**, no. 5, 1870–1900.

[77] BLÖMKER D., FLANDOLI F. & ROMITO M. (2009) Markovianity and ergodicity for a surface growth PDE, *Ann. Probab.*, **37**, no. 1, 275–313.

[78] BLÖMKER D. & GUGG C. (2002) On the existence of solutions for amorphous molecular beam epitaxy, *Nonlinear Anal.*, **3**, no. 1, 61–73.

[79] BLÖMKER D. & GUGG C. (2004) Thin-film-growth-models: on local solutions. In *Recent Developments in Stochastic Analysis and Related Topics*, World Scientific, 66–77.

[80] BLÖMKER, D., GUGG C. & RAIBLE M. (2002) Thin-film-growth models: Roughness and correlation functions, *Eur. J. Appl. Math.*, **13**, no. 4, 385–402.

[81] BLÖMKER D. & HAIRER M. (2004) Multiscale expansion of invariant measures for SPDEs, *Commun. Math. Phys.* **251**, no. 3, 515–555.

[82] BLÖMKER D., HAIRER M. & PAVLIOTIS G. A. (2005) Modulation equations: stochastic bifurcation in large domains, *Commun. Math. Phys.* **258**, no. 2, 479–512.

[83] BLÖMKER D., HAIRER M. & PAVLIOTIS G. A. (2007) Multiscale analysis for stochastic partial differential equations with quadratic nonlinearities, *Nonlinearity*, **20**, no. 7, 1721–1744.

[84] BLOWEY J. F. & ELLIOTT C. M. (1994) A phase-field model with a double obstacle potential. In *Motion by Mean Curvature*, G. Buttazzo and A. Visintin (eds.), De Gruyter.

[85] BOGACHEV V., DA PRATO G. & RÖCKNER M. (2008) On parabolic equations for measures, *Commun. Partial Differential Equations*, **33**, no. 1–3, 397–418.

[86] BOGACHEV V., DA PRATO G. & RÖCKNER M. (2008) Parabolic equations for measures on infinite-dimensional spaces, *Dokl. Math.*, **78**, no. 1, 544–549.

[87] BOGACHEV V., DA PRATO G. & RÖCKNER M. (2009) Fokker–Planck equations and maximal dissipativity for Kolmogorov operators with time dependent singular drifts in Hilbert spaces, *J. Funct. Anal.,* **256**, no. 4, 1269–1298.

[88] BOGACHEV V., DA PRATO G. & RÖCKNER M. (2010) Existence and uniqueness of solutions for Fokker–Planck equations on Hilbert spaces, *J. Evol. Equations*, **10**, no. 3, 487–509

[89] BOGACHEV V., DA PRATO G. & RÖCKNER M. (2011) Uniqueness for solutions of Fokker–Planck equations on infinite dimensional spaces, *Commun. Partial Differential Equations*, **36**, no. 6, 925–939.

[90] BOGACHEV V., DA PRATO G., RÖCKNER M. & STANNAT W. (2007) Uniqueness of solutions to weak parabolic equations for measures, *Bull. London Math. Soc.*, no. 4, 631–640.

[91] BOJDECKI T. & GOROSTIZA L. G. (1986) Langevin equations for S'-valued Gaussian processes and fluctuations limits of infinite particle systems, *Probab. Theory Relat. Fields*, **73**, no. 2, 227–244.

[92] BORKAR V. S, CHARI R. T. & MITTER S. K. (1988) Stochastic quantization field theory in finite and infinite volume, *J. Funct. Anal.*, **81**, no. 1, 184–206.

[93] BOURGAIN J. (1993) Fourier restriction phenomena for certain lattice subsets and applications to nonlinear evolution equations, Part II. The KdV-equation, *Geom. Funct. Anal.*, **3**, no. 3, 209–262.

[94] BRIAND Ph. & CONFORTOLA F. (2008) Differentiability of backward stochastic differential equations in Hilbert spaces with monotone generators, *Appl. Math. Optim.*, **57**, no. 2, 149–176.

[95] BRIAND Ph. & CONFORTOLA F. (2008) BSDEs with stochastic Lipschitz condition and quadratic PDEs in Hilbert spaces, *Stochastic Process. Appl.*, **118**, no. 5, 818–838.

[96] BRIAND Ph. & CONFORTOLA F. (2008) Quadratic BSDEs with random terminal time and elliptic PDEs in infinite dimension, *Electron. J. Probab.*, **13**, no. 54, 1529–1561.

[97] BRIAND Ph., DELYON B., HU Y., PARDOUX E. & STOICA L. (2003), L^p solutions of backward stochastic differential equations, *Stochastic Process. Appl.*, **108**, no. 1, 109–129.

[98] BRIAND Ph. & HU Y. (2006) BSDE with quadratic growth and unbounded terminal value, *Probab. Theory Relat. Fields*, **136**, no. 4, 604–618.

[99] BRICMONT J., KUPIAINEN A., & LEFEVERE R. (2002) Exponential mixing for the 2D stochastic Navier–Stokes dynamics, *Commun. Math. Phys.*, **230**, no. 1, 87–132.

[100] BRZEŹNIAK Z. (1995) Stochastic partial differential equations in M-type 2 Banach spaces, *Potential Anal.*, **4**, no. 1, 1–45.

[101] BRZEŹNIAK Z. (1997) On stochastic convolution in Banach spaces and applications, *Stochastics Stochastics Rep.*, **61**, 245–295.

[102] BRZEZNIAK Z., CAPINSKI M. & FLANDOLI F. (1988) A convergence result for stochastic partial differential equations, *Stochastics*, **24**, no. 4, 423–445.

[103] BRZEZNIAK Z., CAPINSKI M. & FLANDOLI F. (1991) Stochastic partial differential equations and turbulence, *Math. Models Methods Appl. Sci.*, **1**, no. 1, 41–59.

[104] BRZEZNIAK Z., CAPINSKI M. & FLANDOLI F. (1992) Stochastic Navier–Stokes equations with multiplicative noise, *Stochastic Anal. Appl.*, **10**, no. 5, 523–532.

[105] BRZEŹNIAK Z. & DEBBI L. (2007) On Stochastic Burgers equation driven by a fractional Laplacian and space-time white noise. In *Stochastic Differential Equations: Theory and Applications, a Volume in Honor of Professor Boris L. Rozovskii*, P. H. Baxendale & S. V. Lototsky (eds.), Interdiscip. Math. Sci., 2, World Scientific, 135–167.

[106] BRZEZNIAK Z. & FLANDOLI F. (1992) Regularity of solutions and random evolution operator for stochastic parabolic equations, *Pitman Res. Notes Math.*, no. 268, G. Da Prato and L. Tubaro (eds.), 54–71.

[107] BRZEŹNIAK Z. & GĄTAREK D. (1999) Martingale solutions and invariant measures for stochastic evolution equations in Banach spaces, *Stochastic Process. Appl.*, **84**, no. 4, 187–226.

[108] BRZEŹNIAK Z., GOLDYS B., IMKELLER P., PESZAT S., PRIOLA E. & ZABCZYK J. (2010) Time irregularity of generalized Ornstein–Uhlenbeck processes, *C. R. Acad. Sci. Paris, Ser. I,* **348**, no. 5–6, 273–276.

[109] BRZEŹNIAK Z., GOLDYS B. & JEGARAJ T. (2013) Weak solutions of the stochastic Landau–Lifshitz–Gilbert equation, *Appl. Math. Res. Express. AMRX*, no. 1, 1–33.

[110] BRZEŹNIAK Z., GOLDYS B. & JEGARAJ T. (2012) Large deviations for a stochastic Landau–Lifschitz equation, extended version, arXiv:1202.0370 [math.PR].

[111] BRZEŹNIAK Z., GOLDYS B. & ONDREJÀT M. (2012) Partial differential equations. In *New Trends in Stochastic Analysis and Related Topics*, 132, Interdiscip. Math. Sci., 12, World Scientific.

[112] BRZEŹNIAK Z. & ONDREJÀT M. (2011) Weak solutions to stochastic wave equations with values in Riemannian manifolds, *Commun. Partial Differential Equations*, **36**, no. 9, 1624–1653.

[113] BRZEŹNIAK Z., GOLDYS B., PESZAT S. & RUSSO F. (2007) PDEs with the noise on the boundary, in preparation.

[114] BRZEŹNIAK Z. & HAUSENBLAS E. (2009) Maximal regularity for stochastic convolutions driven by Lévy processes, *Probab. Theory Relat. Fields*, **145**, 615–637.

[115] BRZEŹNIAK Z. & LI Y. (2006) Asymptotical compactness of 2D stochastic Navier–Stokes equations on some unbounded domains, *Trans. Am. Math. Soc.*, **358**, no. 12, 5587–5629.

[116] BRZEZNIAK Z., LONG H. & SIMAO I. (2010) Invariant measures for stochastic evolution equations in M-type 2 Banach spaces, *J. Evol. Equations*, **10**, no. 4, 785–810.

[117] BRZEŹNIAK Z., MASLOWSKI B. & SEIDLER J. (2005) Stochastic nonlinear beam equations, *Probab. Theory Relat. Fields*, **132**, no. 1, 119–149.

[118] BRZEZNIAK Z. & ONDREJÁT M. (2007) Strong solutions to stochastic wave equations with values in Riemannian manifolds, *J. Funct. Anal.*, **253**, no. 2, 449–481.

[119] BRZEŹNIAK Z., ONDREJÁT M. & SEIDLER J. (2012) Invariant measures for stochastic nonlinear beam and wave equations, preprint.

[120] BRZEZNIAK Z. & PESZAT S. (1999) Space-time continuous solutions to SPDEs driven by a homogeneous Wiener process, *Studia Math.*, **137**, no. 3, 261–299.

[121] BRZEZNIAK Z. & PESZAT S. (2000) Maximal inequalities and exponential estimates for stochastic convolutions in Banach spaces. In *Stochastic Processes, Physics and Geometry: New Interplays, I (Leipzig)*, CMS Conf. Proc., Am. Math. Soc., Providence, RI, 28, 55–64.

[122] BRZEZNIAK Z. & PESZAT S. (2001) Stochastic two dimensional Euler equations, *Ann. Probab.*, **29**, no. 4, 1796–1832.

[123] BRZEŹNIAK Z. & PESZAT S. (2010) Hyperbolic equations with random boundary conditions. In *Recent Development in Stochastic Dynamics and Stochastic Analysis*, J. Duan, S. Luo and C. Wang (eds.), Interdisciplinary Mathematical Sciences, 8, World Scientific, 1–22.

[124] BRZEZNIAK Z., PESZAT S. & ZABCZYK J. (2001) Continuity of stochastic convolutions, *Czech. Math. J.*, **51**, no. 126, 679–684.

[125] BRZEŹNIAK Z. & VAN NEERVEN J. (2003) Space-time regularity for linear stochastic evolution equations driven by spatially homogeneous noise, *J. Math. Kyoto Univ.*, **43**, no. 2, 261–303.

[126] BRZEZNIAK Z., VAN NEERVEN J., VERAAR M. & WEIS L. (2008) Itô's formula in UMD Banach spaces and regularity of solutions of the Zakai equation, *J. Differential Equations*, **245**, no. 1, 30–58.

[127] BRZEŹNIAK Z. & ZABCZYK J. (2010) Regularity of Ornstein–Uhlenbeck processes driven by a Lévy white noise, *Potential Anal.*, **32**, no. 2, 153–188.

[128] BUCKDAHN R. & PARDOUX E. (1990) Monotonicity methods for white noise driven quasi linear SPDE's. In *Diffusion Processes and Related Problems in Analysis*, M. Pinsky (ed.), Progress in Probability, 22, Birkhäuser, 219–233.

[129] BUDHIRAJA A., DUPUIS P. & MAROULAS V. (2008) Large deviations for infinite dimensional stochastic dynamical systems, *Ann. Probab.*, **36**, no. 4, 1390–1420.

[130] BURGERS J. M. (1939) Mathematical examples illustrating relations occuring in the theory of turbulent fluid motion, *Verh. Ned. Akad. Wetensch. Afd. Natuurk.*, **17**, no. 2, 1–53.

[131] BUTZER P. L. & BERENS H. (1967) *Semigroups of Operators and Approximations*, Springer-Verlag.

[132] CABANA E. (1970) The vibrating string forced by white noise, *Z. Wahrsch. Verw. Gebiete*, **15**, 111–130.

[133] CAHN J. W. & HILLIARD J. E. (1958) Free energy for a non-uniform system I. Interfacial free energy, *J. Chem. Phys.*, **2**, 258–267.

[134] CANNARSA P. & DA PRATO G. (1991) A semigroup approach to Kolmogoroff equations in Hilbert spaces, *Appl. Math. Lett.*, **4**, no. 1, 49–52.

[135] CANNARSA P. & DA PRATO G. (1991) Second-order Hamilton–Jacobi equations in infinite dimensions, *SIAM J. Control Optim.*, **29**, no. 2, 474–492.

[136] CANNARSA P. & VESPRI V. (1985) Existence and uniqueness of solutions to a class of stochastic partial differential equations, *Stochastic Anal. Appl.*, **3**, no. 3, 315–339.

[137] CANNARSA P. & VESPRI V. (1985) Generation of analytic semigroups by elliptic operators with unbounded coefficients, *SIAM J. Math. Anal.*, **18**, no. 3, 857–872.

[138] CANNARSA P. & VESPRI V. (1987) Existence and uniqueness results for a nonlinear stochastic partial differential equation. In *Stochastic Partial Differential Equations and Applications*, G. Da Prato and L. Tubaro (eds.), Lecture Notes in Mathematics, no. 1236, Springer-Verlag, 1–24.

[139] CAPINSKI M. & CUTLAND N. (1994) Statistical solutions of stochastic Navier–Stokes equations, *Indiana Univ. Math. J.*, **43**, no. 3, 927–940.

[140] CAPINSKI M. & CUTLAND N. (1999) Stochastic Euler equations on the torus, *Ann. Appl. Probab.*, **9**, no. 3, 688–705.

[141] CAPINSKI M. & GĄTAREK D. (1994) Stochastic equations in Hilbert spaces with applications to Navier–Stokes equations in any dimension, *J. Funct. Anal.*, **126**, no. 1, 26–35.

[142] CARDON-WEBER C. (1999) Large deviations for a Burgers'-type SPDE, *Stochastic Process. Appl.*, **84**, no. 1, 53–70.

[143] CARELLI E., HAUSENBLAS E. & PROHL A. (2009) Time-splitting methods to solve the stochastic incompressible Stokes equation. Preprint, Universität Tübingen, Numerical Analysis Group.

[144] CARUANA M. & FRIZ P. (2009) Partial differential equations driven by rough paths, *J. Differential Equations*, **247**, no. 1, 140–173.

[145] CARUANA M., FRIZ P. & OBERHAUSER H. (2011) A (rough) pathwise approach to a class of non-linear stochastic partial differential equations, *Ann. Inst. H. Poincaré, Anal. Non Linéaire*, **28**, 1294–1449.

[146] CARVERHILL A. P. & ELWORTHY K. D. (1983) Flow of stochastic dynamical systems: the Functional Analysis approach, *Z. Wahrsch. Verw. Gebiete*, **65**, no. 2, 245–267.

[147] CAZENAVE T. (2003) *Semilinear Schrödinger Equations,* Courant Lecture Notes in Mathematics, no. 10. Am. Math. Soc., Providence, RI.

[148] CÉPA E. (1998) Problème de Skorohod multivoque, *Ann. Probab.*, **26**, no. 2, 500–532.

[149] CERRAI S. (2001). *Second Order PDEs in Finite and Infinite Dimensions. A Probabilistic Approach*, Lecture Notes in Mathematics, no. 1762, Springer-Verlag.

[150] CERRAI S. (2003) Stochastic reaction-diffusion systems with multiplicative noise and non-Lipschitz reaction term, *Probab. Theory Relat. Fields*, **125**, no. 2, 271–304.

[151] CERRAI S. (2009) Normal deviations from the averaged motion for some reaction-diffusion equations with fast oscillating perturbation, *J. Math. Pures Appl.*, **91**, no. 6, 614–647.

[152] CERRAI S. (2009) A Khasminskii type averaging principle for stochastic reaction-diffusion equations, *Ann. Appl. Probab.*, **19**, no. 3, 899–948.

[153] CERRAI S., DA PRATO G. & FLANDOLI F. Pathwise uniqueness for stochastic reaction-diffusion equations in Banach spaces with an Hölder drift component, *Ann. Probab.* (to appear), arXiv:1212.5376.

[154] CERRAI S. & FREIDLIN M. (2006) On the Smoluchowski–Kramers approximation for a system with an infinite number of degrees of freedom, *Probab. Theory Relat. Fields*, **135**, no. 3, 363–394.

[155] CERRAI S. & FREIDLIN M. (2006) Smoluchowski–Kramers approximation for a general class of SPDEs, *J. Evol. Equations*, **6**, no. 4, 657–689.

[156] CERRAI S. & FREIDLIN M. (2009) Averaging principle for a class of stochastic reaction-diffusion equations, *Probab. Theory Relat. Fields*, **144**, no. 1–2, 137–177.

[157] CERRAI S. & RÖCKNER M. (2004) Large deviations for stochastic reaction-diffusion systems with multiplicative noise and non-Lipschitz reaction term, *Ann. Probab.*, **32**, no. 1B, 1100–1139.

[158] CERRAI S. & RÖCKNER M. (2005) Large deviations for invariant measures of stochastic reaction-diffusion systems with multiplicative noise and non-Lipschitz reaction term, *Ann. Inst. H. Poincaré Probab. Statist.*, **41**, no. 1, 69–105.

[159] CHALEYAT-MAUREL M. (1987) Continuity in nonlinear filtering. Some different approaches. In *Stochastic Partial Differential Equations and Applications*, G. Da Prato and L. Tubaro (eds.), Lecture Notes in Mathematics, no. 1236, Springer-Verlag, 25–39.

[160] CHALEYAT-MAUREL M. & MICHEL D. (1989) The support of the density of a filter in the uncollerated case. In *Stochastic Partial Differential Equations and Applications II*, G. Da Prato and L. Tubaro (eds.), Lecture Notes in Mathematics, no. 1390, Springer-Verlag, 33–41.

[161] CHANG M.-H. (1996) Large deviations for Navier–Stokes equations with small stochastic perturbations, *Appl. Math. Comput.*, **76**, 65–93.

[162] CHENAL F. & MILLET A. (1997) Uniform large deviations for parabolic SPDEs and applications, *Stochastic Process. Appl.*, **72**, no. 2, 161–186.

[163] CHOJNOWSKA-MICHALIK A. (1977) *Stochastic differential equations in Hilbert spaces and some of their applications*, Thesis, Institute of Mathematics, Polish Academy of Sciences.

[164] CHOJNOWSKA-MICHALIK A. (1978) Representation theorem for general stochastic delay equations, *Bull. Acad. Pol. Sci. Ser. Sci. Math.*, **26**, 7, 634–641.

[165] CHOJNOWSKA-MICHALIK A. (1987) On processes of Ornstein–Uhlenbeck type in Hilbert spaces, *Stochastics*, **21**, no. 3, 251–286.

[166] CHOW P. L. (1978) Stochastic partial differential equations in turbulence. In A. T. Bharucha-Reid (ed.), *Probabilistic Analysis and Related Topics*, vol. 1, Academic Press, 1–43.

[167] CHOW P. L. (1987) Expectation functionals associated with some stochastic evolution equations. In *Stochastic Partial Differential Equations and Applications*, G. Da Prato and L. Tubaro (eds.), Lecture Notes in Mathematics, no. 1236, Springer-Verlag, 40–56.

[168] CHOW P. L. (1992) Large deviation problem for some parabolic Itô equations, *Commun. Pure Appl. Math.*, **45**, no. 1, 97–120.

[169] CHOW P. L. (2007) *Stochastic Partial Differential Equations,* Applied Mathematics and Nonlinear Science Series, Chapman & Hall/CRC.

[170] CHOW P. L. & MENALDI J. L. (1989) Variational inequalities for the control of stochastic partial differential equations. In *Stochastic Partial Differential Equations and Applications II*, G. Da Prato and L. Tubaro (eds.), Lecture Notes in Mathematics, no. 1390, Springer-Verlag, 42–52.

[171] CHUESHOV I. & MILLET A. (2010) Stochastic 2D hydrodynamical type systems: well posedness and large deviations, *Appl. Math. Optim.*, **61**, no. 3, 379–420.

[172] CONFORTOLA F. (2007) Dissipative backward stochastic differential equations with locally Lipschitz nonlinearity, *Stochastic Process. Appl.*, **117**, no. 5, 613–628.

[173] CONSTANTIN P. & FOIAS C. (1988) *Navier–Stokes Equations*, Chicago Lectures in Mathematics. University of Chicago Press.

[174] CRISAN D. & LYONS T. (2002) Minimal entropy approximations and optimal algorithms, *Monte Carlo Methods Appl.*, **8**, no. 4, 343–355.

[175] CUERNO R., MAKSE H. A., TOMASSONE S., HARRINGTON S. T. & STANLEY H. E. (1995) Stochastic model for surface erosion via ion sputtering: dynamical evolution from ripple morphology to rough morphology, *Phys. Rev. Lett.*, **75**, 4464–4467.

[176] CURTAIN R. F. & FALB P. (1971) Stochastic differential equations in Hilbert spaces, *J. Differential Equations,* **10**, 412–430.

[177] CURTAIN R. F. & PRITCHARD A. J. (1978) *Infinite Dimensional Linear Systems Theory*, Lecture Notes in Control and Information Sciences, no. 8, Springer-Verlag.

[178] CRUZEIRO A. B., FLANDOLI F. & MALLIAVIN P. (2007) Brownian motion on volume preserving diffeomorphisms group and existence of global solutions of 2D stochastic Euler equation, *J. Funct. Anal.*, **242**, no. 1, 304–326.

[179] DALANG R. C. & FRANGOS N. (1998) The stochastic wave equation in two spatial dimensions, *Ann. Probab.*, **26**, no. 1, 187–212.

[180] DALANG R. C., KHOSHNEVISAN D., MUELLER C., NUALART D. & XIAO Y. (2009) *A Minicourse on Stochastic Partial Differential Equations*, Lecture Notes in Mathematics, no. 1962, Springer-Verlag.

[181] DALANG R. C., MUELLER C. & ZAMBOTTI L. (2006) Hitting properties of parabolic s.p.d.e.'s with reflection, *Ann. Probab.*, **34**, no. 4, 1423–1450.

[182] DALANG R. C. & SANZ-SOLÉ M. (2009) *Hölder–Sobolev Regularity of the Solution to the Stochastic Wave Equation in Dimension Three*, Mem. Am. Math. Soc., **199**, no. 931.

[183] DALECKII Yu. L. (1966) Differential equations with functional derivatives and stochastic equations for generalized random processes, *Dokl. Akad. Nauk SSSR*, **166**, 1035–1038.

[184] DALECKIJ Yu. L. & FOMIN S. V. (1991) *Measures and Differential Equations in Infinite-Dimensional Space.* Translated from the Russian. With

additional material by V. R. Steblovskaya, Yu. V. Bogdansky and N. Yu. Goncharuk. Mathematics and its Applications (Soviet Series), no. 76, Kluwer.

[185] DA PRATO G. (1976) *Applications Croissantes et Équations d'Évolution dans les Espaces de Banach*, Academic Press.

[186] DA PRATO G. (1982) Regularity results of a convolution stochastic integral and applications to parabolic stochastic equations in a Hilbert space, *Conferenze del Seminario Matematico dell'Universitá di Bari*, no. 182, Laterza.

[187] DA PRATO G. (1983) Some results on linear stochastic differential equations in Hilbert spaces by semi-groups methods, *Stochastic Anal. Appl.*, **1**, 57–88.

[188] DA PRATO G. & DEBUSSCHE A. (1996) Stochastic Cahn–Hilliard equation, *Nonlinear Anal.*, **26**, no. 2, 241–263.

[189] DA PRATO G. & DEBUSSCHE A. (2002) 2D Navier–Stokes equations driven by a space-time white noise, *J. Funct. Anal.*, **196**, no. 1, 180–210.

[190] DA PRATO G. & DEBUSSCHE A. (2003) Ergodicity for the 3D stochastic Navier–Stokes equations, *J. Math. Pures Appl.*, **82**, 877–947.

[191] DA PRATO G. & DEBUSSCHE A. (2003) Strong solutions to the stochastic quantization equations, *Ann. Probab.,* **31**, no. 4, 1900–1916.

[192] DA PRATO G. & DEBUSSCHE A. (2008) 2D stochastic Navier–Stokes equations with a time-periodic forcing term, *J. Dynam. Differential Equations*, **20**, no. 2, 301–335.

[193] DA PRATO G. & DEBUSSCHE A. (2008) On the martingale problem associated to the 2D and 3D stochastic Navier–Stokes equations, *Rend. Lincei Math. Appl.*, **19**, no. 3, 247–264.

[194] DA PRATO G., DEBUSSCHE A. & TEMAM R. (1994) Stochastic Burgers' equation, *Nonlinear Differential Equations Appl.*, **1**, no. 4, 389–402.

[195] DA PRATO G., DEBUSSCHE A. & TUBARO L. (2004) Irregular semi-convex gradient systems perturbed by noise and application to the stochastic Cahn–Hilliard equation, *Ann. Inst. H. Poincarè Probab. Statist.*, **40**, no. 1, 73–88.

[196] DA PRATO G., DEBUSSCHE A. & TUBARO L. (2005) Coupling for some partial differential equations driven by white noise, *Stochastic Process. Appl.*, **115**, 1384–1407.

[197] DA PRATO G., DEBUSSCHE A. & TUBARO L. (2007) A modified Kardar–Parisi–Zhang model, *Electron. Commun. Probab.*, **12**, 442–453.

[198] DA PRATO G. & FLANDOLI F. (2010) Pathwise uniqueness for a class of SDE in Hilbert spaces and applications, *J. Funct. Anal.*, **259**, no. 1, 243–267.

[199] DA PRATO G. & GĄTAREK D. (1995) Stochastic Burgers equation with correlated noise, *Stochastics Stochastics Rep.*, **52**, no. 1–2, 29–41.

[200] DA PRATO G., GĄTAREK D. & ZABCZYK J. (1992) Invariant measures for semilinear stochastic equations, *Stochastic Anal. Appl.*, **10**, no. 4, 387–408.

[201] DA PRATO G. & GRISVARD P. (1979) Équations d'évolution abstraites non linéaires de type parabolique, *Ann. Mat. Pura Appl.*, **120**, no. 4, 329–396.

[202] DA PRATO G., IANNELLI M. & TUBARO L. (1981–1982) Some results on linear stochastic differential equations in Hilbert spaces, *Stochastics*, **6**, no. 2, 105–116.

[203] DA PRATO G., IANNELLI M. & TUBARO L. (1981–1982) On the path regularity of a stochastic process on a Hilbert space, *Stochastics*, **6**, no. 3–4, 315–322.

[204] DA PRATO G. & ICHIKAWA A. (1985) Stability and quadratic control for linear stochastic equations with unbounded coefficients, *Boll. Unione Mat. Ital. B,* **4**, no. 3, 987–1001.

[205] DA PRATO G., KWAPIÉN. S & ZABCZYK J. (1987) Regularity of solutions of linear stochastic equations in Hilbert Spaces, *Stochastics*, **23**, no. 1, 1–23.

[206] DA PRATO G. & LUNARDI A. (1998) Maximal regularity for stochastic convolutions in L^p spaces, *Rend. Mat. Acc. Lincei*, **9**, no. 4, 25–29.

[207] DA PRATO G., PRITCHARD A. J. & ZABCZYK J. (1991) On minimum energy problems, *SIAM J. Control Optim.,* **29**, no. 1, 209–221.

[208] DA PRATO G. & RÖCKNER M. (2002) Singular dissipative stochastic equations in Hilbert spaces, *Probab. Theory Relat. Fields*, **124**, no. 2, 261–303.

[209] DA PRATO G., RÖCKNER M., ROZOVSKII B. L. & WANG. F. (2006) Strong solutions of stochastic generalized porous media equations: existence, uniqueness, and ergodicity, *Commun. Partial Differential Equations*, **31**, no. 1–3, 277–291.

[210] DA PRATO G., RÖCKNER M. & WANG F. Y. (2009) Singular stochastic equations on Hilbert spaces: Harnack inequalities for their transition semigroups, *J. Funct. Anal.*, **257**, no. 4, 992–1017.

[211] DA PRATO G. & TUBARO L. (1985) Some results on semilinear stochastic differential equations in Hilbert spaces, *Stochastics*, **15**, no. 4, 271–281.

[212] DA PRATO G. & TUBARO L. (2000) Self-adjointness of some infinite-dimensional elliptic operators and application to stochastic quantization, *Probab. Theory Relat. Fields*, **118**, no. 1, 131–145.

[213] DA PRATO G. & ZABCZYK J. (1988) A note on semilinear stochastic equations, *Differential Integral Equations*, **1**, no. 2, 143–155.

[214] DA PRATO G. & ZABCZYK J. (1991) Smoothing properties of transition semigroups in Hilbert spaces, *Stochastics,* **35**, no. 2, 63–77.

[215] DA PRATO G. & ZABCZYK J. (1992) Non explosion, boundedness and ergodicity for stochastic semilinear equations, *J. Differential Equations*, **98**, no. 1, 181–195.

[216] DA PRATO G. & ZABCZYK J. (1992) *Stochastic Equations in Infinite Dimensions,* Encyclopedia of Mathematics and its Applications, Cambridge University Press.

[217] DA PRATO G. & ZABCZYK J. (1992) A note on stochastic convolution, *Stochastic Anal. Appl.*, **10**, no. 2, 143–153.

[218] DA PRATO G. & ZABCZYK J. (1993). Evolution equations with white-noise boundary conditions, *Stochastics Stochastics Rep.*, **42**, no. 3–4, 167–182.

[219] DA PRATO G. & ZABCZYK J. (1995) Convergence to equilibrium for classical and quantum spin systems, *Probab. Theory Relat. Fields*, **103**, 529–552.

[220] DA PRATO G. & ZABCZYK J. (1996) *Ergodicity for Infinite Dimensional Systems,* London Mathematical Society Lecture Notes, no. 229, Cambridge University Press.

[221] DA PRATO G. & ZABCZYK J. (1997) Differentiability of the Feynman–Kac semigroup and a control application, *Rend. Lincei Math. Appl.*, **8**, no. 3, 183–188.

[222] DA PRATO G. & ZABCZYK J. (2002) *Second Order Partial Differential Equations in Hilbert spaces,* London Mathematical Society Lecture Notes, no. 293, Cambridge University Press.

[223] DATKO R. (1970) Extending a theorem of A. M. Liapunov to Hilbert space, *J. Math. Anal. Appl.*, **32**, 610–616.

[224] DAVIES E. B. (1980) *One-Parameter Semigroups,* London Mathematical Society Monographs, no. 15, Academic Press.

[225] DAWSON D. A. (1972) Stochastic evolution equations, *Math. Biosci.,* **154**, no. 3–4, 187–316.

[226] DAWSON D. A. (1975) Stochastic evolution equations and related measures processes, *J. Multivariate Anal.*, **5**, 1–52.

[227] DAWSON D. A. & GOROSTIZA L. G. (1989) Generalized solutions of stochastic evolution equations. In *Stochastic Partial Differential Equations and Applications II*, G. Da Prato and L. Tubaro (eds.), Lecture Notes in Mathematics, no. 1390, Springer-Verlag, 53–64.

[228] DAWSON D. A. & SALEHI H. (1980) Spatially homogeneous random evolutions, *J. Multivariate Anal.*, **10**, no. 2, 141–180.

[229] DE ACOSTA A. (2000) A general non-convex large deviation result with applications to stochastic equations, *Probab. Theory Relat. Fields*, **118**, no. 4, 483–521.

[230] DE BOUARD A. & DEBUSSCHE A. (1998) On the stochastic Korteweg–de Vries equation, *J. Funct. Anal.*, **154**, no. 1, 215–251.

[231] DE BOUARD A. & DEBUSSCHE A. (1999) A stochastic nonlinear Schrödinger equation with multiplicative noise, *Commun. Math. Phys.*, **205**, no. 1, 161–181.

[232] DE BOUARD A. & DEBUSSCHE A. (2002) On the effect of a noise on the solutions of the focusing supercritical nonlinear Schrödinger equation, *Probab. Theory Relat. Fields*, **123**, no. 1, 76–96.

[233] DE BOUARD A. & DEBUSSCHE A. (2003) The stochastic nonlinear Schrödinger equation in H^1, *Stochastic Anal. Appl.*, **21**, no. 1, 97–126.

[234] DE BOUARD A. & DEBUSSCHE A. (2004) A semi-discrete scheme for the stochastic nonlinear Schrödinger equation, *Numer. Math.*, **96**, no. 4, 733–770.

[235] DE BOUARD A. & DEBUSSCHE A. (2005) Blow-up for the stochastic nonlinear Schrödinger equation with multiplicative noise, *Ann. Probab.*, **33**, no. 3, 1078–1110.

[236] DE BOUARD A. & DEBUSSCHE A. (2006) Weak and strong order of convergence of a semi discrete scheme for the stochastic nonlinear Schrodinger equation, *Appl. Math. Optim.*, **54**, no. 3, 369–399.

[237] DE BOUARD A. & DEBUSSCHE A. (2007) The Korteweg–de Vries equation with multiplicative homogeneous noise. In *Stochastic Differential Equations – Theory and Applications*, P. Baxendale and S. Lototsky (eds.), World Scientific.

[238] DE BOUARD A. & DEBUSSCHE A. (2007) Random modulation of solitons for the stochastic Korteweg–de Vries equation, *Ann. Inst. H. Poincaré, Anal. Non Linéaire*, **24**, no. 2, 251–278.

[239] DE BOUARD A. & DEBUSSCHE A. (2009) Soliton dynamics for the Korteweg–de Vries equation with multiplicative homogeneous noise, *Electron. J. Probab.*, **14**, no. 58, 1727–1744.

[240] DE BOUARD A. & DEBUSSCHE A. (2010) The nonlinear Schrodinger equation with white noise dispersion, *J. Funct. Anal.*, **259**, no. 5, 1300–1321.

[241] DE BOUARD A., DEBUSSCHE A & TSUTSUMI Y. (1999) White noise driven Korteweg–de Vries equation, *J. Funct. Anal.*, **169**, no. 2, 532–558.

[242] DE BOUARD A., DEBUSSCHE A & TSUTSUMI Y. (2004–2005) Periodic solutions of the Korteweg–de Vries equation driven by white noise, *SIAM J. Math. Anal.*, **36**, no. 3, 815–855 (electronic).

[243] DEBUSSCHE A. (2011) Weak approximation of stochastic partial differential equations: the nonlinear case, *Math. Comput.*, **80**, no. 273, 89–117.

[244] DEBUSSCHE A. & DETTORI L. (1995) On the Cahn–Hilliard equation with a logarithmic free energy, *Nonlinear Anal.*, **24**, no. 10, 1491–1514.

[245] DEBUSSCHE A. & DI MENZA L. (2002) Numerical resolution of stochastic focusing NLS equations, *Appl. Math. Lett.*, **15**, no. 6, 661–669.

[246] DEBUSSCHE A. & GAUTIER E. (2008) Small noise asymptotic of the timing jitter in soliton transmission, *Ann. Appl. Probab.*, **18**, no. 1, 178–208.

[247] DEBUSSCHE A., GLATT-HOLTZ N. & TEMAM R. (2010) Local martingale and pathwise solutions for an abstract fluids model, arXiv:1007.2831.

[248] DEBUSSCHE A., GLATT-HOLTZ N., TEMAM R. & ZIANE M. (2012) Global existence and regularity for the 3D stochastic primitive equations of the ocean and atmosphere with multiplicative white noise, *Nonlinearity*, **25**, 2093–2118.

[249] DEBUSSCHE A. & GOUDENÈGE L. (2009) Stochastic Cahn–Hilliard equation with double singular nonlinearities and two reflections, Prepublication, IRMAR 2009-29.

[250] DEBUSSCHE A., HU Y, & TESSITORE G. (2011) Ergodic BSDEs under weak dissipative assumptions, *Stochastic Process. Appl.,* **121**, no. 3, 407–426.

[251] DEBUSSCHE A. & ODASSO C. (2006) Markov solutions for the 3D stochastic Navier–Stokes equations with state dependent noise, *J. Evol. Equations*, **6**, no. 2, 305–324.

[252] DEBUSSCHE A. & PRINTEMS J. (2009) Weak order for the discretization of the stochastic heat equation, *Math. Comput.*, **78**, no. 266, 845–863.

[253] DEBUSSCHE A. & TSUTSUMI Y. (2011) 1D quintic nonlinear Schrödinger equation with white noise dispersion, *Journal Math. Pures Appl.*, **96**, no. 4, 363–376.

[254] DEBUSSCHE A. & VOVELLE J. (2010) Scalar conservation laws with stochastic forcing, *J. Funct. Anal.*, **259**, no. 4, 1014–1042.

[255] DEBUSSCHE A. & ZAMBOTTI L. (2007) Conservative stochastic Cahn–Hilliard equation with reflection, *Ann. Probab.*, **35**, no. 5, 1706–1739.

[256] DELFOUR M. C. (1980) The largest class of hereditary systems defining a C_0 semigroup on the product space, *Can. J. Math.*, **32**, no. 4, 969–978.

[257] DELFOUR M. C. & KARRAKCHOU J. (1987) State space theory of linear time invariant systems with delays in state, control and observation variables, I, II, *J. Math. Anal. Appl.*, **125**, no. 2, 361–450.

[258] DELFOUR M. C. & MITTER S. K. (1972) Hereditary differential systems with constant delay, *J. Differential Equations*, **12**, 213–235; erratum, (1973), 397.

[259] DETTWEILER E. (1989) On the martingale problem for Banach space valued SDE, *J. Theor. Probab.*, **2**, no. 2 159–197.

[260] DI BLASIO G. (1993) Holomorphic semigroups in interpolation and extrapolation spaces, *Semigroup Forum*, **47**, no. 1, 105–114.

[261] DIEUDONNÉ J. (1960) *Foundation of Modern Analysis*, Academic Press.

[262] DI PERNA R. J. & LIONS P. L. (1989) Ordinary differential equations, transport theory and Sobolev spaces, *Invent. Math.*, **98**, no. 3, 511–547.

[263] DONATI-MARTIN C. & PARDOUX E. (1993) White noise driven SPDEs with reflection, *Probab. Theory Relat. Fields*, **95**, no. 1, 413–425.

[264] DOOB J. L. (1948) Asymptotic properties of Markoff transition probabilities, *Trans. Am. Math. Soc.*, **63**, 394–421.

[265] DOSS H. & ROYER G. (1978–79) Processus de diffusion associé aux mesures de Gibbs sur $\mathbb{R}^{\mathbb{Z}^d}$, *Z. Wahrsch. Verw. Gebiete*, **46**, no. 1, 107–124.

[266] DU K., QIU J. & TANG S. (2012) L^p theory for super-parabolic backward stochastic partial differential equations in the whole space, *Appl. Math. Optim.*, **65**, no. 2, 175–219.

[267] DUAN J. & ERVIN V. J. (2001) On the stochastic Kuramoto–Sivashinsky equation, *Nonlinear Anal.*, **44**, no. 2, 205–216.

[268] DUAN J. & MILLET A. (2009) Large deviations for the Boussinesq equation under random influences, *Stochastic Process. Appl.*, **119**, no. 6, 2052–2081.

[269] DUNCAN T. E., MASLOWSKI B. & PASIK-DUNCAN B. (2005) Stochastic equations in Hilbert space with a multiplicative fractional Gaussian noise, *Stochastic Process. Appl.*, **115**, no. 8, 1357–1383.

[270] DUNCAN T. E., PASIK-DUNCAN B. & MASLOWSKI B. (2002) Fractional Brownian motion and stochastic equations in Hilbert spaces *Stoch. Dyn.*, **2**, no. 2, 225–250.

[271] DUNFORD N. & SCHWARTZ J. T. (1963) *Linear Operators. Part II: Spectral Theory. Self Adjoint Operators in Hilbert Space* (with the assistance of William G. Bade and Robert G. Bartle), Interscience Publishers John Wiley & Sons.

[272] DUNFORD N. & SCHWARTZ J. T. (1988) *Linear Operators. Part I: General Theory* (with the assistance of William G. Bade and Robert G. Bartle), reprint of the 1958 original, Interscience Publishers John Wiley & Sons.

[273] DUNST T., HAUSEMBLAS E. & PROHL A. Approximate Euler method for parabolic stochastic partial differential equations driven by space-time Lévy noise, preprint.

[274] DYNKIN E. B. (1965) *Markov Processes, Vols. I, II,* translated with the authorization and assistance of the author by J. Fabius, V. Greenberg, A. Maitra, G. Majone. Die Grundlehren der Mathematischen Wissenschaften, Bnde 121, 122, Academic Press.

[275] E W., KHANIN K., MAZEL A. & SINAI Ya. (2000) Invariant measures for Burgers equation with stochastic forcing, *Ann. Math.*, **151**, no. 3, 877–960.

[276] E W., MATTINGLY J. C. & SINAI Ya. (2001) Gibbsian dynamics and ergodicity for the stochastically forced Navier–Stokes equation, *Commun. Math. Phys.*, **224**, no. 1, 83–106.

[277] ELEZOVIÈ N. & MIKELIÈ A. (1991) On the stochastic Cahn–Hilliard equation, *Nonlinear Anal.,* **16**, no. 12, 1169–1200.

[278] El KAROUI N., PENG S. & QUENEZ M. C. (1997) Backward stochastic differential equations in finance, *Math. Finance*, **7**, no. 1, 1–71.

[279] ELLIOT C. M. & OCKENDON J. R. (1982) *Weak and Variational Methods for Moving Boundary Problems*, Pitman Research Notes in Mathematics, no. 59, Pitman Publishing.

[280] ELWORTHY K. D (1982) *Stochastic Differential Equations on Manifolds*, LMS Lecture Notes Series, no. 70, Cambridge University Press.

[281] ELWORTHY K. D (1992) Stochastic flows on Riemannian manifolds. In *Diffusion Processes and Related Problems in Analysis*, Vol. II, M. A. Pinsky and V. Wihstutz (eds.), Birkhäuser, 33–72.

[282] ENGEL K.-J. & NAGEL R. (2000) *One-Parameter Semigroups for Linear Evolution Equations*, Springer–Verlag.

[283] ES-SARHIR A. & VON RENESSE M. (2012) Ergodicity of stochastic curve shortening flow in the plane, *SIAM J. Math. Anal.*, **44**, no. 1, 224–244.

[284] ES-SARHIR A., VON RENESSE M. & STANNAT W. (2012) Estimates for the ergodic measure and polynomial stability of plane stochastic curve shortening flow, *Nonlinear Differential Equations Appl.*, **19**, no. 6, 663–675.

[285] EVANS L. C. (1998). *Partial Differential Equations*, Am. Math. Soc.

[286] EWALD B., PETCU M. & TEMAM R. (2007) Stochastic solutions of the two-dimensional primitive equations of the ocean and atmosphere with an additive noise, *Anal. Appl. (Singapore)*, **5**, no. 2, 183–198.

[287] FANG S. & LUO D. (2010) Transport equations and quasi-invariant flows on the Wiener space, *Bull. Sci. Math.*, **134**, no. 3, 295–328.

[288] FARIS W. G. & JONA-LASINIO G. (1982) Large fluctuations for a nonlinear heat equation with noise, *J. Phys. A: Math. Gen.*, **15**, 3025–3055.

[289] FENG J. & T. KURTZ T. (2006) *Large Deviations for Stochastic Processes,* Mathematical Surveys and Monographs, no. 131, Am. Math. Soc., Providence, RI.

[290] FENG J. & NUALART D. (2008) Stochastic scalar conservation laws, *J. Funct. Anal.*, **255**, no. 2, 313–373.

[291] FERNIQUE X. (1975) Regularité des trajectoires des fonctions aleatoires Gaussiennes, *École d'Été de Probabilités de Saint-Flour IV–1974*, P. Hennequin (ed.), Lecture Notes in Mathematics, no. 480, Springer-Verlag, 2–96.

[292] FERRARIO, B. (2008) Invariant mesures for a stochastic Kuramoto–Sivashinsky equation, *Stochastic Anal. Appl.*, **26**, no. 2, 379–407.

[293] FILIPOVIĆ D. (2001) *Consistency Problems for Heath–Jarrow–Morton Interest Rate Models,* Lecture Notes in Mathematics, no. 1760, Springer-Verlag.

[294] FILIPOVIĆ D., TAPPE S. & TEICHMAN J. (2010) Term structure models driven by Wiener processes and Poisson measures: existence and positivity, *SIAM J. Financial Math.*, **1**, 523–554.

[295] FLANDOLI F. (1990) Solution and control of a bilinear stochastic delay equation, *SIAM J. Control Optim.*, **28**, no. 4, 936–949.

[296] FLANDOLI F. (1990) Dirichlet boundary value problem for stochastic parabolic equations: compatibility relations and regularity of solutions, *Stochastics Stochastics Rep.,* **29**, no. 3, 331–357.

[297] FLANDOLI F. (1991) Stochastic flow and Liapunov exponents for abstract stochastic PDEs of parabolic type, *Lyapunov Exponents (Oberwolfach, 1990)*, Lecture Notes in Mathematics, no. 1486, Springer–Verlag, 196–205.

[298] FLANDOLI F. (1991) A stochastic reaction-diffusion equation with multiplicative noise, *Appl. Math. Lett.,* **4**, 45–48.

[299] FLANDOLI F. (1992) On the semi-group approach to stochastic evolution equations, *Stochastic Anal. Appl.*, **10**, no. 2, 181–203.

[300] FLANDOLI F. (1992) Stochastic evolution equations with non coercive monotone operators. In *Stochastic Partial Differential Equations and their Applications (Charlotte, NC, 1991)*, Lecture Notes in Control and Inform. Sci., no. 176, Springer-Verlag, 70–80.

[301] FLANDOLI F. (1994) Dissipativity and invariant measures for stochastic Navier–Stokes equations, *Nonlinear Differential Equations Appl.*, **1**, no. 4, 403–423.

[302] FLANDOLI F. (2008) An introduction to 3D stochastic Fluid Dynamics. In *SPDE in Hydrodynamics: Recent Progress and Prospects*, G. Da Prato and M. Röckner (eds.), CIME Lecture notes in Mathematics, no. 1942, Springer-Verlag.

[303] FLANDOLI F. (2011) Random Perturbations of PDEs and Fluid Dynamics Models, *École d'Été de Probabilités de Saint-Flour, XL–2010*, Springer-Verlag.

[304] FLANDOLI F. & GĄTAREK D. (1995) Martingale and stationary solutions for stochastic Navier–Stokes equations, *Probab. Theory Relat. Fields*, **102**, no. 3, 367–391.

[305] FLANDOLI F., GUBINELLI M. & PRIOLA E. (2010) Well-posedness of the transport equation by stochastic perturbation, *Invent. Math.*, **180**, no. 1, 1–53.

[306] FLANDOLI F. & MASLOWSKI B. (1995) Ergodicity of the 2D Navier–Stokes equation under random perturbations, *Commun. Math. Phys.*, **172**, no. 1, 119–141.

[307] FLANDOLI F. & ROMITO M. (2002) Partial regularity for the stochastic Navier–Stokes equations, *Trans. Am. Math. Soc.*, **354**, no. 6, 2207–2241.

[308] FLANDOLI F. & ROMITO M. (2008) Selections for the 3D stochastic Navier–Stokes equations, *Probab. Theory Relat. Fields*, **140**, no. 3–4, 407–458.

[309] FLANDOLI F. & SCHAUMLÖFFEL K. U. (1990) Stochastic parabolic equations in bounded domains: random evolution operator and Lyapunov exponents, *Stochastics Stochastics Rep.*, **29**, no. 4, 461–485.

[310] FLANDOLI F. & SCHAUMLÖFFEL K. U. (1991) A multiplicative ergodic theorem with applications to a first order stochastic hyperbolic equation in a bounded domain, *Stochastics Stochastics Rep.*, **34**, no. 3–4, 241–255.

[311] FLEMING W. H. (1975) Distributed parameter stochastic systems in population biology. In *Control Theory, Numerical Methods and Computer Systems Modeling*, A. Bensoussan and J. L. Lions (eds.), Lecture Notes in Economics & Mathematical Systems, no. 107, Springer-Verlag, 179–191.

[312] FLEMING W. H. & MITTER S. K. (1982) Optimal control and nonlinear filtering for nondegenerated diffusion processes, *Stochastics*, **8**, no. 1, 63–77.

[313] FÖLLMER H. (1990) Martin boundaries on Wiener spaces. In *Diffusion Processes and Related Problems in Analysis*, M. Pinsky (ed.), Birkhäuser.

[314] FÖLLMER H. & WAKOLBINGER A. (1986) Time reversal of infinite dimensional diffusions, *Stochastic Process. Appl.*, **22**, no. 1, 59–77.

[315] FREIDLIN M. I. (1988) Random perturbations of reaction-diffusion equations: the quasi-deterministic approximation, *Trans. Am. Math. Soc.*, **305**, no. 2, 665–697.

[316] FREIDLIN M. I. & VENTZELL A. (1984) *Random Perturbations of Dynamical Systems*, Springer-Verlag.

[317] FREIDLIN M. I. & VENTZELL A. (1992) Reaction-diffusion equation with randomly perturbed boundary condition, *Ann. Probab.*, **20**, no. 2, 963–986.

[318] FRISCH U. (1968) Wave propagation in random media. In *Probabilistic Methods in Applied Mathematics*, A. T. Bharucha-Reid (ed.), Academic Press.

[319] FRITZ J. (1982) Infinite lattice systems of interacting diffusion processes, existence and regularity properties, *Z. Wahrsch. Verw. Gebiete*, **59**, no. 3, 291–309.

[320] FRIZ P. & OBERHAUSER H. (2011) On the splitting-up method for rough (partial) differential equations, *J. Differential Equations*, **251**, no. 2, 316–338.

[321] FRIZ P. & OBERHAUSER H. (2010) Rough path stability of SPDEs arising in non-linear filtering, arXiv:1005.1781.

[322] FRIZ P. & VICTOIR N. (2010) *Multidimensional Stochastic Processes as Rough Paths. Theory and Applications*, Cambridge Stud. Adv. Math., 120, Cambridge University Press.

[323] FUHRMAN M. & HU Y. (2007) Backward stochastic differential equations in infinite dimensions with continuous driver and applications, *Appl. Math. Optim.*, **56**, no. 2, 265–302.

[324] FUHRMAN M., MASIERO F. & TESSITORE G. (2010) Stochastic equations with delay: optimal control via BSDEs and regular solutions of Hamilton–Jacobi–Bellman equations, *SIAM J. Control Optim.,* **48**, no. 7, 4624–4651.

[325] FUHRMAN M. & TESSITORE G. (2002) Nonlinear Kolmogorov equations in infinite dimensional spaces: the backward stochastic differential equations approach and applications to optimal control, *Ann. Probab.,* **30**, no. 3, 1397–1465.

[326] FUHRMAN M. & TESSITORE G. (2002) The Bismut–Elworthy formula for backward SDEs and applications to nonlinear Kolmogorov equations and control in infinite dimensional spaces, *Stochastics Stochastics Rep.*, **74**, no. 1–2, 429–464.

[327] FUHRMAN M. & TESSITORE G. (2004). Infinite horizon backward stochastic differential equations and elliptic equations in Hilbert spaces, *Ann. Probab.*, **32**, no. 1B, 607–660.

[328] FUHRMAN M. & TESSITORE G. (2005) Generalized directional gradients, backward stochastic differential equations and mild solutions of semilinear parabolic equations, *Appl. Math. Optim.*, **51**, no. 3, 279–332.

[329] FUHRMAN M., TESSITORE G. & HU Y. (2009) Ergodic BSDES and optimal ergodic control in Banach spaces, *SIAM J. Control Optim.,* **48**, no. 3, 1542–1566.

[330] FUJISAKI M., KALLIANPUR G. & KUNITA H. (1972) Stochastic differential equations for the nonlinear filtering problem, *Osaka J. Math.*, **9**, 19–40.

[331] FUNAKI T. (1983) Random motions of string and related stochastic evolution equations, *Nagoya Math. J.*, **89**, 129–193.

[332] FUNAKI T. (1992) A stochastic partial differential equation with values in a manifold, *J. Funct. Anal.*, **109**, no. 2, 257–288.

[333] FUNAKI T. (1999) Singular limit for stochastic reaction-diffusion equation and generation of random interfaces, *Acta Math. Sin. (Engl. Ser.)*, **15**, no. 3, 407–438.

[334] FUNAKI T. & OLLA S. (2001) Fluctuations for $\nabla\phi$ interface model on a wall, *Stochastic Process. Appl.*, **94**, no. 1, 1–27.

[335] GARSIA A. M., RADEMICH E. & RUMSAY H. Jr. (1970–1971) A real variable lemma and the continuity of paths of some Gaussian process, *Indiana Univ. Math. J.*, **20**, 565–578.

[336] GĄTAREK D. & GOLDYS B. (1992) On solving stochastic evolution equations by the change of drift with applications to optimal control. In *Stochastic Partial Differential Equations and Applications (Trento 1990)*, *Pitman Res. Notes Math.*, no. 268, G. Da Prato and L. Tubaro (eds.), 180–190.

[337] GĄTAREK D. & GOLDYS B. (1994) On weak solutions of stochastic equations in Hilbert spaces, *Stochastics Stochastics Rep.*, **46**, no. 1–2, 41–51.

[338] GĄTAREK D. & GOLDYS B. (1996) Existence, uniqueness and ergodicity for the stochastic quantization equation, *Studia Math.*, **119**, no. 2, 179–193.

[339] GAUTIER E. (2005) Large deviations and support results for nonlinear Schrödinger equations with additive noise and applications, *ESAIM Probab. Stat.*, **9**, 74–97 (electronic).

[340] GAVARECKI L. & MANDREKAR V. (2010) *Stochastic Differential Equations in Infinite Dimensions*, Springer–Verlag.

[341] GAVEAU B. (1981) Noyau de probabilité de transition de certains opérateurs d'Ornstein Uhlenbeck dans les espaces de Hilbert, *C. R. Acad. Sci. Paris Sér. I, Math.*, **293**, no. 9, 469–472.

[342] GAVEAU B. & MOULINIER J. M. (1985) Régularité des mesures et perturbations stochastiques de champs de dimension infinie, *Pub. Res. Inst. Math. Sci., Kyoto Univ.*, **21**, no. 3, 593–616.

[343] GEISSERT M., KOVACS M. & LARSSON S. (2009) Rate of weak convergence of the finite element method for the stochastic heat equation with additive noise, *BIT 49*, no. 2, 343–356.

[344] GEL'FAND M. & SHILOV G. (1964) *Generalized Functions I. Properties and Operations*, Academic Press.

[345] GEL'FAND M. & VILENKIN N. (1964) *Generalized Functions IV. Applications of Harmonic Analysis*, Academic Press.

[346] GESS B., LIU W. & RÖCKNER M. (2011) Random attractors for a class of stochastic partial differential equations driven by general additive noise, *J. Differential Equations*, **251**, no. 4–5, 1225–1253.

[347] GIKHMAN I. I (1946) A method of constructing random processes, *Dokl. Acad. Nauk, SSSR*, **58**, 961–964.

[348] GIKHMAN I. I & SKOROKHOD A. V. (1972) *Stochastic Differential Equations*, Springer-Verlag.

[349] GLATT-HOLTZ N. & TEMAM R. (2011) Pathwise solutions of the 2D stochastic primitive equations, *Appl. Anal.*, **90**, no. 1, 85–102.

[350] GLATT-HOLTZ N. & ZIANE M. (2008) The stochastic primitive equations in two space dimensions with multiplicative noise, *Discrete Contin. Dyn. Syst. Ser. B*, **10**, no. 4, 801–822.

[351] GLIMM J. & JAFFE, A. (1981) *Quantum Physics. A Functional Integral Point of View*, Springer-Verlag.

[352] GOBET E., PAGÉS G., PHAM H. & PRINTEM J. (2006) Discretization and simulation of the Zakai equation, *SIAM J. Numer. Anal.*, **44**, no. 6, 2505–2538 (electronic).

[353] GODUNOV A. N. (1975), Peano's theorem in Banach spaces, *Functional Anal. Appl.*, **9**, 53–55.

[354] GOLDYS B. & MASLOWSKI B. (2005) Exponential ergodicity for stochastic Burgers and 2D Navier–Stokes equations, *J. Funct. Anal.*, **226**, no. 1, 230–255.

[355] GOLDYS B. & MASLOWSKI B. (2006) Lower estimates of transition densities and bounds on exponential ergodicity for stochastic PDE's, *Ann. Probab.*, **34**, no. 4, 1451–1496.

[356] GOLDYS B. & MASLOWSKI B. (2006) Exponential ergodicity for stochastic reaction-diffusion equations. In *Stochastic Partial Differential Equations and Applications VII*, Lecture Notes Pure Appl. Math., no. 245, G. Da Prato and L. Tubaro (eds.), Chapman & Hall/CRC, 115–131.

[357] GOLDYS B., RÖCKNER M. & ZHANG (2009) Martingale solutions and Markov selections for stochastic partial differential equations, *Stochastic Process. Appl.*, **119**, no. 5, 1725–1764.

[358] GOROSTIZA L. & LEÒN J. A. (1991) A stochastic Fubini theorem and equivalence of extended solutions of stochastic evolution equations in Hilbert spaces. In *Random Partial Differential Equations (Oberwolfach, 1989)*, 85–94, Int. Ser. Numer. Math., 102, Birkhäuser.

[359] GOUDENÈGE L. L. (2009) Stochastic Cahn–Hilliard equation with singular nonlinearity and reflection, *Stochastic Process. Appl.*, **119**, no. 10, 3516–3548.

[360] GRECKSCH W. & TUDOR C. (1995) *Stochastic Evolution Equations. A Hilbert Space Approach*, Akademie-Verlag, Berlin.

[361] GRIESER D. (2002) Uniform bounds for eigenfunctions of the Laplacian on manifolds with boundary, *Commun. Partial Differential Equations*, **27**, no. 7–8, 1283–1299.

[362] GRISVARD P. (1966) Commutativité de deux foncteurs d'interpolation et applications, *J. Math. Pures Appl.*, **45**, 143–290.

[363] GROSS L. (1967) Potential theory in Hilbert spaces, *J. Funct. Anal.*, **1**, 123–181.

[364] GUATTERI G. (2007) On a class of forward-backward stochastic differential systems in infinite dimensions, *J. Appl. Math. Stoch. Anal.*, article 42640.

[365] GUATTERI G. (2011) Stochastic maximum principle for SPDEs with noise and control on the boundary, *Systems Control Lett.*, **60**, no. 3, 198–204.

[366] GUBINELLI M. (2004) Controlling rough paths, *J. Func. Anal.*, **216**, no. 1, 86–140.

[367] GUBINELLI M. (2010) Ramification of rough paths, *J. Differential Equations*, **248**, no. 4, 693–721.

[368] GUBINELLI M. (2012) Rough solutions of the periodic Korteweg–de Vries equation, *Commun. Pure Appl. Anal.*, **11**, no. 2, 709–733.

[369] GUBINELLI M. & TINDEL S. (to appear) Non-linear rough heat equations, *Probab. Theory Relat. Fields*.

[370] GYONGY I. (1988) On the approximation of stochastic partial differential equations, *Stochastics*, **25**, no. 2, 59–86.

[371] GYONGY I. (1989) On the approximation of stochastic partial differential equations, *Stochastics*, **26**, no. 3, 129–164.

[372] GYONGY I. (1989) The stability of stochastic partial differential equations and applications. In *Stochastic Partial Differential Equations and Applications II*, G. Da Prato and L. Tubaro (eds.), Lecture Notes in Mathematics, no. 1390, Springer-Verlag, 91–118.

[373] GYÖNGY I. (1995) On non-degenerate quasi-linear stochastic parabolic partial differential equations, *Potential Anal.*, **4**, no. 2, 157–171.

[374] GYÖNGY I. (1998) Existence and uniqueness results for semilinear stochastic partial differential equations, *Stochastic Process. Appl.*, **73**, no. 2, 271–299.

[375] GYÖNGY I. (1998) Lattice approximations for stochastic quasi-linear parabolic partial differential equations driven by space-time white noise. I, *Potential Anal.*, **9**, no. 1, 1–25.

[376] GYÖNGY I. (1999) Lattice approximations for stochastic quasi-linear parabolic partial differential equations driven by space-time white noise. II, *Potential Anal.*, **11**, no. 1, 1–37.

[377] GYONGY I. & KRYLOV N. V. (1981–1982) On stochastic equations with respect to semimartingales II. Ito formula in Banach spaces, *Stochastics,* **6**, no. 3–4, 153–163.

[378] GYÖNGY I. & MARTINEZ T. (2001) On stochastic differential equations with locally unbounded drift, *Czech. Math. J.*, **51**, no. 4, 763–783.

[379] GYÖNGY I. & MILLET A. (2005) Implicit scheme for stochastic parabolic partial differential equations driven by space-time white noise, *Potential Anal.*, **7**, no. 4, 725–757.

[380] GYÖNGY I. & MILLET A. (2009) Rate of convergence of space time approximations for stochastic evolution equations, *Potential Anal.*, **30**, no. 1, 29–64.

[381] GYÖNGY I. & NUALART D. (1997) Implicit scheme for stochastic parabolic partial differential equations driven by space-time white noise, *Potential Anal.* **7**, no. 4, 725–757.

[382] GYÖNGY I. & NUALART D. (1999) On the stochastic Burgers' equation in the real line, *Ann. Probab.*, **27**, no. 2, 782–802.

[383] GYÖNGY I. & PARDOUX E. (1993) On the regularization effect of space-time white noise on quasi-linear parabolic partial differential equations, *Probab. Theory Relat. Fields*, **97**, no. 1–2, 211–229.

[384] HAIRER M. (2005) Coupling of stochastic PDEs, *XIVth International Congress on Mathematical Physics*, World Scientific, Hackensack, NJ, 281–289.

[385] HAIRER M. (2009) An introduction to stochastic PDEs, arXiv:0907.4178.

[386] HAIRER M. (2011) Rough stochastic PDEs, *Commun. Pure Appl. Math.*, **64**, no. 11, 1547–1585.

[387] HAIRER M. (2012) Solving the KPZ equation, arXiv:1109.6811v1.

[388] HAIRER M. & MATTINGLY J. (2006) Ergodicity of the 2D Navier–Stokes equations with degenerate stochastic forcing, *Ann. Math.* **164**, no. 3, 993–1032.

[389] HAIRER M. & MATTINGLY J. (2008) Spectral gaps in Wasserstein distances and the 2D stochastic Navier–Stokes equations, *Ann. Probab.*, **36**, 2050–2091.

[390] HAIRER M., MATTINGLY J. & PARDOUX É. (2004) Calcul de Malliavin pour les Équations de Navier–Stokes 2D stochastiques, hautement dégénérées, *C. R. Acad. Sci. Paris*, **339**, no. 11, 793–796.

[391] HAIRER M., MATTINGLY J. & SCHEUTZOW M. (2011) Asymptotic coupling and a general form of Harris theorem with applications to stochastic delay equations, *Probab. Theory Relat. Fields*, **149**, no. 1–2, 223–259.

[392] HALE J. (1977) *Theory of Functional Differential Equations,* Springer-Verlag.

[393] HALMOS P. R. (1950) *Measure Theory*, Van Nostrand.

[394] HAS'MINSKII R. Z. (1980) *Stochastic Stability of Differential Equations,* Sijthoff & Noordhoff.

[395] HAUSEMBLAS E. (2002) Numerical analysis of semilinear stochastic evolution equations in Banach spaces, *J. Comput. Appl. Math.*, **147**, no. 2, 485–516.

[396] HAUSEMBLAS E. (2003) Approximation for semilinear stochastic evolution equations, *Potential Anal.*, **18**, no. 2, 141–186.

[397] HAUSEMBLAS E. & SEIDLER J. (2001) A note on maximal inequality for stochastic convolutions, *Czech. Math. J.*, **51(126)**, no. 4, 785–790.

[398] HAUSSMAN U. G. (1978) Asymptotic stability of the linear Ito equation in infinite dimensions, *J. Math. Anal. Appl.*, **65**, no. 1, 219–235.

[399] HAUSSMAN U. G. & PARDOUX E. (1989) Stochastic variational inequalities of parabolic type, *Appl. Math. Optim.*, **20**, no. 2, 163–192.

[400] HEATH D., JARROW R. & MORTON A. (1992) Bond pricing and the term structure of internal rates: a new methodology for contingent claims valuation, *Econometica*, **60**, 77–105.

[401] HENRY D. (1981) *Geometric Theory of Semilinear Parabolic Equations*, Springer-Verlag.

[402] HIDA T., KUO H., POTTHOFF J. & STREIT L. (1993) *White Noise. An Infinite-Dimensional Calculus*, Kluwer, Dordrecht.

[403] HIDA T. & STREIT L. (1977) On quantum theory in terms of white noise, *Nagoya Math. J.*, **68**, 21–34.

[404] HINO M. (2002) On short time asymptotic behavior of some symmetric diffusions on general state spaces. *Potential Anal.*, **16**, no. 3, 249–264.

[405] HINO M. & RAMÌREZ J. (2003) Small-time Gaussian behavior of symmetric diffusion semigroups, *Ann. Probab.*, **31**, no. 3, 1254–1295.

[406] HÖGELE M. A. (2010) *Metastability of the Chafee–Infante equation with small heavy-tailed Lévy noise*, PhD Dissertation, Humboldt University.

[407] HOLDEN H., ØKSENDAL B., UBØE J. & ZHANG T. (2010) *Stochastic Partial Differential Equations. A Modeling, White Noise Functional Approach*, Universitext, Springer, New York.

[408] HU Y., KALLIANPUR G. & XIONG J. (2002) An approximation of the Zakai equation, *Appl. Math. Optim.*, **45**, no. 1, 23–44.

[409] HU Y., MA J. & YONG J. (2002) On semi-linear degenerate backward stochastic partial differential equations, *Probab. Theory Relat. Fields*, **123**, no. 3, 381–411.

[410] HU Y. & PENG S. (1990) Maximum principle for semilinear stochastic evolution control systems, *Stochastics Stochastics Rep.*, **33**, no. 3–4, 159–180.

[411] HU Y. & PENG S. (1991) Adapted solution of a backward semilinear stochastic evolution equation, *Stochastic Anal. Appl.*, **9**, no. 4, 445–459.

[412] ICHIKAWA A. (1978) Linear stochastic evolution equations in Hilbert spaces, *J. Differential Equations*, **28**, no. 2, 266–283.

[413] ICHIKAWA A. (1979) Dynamic programming approach to stochastic evolution equations, *SIAM J. Control Optim.*, **17**, no. 1, 152–174.

[414] ICHIKAWA A. (1982) Stability of semilinear stochastic evolution equations, *J. Math. Anal. Appl.*, **90**, no. 1, 12–44.

[415] ICHIKAWA A. (1984) Semilinear stochastic evolution equations. Boundedness, stability and invariant measure, *Stochastics*, **12**, no. 1, 1–39.

[416] ICHIKAWA A. (1986) Some inequalities for martingales and stochastic convolutions, *Stochastic Anal. Appl.*, **4**, 329–39.

[417] ICHIKAWA A. (1987) The separation principle for stochastic differential equations with unbounded coefficients. In *Stochastic Partial Differential Equations and Applications*, G. Da Prato and L. Tubaro (eds.), Lecture Notes in Mathematics, no. 1236, Springer-Verlag, 164–171.

[418] IKEDA N. & WATANABE S. (1981) *Stochastic Differential Equations and Diffusion Processes*, North-Holland.

[419] IL'IN A.M. & HAS'MINSKII R.Z. (1963) Asymptotic behavior of solutions of parabolic equations and an ergodic property of non-homogeneous diffusion processes, *Mat. Sb. (Russian)*, **60**, 366–392.

[420] ILJASOV J. S. & KOMIEC A. I. (1987) Girsanov theorem and ergodic properties of stochastic solutions of nonlinear parabolic equations, *Trudy Sem. Petrovsk.(Russian)*, **12**, 88–117.

[421] IMAJKIN V. M. & KOMIEC A. I. (1988) On large deviations for solutions of nonlinear stochastic equations, *Trudy Sem. Petrovsk.(Russian)*, **13**, 177–196.

[422] ISCOE I., MARCUS M. B., McDONALD D., TALAGRAND M. & ZINN J. (1990) Continuity of ℓ^2-valued Ornstein–Uhlenbeck processes, *Ann. Probab.*, **18**, no. 1, 68–84.

[423] ITÔ K. (1942) Differential equations determining Markov processes, *Zenkoku Shijo Sugaku Danwakai*, no. 1077, 1352–1400.

[424] ITÔ K. (1982) Infinite dimensional Ornstein–Uhlenbeck processes, *Taniguchi Symposium, FA, Katata*, 197–224.

[425] ITÔ K. (1984) *Foundations of Stochastic Differential Equations in Infinite Dimensional Spaces,* CBMS Notes, Baton Rouge 1983, SIAM.

[426] ITURRIAGA R. & KHANIN K. (2003) Burgers turbulence and random Lagrangian systems, *Commun. Math. Phys.*, **232**, no. 3, 377–428.

[427] IWATA K. (1987) An infinite dimensional stochastic equation with state space $C(R)$, *Probab. Theory Relat. Fields,* **74**, no. 1, 141–159.

[428] JACOD J. (1979) *Calcul Stochastique et Problème de Martingales*, Lecture Notes in Mathematics, no. 714, Springer-Verlag.

[429] JENTZEN A. (2009) Pathwise numerical approximation of SPDEs with additive noise under non-global Lipschitz coefficients, *Potential Anal.*, **31**, no. 4, 375–404.

[430] JENTZEN A. (2010) Taylor expansions of solutions of stochastic partial differential equations, *Discrete Contin. Dyn. Syst., Ser. B*, **14**, no. 2, 515–557.

[431] JENTZEN A. & KLOEDEN P. (2010) Taylor expansions of solutions of partial differential equations with additive noise, *Ann. Probab.*, **38**, no. 2, 532–569.

[432] JONA LASINIO G. & MITTER S. K. (1985) On the stochastic quantization of field theory, *Commun. Math. Phys.*, **101**, no. 3, 409–436.

[433] KALLENBERG O. (2001) *Foundations of Modern Probability,* Springer-Verlag.

[434] KALLIANPUR G. & ODAIRA H. (1978) Freidlin–Ventzell type estimates for abstract linear spaces, *Sankhya: Indian J. Statistics*, **40**, no. 2, 116–135.

[435] KALLIANPUR G. & PÉREZ-ABREU V. (1988) Stochastic evolution equations driven by nuclear space valued martingales, *Appl. Math. Optim.*, **17**, no. 3, 237–272.

[436] KALLIANPUR G. & PÉREZ-ABREU V. (1989) Weak convergence of solutions of stochastic evolution equations. In *Stochastic Partial Differential Equations and Applications II*, G. Da Prato and L. Tubaro (eds.), Lecture Notes in Mathematics, no. 1390, Springer-Verlag, 119–131.

[437] KALLIANPUR G. & VOLPERT R. (1984) Infinite dimensional stochastic differential equations modelled for spatially distributed neurons, *Appl. Math. Optim.*, **12**, no. 2, 125–172.

[438] KALLIANPUR G. & XIONG J. (1996), Large deviations for a class of stochastic partial differential equations, *Ann. Probab.*, **24**, no. 1, 320–345.

[439] KALPINELLI E. A., FRANGOS N. E. & YANNACOPOULOS A. N. (2011) A Wiener chaos approach to hyperbolic spdes, *Stochastic Anal. Appl.*, **29**, no. 2, 237-258.

[440] KARCZEWSKA A. & ZABCZYK J. (2000) Stochastic PDEs with function-valued solutions. In *Infinite Dimensional Stochastic Analysis, Amsterdam 1999*, Ph. Clément, F. den Hollander, J. van Neerven and B. de Pagter (eds.), Royal Netherlands Academy of Arts and Sciences, Amsterdam, 197–216.

[441] KARCZEWSKA A. & ZABCZYK J. (2001) A note on stochastic wave equations. In *Proc. Conf. on Evolution Equations and their Applications in Physical and Life Sciences (Bad Herrenhalb 1998)*, G. Lumer and L. Weis (eds.), Dekker, 501–511.

[442] KARDAR M., PARISI G. & ZHANG J. C. (1986) Dynamical scaling of growing interfaces, *Phys. Rev. Lett.*, **56**, 889–892.

[443] KATO T. (1962) Fractional powers of dissipative operators II, *J. Math. Soc. Jpn.*, **14**, 242–248.

[444] KATO T. (1984) Strong L^p-solutions of the Navier–Stokes equation in $\mathbb{R}^m$ with applications to weak solutions, *Math. Z.*, **187**, no. 4, 471–480.

[445] KELLER J. B. (1964) Stochastic equations and wave propagation in random media, *Proc. Symp. Appl. Math.*, no. 16, 145–170.

[446] KENIG C. E. , PONCE G. & VEGA L. (2003) On unique continuation for nonlinear Schrödinger equations, *Commun. Pure Appl. Math.*, **56**, no. 9, 1247–1262.

[447] KIM J. U. (2003) On a stochastic scalar conservation law, *Indiana Univ. Math. J.* **52**, no. 1, 227–256.

[448] KIM J. U. (2005) On the stochastic Burgers equations with a polynomial nonlinearity in the real line, *Discrete Contin. Dyn. Syst, Ser. B*, **6**, no. 4, 835–866.

[449] KIM J. U. (2011) On the stochastic quasi-linear symmetric hyperbolic system, *J. Differential Equations*, **250**, no. 3, 1650–1684.

[450] KOBILANSKI M. (2000) Backward stochastic differential equations and partial differential equations with quadratic growth, *Ann. Probab.*, **28**, no. 2, 55–602.

[451] KOMOROWSKI T., PESZAT S. & SZAREK T. (2010) On ergodicity of some Markov processes, *Ann. Probab.*, **38**, no. 4, 1401–1443.

[452] KOSKI T. & LOGES W. (1985) Asymptotic statistical inference for stochastic heat flow problem, *Statist. Probab. Lett.*, **3**, no. 4, 185–189.

[453] KOTELENEZ P. (1982) A submartingale type inequality with applications to stochastic evolution equations, *Stochastics,* **8**, no. 2, 139–151.

[454] KOTELENEZ P. (1984) A stopped Doob inequality for stochastic convolution integrals and stochastic evolution equations, *Stochastic Anal. Appl.*, **2**, no. 3, 245–265.

[455] KOTELENEZ P. (1985) On the semigroup approach to stochastic evolution equations. In *Stochastic Space Time Models and Limit Theorems*, L. Arnold and P. Kotelenez (eds.), Reidel, 95–139.

[456] KOTELENEZ P. (1987) A maximal inequality for stochastic convolution integrals on Hilbert space and space–time regularity of linear stochastic partial differential equations, *Stochastics,* **21**, no. 4, 345–458.

[457] KOTELENEZ P. (1989) A stochastic reaction–diffusion model. In *Stochastic Partial Differential Equations and Applications II*, G. Da Prato and L. Tubaro (eds.), Lecture Notes in Mathematics, no. 1390, Springer-Verlag, 132–137.

[458] KOTELENEZ P. (2008) *Stochastic Ordinary and Stochastic Partial Differential Equations,* Springer-Verlag.

[459] KOVACS M., LARSSON S. & SAEDPANAH F. (2010) Finite element approximation of the linear stochastic wave equation with additive noise, *SIAM J. Numer. Anal.*, **48**, no. 2, 408–427.

[460] KOZLOV S. M. (1977) Equivalence of measures for linear stochastic Ito equations with partial derivatives, *West. Moscow University, Sov. Math. Mech.*, **4**, 47–52 (in Russian).

[461] KOZLOV S. M. (1978) Some questions of stochastic equations with partial derivatives, *Trudy Sem. Petrovsk*, **4**, 147–172, (in Russian).

[462] KREIN S. G. (1971) Linear differential equations in Banach spaces, *Trans. Am. Math. Soc.*, **29**.

[463] KRUŽKOV S. (1972) First order quasilinear equations in several independent variables, *Math. USSR Sb.*, **10**, 217–243.

[464] KRYLOV N. & BOGOLIUBOV N. (1937), La théorie générale de la mesure dans son application à l'étude des systèmes de la mécanique nonlinéaire, *Ann. Math.*, **38**, 65–113.

[465] KRYLOV N. V. (1973) The selection of a Markov process from a Markov system of processes, and the construction of quasidiffusion processes, *Izv. Akad. Nauk SSSR Ser. Mat.*, **37**, 691–708.

[466] KRYLOV N. V. (1996) On L_p theory of stochastic partial differential equations in the whole space, *SIAM J. Math. Anal.*, **27**, no. 2, 313–340.

[467] KRYLOV N. V. (1999) An analytic approach to SPDE's. In *Stochastic Partial Differential Equations: Six Perspectives*, R. A. Carmona and B. Rozoskii (eds.), Mathematical Surveys and Monograph no. 64, Am. Math. Soc., 183–242.

[468] KRYLOV N. V. & RÖCKNER M. (2005) Strong solutions to stochastic equations with singular time dependent drift, *Probab. Theory Relat. Fields*, **131**, no. 2, 154–196.

[469] KRYLOV N. V. & ROZOVSKII B. L. (1981) Stochastic evolution equations, translated from *Itogi Naukii Tekhniki, Seriya Sovremennye Problemy Matematiki*, **14** (1979), 71–146, Plenum Publishing Corp.

[470] KUKSIN S. (2004) The Eulerian limit for 2D statistical hydrodynamics, *J. Statist. Phys.*, **115**, no. 1–2, 469–492.

[471] KUKSIN S. (2008) On distribution of energy and vorticity for solutions of 2D Navier–Stokes equation with small viscosity, *Commun. Math. Phys.*, **284**, no. 2, 407–424.

[472] KUKSIN S., PIATNITSKI A. & SHIRIKYAN A. (2002) A coupling approach to randomly forced nonlinear PDE's II, *Commun. Math. Phys.*, **230**, no. 1, 81–85.

[473] KUKSIN S. & SHIRIKYAN A. (2001) A coupling approach to randomly forced nonlinear PDE's I, *Commun. Math. Phys.*, **221**, no. 2, 351–366.

[474] KUKSIN S. B. & SHIRIKYAN A. (2012) *Mathematics of Two-Dimensional Turbulence*, Cambridge Tracts in Mathematics, Cambridge University Press.

[475] KUNITA H. (1981) Densities of a measure valued process governed by a stochastic partial differential equation, *Systems Control Lett.*, **1**, no. 1, 37–41, **1**, no. 2, 100–104.

[476] KUNITA H. (1990) *Stochastic Flows and Stochastic Differential Equations,* Cambridge University Press.

[477] KUO H. H. (1965) *Gaussian Measures in Banach Spaces*, Springer-Verlag.

[478] KUO H. H. (1980) Integration in Banach spaces. In *Notes in Banach Spaces*, H. Elton Lacey (ed.), University of Texas Press, 1–38.

[479] KUO H. H. (1989) Stochastic partial differential equations of generalized Brownian functionals. In *Stochastic Partial Differential Equations and Applications II*, G. Da Prato and L. Tubaro (eds.), Lecture Notes in Mathematics, no. 1390, Springer-Verlag, 138–146.

[480] KURATOWSKI K. & RYLL-NARDZEWSKI C. (1965) A general theorem on selectors, *Bull. Acad. Pol. Sci.*, **13**, 397–403.

[481] KUSHNER H. J. (1978) On the optimal control of a system governed by a linear parabolic equation with white noise inputs, *SIAM J. Control Optim.*, **6**, no. 4, 596–614.

[482] LADYZHENSKAYA, O. A. (1969) *The Mathematical Theory of Viscous Incompressible Flow*, Second English edition, revised and enlarged. Translated from the Russian by Richard A. Silverman and John Chu. Mathematics and its Applications, Vol. 2, Gordon and Breach.

[483] LANDAU L. & LIFSHITZ E. (1969) *Electrodynamique des Milieux Continus*, *Physique Théorique*, Vol. VIII, MIR, Moscow.

[484] LANDKOFF N. (1972) *Foundations of Modern Potential Theory*, Springer-Verlag, Berlin.

[485] LARSSON S. & MESFOROUSH A. (2011) Finite element approximation of the linearized Cahn–Hilliard–Cook equation, *IMA J. Numer. Anal.*, **31**, no. 4, 1315–1333.

[486] LASIECKA I. & TRIGGIANI R. (1983) Regularity of hyperbolic equations under $L_2(0, T; L^2(\Gamma))$-Dirichlet boundary terms, *Appl. Math. Optim,* **10**, no. 3, 275–286.

[487] LAURITSEN K. B., CUERNO R. & MAKSE H. A. (1996) Noisy Kuramoto–Sivashinsky equation for an erosion model, *Phys. Rev. E*, **54**.

[488] LEE Y. J. (1986) *Sharp Inequalities and Regularity of Heat Semigroup on Infinite Dimensional Spaces*, IMA Preprint Series 217, University of Minnesota.

[489] LEHA G. & RITTER G. (1984) On diffusion processes and their semigroups in Hilbert spaces with an application to interacting stochastic systems, *Ann. Probab.*, **12**, no. 4, 1077–1112.

[490] LENGLART E., LEPINGLE D. & PRATELLI M. (1981) Presentation unifiée de certaines inegalitées de la théorie des martingales, *Séminaire des Probabilités 14*, Lecture Notes in Mathematics, no. 784, Springer-Verlag, 110–119.

[491] LEÒN J. A. (1989) Stochastic evolution equations with respect to semimartingales in Hilbert spaces, *Stochastics Stochastics Rep.,* **27**, no. 1, 1–21.

[492] LEÒN J. A. (1990) Stochastic Fubini theorem for semi-martingales in Hilbert spaces, *Can. J. Math.,* **52**, no. 5, 890–901.

[493] LERAY J. (1934) Sur le mouvement d'un liquide visqueux remplissant l'espace, *Acta Mathematica*, **63**, 193–248.

[494] LEVIEUX F. (1973) Un théorème d'éxistence et unicité de la solution d'une équation integro–différentielle stochastique, *C. R. Acad. Sci. Paris, Ser.* A-B, **277**, 281–284.

[495] LI X.-M. (1992) *Stochastic flows on noncompact manifolds*, Ph.D. Thesis, Warwick University.

[496] LI X. & TANG S. (1994) Maximum principle for optimal control of distributed parameter stochastic systems with random jumps. In *Differential Equations, Dynamical Systems, and Control Science*, Lecture Notes in Pure and Applied Mathematics, no. 152, Dekker, 867–890.

[497] LINDVALL T. (1992) *Lectures on the Coupling Method*, John Wiley & Sons.

[498] LIONS J. L. & MAGENES E. (1968) *Problèmes aux Limites Nonhomogènes et Applications*, Dunod.

[499] LIONS J. L. & PEETRE J. (1964) *Sur une Classe d'Espaces d'Interpolation*, Institut des Hautes Études Scientifiques, Publications Mathématiques no. 19, 5–68.

[500] LIONS J. L. & STRAUSS W. (1965) Some nonlinear evolution equations, *Bull. Soc. Math. Fr.*, **93**, 43–96.

[501] LIPTSER R. S. & SHIRYAYEV A. N. (2001) *Statistics of Random Processes. I. General Theory*. Translated from the 1974 Russian original by A. B. Aries. Second, revised and expanded edition. Applications of Mathematics (New York), 5. Stochastic Modelling and Applied Probability. Springer-Verlag, Berlin.

[502] LISKEVICH V. & RÖCKNER M. (1998) Strong uniqueness for a class of infinite dimensional Dirichlet operators and application to stochastic quantization, *Ann. Scuola Norm. Sup. Pisa*, **27**, no. 4, 69–91.

[503] LIU W. (2011) Ergodicity of transition semigroups for stochastic fast diffusion equations, *Front. Math. China*, **6**, no. 3, 449–472.

[504] LIU W., RÖCKNER M. & ZHU X. (2013) Large deviation principles for the stochastic quasi-geostrophic equation, *Stochastic Process Appl.*, **123**, no. 8, 3299–3327.

[505] LORD G. & SHARDLOW T. (2007) Postprocessing for stochastic parabolic partial differential equations, *SIAM J. Numer. Anal.*, **45**, no. 2, 870–889.

[506] LOTOTSKY S. V. (2006) Wiener chaos and nonlinear filtering, *Appl. Math. Optim.*, **54**, no. 3, 265–291.

[507] LOTOTSKY S. V. (2009) Stochastic partial differential equations driven by purely spatial noise, *SIAM J. Math. Anal.*, **41**, no. 4, 1295–1322.

[508] LOTOTSKY S. V. & ROZOVSKII B. (2006) Wiener chaos solutions of linear stochastic evolution equations, *Ann. Probab.*, **34**, no. 2, 638–662.

[509] LOTOTSKY S. V. & ROZOVSKII B. (2006) Stochastic differential equations: a Wiener chaos approach. In *From Stochastic Calculus to Mathematical Finance*, Springer, 433–506.

[510] LUNARDI A. (1985) Interpolation spaces between domains of elliptic operators and spaces of continuous functions with applications to nonlinear parabolic equations, *Math. Nachr.*, **121**, 323–349.

[511] LUNARDI A. (1987) On the evolution operator for abstract parabolic equations, *Isr. J. Math.*, **60**, no. 3, 281–314.

[512] LUNARDI A. (1995) *Analytic Semigroups and Optimal Regularity in Parabolic Problems*, Birkhäuser.

[513] LYONS T. (1998) Differential equations driven by rough signals, *Rev. Mat. Iberoamericana*, **14**, no. 2, 215–310.

[514] LYONS T., CARUANA M. & LÉVY T. (2007) *Differential Equations Driven by Rough Paths: École d'Eté de Probabilités De Saint-Flour XXXIV-2004*, Springer-Verlag.

[515] LYONS T. & QIAN Z. (2002) *System Control and Rough Paths*, Oxford Mathematical Monographs, Oxford University Press.

[516] MA Z. & RÖCKNER M. (1992) *Introduction to the (Non Symmetric) Theory of Dirichlet Forms*, Springer-Verlag.

[517] MA J. & YONG J. (1999) *Forward-Backward Stochastic Differential Equations and their Applications*, Lecture Notes in Mathematics, no. 1702, Springer-Verlag.

[518] MALLIAVIN P. (1978) Stochastic calculus of variations and hypoelliptic operators. In *Proceedings of International Symposium SDE, Kyoto*, K. Ito (ed.), Kinokuniya, Tokyo, 195–263.

[519] MALLIAVIN P. (1997) *Stochastic Analysis*, Springer-Verlag.

[520] MANTHEY R. (1986) Existence and uniqueness of a solution of a reaction–diffusion equation with polynomial nonlinearity and white noise disturbance, *Math. Nachr.,* **125**, 121–133.

[521] MANTHEY R. (1988) On the Cauchy problem for reaction–diffusion equations with white noise, *Math. Nachr.,* **136**, 209–228.

[522] MARCUS R. (1974) Parabolic Itô equations, *Trans. Am. Math. Soc.*, **198,** 177–190.

[523] MARCUS R. (1974) Parabolic Itô equations with monotone nonlinearities, *J. Funct. Anal.*, **29**, no. 3, 257–287.

[524] MARIANI M. (2010) Large deviation principles for stochastic scalar conservation laws, *Probab. Theory Relat. Fields*, **147**, no. 3–4, 607–648.

[525] MARINELLI C. (2010) Local well-posedness of Musiela's SPDEs with Lévy noise, *Math. Finance*, **20**, no. 3, 341–362.

[526] MARINELLI C., PREVOT C. & RÖCKNER M. (2010) Regular dependence on initial data for stochastic evolution equations with multiplicative Poisson noise, *J. Funct. Anal.*, **258**, no. 2, 616–649.

[527] MARINOSCHI G. (2006) *Functional Approach to Nonlinear Models of Water Flow in Soils*, Springer.

[528] MARTIN R. (1970) A global existence theorem for autonomous differential equations in a Banach spaces, *Proc. Am. Math. Soc.*, **26**, 307–314.

[529] MASIERO F. (2008) Stochastic optimal control problems and parabolic equations in Banach spaces, *SIAM J. Control Optim.*, **47**, no. 1, 25–300.

[530] MASIERO F. (2012) Hamilton Jacobi Bellman equations in infinite dimensions with quadratic and superquadratic Hamiltonian, *Discrete Contin. Dyn. Syst.*, **32**, no. 1, 223–263.

[531] MASLOWSKI B. (1989) Strong Feller property for semilinear stochastic evolution equations. In *Stochastic Partial Differential Equations and Applications*, G. Da Prato and L. Tubaro (eds.), Lecture Notes in Mathematics, no. 1236, Springer-Verlag, 210–224.

[532] MASLOWSKI B. (1989) Uniqueness and stability of invariant measures for stochastic differential equations in Hilbert spaces, *Stochastics Stochastics Rep.,* **28**, no. 2, 85–114.

[533] MASLOWSKI B. (1991) On ergodic behaviour of solutions to systems of stochastic reaction-diffusion equations with correlated noise. In *Stochastic Processes and Related Topics (Georgenthal, 1990)*, Math. Res., no. 61, Akademie-Verlag, 93–102.

[534] MASLOWSKI B. & NUALART D. (2003) Evolution equations driven by a fractional Brownian motion, *J. Funct. Anal.*, **202**, no. 1, 277–305.

[535] MASLOWSKI B. & POSPIŠIL J. (2008) Ergodicity and parameter estimates for infinite-dimensional fractional Ornstein–Uhlenbeck process, *Appl. Math. Optim.*, **57**, no. 3, 401–429.

[536] MASLOWSKI B. & SEIDLER J. (1998) Invariant measures for nonlinear SPDE's: uniqueness and stability, *Arch. Math. (Brno)*, **34**, no. 1, 153–172.

[537] MASLOWSKI B. & SEIDLER J. (2001) Strong Feller solutions to SPDE's are strong Feller in the weak topology, *Studia Math.*, **148**, no. 2, 111–129.

[538] MASLOWSKI B. & SEIDLER J. (2002) Strong Feller infinite-dimensional diffusions. In *Stochastic Partial Differential Equations and Applications*, G. Da Prato and L. Tubaro (eds.), Lecture Notes in Pure and Applied Mathematics, Dekker, 373–387.

[539] McKEAN H. P. Jr. (1969) *Stochastic Integrals,* Academic Press.

[540] McKEAN H. P. Jr. (1970) Nagumo's equation, *Adv. Math.* **4**, 209–423.

[541] MEDJO T. (2011) The exponential behavior of the stochastic three-dimensional primitive equations with multiplicative noise, *Nonlinear Anal.,* **12**, no. 2, 799–810.

[542] MÉTIVIER M. (1988) *Stochastic Partial Differential Equations in Infinite Dimensional Spaces*, Quaderni Scuola Normale Superiore di Pisa.

[543] MÉTIVIER M. & PELLAUMAIL J. (1980) *Stochastic Integration*, Academic Press.

[544] MÉTIVIER M. & PISTONE G. (1975) Une formule d'isometrie pour l'integral stochastique Hilbertienne et équations d'évolution linéaires stochastiques, *Z. Wahrsch. Verw. Gebiete,* **33**, no. 1, 1–11.

[545] MIKULEVICIUS R. & ROZOVSKII B. (1999) Martingale problems for stochastic PDE's. In *Stochastic Partial Differential Equations: Six Perspectives*, R. A. Carmona and B. Rozoskii (eds.), Mathematical Surveys and Monographs, no. 64, Am. Math. Soc., 243–325.

[546] MIKULEVICIUS R. & ROZOVSKII B. (2004) Stochastic Navier–Stokes equations for turbulent flows, *SIAM J. Math. Anal.*, **35**, no. 5, 1250–1310.

[547] MILLET A. & SANZ SOLÉ M. (1994) The support of an hyperbolic stochastic partial differential equation, *Probab. Theory Relat. Fields,* **98**, no. 3, 361–387.

[548] MILLET A. & SANZ SOLÉ M. (2000) Approximation and support theorem for a wave equation in two space dimensions, *Bernoulli*, **6**, no. 5, 887–915.

[549] MILLET A. & SMOLENSKI W. (1992) On the continuity of Ornstein-Uhlenbeck processes in infinite dimensions, *Probab. Theory Relat. Fields,* **92**, no. 4, 529–547.

[550] MIZOHATA S. (1973) *The Theory of Partial Differential Equations*, Cambridge University Press.

[551] MOHAMMED S. A. E. (1984) *Stochastic Functional Differential Equations*, Pitman.

[552] MORTON M. J. (1989) *Arbitrage and martingales*, PhD Thesis, Cornell University.

[553] MUELLER C. & TRIBE R. (1995) Stochastic P.D.E.'s arising from the long range contact and long range voter processes, *Probab. Theory Relat. Fields*, **102**, no. 4, 519–545.

[554] MUELLER C. (1998) The heat equation with Lévy noise, *Stochastic Process. Appl.*, **74**, no. 1, 67–82.

[555] MUSIELA M. (1993) *Stochastic PDEs and Term Structure Models,* Journées Internationales de Finance, Association Francaise de Finance, IGR-AFFI, vol. 1, La Baule.

[556] NAGASE N. (1992) Note on stochastic partial differential equations, *Sci. Rep. Hirosaki Univ.*, **39**, no. 2, 73–87.

[557] NAGASE N. & NISIO M. (1990) Optimal controls for stochastic partial differential equations, *SIAM J. Control Optim.,* **28**, no. 1, 186–213.

[558] NEIDHART A.L. (1978) *Stochastic integrals in 2-uniformly smooth Banach spaces*, Ph.D. Thesis, University of Winsconsin.

[559] NEVEU J. (1970) *Bases Mathematiques du Calcul des Probabilités*, Masson.

[560] NOVIKOV E. A. (1965) Functionals and the random-force method in turbulence theory, *Sov. Phys. JETP*, **20**, 1290–1294.

[561] NUALART D. & PARDOUX E. (1988) Stochastic calculus with anticipating integrands, *Probab. Theory Relat. Fields*, **78**, no. 4, 535–581.

[562] NUALART D. & PARDOUX E. (1992) White noise driven by quasilinear SPDE's with reflection, *Probab. Theory Relat. Fields*, **93**, no. 1, 77–89.

[563] NUALART D. & ÜSTUNEL A. S. (1989) Mesures cylindriques et distributions sur l'espace de Wiener. In *Stochastic Partial Differential Equations and Applications II*, G. Da Prato and L. Tubaro (eds.), Lecture Notes in Mathematics, no. 1390, Springer-Verlag, 186–191.

[564] NUALART D. & ZAKAI M. (1986) Generalized stochastic integrals and the Malliavin calculus, *Probab. Theory Relat. Fields*, **73**, no. 2, 255–280.

[565] OCONE D. (1984) Malliavin calculus and stochastic integral representation of diffusion processes, *Stochastics*, **12**, no. 3–4, 161–185.

[566] OGAWA S. (1973) A partial differential equation with the white noise as a coefficient, *Z. Wahrsch. Verw. Gebiete,* **28**, 53–71.

[567] ØKSENDAL B. (1998) *Stochastic Differential Equations. An Introduction with Applications*, Fifth edition, Universitext, Springer-Verlag.

[568] ONDREJÁT M. (2004) Existence of global mild and strong solutions to stochastic hyperbolic evolution equations driven by a spatially homogeneous Wiener process, *J. Evol. Equations*, **4**, no. 2, 169–191.

[569] ONDREJÁT M. (2004) *Uniqueness for stochastic evolution equations in Banach spaces,* Dissertationes Math. (Rozprawy Mat.), no. 426.

[570] ONDREJÁT M. (2006) Existence of global martingale solutions to stochastic hyperbolic equations driven by a spatially homogeneous Wiener process, *Stoch. Dyn.*, **6**, no. 1, 23–52.

[571] ONDREJÁT M. (2007) Uniqueness for stochastic non-linear wave equations, *Nonlinear Anal.*, **67**, no. 12, 3287–3310.

[572] ONDREJÁT M. (2010) Stochastic nonlinear wave equations in local Sobolev spaces, *Electron. J. Probab.*, **15**, no. 33, 1041–1091.

[573] ONDREJÁT M. (2010) Stochastic wave equation with critical nonlinearities: temporal regularity and uniqueness, *J. Differential Equations*, **248**, no. 7, 1579–1602.

[574] ORTIZ-LOPEZ V. AND SANZ-SOLÉ M. (2011) Laplace principle for a stochastic wave equation in spatial dimension three, *Stochastic Analysis*, 31–49.

[575] PARDOUX E. (1975) *Equations aux derivées partielles stochastiques non-linéaires monotones*, Thèse, Université Paris XI.

[576] PARDOUX E. (1976) *Integrales Stochastiques Hilbertiennes*, Cahiers Mathématiques de la Decision, no. 7617, Université Paris Dauphine.

[577] PARDOUX E. (1979) Stochastic partial differential equations and filtering of diffusion processes, *Stochastics,* **3**, 127–167.

[578] PARDOUX E. (1982) Equations du filtrage nonlinéaire, de la prédiction et du lissage, *Stochastics,* **6**, no. 3–4, 193–131.

[579] PARDOUX E. (1985) Asymptotic analysis of a semilinear PDE with wide-band noise disturbances. In *Stochastic Space Time Models and Limit Theorems*, L. Arnold and P. Kotelenez (eds.), Reidel, 95–139.

[580] PARDOUX E. (1987) Two-sided stochastic calculus for SPDEs. In *Stochastic Partial Differential Equations and Applications*, G. Da Prato and L. Tubaro (eds.), Lecture Notes in Mathematics, no. 1236, Springer-Verlag, 200–207.

[581] PARDOUX E. (1998) BSDE's and semilinear PDE's. Stochastic Analysis and Related Topics VI. In *The Geilo Workshop 1996*, Progress in Probability, Vol. 42, Birkhauser, 79–127.

[582] PARDOUX E. (1999) BSDE's, weak convergence and homogenization of semilinear PDE's. In *Nonlinear Analysis, Differential Equations and Control*, F. H. Clarke and R. J. Stern (eds.), Kluwer, 503–549.

[583] PARDOUX E. & PENG S. (1992) Backward SDEs and quasilinear PDEs. In *Stochastic Partial Differential Equations and Their Applications*, B. L. Rozovskii and R. B. Sowers (eds.), Lecture Notes in Control and Information Science, no. 176, Springer-Verlag.

[584] PARDOUX E. & RASCANU A. (1999) Backward stochastic variational inequalities, *Stochastics Stochastic Rep.,* **67**, no. 3–4, 159–167.

[585] PARISI G. & WU Y. S. (1981), Perturbation theory without gauge fixing, *Sci. Sinica,* **24**, no. 4, 483–496.

[586] PARTASARATHY K. R. (1967) *Probability Measures in Metric Spaces*, Academic Press.

[587] PAZY A. (1983) *Semigroups of Linear Operators and Applications to Partial Differential Equations*, Springer-Verlag.

[588] PENG S. (1991) Probabilistic interpretation for systems of quasilinear parabolic partial differential equations, *Stochastics Stochastics Rep.,* **37**, no. 1–2, 61–74.

[589] PENG S. (1993) Backward stochastic differential equations and applications to optimal control, *Appl. Math. Optim.,* **27**, no. 2, 125–144.

[590] PESZAT S. (1992) Equivalence of distribution of some Ornstein–Uhlenbeck processes taking values in Hilbert space, *Probab. Mat. Statis.*, **13**, no. 1, 7–17.

[591] PESZAT S. (1992) Large deviation principle for stochastic evolution equations, *Probab. Theory Relat. Fields,* **98**, no. 1, 113–136.

[592] PESZAT S. (1992) Law equivalence of solutions of linear stochastic equations driven by cylindrical Wiener process, *Studia Math.*, **101**, no. 3, 269–284.

[593] PESZAT S. (1995) Existence and uniqueness of the solution for stochastic equations on Banach spaces, *Stochastics Stochastics Rep.*, **55**, no. 3–4, 167–193.

[594] PESZAT S. (2002) The Cauchy problem for a nonlinear stochastic wave equation in any dimension, *J. Evol. Equations*, **2**, no. 3, 383–394.

[595] PESZAT S. & TINDEL S. (2010) Stochastic heat and wave equations on a Lie group, *Stochastic Anal. Appl.*, **28**, no. 4, 662–695.

[596] PESZAT S. & ZABCZYK J. (1995) Strong Feller property and irreducibility for diffusions on Hilbert spaces, *Ann. Probab.*, **23**, no. 1, 157–172.

[597] PESZAT S. & ZABCZYK J. (1997) Stochastic evolution equations with a spatially homogeneous Wiener process, *Stochastic Process. Appl.*, **72**, no. 2, 187–204.

[598] PESZAT S. & ZABCZYK J. (2000) Nonlinear stochastic wave and heat equations, *Probab. Theory Relat. Fields*, **116**, no. 3, 421–443.

[599] PESZAT S. & ZABCZYK J. (2007) *Stochastic Partial Differential Equations with Lévy Noise*, Cambridge University Press.

[600] POLYANIN A. D. & ZAITSEV V. F. (2004) *Handbook of Nonlinear Partial Differential Equations*, Chapman and Hall.

[601] PRATELLI M. (1988) Intégration stochastique et géometrie des espaces de Banach. In *Séminaire des Probabilitées 22*, in Mathematics, no. 1321, Springer-Verlag, 129–137.

[602] PRÉVOT C. & RÖCKNER M. (2007) *A Concise Course on Stochastic Partial Differential Equations*, Lecture Notes in Mathematics, no. 1905, Springer-Verlag.

[603] PRINTEMS J. (1999) The stochastic Korteweg–de Vries equation in $L^2(\mathbb{R})$, *J. Differential Equations*, **153**, no. 2, 338–373.

[604] PRINTEMS J. (2001) On the discretization in time of parabolic stochastic partial differential equations, *M2AN Math. Model. Numer. Anal.*, **35**, no. 6, 1055–1078.

[605] PRIOLA E., SHIRIKYAN L., XU L. & ZABCZYK J. (2012) Exponential ergodicity in variation distance for equations with Lèvy noise, *Stochastic Process. Appl.*, **122**, no. 1, 106–133.

[606] PRIOLA E., XU L. & ZABCZYK J. (2011) Exponential mixing for some SPDEs with Lèvy noise, *Stoch. Dyn.*, **11**, no. 2–3, 521–534.

[607] PRIOLA E. & ZABCZYK J. (2004) Liouville theorems for non-local operators, *J. Funct. Anal.*, **216**, no. 2, 455–490.

[608] PRIOLA E. & ZABCZYK J. (2011) Structural properties of semilinear SPDEs driven by cylindrical stable processes, *Probab. Theory Relat. Fields*, **149**, no. 1–2, 97–137.

[609] PROTTER P. (1992) *Stochastic Integration and Differential Equations*, Springer-Verlag.

[610] RAIBLE M., LINZ S. J. & HÄNGGI P. (1961) Amorphous thin film growth: minimal deposition equation, *Phys. Rev. E*, **62**, 1691–1705.

[611] RASCANU A. (1981) Existence for a class of stochastic parabolic variational inequalities, *Stochastics*, **5**, no. 3, 201–239.

[612] RAMÌREZ J. (2001) Short-time asymptotics in Dirichlet spaces., *Commun. Pure Appl. Math.*, **54**, no. 3, 259–293.

[613] REN J., RÖCKNER M. & WANG F. Y. (2007) Stochastic generalized porous media and fast diffusions equations, *J. Differential Equations*, **238**, no. 11, 118–152.

[614] RIEDLE M. (2006) Lyapunov exponents for linear delay equations in arbitrary phase spaces, *Integral Equations Oper. Theory*, **54**, no. 2, 259–278.

[615] RIEDLE M. (2008) Solutions of affine stochastic functional differential equations in the state space, *J. Evol. Equations*, **8**, no. 1, 71–97.

[616] RÖCKNER M., SCHMULAND B. & ZHANG X. (2008) The Yamada–Watanabe theorem for stochastic evolution equations in infinite dimensions, *Commun. Mat. Phys.*, **11**, no. 2, 247–259.

[617] RÖCKNER M. & SOBOL Z. (2006) Kolmogorov equations in infinite dimensions: well-posedness and regularity of solutions, with applications to stochastic generalized Burgers equations, *Ann. Probab.*, **34**, no. 2, 663–727.

[618] RÖCKNER M. & SOBOL Z. (2007) A new approach to Kolmogorov equations in infinite dimensions and applications to the stochastic 2D Navier–Stokes equation, *C. R. Math. Acad. Sci. Paris*, **345**, no. 5, 289–292.

[619] RÖCKNER M. & WANG F. Y. (2003) Supercontractivity and ultracontractivity for (non-symmetric) diffusion semigroups on manifolds, *Forum Math.*, **15**, no. 6, 893–921.

[620] RÖCKNER M. & WANG F. Y. (2008) On monotone stochastic generalized porous media equations, *J. Differential Equations*, **245**, no. 12, 3898–3935.

[621] RÖCKNER M. & WANG F. Y. (2013) General extinction results for stochastic partial differential equations and applications, *J. London Math. Soc.*, **87**, no. 2, 545–560.

[622] RÖCKNER M. & ZHANG T. (2007) Stochastic evolution equations of jump type: existence, uniqueness and large deviation principles, *Potential Anal.*, **26**, no. 3, 255–279.

[623] RÖCKNER M. & ZHANG X. (2009) Stochastic tamed 3D Navier–Stokes equations: existence, uniqueness and ergodicity, *Probab. Theory Relat. Fields*, **145**, no. 1–2, 211–267.

[624] RÖCKNER M. & ZHANG X. (2009) Tamed 3D Navier–Stokes equation: existence, uniqueness and regularity, *Infin. Dimens. Anal. Quantum Probab. Relat. Topics*, **12**, no. 4, 525–549.

[625] RÖCKNER M., ZHANG T., & ZHANG X. (2010) Large deviations for stochastic tamed 3D Navier–Stokes equations, *Appl. Math. Optim.*, **61**, no. 2, 267–285.

[626] RÖCKNER M., ZHU R. & ZHU X. (2011) The stochastic reflection problem on an infinite dimensional convex set and BV functions in a Gelfand triple, arXiv:1011.3996.

[627] RÖCKNER M., ZHU R. & ZHU X. (2011) Stochastic quasi-geostrophic equation, arXiv:1108.4896.

[628] RÖCKNER M., ZHU R. & ZHU X. (2012) BV functions in a Gelfand triple for differentiable measure and its applications, arXiv:1011.3996.

[629] ROMITO M. (2008) Analysis of equilibrium states of Markov solutions of 3D-Navier–Stokes equations driven by additive noise, *J. Statist. Phys.*, **131**, no. 3, 415–444.

[630] ROSEMBLATT M. (1971) *Markov Processes. Structure and Asymptotic Behaviour*, Springer-Verlag.

[631] ROYER G. (1979) Processus de diffusion associées a certains models d'Ising à spin continu, *Z. Wahrsch. Verw. Gebiete*, **59**, 479–490.

[632] ROYER M. (2004) BSDEs with a random terminal time driven by a monotone generator and their links with PDEs, *Stochastics Stochastics Rep.*, **76**, no. 4, 281–307.

[633] ROZOVSKII B. L. (1990) *Stochastic Evolution Equations. Linear Theory and Applications to Non-linear Filtering*, Kluwer.

[634] ROZOVSKII B. L. & SHIMIZU (1981) Smoothness of solutions of stochastic evolution equations and the existence of a filtering transition density, *Nagoya Math. J.*, **84**, 195–208.

[635] RUSINEK A. (2010) Mean reversion for HJMM forward rate models, *Adv. Appl. Probab.*, **42**, no. 2, 371–391.

[636] SAMARSKIJ A. and TYCHONOFF A. (1963) *Equations of Mathematical Physics*. Translated by A. R. M. Robson and P. Basu, translation edited by D. M. Brink, Pergamon Press.

[637] SANZ-SOLÉ M. (2005) *Malliavin Calculus with Applications to Stochastic Partial Differential Equations*, Fundamental Sciences, Mathematics, EPFL Press.

[638] SANZ-SOLÉ M. & SARRÁ M. (2002) Hölder continuity for the stochastic heat equation with spatially correlated noise. In *Stochastic Analysis, Random Fields and Applications*, R. C. Dalang, M. Dozzi and F. Russo (eds.), Progress in Probability, no. 52, 259–268.

[639] SCHAUMLÖFFEL K. U. (1989) White noise in space and time as the time derivative of a cylindrical Wiener process. In *Stochastic Partial Differential Equations and Applications II*, G. Da Prato and L. Tubaro (eds.), Lecture Notes in Mathematics, no. 1390, Springer-Verlag, 225–229.

[640] SCHAUMLÖFFEL K. U. (1990) *Zufällige Evolutionsoperatoren für stochastische partielle Differentialgleichungen,* Ph.D. Thesis, Universität Bremen.

[641] SEIDLER J. (2010) Exponential estimates for stochastic convolutions in 2-smooth Banach spaces, *Electron. J. Probab.*, **15**, no. 50, 1556–1573.

[642] SEIDLER J. & VRKOC I. (1990) An averaging principle for stochastic evolution equations, *Časopis Pěst. Mat.*, **115**, no. 3, 240–263.

[643] SELL R. & YOU Y. (2002) *Dynamics of Evolutionary Equations*, Applied Mathematical Sciences, no. 143, Springer-Verlag.

[644] SERMANGE M. & TEMAM R. (1983) Some mathematical questions related to the MHD equations, *Commun. Pure Appl. Math.*, **36**, no. 5, 635–664.

[645] SHIGEKAWA I. (1987) Existence of invariant measures of diffusions on an abstract Wiener space, *Osaka J. Math.,* **24**, no. 1, 37–59.

[646] SIMON B. (1974) *The $P(\phi)_2$ Euclidean (Quantum) Field Theory,* Princeton University Press.

[647] SINESTRARI E. (1976) Accretive differential operators, *Boll. Unione Mat. Ital. B,* **13**, no. 1, 19–31.

[648] SINESTRARI E. (1985) On the abstract Cauchy problem of parabolic type in spaces of continuous functions, *J. Math. Anal. Appl.*, **107**, no. 1, 16–66.

[649] SKOROHOD A. V. (1961) Stochastic equations for diffusions in a bounded region, *Theory Probab. Appl.*, **6**, 264–274.

[650] SKOROHOD A. V. (1984) *Random Linear Operators*, Reidel.

[651] SMOLENSKI W., SZTENCEL R. & ZABCZYK J. (1986) Large deviation estimates for semilinear stochastic equations . In *Stochastic Differential Systems*, H. J. Engelbert and W. Schmidt (eds.), Lecture Notes in Control and Information Sciences, no. 96, Springer-Verlag, 218–231.

[652] SOUGANIDIS P. E. & YIP N. K. (2004) Uniqueness of motion by mean curvature perturbed by stochastic noise, *Ann. Inst. H. Poincaré, Anal. Non Linéaire*, **21**, no. 1, 1–23.

[653] SOWERS R. B. (1991) *New asymptotic results for stochastic partial differential equations,* Ph.D. Dissertation, University of Maryland.

[654] SOWERS R. (1994) Multi-dimensional reaction-diffusion equations with white noise boundary perturbations, *Ann. Probab.*, **22**, no. 4, 2071–2121.

[655] SPOHN H. (1993) Interface motion in models with stochastic dynamics, *J. Statist. Phys.*, **71**, no. 5–6, 1081–1132.

[656] SRITHARAN S. S. & SUNDAR P. (2006) Large deviations for the two-dimensional Navier–Stokes equations with multiplicative noise, *Stochastic Process. Appl.*, **116**, no. 11, 1636–1659.

[657] STANNAT W. (2003) L^1-Uniqueness of regularized 2D-Euler and stochastic Navier–Stokes equations, *J. Funct. Anal.*, **200**, no. 1, 101–117.

[658] STASI R. (2007) Maximum principle for infinite dimensional diffusion equations, *Potential Anal.*, **26**, no. 3, 213–224.

[659] STEIN O. & WINKLER M. (2005) Amorphous molecular beam epitaxy: global solutions and absorbing sets, *Eur. J. Appl. Math.*, **16**, no. 6, 767–798.

[660] STEWART H. B. (1974) Generation of an analytic semigroup by strongly elliptic operators, *Trans. Am. Math. Soc.,* **199**, 141–162.

[661] STEWART H. B. (1980) Generation of analytic semigroups by strongly elliptic operators under general boundary conditions, *Trans. Am. Math. Soc.,* **259**, no. 1, 299–310.

[662] STRAUSS W. (1966) On continuity of functions with values in various Banach spaces, *Pac. J. Math.*, **19**, 543–551.

[663] STROOCK D. W. (1981) The Malliavin calculus, a functional analytic approach, *J. Funct. Anal.,* **44**, no. 2, 212–257.

[664] STROOCK D.W. (1983) Some application of stochastic calculus to partial differential equations. In *École d'Eté de Probabilité de Saint Flour*, Lecture Notes in Mathematics, no. 976, 267–382.

[665] STROOCK D. W. & VARHADAN S. R. S (1979) *Multidimensional Diffusion Processes*, Springer-Verlag.

[666] SWIECH A. (2009), A PDE approach to large deviations in Hilbert spaces, *Stochastic Process. Appl.*, **119**, no. 4, 1081–1123.

[667] SWIECH A. & ZABCZYK J. (2011), Large deviations for stochastic PDEs with Lèvy noise, *J. Funct. Anal.*, **260**, no. 3, 674–723.

[668] TANABE H. (1979) *Equations of Evolution,* Pitman.

[669] TANAKA H. (1979) Stochastic differential equations with reflecting boundary conditions in convex regions, *Hiroshima Math J.*, **9**, no. 1, 163–177.

[670] TEICHMANN J. (2011) Another approach to some rough and stochastic partial differential equations, *Stoch. Dyn.*, **11**, no. 2–3, 535–550.

[671] TEMAM R. (1977) *Navier–Stokes Equations,* North-Holland.

[672] TESSITORE G. (1996) Existence, uniqueness and space regularity of the adapted solutions of a backward SPDE, *Stochastic Anal. Appl.*, **14**, no. 4, 461–486.

[673] TESSITORE G. & ZABCZYK J. (1998) Invariant measures for stochastic heat equations, *Probab. Math. Statist.*, **18**, no. 2, 271–287.

[674] TESSITORE G. & ZABCZYK J. (1998) Strict positivity for stochastic heat equations, *Stochastic Process. Appl.*, **77**, no. 1, 83–98.

[675] TESSITORE G. & ZABCZYK J. (2006) Wong–Zakai approximations of stochastic evolution equations, *J. Evol. Equations*, **6**, no. 4, 621–655.

[676] TINDEL S. & VIENS F. (1999) On space-time regularity for the stochastic heat equation on Lie groups, *J. Funct. Anal.*, **169**, no. 2, 559–603.

[677] TRIEBEL H. (1978) *Interpolation Theory, Function Spaces, Differential Operators*, North-Holland, Amsterdam.

[678] TUBARO L. (1983) On abstract differential stochastic equation in Hilbert spaces with dissipative drift, *Stochastic Anal. Appl.*, **1**, no. 2, 205–214.

[679] TUBARO L. (1984) An estimate of Burkholder type for stochastic processes defined by the stochastic integral, *Stochastic Anal. Appl.*, **2**, no. 2, 187–192.

[680] TUBARO L. (1988) Some results on stochastic partial differential equations by the stochastic characteristic method, *Stochastic Anal. Appl.*, **6**, no. 2, 217–230.

[681] TWARDOWSKA K. (1992) An extension of the Wong–Zakai theorem for stochastic equations in Hilbert spaces, *Stochastic Anal. Appl.*, **10**, no. 4, 471–500.

[682] TWARDOWSKA K. & ZABCZYK J. (2004) A note on stochastic Burgers' system of equations, *Stochastic Anal. Appl.*, **22**, no. 6, 1641–1670.

[683] TWARDOWSKA K. & ZABCZYK J. (2006). Qualitative properties of solutions to stochastic Burgers's system of equations. In *Stochastic Partial Differential Equations and Applications VII*, G. Da Prato and L. Tubaro (eds.), Lecture Notes in Pure and Applied Mathematics, no. 245, Chapman & Hall/CRC, 311–322.

[684] VAKHANIA N. N., TASIELADRE V. I. & CHOBANYAN S. (1987) *Probability Distributions on Banach Spaces*, Reidel.

[685] VALLET G. & WITTBOLD P. (2009) On a stochastic first-order hyperbolic equation in a bounded domain, *Infin. Dimens. Anal. Quantum Probab. Relat. Topics*, **12**, no. 4, 613–651.

[686] VAN NEERVEN J. (2010) Stochastic Evolution Equations, Internet Seminar, http://fa.its.tudelft.nl/isemwiki.

[687] VAN NEERVEN J. & RIEDLE M. (2007) A semigroup approach to stochastic delay equations in spaces of continuous functions, *Semigroup Forum*, **74**, no. 2, 227–239.

[688] VAN NEERVEN J. & VERAAR M. (2006) On the stochastic Fubini theorem in infinite dimensions. In *Stochastic Partial Differential Equations and Applications VII*, G. Da Prato and L. Tubaro (eds.), Lecture Notes in Pure and Applied Mathematics, no. 245, Chapman & Hall/CRC, 323–336.

[689] VAN NEERVEN J., VERAAR M. & WEIS L. (2007) Stochastic integration in UMD Banach spaces, *Ann. Probab.*, **35**, no. 4, 1438–1478.

[690] VAN NEERVEN J., VERAAR M. & WEIS L. (2008) Stochastic evolution equations in UMD Banach spaces, *J. Funct. Anal.*, **255**, no 4, 940–993.

[691] VARADHAN S. S. (1968), *Stochastic Processes*, Courant Institute, New York University.

[692] VARADHAN S. S. (1983) *Large Deviations and Applications*, Courant Institute, New York University.

[693] VERAAR M. (2006) *Stochastic integration in Banach spaces and applications to parabolic evolution equations*, Ph.D. Thesis, Thomas Stieltjes Institute for Mathematics.

[694] VESPRI V. (1989) Linear stochastic integrodifferential equations. In *Volterra Integro-Differential Equations in Banach Spaces and Applications*, G. Da Prato and M. Iannelli (eds.), Pitman Research Notes in Mathematics Series, no. 190, Pitman, 387–404.

[695] VIOT M. (1976) *Solution faibles d'équations aux dérivées partielles non-linéaires,* Thèse, Université Pierre et Marie Curie, Paris.

[696] VIOT M. (1974) Solution en lois d'une équation aux derivées partielles non linéaire: methodes de compacité, *C. R. Acad. Sci. Paris, Sér. A*, **278**, 1185–1188.

[697] VIOT M. (1974) Solution en lois d' une equations aux derivées partielles non linéaire: methode de monotonie, *C. R. Acad. Sci. Paris, Sér. A*, **278**, 1405–1408.

[698] VINTER R. B. (1975) A representation of solution to stochastic delay equations, Imperial College Department, Computing and Control, Report.

[699] VINTER R. B. & R. H. KWONG (1981) The infinite time quadratic control problem for linear systems with state and control delays: an evolution equation approach , *SIAM J. Control Optim.*, **19**, no. 1, 139–53.

[700] VISHIK M. I. & FURSIKOV A. (1979) *Mathematical Problems of Statistical Hydromechanics*, Kluwer Academic Press.

[701] VON WAHL W. (1985) *The Equations of Navier–Stokes and Abstract Parabolic Equations*, Fried. Vieweg & Sohn.

[702] WALSH J. B. (1984) An introduction to stochastic partial differential equations. In *École d'Eté de Probabilité de Saint Flour XIV 1984*, P. L. Hennequin (ed.), Lecture Notes in Mathematics, no. 1180, 265–439.

[703] WALSH J. B. (2005) Finite element methods for parabolic stochastic PDE's, *Potential Anal.*, **23**, no. 1, 1–43.

[704] WANG F. Y. (1997), Logarithmic Sobolev inequalities on noncompact Riemannian manifolds, *Probab. Theory Relat. Fields*, **109**, no. 3, 417–424.

[705] WANG F. Y. (2004) *Functional Inequalities, Markov Semigroups and Spectral Theory*, Science Press.

[706] WANG F. Y. (2007) Harnack inequalities and applications for stochastic generalized porous media equations, *Ann. Probab.*, **35**, no. 4, 1333–1350.

[707] WANG P. K. C. (1966) On the almost sure stability of linear stochastic distributed parameter dynamical systems, *ASME Trans. J. Appl. Mech.,* 182–186.

[708] WEBB G. F. (1972) Continuous nonlinear perturbations of linear accretive operators, *J. Funct. Anal.,* **10**, 191–203.

[709] WEBER H. (2010) On the short time asymptotic of the stochastic Allen–Cahn equation, *Ann. Inst. Henri Poincaré, Probab. Stat.*, **46**, no. 4, 965–975.

[710] WEBER H. (2010) Sharp interface limit for invariant measures of a stochastic Allen–Cahn equation, *Commun. Pure Appl. Math.*, **63**, no. 8, 1071–1109.

[711] WILLEMS J. C. (1981) Stochastic systems: the mathematics of filtering and identification and applications. In *NATO Advanced Study Institute, Les Arcs, France, 1980*, M. Hazewinkel and J. C. Willems (eds.), Reidel, 111–134.

[712] XIA D.-X. (1972) *Measure and Integration Theory on Infinite-Dimensional Spaces*, Academic Press.

[713] XU L. & ZEGARLINSKI B. (2009) Ergodicity of the finite and infinite dimensional α-stable systems, *Stochastic Anal. Appl.*, **27**, no. 4, 797–824.

[714] YAGI A. (1984) Coincidence entre des espaces d'interpolation et des domaines de puissances fractionnaires d'opérateurs, *C. R. Acad. Sci. Paris, Sér. I*, **299**, no. 6. 173–176.

[715] YANG D. (to appear) The Kolmogorov equation associated to a stochastic Kuramoto–Sivashinsky equation, *J. Funct. Anal.*

[716] YANNACOPOULOS A. N., FRANGOS N. E. & KARATZAS I. (2011) Wiener chaos solutions for linear backward stochastic evolution equations, *SIAM J. Math. Anal.*, **43**, no. 1, 68–113.

[717] YIP N. K. (1998) Stochastic motion by mean curvature, *Arch. Rational Mech. Anal.*, **144**, no. 4, 313–355.

[718] YONG J. & ZHOU K. (1999), *Stochastic Controls. Hamiltonian Systems and HJB Equations,* Applications of Mathematics, no. 43, Springer-Verlag, New York.

[719] YOR M. (1974) Existence et unicité de diffusions á valeurs dans un espace de Hilbert, *Ann. Inst. H. Poincaré, Sec. B*, **10**, 55–88.

[720] YOR M. (1974) Sur les intégral stochastiques à valeurs dans un espace de Banach, *Ann. Inst. H. Poincaré Sect. B*, **10**, 31–36.

[721] YOSIDA K. (1965) *Functional Analysis,* Springer-Verlag.

[722] YUBIN Y. (2004) Semidiscrete Galerkin approximation for a linear stochastic parabolic partial differential equation driven by an additive noise, *BIT Numer. Math.*, **44**, no. 4, 829–847.

[723] YUBIN Y. (2005) Smoothing properties in multistep backward difference method and time derivative approximation for linear parabolic equations, *Int. J. Math. Math. Sci.*, **4**, 523–536.

[724] YUBIN Y. (2006) Galerkin finite element methods for stochastic parabolic partial differential equations, *SIAM J. Numer. Anal.*, **43**, no. 4, 1363–1384.

[725] ZABCZYK J. (1975) Remarks on the algebraic Riccati equation in Hilbert spaces, *J. Appl. Math. Optim.,* **2**, no. 3, 251–58.

[726] ZABCZYK J. (1979) *On the Stability of Infinite-Dimensional Linear Stochastic Systems,* Banach Center Publications, vol 5, PWN-Polish Scientific Publishers.

[727] ZABCZYK J. (1981) *Linear stochastic systems in Hilbert spaces; spectral properties and limit behaviour,* Institute of Mathematics, Polish Academy of Sciences, report no. 236. Also in Banach Center Publications, vol. 41 (1985), 591–609.

[728] ZABCZYK J. (1985) Exit problem and control theory, *Systems Control Lett.*, **6**, no. 3, 165–172.

[729] ZABCZYK J. (1987) Exit problem for infinite dimensional systems. In *Stochastic Partial Differential Equations and Applications*, G. Da Prato and

L. Tubaro (eds.), Lecture Notes in Mathematics, no. 1236, Springer-Verlag, 239–257.

[730] ZABCZYK J. (1987) Stable dynamical systems under small perturbations, *J. Math. Anal. Appl.*, **125**, no. 2, 568–588.

[731] ZABCZYK J. (1988) *Topics in Stochastic Systems, Asymptotics and Regularity,* Flight Systems Research Laboratory, University of California, Los Angeles, TR No. 1-Z-4015-88.

[732] ZABCZYK J. (1989) Symmetric solutions of semilinear stochastic equations. In *Stochastic Partial Differential Equations and Applications II*, G. Da Prato and L. Tubaro (eds.), Lecture Notes in Mathematics, no. 1390, Springer–Verlag, 237–256.

[733] ZABCZYK J. (1989) Law equivalence of Ornstein–Uhlenbeck processes and control equivalence of linear systems, IMPAN preprint, no. 457.

[734] ZABCZYK J. (1991) Law equivalence of Ornstein–Uhlenbeck processes, *Gaussian Random Fields (Nagoya, 1990)*, K. Ito and T. Hida (eds.), Ser. Probab. Statist., 1, World Scientific, River Edge, NJ, 420–432.

[735] ZABCZYK J. (2000) Stochastic invariance and consistency of financial models, *Rend. Lincei Math. Appl.*, **11**, no. 2, 67–80.

[736] ZABCZYK J. (2001) A mini course on stochastic partial differential equations, *Progr. Probab.*, **49**, 257–284.

[737] ZAKAÏ M. (1969) On the optimal filtering of diffusion processes, *Z. Wahrsch. Verw. Gebiete*, **11**, 230–243.

[738] ZAMBOTTI L. (2002) Integration by parts formulae on convex sets of paths and applications to SPDEs with reflection, *Probab. Theory Relat. Fields*, **123**, no. 4, 579–600.

[739] ZAMBOTTI L. (2004) Occupation densities for SPDEs with reflection, *Ann. Probab.*, **32**, no. 1A, 191–215.

[740] ZHANG X. (2010) Stochastic Volterra equations in Banach spaces and stochastic partial differential equation, *J. Funct. Anal.*, **258**, no. 4, 1361–1425.

[741] ZHOU J. & LIU B. (2010) Optimal control problem for stochastic evolution equations in Hilbert spaces, *Int. J. Control,* **83**, no. 9, 1771–1784.

[742] ZHOU J. & ZHANG Z. (2011) Optimal control problems for stochastic delay evolution equations in Banach spaces, *Int. J. Control*, **84**, no. 8, 1295–1309.

[743] ZHOU X. (1993) On the necessary conditions of optimal controls for stochastic partial differential equations, *SIAM J. Control Optim.*, **31**, no. 6, 1462–1478.

Index